THE IN
AN OU
ENTOMOLOGY

2

THE INSECTS:
AN OUTLINE OF
ENTOMOLOGY

P.J. GULLAN

Division of Botany and Zoology
The Australian National University
Canberra
Australia

and

P.S. CRANSTON

Division of Entomology
CSIRO
Canberra
Australia

With illustrations by
K. Hansen McInnes

CHAPMAN & HALL

London · Glasgow · Weinheim · New York · Tokyo · Melbourne · Madras

Published by Chapman & Hall, 2-6 Boundary Row, London SE1 8HN

Chapman & Hall, 2-6 Boundary Row, London SE1 8HN, UK

Blackie Academic & Professional, Wester Cleddens Road, Bishopbriggs, Glasgow G64 2NZ, UK

Chapman & Hall GmbH, Pappelallee 3, 69469 Weinheim, Germany

Chapman & Hall USA, One Penn Plaza, 41st Floor, New York, NY10119, USA

Chapman & Hall Japan, ITP - Japan, Kyowa Building, 3F, 2-2-1 Hirakawacho, Chiyoda-ku, Tokyo 102, Japan

Chapman & Hall Australia, Thomas Nelson Australia, 102 Dodds Street, South Melbourne, Victoria 3205, Australia

Chapman & Hall India, R. Seshadri, 32 Second Main Road, CIT East, Madras 600 035, India

First edition 1994
Reprinted 1995

© 1994 P.J. Gullan and P.S. Cranston

Typeset in 10/12pt Palatino by Type Study, Scarborough
Printed in Great Britain by Alden Press, Osney Mead, Oxford

ISBN 0 412 49360 8

A Catalogue record for this book is available from the British Library

Library of Congress Catalog Card Number: 93-74492

Printed on acid-free text paper, manufactured in accordance with ANSI/NISO Z39.48-1992 (Permanence of Paper).

CONTENTS

PREFACE

Insects are extremely successful animals and they affect many aspects of our lives, despite their small size. All kinds of natural and modified ecosystems, both terrestrial and aquatic, support communities of insects that present a bewildering variety of lifestyles, forms and functions. The recent high profile of biodiversity as a scientific issue is leading to increasing interest in insects because of their astonishingly high species richness.

Students of biodiversity need to be versed in entomology, which covers not only the classification, evolutionary relationships and natural history of insects, but also how they interact with each other and the environment. The effects of insects on us, our crops and domestic stock, and how insect activities (both deleterious and beneficial) might be modified or controlled, are amongst the concerns of entomologists.

We, the authors, are systematic entomologists teaching and researching insect identification, distribution, evolution and ecology. Our study insects belong to two groups – scale insects and midges – and we make no apologies for using these, our favourite organisms, to illustrate some points in this book.

This book is not an identification guide, but addresses entomological issues of a more general nature. We commence with the significance of insects, their internal and external structure, and how they sense their environment, followed by their modes of reproduction and development. Succeeding chapters are based on major themes in insect biology: the ecology of ground-dwelling, aquatic and plant-feeding insects, and the behaviours of sociality, predation and parasitism, and defence. Finally, aspects of medical and veterinary entomology and the management of insect pests are considered.

Those to whom this book is addressed, namely students contemplating entomology as a career, or studying insects as a subsidiary to specialized disciplines such as agricultural science, forestry, medicine or veterinary science, ought to know something about insect systematics – this is the framework for scientific observations. However, we depart from the traditional order-by-order systematic arrangement seen in many entomological text books. The systematics of each insect order are presented in separate sections following the ecological/behavioural chapter appropriate to the predominant biology of the order. We have attempted to keep a phylogenetic perspective throughout, and one complete chapter is devoted to insect phylogeny, including examination of the evolution of several key features.

We believe that a picture is worth a thousand words. Much effort has been made to enable Karina Hansen McInnes to draw insects as seen under natural conditions. We hope these illustrations, together with the text, will encourage the rewarding study of insects, whether as a hobby or as a vocation.

Penny Gullan and Peter Cranston

ACKNOWLEDGEMENTS

All illustrations were drawn by Karina Hansen McInnes, who holds an Honours degree in Zoology from The Australian National University, Canberra. We are delighted with her artwork and are grateful for her hours of effort, attention to detail and skill in depicting the essence of the many subjects that are figured in the following pages. Thank you Karina.

This book would still be on the computer without the efforts of John Trueman, who job-shared with Penny in second semester 1992. John delivered invertebrate zoology lectures and ran lab classes while Penny revelled in valuable writing time, free from undergraduate teaching. Aimorn Stewart also assisted Penny by keeping her research activities alive during book preparation and by helping with labelling of figures. Eva Bugledich acted as a library courier and brewed hundreds of cups of coffee.

The following people generously reviewed one or more chapters for us: Andy Austin, Tom Bellas, Keith Binnington, Ian Clark, Geoff Clarke, Paul Cooper, Kendi Davies, Don Edward, Penny Greenslade, Terry Hillman, Dave McCorquodale, Rod Mahon, Dick Norris, Chris Reid, Steve Shattuck, John Trueman and Phil Weinstein. We also enjoyed many discussions on hymenopteran phylogeny and biology with Andy. Tom sorted out our chemistry and Keith gave expert advice on insect cuticle. Paul's broad knowledge of insect physiology was absolutely invaluable. Penny put us straight with springtail facts. Chris's entomological knowledge, especially on beetles, was a constant source of information. Steve patiently answered our endless questions on ants. Numerous other people read and commented on sections of chapters or provided advice or helpful discussion on particular entomological topics. These people included John Balderson, Mary Carver, Lyn Cook, Jane Elek, Adrian Gibbs, Ken Hill, John Lawrence, Chris Lyal, Patrice Morrow, Dave Rentz, Eric Rumbo, Vivienne Turner, John Vranjic and Tony Watson. Mike Crisp assisted with checking on current host-plant names. Sandra McDougall inspired part of Chapter 15. Thank you everyone for your many comments which we have endeavoured to incorporate as far as possible, for your criticisms which we hope we have answered, and for your encouragement.

We benefited from discussions concerning published and unpublished views on insect phylogeny (and fossils), particularly with Jim Carpenter, Mary Carver, Niels Kristensen, Jarmila Kukulová-Peck and John Trueman. Our views are summarized in the phylogenies shown in this book and do not necessarily reflect a consensus of our discussants' views (this was unattainable).

Our writing was assisted by the Commonwealth Scientific and Industrial Research Organisation (CSIRO) providing somewhere for both of us to work during the many weekdays, nights and weekends during which this book was prepared. In particular, Penny managed to escape from the distractions of her university position by working in CSIRO. Eventually, however, everyone discovered her whereabouts. The Division of Entomology of CSIRO provided generous support: Carl Davies gave us driving lessons on the machine that produced reductions of the figures and Sandy Smith advised us on labelling. The Division of Botany and Zoology of the Australian National University also provided assistance in aspects of the book production: Aimorn Stewart prepared the SEMs from which Fig. 4.7 was drawn, and Judy Robson typed the labels for some of the figures.

We are especially grateful to everyone at Chapman & Hall who helped with the production of the book.

1

THE IMPORTANCE AND DIVERSITY OF INSECTS

Bees representing three genera (from left): Bombus, Apis *and* Trigona

There are millions of kinds of insects, though we do not know exactly (or even approximately) how many there are. Our ignorance of how many organisms we share our planet with is surprising considering that comparable millions of stars and other galactic objects are listed, mapped and uniquely named or numbered. Some estimates (discussed later) imply that the species richness of insects is so great that, to a near approximation, all organisms can be considered to be insects. The questions raised by this insect diversity are legion. For example, the vignette for this chapter shows bees belonging to three genera, from left, *Bombus*, *Apis* and *Trigona*. Although they superficially resemble one another in morphology, each of these species differs from the others in size and ecology, not least in their pollination preferences.

Insect dominance applies to terrestrial and freshwater ecosystems, but extends no further into the marine environment than the high tidemark. Scientific curiosity concerning the identities and lifestyles of the fellow inhabitants of our planet is justification enough for their study. Furthermore, when economic considerations are taken into account, the effects of some insects are enormous. Clearly, we ought to understand these pervasive animals.

In this opening chapter, we outline the ecological importance and human uses of insects and discuss insect diversity, distribution and classification. We start by introducing the ecological functions of insects. Next we cover a little appreciated contribution of insects to our livelihoods – their use as human food. We explore the species richness and occurrence of insects and proceed to a discussion of naming and classification systems. The final section of the chapter is a tabular summary of the characteristics of each insect order.

1.1 WHAT IS ENTOMOLOGY?

Entomology is the study of insects and entomologists are people who study them. Worldwide there are countless thousands of entomologists observing, collecting, rearing and experimenting with insects. The researches undertaken by entomologists cover the total range of biological disciplines, including evolution, ecology, behaviour, anatomy, physiology, biochemistry and genetics. The unifying feature is that the study organisms are insects. Although many people working with insects hold professional positions, entomology can involve amateurs, though there are fewer than in bird-watching.

Biologists choose to work with insects for many reasons: the ease of maintaining laboratory cultures, short life-cycle and availability of many individuals are important factors. The few ethical problems concerning the responsible experimental use of insects, as compared with vertebrates, is another important consideration. It is a popular perception that entomologists kill or at least control insects but entomology includes many positive aspects of insects since their benefits to us outweigh their harm.

1.2 THE IMPORTANCE OF INSECTS

We should study insects for many reasons. Their ecologies are highly diverse and often they dominate food chains and food webs in biomass and species richness. The feeding specializations of different insect groups range over detritivory, saprophagy, xylophagy and fungivory (Chapter 8), aquatic filter feeding and grazing (Chapter 9), phytophagy (herbivory), including sap feeding (Chapter 10), and carnivory (predation and parasitism) (Chapter 12). Other insect groups show various combinations of these major feeding types. Insects may be aquatic (Chapter 9) or terrestrial throughout, or during part of, their lives. Their lifestyles encompass solitary, gregarious, subsocial and highly social modes (Chapter 11). They may be conspicuous or concealed (Chapter 13) and active by day or night. Insect life-cycles are adapted to a variety of abiotic conditions, including seasonal extremes of heat and cold, wet and dry, and notably to unpredictable climate. Insects are essential in the following roles within ecosystems:

- Nutrient recycling via leaf-litter and wood degradation, carrion and dung disposal, and soil turnover.
- Plant pollination and sometimes seed dispersal.
- Maintenance of plant community composition and structure via phytophagy, including seed feeding.
- Supporting insectivorous animals, such as many birds, mammals, reptiles and fish.

Each insect species is part of a wider community and, if lost, the complexities and abundance of other life will be affected.

From an anthropocentric point of view, some insects damage the health of domestic animals and ourselves (Chapter 14) and adversely affect agriculture and horticulture (Chapter 15). Other insects greatly benefit human society, either by providing us with food directly or by contributing to materials that we use or the food that we eat. For example, bees provide us with honey but honey bees also are valuable agricultural pollinators, worth an estimated US$1.6 5.7 billion annually in the United States alone. Furthermore, the services of predatory beetles and bugs or parasitic wasps that control pests often go unrecognized, especially by city-dwellers.

Insects also contain a vast array of chemical compounds, some of which can be collected, extracted or synthesized and used by ourselves. Silk from the cocoons of silkworm moths, *Bombyx mori*, has been used for fabric for centuries. The red dye, cochineal, is obtained commercially from scale insects of *Dactylopius coccus* cultured on *Opuntia* cacti. Another scale insect, the lac insect *Kerria lacca*, is a source of a commercial varnish called shellac. Chitin, a component of insect cuticle, or a derivative of chitin, can act as an anticoagulant or a haemostatic agent for tissue repair

in humans, enhancing the healing of wounds and burns, reducing serum cholesterol, serving as a non-allergenic drug carrier, providing a biodegradable plastic of high-tensile strength, and enhancing the removal of pollutants from waste water, to mention just a few of its possible applications.

Benefits from insects are more than economic or environmental; characteristics of certain insects make them useful models for understanding biological processes in general. For instance, the now widespread vinegar fly, *Drosophila melanogaster*, has a short generation time, high fecundity and ease of laboratory rearing and manipulation, making it ideal for genetic and cytological research. Aesthetically, the enormous variety of structure and colour in insects is worthy of admiration, by collection or depiction in drawings or photographs. Lastly and perhaps most importantly, the sheer number of insects means that their impact upon the environment, and hence our lives, is highly significant. Insects are the major component of biodiversity and, if only for this reason, we should try to understand them better.

1.2.1 Insects as human food: entomophagy

The study of insects as human food is an increasingly popular field of entomology. About 500 species of insects in more than 260 genera and 70 families are used for food somewhere in the world, especially in central and southern Africa, Asia, Australia and Latin America. However, western society has largely overlooked entomological cuisine. Insects are high in protein, energy and various vitamins and minerals: they can form 5–10% of the annual animal protein consumed by some indigenous peoples. Food insects generally feed on either living or dead plant matter and chemically-protected species are avoided. Termites, crickets, grasshoppers, locusts, beetles and moth larvae are the most frequently consumed insects.

Table 1.1   Proximate, mineral and vitamin analyses of four edible Angolan insects (percentages of daily human dietary requirements/100 g of insects as consumed) (after Santos Oliveira *et al.*, 1976, as adapted by DeFoliart, 1989)

Nutrient	Requirement per capita (reference person)	Macrotermes subhyalinus (Termitidae)	Imbrasia ertli (Saturniidae)	Usta terpsichore (Saturniidae)	Rhynchophorus phoenicis (Curculionidae)
Energy	2850 kcal	21.5%	13.2%	13.0%	19.7%
Protein	37 g	38.4	26.3	76.3	18.1
Calcium	1 g	4.0	5.0	35.5	18.6
Phosphorus	1 g	43.8	54.6	69.5	31.4
Magnesium	400 mg	104.2	57.8	13.5	7.5
Iron	18 mg	41.7	10.6	197.2	72.8
Copper	2 mg	680.0	70.0	120.0	70.0
Zinc	15 mg	–	–	153.3	158.0
Thiamine	1.5 mg	8.7	–	244.7	201.3
Riboflavin	1.7 mg	67.4	–	112.2	131.7
Niacin	20 mg	47.7	–	26.0	38.9

The typical 'western' repugnance of entomophagy is cultural rather than scientific or rational. After all, other invertebrates such as certain crustaceans and molluscs are favoured culinary items. Objections to eating insects cannot be justified on the grounds of taste or food value. Many are reported to have a nutty flavour and studies report favourably on the nutritional content of insects, although their amino-acid composition is not ideal and needs to be balanced with suitable plant protein. Table 1.1 presents the nutritional values obtained from analyses conducted on samples of four species of insects cooked according to traditional methods in central Angola, Africa. The insects concerned are: reproductive individuals of a termite, *Macrotermes subhyalinus* (Isoptera: Termitidae), which are de-winged and fried in palm oil; the large caterpillars of two species of moth, *Imbrasia ertli* and *Usta terpsichore* (Lepidoptera: Saturniidae), which are de-gutted and either cooked in water, roasted or sun-dried; and the larvae of the palm weevil, *Rhynchophorus phoenicis* (Coleoptera: Curculionidae), which are slit open and then fried whole in oil.

Mature larvae of *Rhynchophorus* species have been appreciated by people in tropical areas of Africa, Asia and the Neotropics for centuries. These fat, legless grubs (Fig. 1.1), often called palmworms, provide one of the richest sources of animal fat with substantial amounts of riboflavin, thiamine, zinc and iron (Table 1.1). Primitive cultivation systems, involving the cutting down of palm trees to provide suitable food for the weevils, are known in Brazil, Colombia, Paraguay and Venezuela. In plantations, however, palmworms are regarded as pests because they can seriously damage living coconut and oil palm trees.

In central Africa, southern Zairean people eat caterpillars belonging to a few dozen species. The calorific value of these caterpillars is high; their protein content ranges from 45% to 80% and they are a rich source of iron. Where there is chronic or seasonal shortage of vertebrate protein reserves elsewhere in sub-Saharan Africa, insect alternatives are often used or even preferred. For instance, caterpillars are the most important source of animal protein in some areas of the Northern Province of Zambia. The edible caterpillars of an emperor moth (Saturniidae), locally called mumpa, are much prized as food. People travel hundreds of kilometres to pick mumpa,

Fig. 1.1 A mature larva of the palm weevil, *Rhynchophorus phoenicis* (Coleoptera: Curculionidae) – a traditional food item in central Angola, Africa. (Larva after Santos Oliveira *et al.*, 1976.)

Fig. 1.2 A delicacy of the Australian Aborigines – a witchety (or witjuti) grub, the caterpillar of a wood moth (Lepidoptera: Cossidae), which feeds on the roots and stems of witjuti bushes (certain *Acacia* species). (After Cherikoff and Isaacs, 1989.)

which provides a highly lucrative market. The caterpillars contain 60–70% protein on a dry-matter basis and offset malnutrition caused by protein deficiency. Mumpa are either fried fresh or they are boiled and sun-dried prior to storage. Further south in Africa, the Pedi people of northern Transvaal much prefer mopanie worms, the larvae of the saturniid *Gonimbrasia belina*, to beef.

In the Philippines, June beetles (melolonthine scarabs), weaver ants (*Oecophylla smaragdina*), mole crickets and locusts are eaten in some regions. Locusts form an important dietary supplement during outbreaks, which apparently have become less common since the widespread use of insecticides. Various species of grasshoppers and locusts were eaten commonly by native tribes in western North America prior to the arrival of Europeans. The number and identity of species used have been poorly documented but species of *Melanoplus* were consumed. To be harvested, the grasshoppers were driven by advancing people or fire into a pit in the ground or were herded into a bed of coals.

Australian Aborigines used a wide range of insect foods, especially moth larvae. The caterpillars of wood or ghost moths (Cossidae and Hepialidae; Fig. 1.2) are called witchety grubs, from an Aboriginal word 'witjuti' for the *Acacia* species (wattles) on which the grubs feed, infesting the roots and stems. Witchety grubs are regarded as a delicacy: they contain 7–9% protein, 14–38% fat and 7–16% sugars and they are good sources of iron and calcium. Adults of the bogong moth, *Agrotis infusa* (Noctuidae), formed another famous Aboriginal food, once collected in their millions from aestivating sites in narrow caves and crevices on mountain summits in south-eastern Australia. Moths cooked in hot ashes provided a rich source of dietary fat.

A favourite source of sugar for Australian aboriginals living in arid regions comes from species of *Melophorus* and *Camponotus*, popularly known as honeypot ants. Specialized workers (called repletes) are fed nectar by other workers to store in their huge distended crops (Fig. 2.4). The repletes serve as food reservoirs for the ant colony and regurgitate part of their crop contents when solicited by another ant. Aborigines dig repletes from their underground nests, an activity most frequently undertaken by women, who sometimes excavate pits to a depth of a metre or more in search of these sweet rewards. Individual nests rarely supply more than 100 g of a honey which is essentially similar in composition to commercial honey. Honeypot ants in the western United States and Mexico belong to a different genus, *Myrmecocystus*. The repletes, a highly valued food, are collected by the rural people of Mexico, a difficult process in the hard soil of the stony ridges where the ants nest.

Even in the urban United States some insects may yet become popular as a food novelty. The millions of 17-year cicadas that periodically plague cities like Chicago are edible. Newly hatched cicadas, called tenerals, are best for eating because their soft body cuticle means that they can be consumed without first removing the legs and wings. These tasty morsels can be marinated or dipped in batter and then deep-fried, boiled and spiced, roasted and ground, or stir-fried with favourite seasonings.

Large-scale harvest or mass production of insects for human consumption brings some practical problems. The small size of most insects presents difficulties in collection or rearing and in processing for sale. The unpredictability of many wild populations needs to be overcome by the development of culture techniques. However, there are several advantages. The encouragement of entomophagy in many rural societies, particularly those with a history of insect use, may help to diversify people's diets. By incorporating mass harvesting of pest insects into control programmes, the use of pesticides can be reduced. Furthermore, cultivating insects for protein should be less environmentally damaging than cattle ranching, which devastates forests and native grasslands. Insect farming

is compatible with low input, sustainable agriculture and most insects have a high food conversion efficiency compared with conventional meat animals. Perhaps the bugburger, cricketburger or beeburger will become acceptable alternatives to the well-known American hamburger.

1.2.2 Insects as feed for domesticated animals

If you do not relish the prospect of eating insects yourself, then perhaps the concept of insects as a protein source for domesticated animals is more acceptable. The nutritive value of insects as feed for fish, poultry, pigs and farm-grown mink certainly is recognized in China, where feeding trials have shown that insect-derived diets can be cost-effective alternatives to more conventional fish-meal diets. The insects involved are primarily the larvae and pupae of house flies (*Musca domestica*), the pupae of silkworms (*Bombyx mori*) and the larvae of mealworms (*Tenebrio molitor*). The same or related insects are being used or investigated elsewhere, particularly as poultry or fish feedstock. Silkworm pupae, a by-product of the silk industry, can be used as a high-protein supplement for chickens. In India, poultry are fed the meal that remains after the oil has been extracted from the pupae. Fly larvae fed to chickens can recycle animal manure and the development of a range of insect recycling systems for converting organic wastes into feed supplements is inevitable, given that most organic substances are fed on by one or more insect species.

Clearly insects can form part of the nutritional base for people and their domesticated animals. Further research is needed and a database with accurate identifications is required to handle biological information. We must know which species we are dealing with in order to make use of information gathered elsewhere on the same or related insects. Data on the nutritional value, seasonal occurrence, host plants or other dietary needs, and rearing or collecting methods must be collated for all actual or potential food insects. Opportunities for insect food enterprises are numerous, given the immense diversity of insects.

1.3 INSECT DIVERSITY

1.3.1 Estimates of insect diversity

Estimates of species richness of insects vary from less than five million to as many as 80 million species. Extrapolation from ratios of temperate:tropical species numbers for well-known groups (such as vertebrates) in the well-known temperate areas to essentially unknown tropical insect faunas provide conservatively low estimates. However, recent sampling emphasizing the previously unknown fauna of the upper layers (canopy) of rainforest has increased enormously the magnitude of previous ideas concerning insect species numbers. Much of this estimated increase comes from arboreal beetles (Coleoptera) but other groups clearly are more numerous than believed previously. Key factors in more recent tropical diversity calculations include identification of the number of beetle species found, allocation to feeding groups, the degree of inferred host-specificity to the surveyed tree species and the proportion of beetles to other arthropods. Certain assumptions can be challenged and some reanalyses incline towards the earlier conservative estimates, whereas others lead to even higher estimates of global insect species numbers.

Does evolution ('God') really show an inordinate 'fondness' for beetles, as J.B.S. Haldane once jested? Generally Coleoptera is claimed to be the most speciose order of insects. This may be true in places such as Australia, Japan and tropical South America but the high relative diversity of beetles may be an artifact of the collection and research biases of entomologists. In some well-studied temperate regions such as Britain and Canada, species of true flies (order Diptera) appear to outnumber beetles. Recent studies of canopy insects of the tropical island of Borneo have shown that both Hymenoptera (wasps, ants and bees) and Diptera can be more diverse than the Coleoptera. Comprehensive regional inventories or credible estimates of insect faunal diversity may eventually tell us which order of insects is globally most diverse.

Whether we estimate 30–80 million species or an order of magnitude less, insects constitute around half of global species diversity (Fig. 1.3). If we consider life on land only, insects comprise an even greater proportion of living species, since the radiation of insects is a predominantly terrestrial phenomenon. Our perception of the relative importance of insects to global diversity will be somewhat lessened if marine diversity, to which insects make a negligible contribution, is shown to be much higher than previously believed. Recent deep-sea surveys, which revealed nearly 800 species of animals living in one small area of ocean floor off the eastern coast of the United States, allow estimates of a total marine diversity of 10 million species, although this extrapolation is disputed.

1.3.2 Some reasons for insect diversity

Whatever the global estimate, the insects constitute a remarkably speciose group of organisms (Fig. 1.3). Their high species diversity has been attributed to several factors. The small size of insects, a limitation imposed by their method of gas exchange via tracheae, is one important determinant of species richness. There are many more niches in any given environment for small organisms than for large organisms. Thus a single acacia tree, which could provide one meal for a giraffe, may support the complete life cycle of dozens of insect species, such as a lycaenid butterfly whose larvae chew the leaves, a bug that sucks the stem sap, a longicorn beetle that bores into the wood, a midge that galls the flower buds, a bruchid beetle that destroys the seeds, a mealybug that sucks the root sap, and

5 Protozoa
6 Porifera (sponges)
7 Cnidaria (jellyfish, corals, etc.)
8 Platyhelminthes (flatworms)
9 Nematoda (roundworms)
10 Annelida (earthworms, leeches, etc.)
11 Mollusca (snails, bivalves, octopus, etc.)
12 Echinodermata (starfish, sea urchins, etc.)
13 Insecta
14 Non-insect Arthropoda
15 Pisces (fish)
16 Amphibia (frogs, salamanders, etc.)
17 Reptilia (snakes, lizards, turtles)
18 Aves (birds)
19 Mammalia (mammals)

1 Monera (bacteria, blue-green algae)
2 Fungi
3 Algae
4 Plantae (multicellular plants)

Fig. 1.3 Speciescape, in which the size of individual organisms is approximately proportional to the number of described species in the higher taxon that it represents. (After Wheeler, 1990.)

several wasp species each of which parasitizes one of these host-specific phytophages. An adjacent acacia of a different species feeds the same giraffe but may have an entirely different suite of phytophagous insects. The environment can be said to be more fine-grained from an insect perspective than from the perspective of a mammal or bird.

Small size alone is insufficient to allow exploitation of this environmental heterogeneity, since organisms must be capable of recognizing and responding to environmental differences. Insects have more highly organized sensory and neuro-motor systems than most other invertebrates, perhaps comparable to those of vertebrate animals. In addition, insects differ from vertebrates not only in size but also in how they respond to environmental change. Generally, vertebrate animals are longer lived than insects and individuals can adapt to change by some degree of learning. Insects, on the other hand, normally respond to or cope with altered conditions (e.g. the application of insecticides to their host plant) by genetic change (e.g. leading to insecticide resistance). High genetic heterogeneity or elasticity within insect species allows persistence in the face of environmental change. Persistence exposes species to the processes that promote speciation, predominantly involving phases of range expansion and subsequent fragmentation. Either selection pressures or stochastic processes may result in relatively rapid genetic alterations that may become fixed in spatially or temporally isolated populations.

All of the above characteristics of insects make possible another potential diversifying influence that may account for the species richness of whole groups of insects. Interactions with other organisms, such as plants in the case of herbivorous insects or hosts for parasitic insects, are thought to promote genetic diversification of eater and eaten. These interactions are often called coevolution and are discussed in more detail in Chapters 7 and 10. The reciprocal nature of such interactions may speed up evolutionary change in one or both partners or sets of partners, perhaps even leading to major radiations in particular groups.

Comparisons among insect groups and between insects and their close relatives can help us understand the reasons for insect diversity. What are the shared characteristics of the most speciose insect orders, the Coleoptera, Hymenoptera, Diptera and Lepidoptera? Which features of insects are lacking in other arthropods, such as arachnids (spiders, mites, scorpions and their allies)? No simple explanation emerges from such comparisons; design features, flexible life-cycle patterns and feeding habits probably all play a part and

some of these factors are explored more fully in Chapter 7. It is notable that arachnids, in contrast to most successful insect groups, lack winged flight, lack the complete transformation of body form during development (metamorphosis) and usually lack dependence on specific food organisms. Mites, the most diverse and abundant of arachnids, are exceptional in having many very specific associations with other living organisms.

High persistence of species or lineages or numerical abundance of individual species all are taken as indicators of insect success. However, insects would be deemed less successful than vertebrates by at least one popular measure of success, that of large body size. The greatest wingspan of a living insect belongs to an dragonfly with a span of 15 cm but fossils show us that in the past some insects were appreciably larger than their extant relatives. For example, an Upper Carboniferous silverfish, *Ramsdelepion schusteri*, had a body length of 6 cm and many winged Carboniferous insects had wingspans exceeding 45 cm. A Permian dragonfly, *Meganeuropsis americana*, had a wingspan of 71 cm. Notably amongst these large insects, the great size comes predominantly with a narrow, elongate body, although the heaviest extant insect, the rhinocerus beetle *Dynastes hercules* (16 cm long), is an exception in having a bulky body. The majority of insects fall into the size range of body lengths of some 1–10 mm but some have succeeded in miniaturization and may be very small: mymarid wasps (parasitic on insect eggs) with a body length of around 0.3 mm are not exceptional.

The barriers to large size include the limitation of the tracheal system to diffuse gases across extended distances from active muscles through the spiracles to the external environment (Box 3.2). Further elaborations of the tracheal system could not be made without jeopardizing water balance in a large insect. Most large insects, being narrow, have not extended greatly the maximum distance between the external oxygen source and the muscular site of gaseous exchange, compared with a smaller insect.

In summary, the radiation of many insect lineages probably depended upon: (a) small size of individuals, combined with (b) short generation time, (c) sensory and neuro-motor sophistication, (d) evolutionary interactions with plants and other organisms, (e) metamorphosis and (f) winged adults. The substantial time since the origin of each of the major insect groups has allowed ample opportunity for lineage diversification (Chapter 7). Present-day species diversity results from either higher rates of speciation (for which there is only limited evidence) and/or lower rates of species extinction (higher persistence) than other organisms.

1.4 INSECT DISTRIBUTIONS

One of the most striking observations is that different plants and animals live in different countries. This is more than a matter of differing climates and ecologies: for example, Australia has suitable trees but no woodpeckers, tropical rainforests without monkeys and prairie grasslands without native ungulates. Non-American deserts lack cacti but have a range of ecological analogues including succulent euphorbs. Biogeography is the study of these distributions and the past historical and current ecological explanations for them. Insects, no less than plants and vertebrates, show patterns of restriction to one geographic area (**endemism**) and entomologists have been and remain amongst the most prominent biogeographers.

Humans, however, have been responsible for the loss of much endemism and the expansion of ranges of species such that many of our most familiar insects are **cosmopolitan** (that is, virtually worldwide) in distribution. There are at least five explanations for this expansion of so many insects of previously restricted distribution into cosmopolitanism.

- Many human-loving (**anthropophilic**) insects such as many cockroaches, silverfish and house flies accompany humans virtually everywhere.
- Humans create disturbed habitats wherever they live and some **synanthropic** (human-associated) insects act rather like weed plants and are able to take advantage of disturbed conditions better than native species can. Synanthropy is a weaker association with humans than anthropophily.
- The insect (and other arthropod) external parasites (**ectoparasites**) and internal parasites (**endoparasites**) of humans and domesticated animals are often cosmopolitan.
- Humans rely on few food crops, most of which are cultivated very widely. Plant-feeding (**phytophagous**) insects associated with plant species once localized but now domesticated and disseminated by humans can follow the crop and, as pests, spread and damage the crop wherever it is cultivated: many have done so.
- Insects have expanded their ranges by deliberate **anthropogenic** (aided by humans) introduction of selected species as **biological control** agents to control pest plants and animals, including other insects.

Attempts are made to restrict the shipment of agricultural and veterinary pests through quarantine regulations, but much of the mixing of insect faunas took place before effective measures were implemented. Thus pest insects tend to be identical throughout climatically similar parts of the world and applied entomologists must take a worldwide perspective in their studies.

1.5 NAMING AND CLASSIFICATION OF INSECTS

The formal naming of insects follows the rules of **nomenclature** developed for all animals (plants have a slightly different system). Formal scientific names are required for unambiguous communication between all scientists no matter what their native language amongst the thousands used worldwide. Vernacular (common) names cannot fulfil this need: the same insects may have different vernacular names even amongst peoples that speak the same language. For instance, the British refer to ladybirds whereas the same coccinellid beetles are known to Americans as ladybugs. Many insects have no vernacular name, or one common name is given to many species as if only one is involved. These difficulties are addressed by the **Linnaean** system, in which every species described is given two names. The first is the **generic** name, used for a usually broader grouping than the second name, which is the **specific** name. These latinized names always are used together and are italicized, as in this book. The combination of the generic and specific names provides a unique name for every organism. Thus the name *Aedes aegypti* is recognised by any medical entomologist, anywhere, whatever the local name (and there are many) for this disease-transmitting mosquito. In scientific publications, the species name often is followed by the name of the original describer of the species and perhaps also the year in which the name was published legally. In this textbook we do not follow this practice but, when particular insects are discussed, we give the order and family names to which the species belongs.

In publications, after first citation of the combination of generic and species names in the text, it is common practice in subsequent citation to abbreviate the genus to the initial letter only (e.g. *A. aegypti*). However, where this might be ambiguous, such as for two mosquito genera *Aedes* and *Anopheles*, the initial two letters *Ae.* and *An.* are used, as in Chapter 14.

Various groups, also called **taxa**, are recognized amongst the insects. As for all other organisms, the basic biological taxon, lying above the individual and population, is the **species**, which is both the fundamental nomenclatural unit in taxonomy and, arguably, a unit of evolution. Multi-species studies

Table 1.2 Taxonomic categories (obligatory categories are shown in bold)

Taxon category	Standard suffix
Order	
Suborder	
Superfamily	-oidea
Family	-idae
Subfamily	-inae
Tribe	-ini
Genus	
Subgenus	
Species	

allow recognition of **genera**, which are more or less discrete higher groups. In similar manner, genera can be grouped into **tribes**, tribes into subfamilies, and subfamilies into **families**. The families of insects are placed in relatively large but easily recognized groups called **orders**. This hierarchy of **ranks** (or categories) thus extends from the species level through a series of 'higher' levels of greater and greater inclusivity until all true insects are included in one **class**, the Insecta. There are standard suffixes for certain ranks in the taxonomic hierarchy, so that the rank of some group names can be recognized by inspection of the ending (Table 1.2).

According to the classification system used, some 26 to 29 orders of Insecta may be recognized. Differences arise principally because there are no hard-and-fast rules for deciding the taxonomic ranks cited above. There is only a general agreement that groups should be monophyletic, comprising all the descendants of a common ancestor (Chapter 7). Orders have been recognized rather arbitrarily in the past two centuries, and the most that can be said is that presently constituted orders contain similar insects sharing major differences from other insect groups. Over time, a relatively stable classification system has developed but there remain differences of opinion as to the boundaries around groups, with 'splitters' recognizing a greater number

Table 1.3 Summary of diagnostic features of the three non-insect hexapod orders and the 29 orders of Insecta

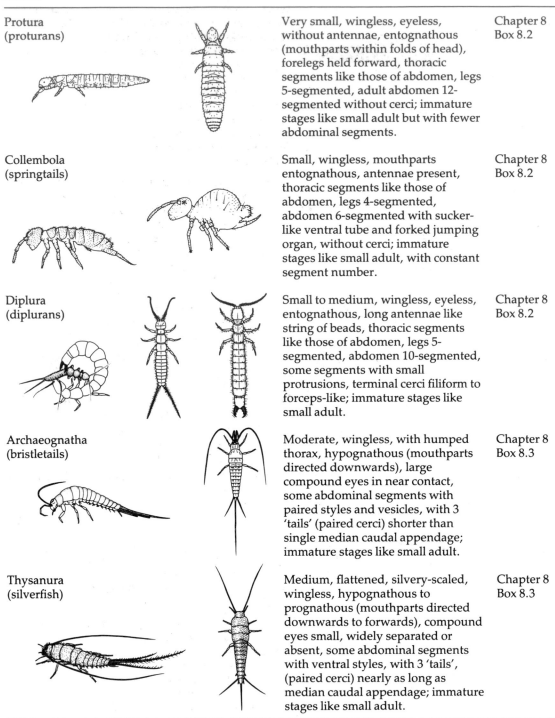

Protura (proturans)	Very small, wingless, eyeless, without antennae, entognathous (mouthparts within folds of head), forelegs held forward, thoracic segments like those of abdomen, legs 5-segmented, adult abdomen 12-segmented without cerci; immature stages like small adult but with fewer abdominal segments.	Chapter 8 Box 8.2
Collembola (springtails)	Small, wingless, mouthparts entognathous, antennae present, thoracic segments like those of abdomen, legs 4-segmented, abdomen 6-segmented with sucker-like ventral tube and forked jumping organ, without cerci; immature stages like small adult, with constant segment number.	Chapter 8 Box 8.2
Diplura (diplurans)	Small to medium, wingless, eyeless, entognathous, long antennae like string of beads, thoracic segments like those of abdomen, legs 5-segmented, abdomen 10-segmented, some segments with small protrusions, terminal cerci filiform to forceps-like; immature stages like small adult.	Chapter 8 Box 8.2
Archaeognatha (bristletails)	Moderate, wingless, with humped thorax, hypognathous (mouthparts directed downwards), large compound eyes in near contact, some abdominal segments with paired styles and vesicles, with 3 'tails' (paired cerci) shorter than single median caudal appendage; immature stages like small adult.	Chapter 8 Box 8.3
Thysanura (silverfish)	Medium, flattened, silvery-scaled, wingless, hypognathous to prognathous (mouthparts directed downwards to forwards), compound eyes small, widely separated or absent, some abdominal segments with ventral styles, with 3 'tails', (paired cerci) nearly as long as median caudal appendage; immature stages like small adult.	Chapter 8 Box 8.3

Table 1.3   Cont'd.

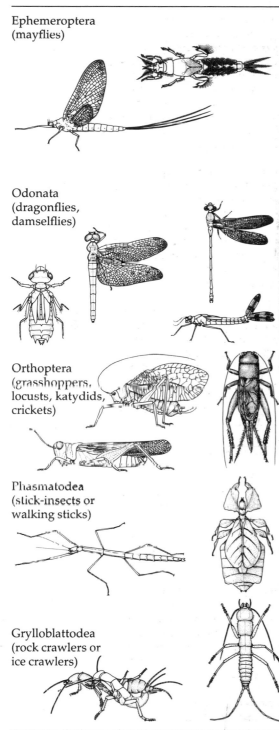

Ephemeroptera (mayflies)	Small to large, winged with large triangular fore wings and smaller hind wings, mouthparts reduced, compound eyes large, long filiform antennae, abdomen slender compared to stout thorax, with 3 'tails' (paired cerci) often as long as median caudal appendage; immature stages (nymphs) aquatic, with 3 'tails' and plate-like abdominal gills, penultimate instar a winged subimago.	Chapter 9 Box 9.1
Odonata (dragonflies, damselflies)	Medium to large, winged, with fore and hind wings equal (Zygoptera) or hind wings wider than fore wings (Anisoptera), head mobile, with large compound eyes separated (Zygoptera) or nearly in contact (Anisoptera), mouthparts mandibulate, antennae short, thorax stout, abdomen slender; immature stages (larvae) aquatic, stout or narrow, with extensible labial 'mask', terminal or rectal gills.	Chapter 9 Boxes 5.3, 9.2
Orthoptera (grasshoppers, locusts, katydids, crickets)	Medium to large, hypognathous, usually winged, fore wings forming leathery tegmina, hind wings broad, at rest pleated beneath tegmina, pronotum curved over pleura, hind legs often enlarged for jumping, cerci 1-segmented; immature stages (nymphs) like small adults.	Chapters 5, 6, 10 Boxes 5.2, 10.5
Phasmatodea (stick-insects or walking sticks)	Medium to large, cylindrical stick-like or flattened leaf-like, prognathous, mandibulate, compound eyes small and laterally placed, fore wings from leathery tegmina, hind wings broad with toughened fore margin, legs elongate for walking, cerci 1-segmented; immature stages (nymphs) like small adults.	Chapters 10, 13 Box 10.6
Grylloblattodea (rock crawlers or ice crawlers)	Medium, soft-bodied, elongate, pale, wingless and often eyeless, prognathous, with stout coxae on legs adapted for running, cerci 5–9-segmented, female with short ovipositor; immature stages (nymphs) like small pale adults.	Chapter 8 Box 8.4

Table 1.3   Cont'd.

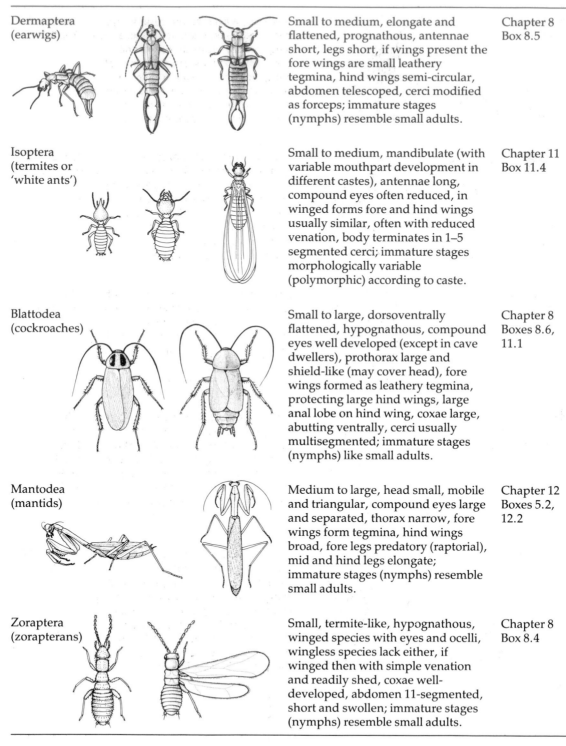

Dermaptera (earwigs)	Small to medium, elongate and flattened, prognathous, antennae short, legs short, if wings present the fore wings are small leathery tegmina, hind wings semi-circular, abdomen telescoped, cerci modified as forceps; immature stages (nymphs) resemble small adults.	Chapter 8 Box 8.5
Isoptera (termites or 'white ants')	Small to medium, mandibulate (with variable mouthpart development in different castes), antennae long, compound eyes often reduced, in winged forms fore and hind wings usually similar, often with reduced venation, body terminates in 1–5 segmented cerci; immature stages morphologically variable (polymorphic) according to caste.	Chapter 11 Box 11.4
Blattodea (cockroaches)	Small to large, dorsoventrally flattened, hypognathous, compound eyes well developed (except in cave dwellers), prothorax large and shield-like (may cover head), fore wings formed as leathery tegmina, protecting large hind wings, large anal lobe on hind wing, coxae large, abutting ventrally, cerci usually multisegmented; immature stages (nymphs) like small adults.	Chapter 8 Boxes 8.6, 11.1
Mantodea (mantids)	Medium to large, head small, mobile and triangular, compound eyes large and separated, thorax narrow, fore wings form tegmina, hind wings broad, fore legs predatory (raptorial), mid and hind legs elongate; immature stages (nymphs) resemble small adults.	Chapter 12 Boxes 5.2, 12.2
Zoraptera (zorapterans)	Small, termite-like, hypognathous, winged species with eyes and ocelli, wingless species lack either, if winged then with simple venation and readily shed, coxae well-developed, abdomen 11-segmented, short and swollen; immature stages (nymphs) resemble small adults.	Chapter 8 Box 8.4

Table 1.3 Cont'd.

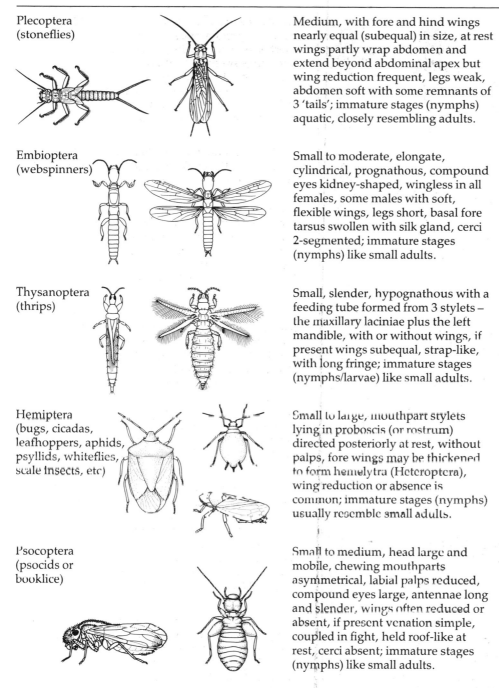

Plecoptera (stoneflies)	Medium, with fore and hind wings nearly equal (subequal) in size, at rest wings partly wrap abdomen and extend beyond abdominal apex but wing reduction frequent, legs weak, abdomen soft with some remnants of 3 'tails'; immature stages (nymphs) aquatic, closely resembling adults.	Chapter 9 Box 9.3
Embioptera (webspinners)	Small to moderate, elongate, cylindrical, prognathous, compound eyes kidney-shaped, wingless in all females, some males with soft, flexible wings, legs short, basal fore tarsus swollen with silk gland, cerci 2-segmented; immature stages (nymphs) like small adults.	Chapter 8 Box 8.4
Thysanoptera (thrips)	Small, slender, hypognathous with a feeding tube formed from 3 stylets – the maxillary laciniae plus the left mandible, with or without wings, if present wings subequal, strap-like, with long fringe; immature stages (nymphs/larvae) like small adults.	Chapter 10 Box 10.7
Hemiptera (bugs, cicadas, leafhoppers, aphids, psyllids, whiteflies, scale insects, etc)	Small to large, mouthpart stylets lying in proboscis (or rostrum) directed posteriorly at rest, without palps, fore wings may be thickened to form hemelytra (Heteroptera), wing reduction or absence is common; immature stages (nymphs) usually resemble small adults.	Chapters 9, 10 Boxes 3.3, 8.1, 9.5, 10.8, 15.4, 15.5
Psocoptera (psocids or booklice)	Small to medium, head large and mobile, chewing mouthparts asymmetrical, labial palps reduced, compound eyes large, antennae long and slender, wings often reduced or absent, if present venation simple, coupled in fight, held roof-like at rest, cerci absent; immature stages (nymphs) like small adults.	Chapter 10 Box 10.9

Table 1.3   Cont'd.

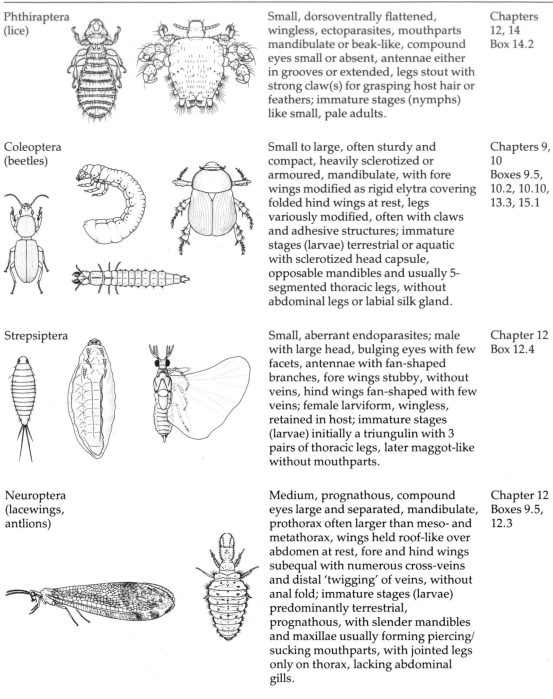

Phthiraptera (lice)	Small, dorsoventrally flattened, wingless, ectoparasites, mouthparts mandibulate or beak-like, compound eyes small or absent, antennae either in grooves or extended, legs stout with strong claw(s) for grasping host hair or feathers; immature stages (nymphs) like small, pale adults.	Chapters 12, 14 Box 14.2
Coleoptera (beetles)	Small to large, often sturdy and compact, heavily sclerotized or armoured, mandibulate, with fore wings modified as rigid elytra covering folded hind wings at rest, legs variously modified, often with claws and adhesive structures; immature stages (larvae) terrestrial or aquatic with sclerotized head capsule, opposable mandibles and usually 5-segmented thoracic legs, without abdominal legs or labial silk gland.	Chapters 9, 10 Boxes 9.5, 10.2, 10.10, 13.3, 15.1
Strepsiptera	Small, aberrant endoparasites; male with large head, bulging eyes with few facets, antennae with fan-shaped branches, fore wings stubby, without veins, hind wings fan-shaped with few veins; female larviform, wingless, retained in host; immature stages (larvae) initially a triungulin with 3 pairs of thoracic legs, later maggot-like without mouthparts.	Chapter 12 Box 12.4
Neuroptera (lacewings, antlions)	Medium, prognathous, compound eyes large and separated, mandibulate, prothorax often larger than meso- and metathorax, wings held roof-like over abdomen at rest, fore and hind wings subequal with numerous cross-veins and distal 'twigging' of veins, without anal fold; immature stages (larvae) predominantly terrestrial, prognathous, with slender mandibles and maxillae usually forming piercing/sucking mouthparts, with jointed legs only on thorax, lacking abdominal gills.	Chapter 12 Boxes 9.5, 12.3

Table 1.3 Cont'd.

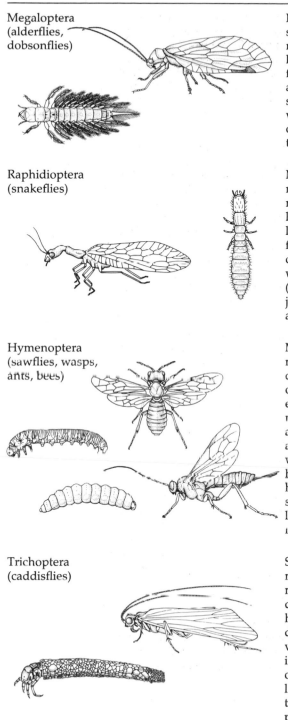

Megaloptera (alderflies, dobsonflies)	Medium, compound eyes large and separated, prognathous, mandibulate, prothorax only slightly longer than meso- and metathorax, fore and hind wings subequal with anal fold in the hind wing; immature stages (larvae) aquatic, prognathous, with stout mandibles, jointed legs only on thorax, with abdominal tracheal gills.	Chapter 12 Boxes 9.5, 12.3
Raphidioptera (snakeflies)	Medium, prognathous, mandibulate, antennae multisegmented, compound eyes large and separated, prothorax much longer than meso- and metathorax, fore wings rather longer than otherwise similar hind wings, without anal fold; immature stages (larvae) terrestrial, prognathous with jointed legs only on thorax, without abdominal gills.	Chapter 12 Box 12.3
Hymenoptera (sawflies, wasps, ants, bees)	Minute to large, mouthparts mandibulate to sucking and chewing, antennae multisegmented often long and held forward, thorax either 3-segmented or forms a mesosoma by incorporation of 1st abdominal segment in which case the abdomen is petiolate (waisted), wings with simple venation, fore and hind wings coupled together by hooks on hind wing; immature stages (larvae) very variable, many lack legs completely, all have distinct mandibles even if head is reduced.	Chapters 10, 11, 12 Boxes 10.3, 10.4, 11.2, 11.3, 12.1
Trichoptera (caddisflies)	Small to large, with long, multisegmented antennae, reduced mouthparts (no proboscis), but well developed maxillary and labial palps, hairy (or rarely scaly) wings, lacking discal cell and with fore wing anal veins looped (see Lepidoptera); immature stages (larvae) aquatic, often case-bearing, but many free-living, with 3 pairs of segmented thoracic legs and lacking abdominal prolegs.	Chapter 9 Box 9.4

Table 1.3   Cont'd.

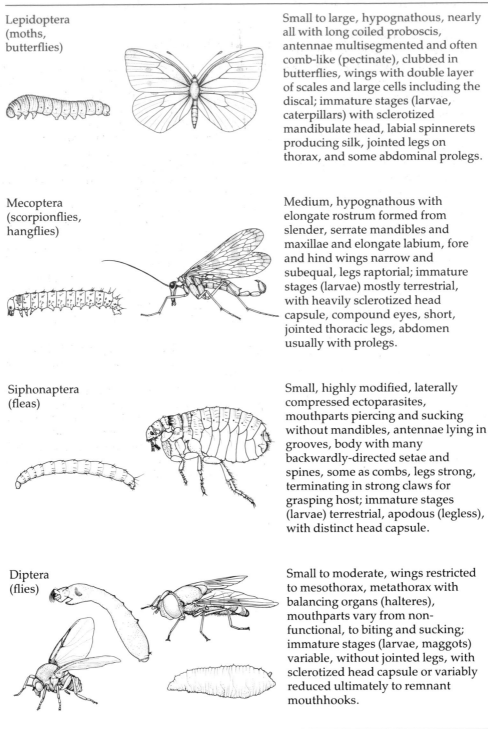

Lepidoptera (moths, butterflies)	Small to large, hypognathous, nearly all with long coiled proboscis, antennae multisegmented and often comb-like (pectinate), clubbed in butterflies, wings with double layer of scales and large cells including the discal; immature stages (larvae, caterpillars) with sclerotized mandibulate head, labial spinnerets producing silk, jointed legs on thorax, and some abdominal prolegs.	Chapters 10, 13 Boxes 10.11, 13.1
Mecoptera (scorpionflies, hangflies)	Medium, hypognathous with elongate rostrum formed from slender, serrate mandibles and maxillae and elongate labium, fore and hind wings narrow and subequal, legs raptorial; immature stages (larvae) mostly terrestrial, with heavily sclerotized head capsule, compound eyes, short, jointed thoracic legs, abdomen usually with prolegs.	Chapters 5, 12 Boxes 5.1, 12.5
Siphonaptera (fleas)	Small, highly modified, laterally compressed ectoparasites, mouthparts piercing and sucking without mandibles, antennae lying in grooves, body with many backwardly-directed setae and spines, some as combs, legs strong, terminating in strong claws for grasping host; immature stages (larvae) terrestrial, apodous (legless), with distinct head capsule.	Chapter 14 Box 14.3
Diptera (flies)	Small to moderate, wings restricted to mesothorax, metathorax with balancing organs (halteres), mouthparts vary from non-functional, to biting and sucking; immature stages (larvae, maggots) variable, without jointed legs, with sclerotized head capsule or variably reduced ultimately to remnant mouthhooks.	Chapters 9, 14 Boxes 5.4, 6.1, 6.2, 9.5, 14.1, 14.4

of groups and 'lumpers' favouring broader categories. For example, some American taxonomists group ('lump') the alderflies, dobsonflies, snakeflies and lacewings into one order, the Neuroptera, whereas others, including ourselves, 'split' the group and recognize three separate (but related) orders, Megaloptera, Raphidioptera and a more narrowly defined Neuroptera. Sometimes the order Hemiptera is split into two orders, Homoptera and Heteroptera, although the validity (monophyly) of the homopteran group is questionable (Box 10.8).

In this book we recognise 29 orders for which the physical characteristics and biologies of their constituent taxa are described, and their relationships considered. Amongst these orders, we distinguish 'major' orders, based upon the numbers of species being appreciably greater in the Coleoptera, Diptera, Lepidoptera, Hymenoptera and Hemiptera than in the remaining 'minor' orders. Furthermore, the minor orders often have quite homogeneous ecologies which can be summarized conveniently in a single descriptive/ecological box following the appropriate ecologically based chapter (Chapters 8–14). The major orders are less readily summarized ecologically and information may appear in two chapters. A summary of the diagnostic features of all 29 insect orders plus the three non-insect hexapod orders (Protura, Collembola and Diplura) and cross-references to fuller identification and ecological information is shown in Table 1.3.

FURTHER READING

Borror, D.J., Triplehorn, C.A. and Johnson, N.F (1989) *An Introduction to the Study of Insects*, 6th edn, Saunders College Publishing, Philadelphia, Fort Worth, Chicago, San Francisco, Montreal, Toronto, London, Sydney, Tokyo.

CSIRO (1991) *The Insects of Australia*, 2nd edn, Melbourne University Press, Carlton.

DeFoliart, G.R. (ed.) (1988–) *The Food Insects Newsletter*, Department of Entomology, University of Wisconsin-Madison, USA

DeFoliart, G.R. (1989) The human use of insects as food and as animal feed. *Bulletin of the Entomological Society of America*, **35**, 22–35.

DeFoliart, G. (1992) Insects as human food. *Crop Protection*, **11**, 395–99.

Gaston, K.J. (1991) The magnitude of global insect species richness. *Conservation Biology*, **5**, 283–96.

Matthews, E.G. and Kitching, R.L. (1984) *Insect Ecology*, 2nd edn, University of Queensland Press, St. Lucia.

May, R.M. (1990) How many species? *Philosophical Transactions of the Royal Society of London*, B**330**, 293–304.

May, R.M. (1992) How many species inhabit the earth? *Scientific American*, **267(4)**, 18–24.

Vane-Wright, R.I. (1991) Why not eat insects? *Bulletin of Entomological Research*, **81**, 1–4.

Wheeler, Q.D. (1990) Insect diversity and cladistic constraints. *Annals of the Entomological Society of America*, **83**, 1031–47.

2
EXTERNAL ANATOMY

'Foot' of a chrysomelid beetle (from SEM)

Insects are segmented invertebrates that possess the articulated external skeleton (exoskeleton) characteristic of all arthropods. Groups are differentiated by various modifications of the exoskeleton and the appendages – for example, the class Hexapoda (to which the Insecta belong) is characterised by having six-legged adults. Many anatomical features of the appendages, especially of the mouthparts, legs and abdominal apex, are important in recognizing the higher groups within the hexapods, including insect orders, families and genera. Differences between species frequently are indicated by anatomical differences. Clearly an understanding of external anatomy is necessary to interpret these various designs and to allow the identification of insects and their hexapod relatives. In this chapter, we describe and discuss the cuticle, body segmentation, the structure of the head, thorax and abdomen and their appendages. However, some basic classification and terminology need to be discussed before we commence.

Adult insects normally have wings (the Pterygota), the structure of which may diagnose orders, but there is a group of primitively wingless insects (the 'apterygotes') (see section 7.5 for defining features). Within the Insecta, three major patterns of development can be recognized; these will be referred to occasionally in this chapter but are discussed in detail in section 6.2. Apterygotes (and non-insect hexapods) develop to adulthood with little change in body form (**ametaboly**), except for sexual maturation through development of gonads and genitalia. All other insects either have a gradual change in body form (**hemimetaboly**) with external wing buds getting larger at each moult, or a drastic change from a wingless immature to winged adult stage that occurs via a pupal stage (**holometaboly**). Immature hemimetabolous insects are called **nymphs**, whereas **larvae** are the equivalent for holometabolous insects.

Anatomical structures of different taxa are **homologous** if they share an evolutionary origin, that is, if the genetic basis is inherited from the most recent common ancestor. For instance, the wings of all insects are believed to be homologous; this means that wings (but not necessarily flight; section 7.4) had a single origin. Generally, homology of structures is inferred by comparison of similarity in **ontogeny** (development from egg to adult), composition (size and detailed appearance) and position (on the same segment and same relative location on that segment). The homology of all insect wings is demonstrated by similarities in venation and articulation – the wings of all insects can be derived from the same basic pattern or groundplan (as explained in section 2.4.2). Sometimes association with other structures of known homologies is helpful in establishing the homology of a structure of uncertain origin. Another sort of homology, called **serial homology**, refers to the same structures on different segments of an individual insect. Thus, the appendages of each body segment are serially homologous, although in living insects those on the head (antennae and mouthparts) are very different in appearance from those on the thorax (walking legs) and abdomen (genitalia and cerci).

2.1 THE CUTICLE

The cuticle is a key contributor to the success of the Insecta. This inert layer provides the strong **exoskeleton** of body and limbs, the **apodemes** (internal supports and muscle attachments) and the material of the wings of pterygote insects, and acts as a barrier between living tissues and the environment. Internally, cuticle lines the tracheal tubes (section 3.5), some gland ducts and the foregut and hindgut of the digestive tract. Cuticle may range from rigid and armour-like, as in most adult beetles, to thin and flexible, as in many immature insects. A critical function of cuticle, the restriction of water loss, is vital to the success of insects on land.

cement layer
wax layer
superficial layer
outer epicuticle

wax filament

inner epicuticle

pore canal

exocuticle

dermal gland
duct opening

epicuticle

exocuticle

procuticle

endocuticle

formation
zone

epidermis

basement membrane

dermal
gland

Fig. 2.1 The general structure of insect cuticle; the enlargement above shows details of the epicuticle. (After Hepburn, 1985; Hadley, 1986; and Binnington, 1993.)

The cuticle is thin but its structure is complex and still the subject of some controversy. A single layer of cells, the **epidermis**, lies beneath and secretes the cuticle, which consists of a thicker **procuticle** overlaid with thin **epicuticle** (Fig. 2.1). The epidermis and cuticle together form an **integument** – the outer covering of the living tissues of an insect.

The epicuticle ranges from 3 μm down to 0.1 μm in thickness, and usually consists of three layers: an **inner epicuticle**, an **outer epicuticle** and a **superficial layer**. In many insects, the superficial layer (probably glycoprotein) is covered by a lipid or wax layer and, external to this, a variably discrete cement layer. The chemistry of the epicuticle and its outer layers is vital in preventing dehydration, a function derived from water-repelling (hydrophobic) lipids, especially hydrocarbons. These compounds include free and protein-bound lipids and the outermost waxy coatings give a bloom to the external surface of some insects. Other cuticular patterns, such as light reflectivity, are produced by various kinds of epicuticular surface microsculpturing, such as close-packed, regular or irregular tubercles, ridges or tiny hairs. Lipid composition can vary and waxiness can increase seasonally or under dry conditions. Besides being water retentive, surface waxes may deter predation, provide patterns for mimicry or camouflage, repel excess rainwater, reflect solar and ultraviolet radiation or give species-specific olfactory cues.

The epicuticle is inextensible and unsupportive. Instead, support is provided by the underlying procuticle, which comprises a thicker **endocuticle** covered by a thinner **exocuticle**, differentiated by sclerotization of the latter. The procuticle is from 10 μm to 0.5 mm thick and consists primarily of chitin complexed with protein.

Chitin is found as a supporting element in fungal cell walls and arthropod exoskeletons, and is especially important in insect extracellular structures. It is an unbranched polymer of high molecular weight – an

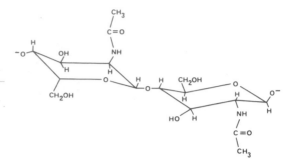

Fig. 2.2 Structure of part of a chitin chain, showing two linked units of *N*-acetyl-D-glucosamine. (After Cohen, 1991.)

amino-sugar polysaccharide composed of β(1–4) linked units of *N*-acetyl-D-glucosamine (Fig. 2.2). Chitin molecules are grouped into bundles and assembled into flexible microfibrils which are embedded in and intimately linked to a protein matrix, giving great tensile strength. The commonest arrangement of chitin microfibrils is in a sheet, in which the microfibrils are in parallel. In the exocuticle, each successive sheet lies in the same plane but may be orientated at a slight angle relative to the previous sheet, such that a thickness of many sheets produces a helicoid arrangement, which in sectioned cuticle appears as alternating light and dark bands (**lamellae**). Thus the parabolic patterns and lamellar arrangement, visible so clearly in sectioned cuticle, represent an optical artifact resulting from microfibrillar orientation (Fig. 2.3). In the endocuticle, alternate stacked or helicoid arrangements of microfibrillar sheets may occur, often giving rise to thicker lamellae than in the exocuticle. Different arrangements may be laid down during darkness compared with daylight, allowing precise age determination in many adult insects.

Much of the strength of cuticle comes from extensive hydrogen bonding of adjacent chitin chains. Additional stiffening comes from **sclerotization**, an irreversible process that darkens the exocuticle and results in the proteins becoming water-insoluble. Sclerotization may result from linkages of adjacent

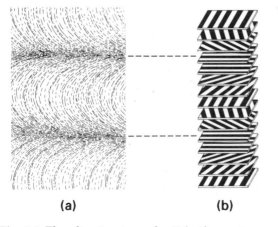

(a)   **(b)**

Fig. 2.3 The ultrastructure of cuticle (from a transmission electron micrograph): (a) the arrangement of chitin microfibrils in a helicoidal array produces characteristic (though artifactual) parabolic patterns; (b) diagram of how the rotation of microfibrils produces a lamellar effect due to microfibrils being either aligned or non-aligned to the plane of sectioning. (After Filshie, 1982.)

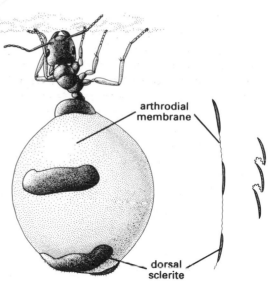

Fig. 2.4 A specialized worker, or replete, of the honeypot ant *Camponotus inflatus* (Hymenoptera: Formicidae), which holds honey in its distensible abdomen and acts as a food store for the colony. The arthrodial membrane between tergal plates is depicted to the right in its unfolded and folded conditions. (After Hadley, 1986; Devitt, 1989.)

protein chains by phenolic bridges (quinone tanning), or from controlled dehydration of the chains, or both. Only exocuticle becomes sclerotized. The deposition of pigment in the cuticle, including deposition of melanin, may be associated with quinones but is additional to sclerotization and not necessarily associated with it.

In contrast to the solid cuticle typical of sclerites and mouthparts such as mandibles, softer, plastic, highly flexible or truly elastic cuticles occur in insects in varying locations and proportions. Where elastic or spring-like movement occurs, such as in wing ligaments or for the jump of a flea, **resilin** – a 'rubber-like' protein – is present. The coiled polypeptide chains of this protein function as a mechanical spring under tension or compression, or in bending.

In soft-bodied larvae and in the membranes between segments, the cuticle must be tough but must have flexibility and extensibility. This 'soft' cuticle, sometimes termed **arthrodial membrane**, is evident in gravid females, for example, in the ovipositing

migratory locust, *Locusta migratoria* (Orthoptera: Acrididae), in which intersegmental membranes may be stretched up to 20-fold for oviposition, and in *Rhodnius* nymphs (Hemiptera: Reduviidae), in which membrane stretching occurs in response to intake of large fluid volume as, for example, in feeding. Similarly, the gross abdominal dilation of gravid queen bees, termites and ants is possible through expansion of this unsclerotized cuticle. The overlying unstretchable epicuticle expands by unfolding from an originally highly folded state, and some new epicuticle is formed. An extreme example of the distensibility of arthrodial membrane is seen in honeypot ants (Fig. 2.4; also section 11.2.3(b)).

Cuticular structural components, waxes, cements, pheromones (Chapter 4) and defensive and other compounds are produced by the epidermis, a virtually continuous single-celled layer beneath the cuticle. Many of these

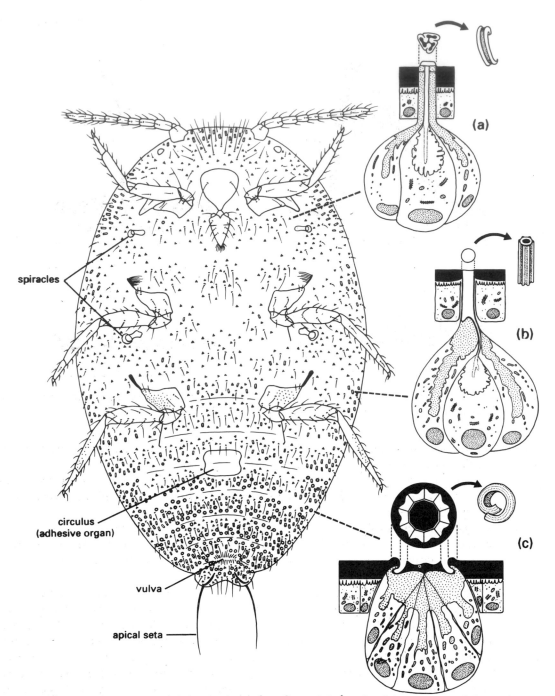

Fig. 2.5 The cuticular pores and ducts on the venter of an adult female of the citrus mealybug, *Planococcus citri* (Hemiptera: Pseudococcidae). Enlargements depict the ultrastructure of the wax glands and the various wax secretions (arrowed) associated with three types of cuticular structure: (a) a trilocular pore; (b) a tubular duct; (c) a multilocular pore. Curled filaments of wax from the trilocular pores form a protective body covering and prevent contamination with their own sugary excreta, or honeydew; long hollow and shorter curled filaments from the tubular ducts and multilocular pores, respectively, form the ovisac. (After Foldi, 1983; Cox, 1987.)

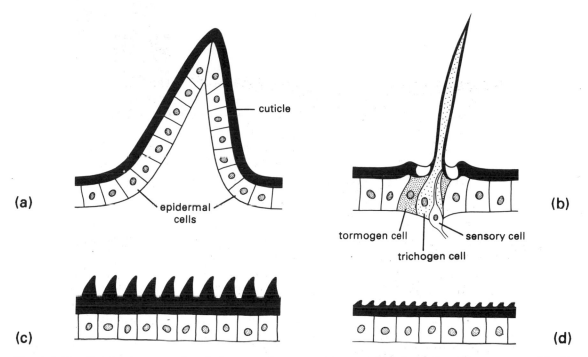

Fig. 2.6 The four basic types of cuticular protuberances: (a) a multicellular spine; (b) a seta, or trichoid sensillum; (c) acanthae; (d) microtrichia. (After Richards and Richards, 1979.)

compounds are secreted to the outside of the insect epicuticle. Numerous fine **pore canals** traverse the procuticle and then branch into numerous finer **wax canals** (or **wax filaments**) within the epicuticle (enlargement in Fig. 2.1); this system transports lipids (waxes) from the epidermis to the epicuticular surface. **Dermal glands**, part of the epidermis, produce cement and/or wax which are transported via larger ducts to the cuticular surface. Wax-secreting glands are particularly well developed in scale insects and mealybugs (Fig. 2.5). The epidermis is closely associated with moulting – the events and processes leading up to and including **ecdysis** (eclosion), the shedding of the old cuticle (Chapter 6).

Insects are well endowed with cuticular extensions, varying from fine and hair-like to robust and spine-like. Four basic types of protuberance (Fig. 2.6), all with sclerotized cuticle, can be recognized on morphological, functional and developmental grounds.

- **Spines** are multicellular with undifferentiated epidermal cells.
- **Setae**, also called **hairs**, **macrotrichia** or **trichoid sensilla**, are multicellular with specialized cells.
- **Acanthae** are unicellular in origin.
- **Microtrichia** are subcellular, with from several to many extensions per cell.

Setae sense much of the insect's tactile environment. If large, these setae may be called bristles or chaetae; the most modified are **scales**, the flattened setae found on butterflies and moths (Lepidoptera) and sporadically elsewhere. Three separate cells form each seta: one for hair formation (**trichogen** cell), one for socket formation (**tormogen** cell) and one sensory cell (Chapter 4, Fig. 4.1).

There is no such cellular differentiation in multicellular spines, unicellular acanthae and subcellular microtrichia. The functions of these types of protuberances are diverse and sometimes debatable but their sensory function appears limited. The production of pattern, including colour, may be significant for some of the microscopic projections. Spines are immovable; if articulated, they are called **spurs**. Both spines and spurs may bear unicellular or subcellular processes.

2.1.1 Colour production

The diverse colours of insects are produced by the interaction of light with cuticle and/or underlying cells or fluid by two different mechanisms. Physical (structural) colours result from light-scattering, interference, and diffraction, whereas pigmentary colours are due to the absorption of visible light by a range of chemicals. Often both mechanisms occur together to produce a colour different from that of either alone.

All physical colours derive from the cuticle and its protuberances. **Interference** colours, such as iridescence and ultraviolet, are produced by refraction from varyingly spaced, close reflective layers produced by microfibrillar orientation within the exocuticle, or, in some beetles, the epicuticle, and by diffraction from regularly textured surfaces such as on many scales. Colours produced by light **scattering** depend on the size of surface irregularities relative to the wavelength of light. Thus, whites are produced by structures larger than the wavelength of light, such that all light is reflected, whereas blues are produced by irregularities that reflect only short wavelengths.

Insect pigments are produced in three ways.

- By the insect's own metabolism.
- By sequestering from a plant source.
- Rarely, by microbial endosymbionts.

Pigments may be located in the cuticle, epidermis, haemolymph or fat body. Cuticular darkening is the most ubiquitous insect colour. This may be due to either sclerotization (unrelated to pigmentation) or the exocuticular deposition of melanins, a heterogeneous group of polymers that may give a black, brown, yellow or red colour. Carotenoids, ommochromes, papiliochromes and pteridines (pterins) mostly produce yellows to reds; flavonoids give yellow; and tetrapyrrholes (including breakdown products of porphyrins such as chlorophyll and haemoglobin) create reds, blues and greens. Quinone pigments occur in scale insects as red and yellow anthraquinones and in aphids as yellow to red to dark blue-green aphins.

Colours have an array of functions in addition to the obvious roles of colour patterns in sexual and defensive display. For example, the ommochromes are the main visual pigments of insect eyes, whereas black melanin, an effective screen for possibly harmful light rays, can convert light energy into heat and may act as a sink for free radicals that could otherwise damage cells. Red haemoglobins (widespread respiratory pigments in vertebrates) occur in a few insects, notably in some midge larvae and a few aquatic bugs, in which they have a similar respiratory function.

2.2 SEGMENTATION AND TAGMOSIS

Metameric segmentation, so distinctive in annelids, is visible only in some unsclerotized larvae (Fig. 2.7(a)). The segmentation seen in the sclerotized adult or nymphal insect is not directly homologous with that of larval insects, since sclerotization extends beyond each primary segment (Fig. 2.7(b),(c)). Each apparent segment represents an area of sclerotization which commences in front of the fold that demarcates the primary segment and extends almost to the rear of that segment, leaving an unsclerotized area of the primary segment, the **conjunctival** or 'intersegmental' membrane. This **secondary** segmentation means that the muscles, which are always inserted on the folds, are attached to solid rather than to soft cuticle. The apparent segments of adult insects, such as on the

Fig. 2.7 Types of body segmentation: (a) primary segmentation, as seen in soft-bodied larvae of some insects; (b) simple secondary segmentation; (c) more derived secondary segmentation; (d) longitudinal section of dorsum of the thorax of winged insects, in which the acrotergites of the second and third segments have enlarged to become the postnota. (After Snodgrass, 1935.)

abdomen, are secondary in origin but they are referred to simply as segments throughout this text.

In adult and nymphal insects, and hexapods in general, one of the most striking external features is the amalgamation of segments into functional units. This process of **tagmosis** has given rise to the familiar **tagmata** (regions) of **head**, **thorax** and **abdomen**. In this process the 20 original segments have been divided into an embryologically detectable 6-segmented head, 3-segmented thorax and 11-segmented abdomen (plus primitively the telson), although varying degrees of fusion mean that the full complement is never visible.

Before discussing the external morphology in more detail, some indication of orientation is required. The bilaterally symmetrical body may be described according to three axes:

- **Longitudinal**, or **anterior** to **posterior**, also termed **cephalic** (head) to **caudal** (tail).
- **Dorsoventral**, or **dorsal** (upper) to **ventral** (lower).
- **Transverse**, or **lateral** (outer) through the longitudinal axis to the opposite lateral (Fig. 2.8).

For appendages, such as legs or wings, **proximal** or **basal** refers to near the body and **distal** or **apical** to distant from the body. In addition, structures may be described as **mesal**, or **medial**, if they are nearer to the **midline** (**median line**), or **lateral** if they are closer to the body margin, relative to other structures.

Four principal regions of the body surface can be recognized: the **dorsum** or upper surface; the **venter** or lower surface; and the two lateral **pleura**, separating the dorsum from the venter and bearing limb bases, if these are present. Sclerotization that takes place in defined areas gives rise to plates called **sclerites**. The major segmental sclerites are the **tergum** (the dorsal plate), the **sternum** (the ventral plate) and the pleuron (the side plate). If a sclerite is a subdivision of the tergum, sternum or pleuron, the diminutive terms **tergite**, **sternite** and **pleurite** apply, although usage may be inconsistent.

The pleura of the abdomen are often at least partly membranous, but on the thorax they are sclerotized and usually linked to the

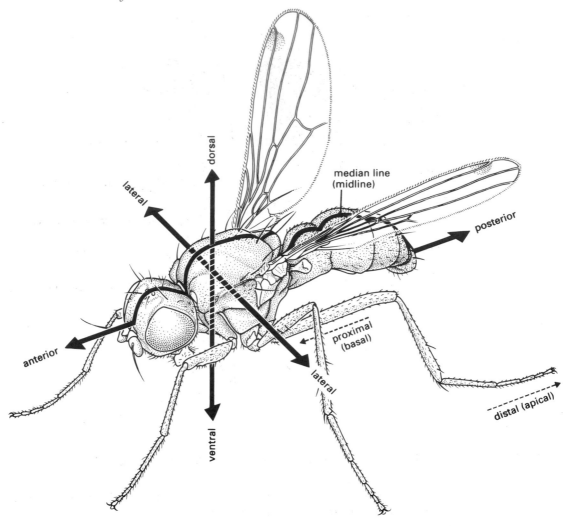

Fig. 2.8 The major body axes and the relationship of parts of the appendages to the body, shown for a sepsid fly. (After McAlpine, 1987.)

tergum and sternum of each segment. This fusion forms a box which contains the leg muscle insertions and, in winged insects, the flight muscles. In the head, all sclerites are fused into a rigid capsule, except in the case of some larvae. In larvae (but not nymphs) the thorax and abdomen may remain membranous and tagmosis may be less apparent (such as in most wasp larvae and fly maggots) and the terga, sterna and pleura are rarely distinct.

2.3 THE HEAD

The rigid cranial capsule has two openings, one posteriorly through the **occipital foramen** to the prothorax, the other to the mouthparts. Typically the mouthparts are directed ventrally (**hypognathous**), although sometimes anteriorly (**prognathous**) as in many beetles, or posteriorly (**opisthognathous**) as in homopterans such as aphids, cicadas and leafhoppers. Several regions can be recognized on

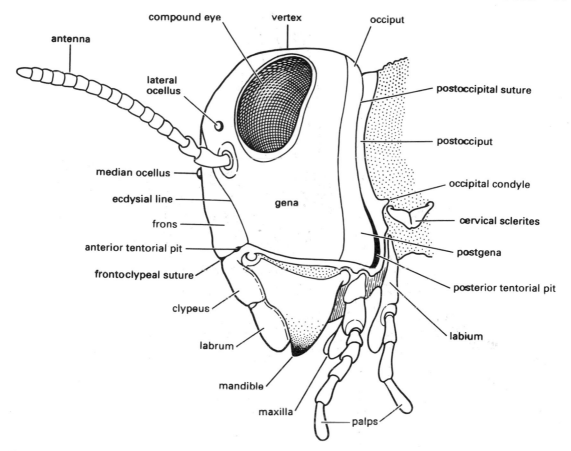

Fig. 2.9 Lateral view of the head of a generalized pterygote insect. (After Snodgrass, 1935.)

the head (Fig. 2.9): the horseshoe-shaped **posterior cranium** (dorsally the **occiput**) contacts the **vertex** dorsally and the **genae** laterally; the vertex merges anteriorly into a **frons** and more anteriorly, a **clypeus**, both of which may be fused into a **frontoclypeus**. In adult and nymphal insects, paired **compound eyes** lie more or less dorsolaterally between the vertex and genae, with a pair of sensory **antennae** placed more medially. In many insects, three light-sensitive 'simple' eyes, or **ocelli** are situated on the anterior vertex, typically arranged in a triangle, and many larvae have stemmatal eyes.

The head regions are often somewhat weakly delimited, with some indications of their extent coming from **sutures** (external grooves or lines on the head). Three sorts may be recognized:

- Remnants of original segmentation, generally restricted to the **postoccipital suture**.
- **Ecdysial** lines of weakness where the head capsule of the immature insect splits at moulting (Chapter 6), including an often prominent inverted 'Y', or **epicranial suture**, on the vertex (Fig. 2.10).
- Grooves, which reflect the underlying internal skeletal ridges, such as the **frontoclypeal** or **epistomal** suture that often delimits the frons from the more anterior clypeus.

The head endoskeleton consists of several invaginated ridges and arms (**apophyses**, or elongate apodemes), the most important of which are the two pairs of **tentorial arms**, one pair being posterior, the other anterior, sometimes with an additional dorsal component. Some of these arms may be absent or, in pterygotes, fused to form the **tentorium**. Pits are discernible on the surface of the cranium at the points where the tentorial arms invaginate. These pits and the sutures may provide prominent landmarks on the head but usually they bear little or no association with the segments.

The segmental origin of the head is most clearly demonstrated by the mouthparts (section 2.3.1). From anterior to posterior, there are six fused head segments:

● The labral segment.
● The antennal segment, with each antenna equivalent to an entire leg.
● The postantennal segment, which is fused with the antennal segment.
● The mandibular segment.
● The maxillary segment.
● The labial segment.

The neck is mainly derived from the first part of the thorax and is not a segment.

2.3.1 Mouthparts

The mouthparts are formed from appendages of all head segments except the second. In omnivorous insects, such as cockroaches, crickets and earwigs, the mouthparts are of a biting and chewing type (**mandibulate**) and resemble the probable basic design of ancestral pterygote insects more closely than the mouthparts of the majority of modern insects. Extreme modifications of basic mouthpart structure, correlated with feeding specializations, occur in most Lepidoptera, Diptera, Hymenoptera, Hemiptera and a number of the smaller orders. Here we first discuss basic mandibulate mouthparts as exemplified by the European earwig, *Forficula auricularia* (Dermaptera: Forficulidae) (Fig. 2.10), and then describe some of the more common

modifications associated with more specialized diets.

There are five basic components of the mouthparts:

● The **labrum**, or 'upper lip', with a ventral surface called the **epipharynx**.
● The **hypopharynx**, a tongue-like structure.
● The **mandibles**, or jaws.
● The **maxillae** (sing. **maxilla**).
● The **labium**, or 'lower lip' (Fig. 2.10).

The labrum forms the roof of the preoral cavity and mouth (Fig. 3.14) and covers the base of the mandibles; it is believed to be formed from fusion of parts of a pair of ancestral appendages. Projecting forward from the back of the preoral cavity is the hypopharynx, a lobe of probable composite origin; in apterygotes, earwigs and nymphal mayflies the hypopharynx bears a pair of lateral lobes, the **superlinguae** (Fig. 2.10). It divides the cavity into a dorsal food pouch, or **cibarium**, and a ventral **salivarium** into which the salivary duct opens (Fig. 3.14). The mandibles, maxillae and labium are the paired appendages of segments 4–6, and their serial homology with walking legs is more apparent than for the labrum and hypopharynx.

The mandibles cut and crush food and may be used for defence; generally they have an apical cutting edge and the more basal molar area grinds the food. They can be extremely hard (approximately 3 on Moh's scale of mineral hardness, or an indentation hardness of about $30\,kg/mm^2$) and thus many termites and beetles have no physical difficulty in boring through foils made from such common metals as copper, lead, tin and zinc. Behind the mandibles lie the maxillae, each consisting of a basal part composed of the proximal **cardo** and the more distal **stipes** and, attached to the stipes, two lobes – the mesal **lacinia** and the lateral **galea** – and a lateral, segmented **maxillary palp**, or **palpus**. Functionally, the maxillae assist the mandibles in processing food; the pointed and sclerotized laciniae hold and macerate the food, whereas the galeae and

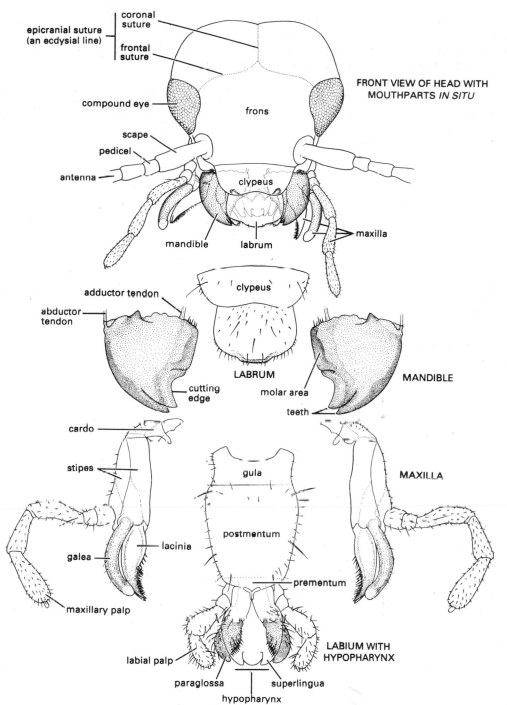

Fig. 2.10 Frontal view of the head and dissected mouthparts of an adult of the European earwig, *Forficula auricularia* (Dermaptera: Forficulidae). Note that the head is prognathous and thus a gular plate, or gula, occurs in the ventral neck region.

palps bear sensory setae (mechanoreceptors) and chemoreceptors which sample items before ingestion. The appendages of the sixth segment of the head are fused with the sternum to form the labium, which is believed to be homologous to the second maxillae of Crustacea. In prognathous insects, such as the earwig, the labium attaches to the ventral surface of the head via a ventromedial sclerotized plate called the **gula** (Fig. 2.10). There are two main parts to the labium – the proximal **postmentum**, closely connected to the poster-oventral surface of the head and sometimes subdivided into a submentum and mentum, and the free distal **prementum**, typically bearing a pair of **labial palps** lateral to two pairs of lobes, the mesal **glossae** and the more lateral **paraglossae**. The glossae and paraglossae, including sometimes the distal part of the prementum to which they attach, are known collectively as the **ligula**; the lobes may be variously fused or reduced as in *Forficula* (Fig. 2.10), in which the glossae are absent. The prementum with its lobes forms the floor of the preoral cavity (functionally a 'lower lip'), whereas the labial palps have a sensory function, similar to that of the maxillary palps.

During insect evolution, an array of different mouthpart types have been derived from the basic design described above. Often feeding structures are characteristic of all members of a genus, family or order of insects, so that knowledge of mouthparts is useful for both taxonomic classification and identification. Mouthpart structure is categorized generally according to feeding method, but mandibles and other components may function in defensive combat or even male–male sexual contests, as in the enlarged mandibles on certain male beetles (Lucanidae). Insect mouthparts have diversified in different orders, with feeding methods that include lapping, suctorial feeding, biting or piercing combined with sucking, and filter feeding, in addition to the basic chewing mode.

The mouthparts of bees are of a chewing and lapping type. Lapping is a mode of

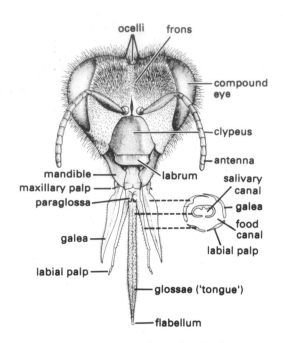

Fig. 2.11 Frontal view of the head of a worker honey bee, *Apis mellifera* (Hymenoptera: Apidae), with transverse section of proboscis showing how the 'tongue' (fused labial glossae) is enclosed within the sucking tube formed from the maxillary galeae and labial palps. (Inset after Wigglesworth, 1964.)

feeding in which liquid or semi-liquid food adhering to a protrusible organ, or 'tongue', is transferred from substrate to mouth. In the honey bee, *Apis mellifera* (Hymenoptera: Apidae), the elongate and fused labial glossae form a hairy tongue, which is surrounded by the maxillary galeae and the labial palps to form a tubular proboscis containing a food canal (Fig. 2.11). In feeding, the tongue is dipped into the nectar or honey, which adheres to the hairs, and then is retracted so that adhering liquid is carried into the space between the galeae and labial palps. This back-and-forth glossal movement occurs repeatedly. Movement of liquid to the mouth apparently results from the action of the cibarial pump, facilitated by each retraction of the tongue pushing liquid up the food canal.

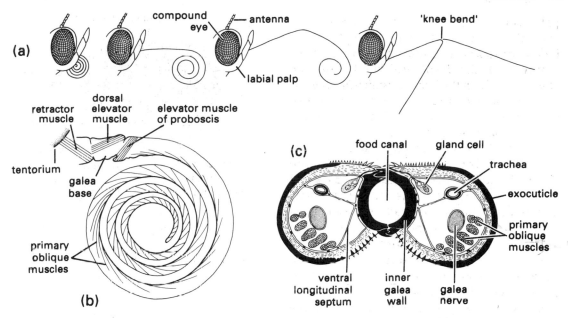

Fig. 2.12 Mouthparts of the cabbage white or cabbage butterfly, *Pieris rapae* (Lepidoptera: Pieridae): (a) positions of the proboscis showing, from left to right, at rest, with proximal region uncoiling, with distal region uncoiling, and fully extended with tip in two of many possible different positions due to flexing at 'knee bend'; (b) lateral view of proboscis musculature; (c) transverse section of the proboscis in the proximal region. (After Eastham and Eassa, 1955.)

The maxillary laciniae and palps are rudimentary and the paraglossae embrace the base of the tongue, directing saliva from the dorsal salivary orifice around into a ventral channel from whence it is transported to the **flabellum**, a small lobe at the glossal tip; saliva may dissolve solid or semi solid sugar. The sclerotized, spoon-shaped mandibles lie at the base of the proboscis and have a variety of functions, including the manipulation of wax and plant resins for nest construction, the feeding of larvae and the queen, grooming, fighting and the removal of nest debris including dead bees.

Most adult Lepidoptera and some adult flies obtain their food solely by sucking up liquids using suctorial (**haustellate**) mouthparts that form a proboscis or rostrum. Pumping of the liquid food is achieved by muscles of the cibarium and/or pharynx. The proboscis of moths and butterflies, formed from the greatly elongated maxillary galeae, is extended (Fig. 2.12(a)) by increases in haemolymph ('blood') pressure. It is loosely coiled by the inherent elasticity of the cuticle, but tight coiling requires contraction of intrinsic muscles (Fig. 2.12(b)). A cross-section of the proboscis (Fig. 2.12(c)) shows how the food canal, which opens basally into the cibarial pump, is formed by apposition and interlocking of the two galeae.

A few moths and many flies combine sucking with piercing or biting. For example, moths that pierce fruit and exceptionally suck blood (species of Noctuidae) have spines and hooks at the tip of their proboscis which are rasped against the skins of either ungulate mammals or fruit. For at least some moths, penetration is effected by the alternate protraction and retraction of the two galeae that slide along each other. Blood-feeding flies have a variety of skin penetration and feeding

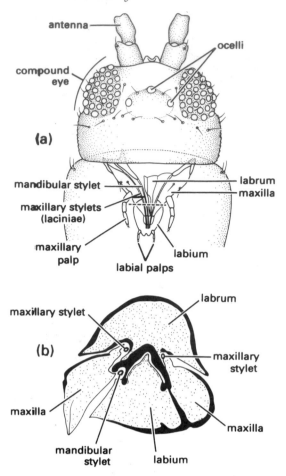

Fig. 2.13 Head and mouthparts of a thrips, *Thrips australis* (Thysanoptera: Thripidae): (a) dorsal view of head showing mouthparts through prothorax; (b) transverse section through proboscis. The plane of the transverse section is indicated by the dashed line in (a). (After CSIRO, 1970; Matsuda, 1965.)

Fig. 2.14 Head and mouthparts of a sucking louse, *Pediculus* (Phthiraptera: Anoplura: Pediculidae): (a) longitudinal section of head (nervous system omitted); (b) transverse section through eversible proboscis. The plane of the transverse section is indicated by the dashed line in (a). (After Snodgrass, 1935.)

mechanisms. In the suborder Nematocera (e.g. mosquitoes, black flies) and the Tabanidae (horse flies, suborder Brachycera), the labium of the adult fly forms a non-piercing sheath for the remainder of the mouthparts which all contribute to the piercing structure, whereas the biting calyptrate dipterans (Brachycera: Calyptratae; e.g. stable flies and tsetse flies) lack mandibles and maxillae and the principal piercing organ is the highly modified labium. The mouthparts of adult Diptera are described in Box 14.4.

Other mouthpart modifications for piercing and sucking are seen in the true bugs (Hemiptera), thrips (Thysanoptera), fleas (Siphonaptera) and sucking lice (Phthiraptera: Anoplura). In each of these insect orders, different mouthpart components form needle-like stylets capable of piercing the plant or animal tissues upon which they feed. Bugs have extremely long, thin paired mandibular and maxillary stylets which fit together to form a flexible stylet-bundle containing a food canal and a salivary canal (Box 10.8).

Thrips have three stylets – paired maxillary stylets (laciniae) plus the left mandibular one (Fig. 2.13). Sucking lice have three stylets – the hypopharyngeal (dorsal), the salivary (median) and the labial (ventral) – lying in a ventral sac of the head and opening at a small eversible proboscis armed with internal teeth that grip the host during blood feeding (Fig. 2.14). Fleas possess a single stylet derived from the epipharynx and the laciniae of the maxillae form two long cutting blades that are ensheathed by the labial palps (Fig. 2.15). The Hemiptera and the Thysanoptera are sister groups and both belong to the same assemblage as the Anoplura (Fig. 7.5), but each group is believed to have had a psocopteroid-like ancestor, presumably with mouthparts of a more generalized mandibulate type. The Siphonaptera are unrelated to the other three taxa; thus similarities in mouthpart structure among these orders result largely from parallel or, in the case of fleas, convergent evolution.

Piercing mouthparts of a slightly different kind are found in antlions and the predatory larvae of other lacewings (Neuroptera). The stylet-like mandible and maxilla on each side of the head fit together to form a sucking tube (Fig. 12.2(c)) and in some families (Chrysopidae, Myrmeleontidae and Osmylidae) there is also a narrow poison channel. Generally, labial palps are present, maxillary palps are absent and the labrum is reduced. Prey is seized by the pointed mandibles and maxillae which are inserted into the victim; its body contents are extra-orally digested and sucked up by the pumping action of the cibarium.

A unique modification of the labium for prey capture occurs in larval damselflies and dragonflies (Odonata). These predators catch other aquatic organisms by rapidly extending their folded labium (or 'mask') and seizing mobile prey with prehensile apical hooks on modified labial palps (Fig. 12.4). The labium is hinged between the prementum and post-mentum and, when folded, covers most of the underside of the head. Labial extension

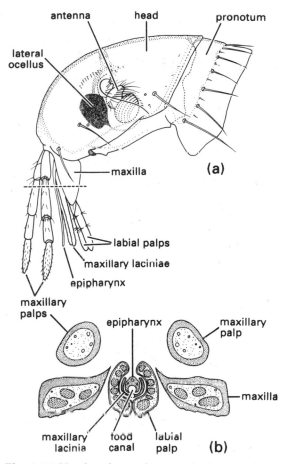

Fig. 2.15 Head and mouthparts of a human flea, *Pulex irritans* (Siphonaptera: Pulicidae): (a) lateral view of head; (b) transverse section through mouthparts. The plane of the transverse section is indicated by the dashed line in (a). (After Snodgrass, 1946; Herms and James, 1961.)

involves the sudden release of energy, produced by increases in blood pressure brought about by the contraction of thoracic and abdominal muscles, and stored elastically in a cuticular click mechanism at the prementum–post-mentum joint. Disengagement of the click mechanism permits the elevated hydraulic pressure to shoot the labium rapidly forwards. Labial retraction then brings the captured prey to the other mouthparts for maceration.

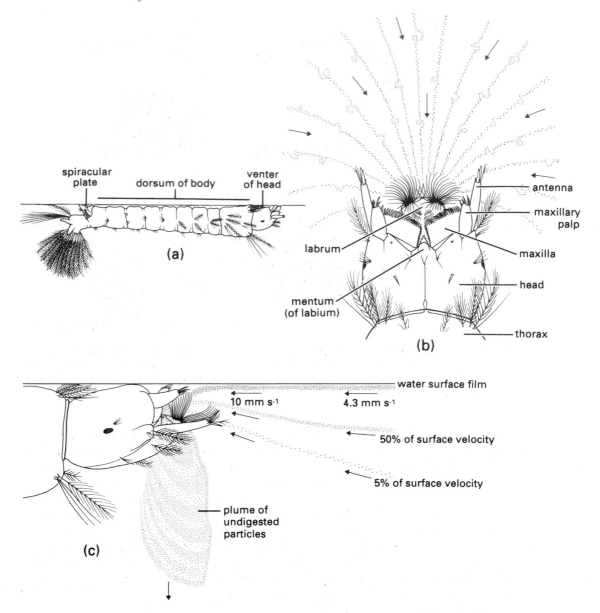

Fig. 2.16 The mouthparts and feeding currents of a mosquito larva of *Anopheles quadrimaculatus* (Diptera: Culicidae): (a) the larva floating just below the water surface, with head rotated through 180 °C relative to body (which is dorsum up so that the spiracular plate near the abdominal apex is in direct contact with the air); (b) viewed from above showing the venter of the head and the feeding current generated by setal brushes on the labrum (direction of water movement and paths taken by surface particles are indicated by arrows and dotted lines, respectively); (c) lateral view showing the particle-rich water being drawn into the preoral cavity between the mandibles and maxillae and its downward expulsion as the outward current. ((b) and (c) after Merritt *et al.*, 1992.)

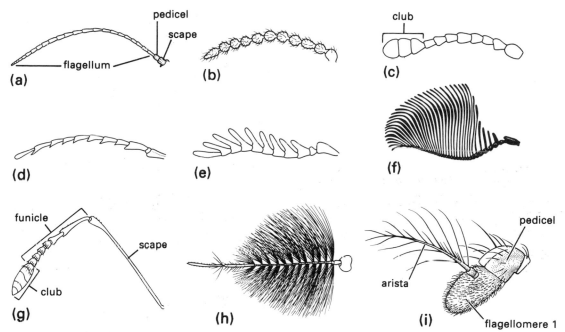

Fig. 2.17 Some types of insect antennae: (a) filiform – linear and slender; (b) moniliform – like a string of beads; (c) clavate or capitate – distinctly clubbed; (d) serrate – saw-like; (e) pectinate – comb-like; (f) flabellate – fan-shaped; (g) geniculate – elbowed; (h) plumose – bearing whorls of setae; (i) aristate – with enlarged third segment bearing a bristle. (After various sources.)

Filter feeding in aquatic insects has been best studied in larval mosquitoes (Diptera: Culicidae), black flies (Diptera: Simuliidae) and net spinning caddisflies (Trichoptera: many Hydropsychoidea and Philopotamoidea), which obtain their food by filtering particles (including bacteria, microscopic algae and detritus) from the water in which they live. The mouthparts of the dipteran larvae have an array of setal 'brushes' or 'fans' which either generate feeding currents or trap particulate matter and then move it to the mouth, whereas the caddisflies spin silk nets that filter particulate matter from flowing water and then use their mouthpart brushes to remove particles from the nets. Thus insect mouthparts are modified for filter feeding chiefly by the elaboration of setae. In mosquito larvae the lateral palatal brushes on the labrum generate the feeding currents (Fig. 2.16); they beat actively, causing particle-rich surface water to flow towards the mouthparts where setae on the mandibles and maxillae help to move particles into the pharynx, where food boluses form at intervals.

In some adult insects, such as mayflies (Ephemeroptera), some Diptera (warble flies), a few moths (Lepidoptera) and male scale insects (Hemiptera: Coccoidea), mouthparts are greatly reduced and non-functional. This mouthpart atrophy is correlated with a short adult life span.

2.3.2 Cephalic sensory structures

The most obvious sensory structures of insects are on the head. Most adults and many nymphs have compound eyes dorsolaterally on head segment 4 and three ocelli on the vertex of the head; the median, or anterior, ocellus lies on segment 1 and is formed from a fused pair; the two lateral ocelli are on segment 2. The only visual structures of larval insects are **stemmata**, or simple eyes, positioned laterally on the head, either singly or in clusters. The structure and functioning of these three types of visual organs are described in detail in section 4.4.

Antennae are mobile, segmented, paired appendages. Primitively, they appear to be 8-segmented in nymphs and adults, but often there are numerous subdivisions, sometimes called **antennomeres**. The entire antenna typically has three main divisions (Fig. 2.17(a)): the first segment, or **scape**, is generally larger than the other segments and is the basal stalk; the second segment, or **pedicel**, nearly always contains a sensory organ known as **Johnston's organ**, which responds to movement of the distal part of the antennae relative to the pedicel; the remainder of the antenna, called the **flagellum**, is often filamentous and multi-segmented (with many **flagellomeres**), but may be reduced or variously modified (Fig. 2.17(b)–(i)). The antennae are reduced or almost absent in some larval insects.

Numerous sensory organs, or **sensilla**, in the form of hairs, pegs, pits or cones, occur on antennae and function as chemoreceptors, mechanoreceptors, thermoreceptors and hygroreceptors (Chapter 4). Antennae of male insects may be more elaborate than those of the corresponding females, increasing the surface area available for detecting female sex pheromones (section 4.3.1).

The mouthparts, other than the mandibles, are well endowed with chemoreceptors and tactile setae. These sensilla are described in detail in Chapter 4.

2.4 THE THORAX

The thorax is composed of three segments: the first or **prothorax**, the second or **mesothorax**, and the third or **metathorax**. Primitively, and in apterygotes and immature insects, these segments are similar in size and structural complexity. In most winged insects the mesothorax and metathorax are enlarged relative to the prothorax and form a **pterothorax**, bearing the wings and associated musculature. Wings occur only on the second and third segments in extant insects but some fossils have prothoracic winglets (Fig. 7.2). Almost all nymphal and adult insects have three pairs of thoracic legs – one pair per segment. Typically the legs are used for walking, although various other functions and associated modifications occur (section 2.4.1). Openings (**spiracles**) of the gas exchange, or tracheal, system (section 3.5) are present laterally on the second and third thoracic segments at most with one pair per segment. However, a secondary condition in some insects is for the mesothoracic spiracles to open on the prothorax.

The tergal plates of the thorax are of simple structure in apterygotes (bristletails and silverfish) and in many immature insects, but are variously modified in winged adults. Thoracic terga are called **nota**, to distinguish them from the abdominal terga. The **pronotum** of the prothorax may be simple in structure and small in comparison to the other nota, but in beetles, mantids, many bugs and some Orthoptera the pronotum is expanded and in cockroaches it forms a shield that covers part of the head and mesothorax. The pterothoracic nota each have two main divisions: the anterior wing-bearing **alinotum** and the posterior phragma-bearing **postnotum** (Fig. 2.18). **Phragmata** are plate-like apodemes which extend inwards below the **antecostal sutures**, marking the primary intersegmental folds between segments; phragmata provide attachment for the longitudinal flight muscles (Fig. 2.7(d)). Each alinotum (sometimes confusingly referred to as a 'notum') may be traversed by sutures that mark the position of internal strengthening ridges and commonly divide the plate into three areas: the anterior **prescutum**, the **scutum** and the smaller posterior **scutellum**.

The lateral pleural sclerites are believed to be derived from the subcoxal segment of the ancestral insect leg (Fig. 7.4(a)). These sclerites may be separate, as in silverfish, or fused

PROTHORAX   MESOTHORAX   METATHORAX

Fig. 2.18 Diagrammatic lateral view of a wing-bearing thoracic segment, showing the typical sclerites and their subdivisions. (After Snodgrass, 1935.)

into an almost continuous sclerotic area, as in most winged insects. In the pterothorax, the pleuron is divided into two main areas – the anterior **episternum** and the posterior **epimeron** – by the internal **pleural ridge** that is visible externally as the **pleural suture** (Fig. 2.18); the ridge runs from the **pleural coxal process** (which articulates with the coxa) to the **pleural wing process** (which articulates with the wing), providing reinforcement for

these articulation points. The **epipleurites** are small sclerites beneath the wing and consist of the **basalaria** anterior to the pleural wing process and the posterior **subalaria**, but usually are reduced to just one basalare and one subalare, which are attachment points for some of the direct flight muscles. The **trochantin** is the small sclerite anterior to the coxa.

The degree of ventral sclerotization on the

thorax varies greatly in different insects. If sternal plates are present, there are typically two per segment: the **eusternum** and the following intersegmental sclerite or **interster-nite** (Fig. 2.7(c)), commonly called the **spinas-ternum** because it usually has an internal apodeme called the **spina**, except that the metasternum never has a spinasternum. The eusterna of the prothorax and mesothorax may fuse with the spinasterna of their segment. Each eusternum may be simple or divided into separate sclerites – typically the **presternum**, **basisternum** and **sternellum**. The eusternum may be fused laterally with one of the pleural sclerites and is then called the **laterosternite**. Often fusion of the sternal and pleural plates results in the formation of **precoxal** and **post-coxal bridges** (Fig. 2.18).

2.4.1 Legs

In most adult and nymphal insects, segmented **fore**, **mid** and **hind legs** occur on the prothorax, mesothorax and metathorax, respectively. Typically, each leg has six segments (Fig. 2.19) and these are, from proximal to distal: **coxa**, **trochanter**, **femur**, **tibia**, **tarsus** and **pretarsus** (or more correctly **post-tarsus**) with claws. Additional segments – the prefemur, patella and basitarsus (Fig. 7.4(a)) – are recognized in many fossil insects and other arthropods, such as arachnids, and one or more of these segments are evident in some Ephemeroptera and Odonata. Primitively, two further segments lie proximal to the coxa and in extant insects one of these, the epicoxa, is associated with the wing articulation or tergum and the other, the subcoxa, with the pleuron (Fig. 7.4(a)).

The tarsus is subdivided into five or fewer components, giving the impression of segmentation; but, because there is only one tarsal muscle, **tarsomere** is a more appropriate term for each 'pseudosegment'. The first tarsomere is sometimes called the basitarsus but should not be confused with the segment called the basitarsus in some fossil insects.

On the underside of the tarsomeres there may be ventral pads, **pulvilli**, also called **euplan-tulae**, which assist in adhesion to surfaces. Terminally on the leg, the small pretarsus (enlargement in Fig. 2.19) bears a pair of lateral **claws** (also called **ungues**) and usually a median lobe, the **arolium**, but in Diptera there may be a central spine-like or pad-like **empodium** (not homologous with the arolium) and a pair of lateral pulvilli. These structures allow flies to walk on walls and ceilings. The pretarsus of Hemiptera may bear a variety of structures, some of which appear to be pulvilli and others have been called empodia or arolia, but the homologies are uncertain. In some beetles such as Coccinellidae, Chrysomelidae and Curculionidae, the ventral surface of some tarsomeres is clothed with adhesive setae that facilitate climbing. The vignette for this chapter shows the underside of the tarsus of *Rhyparida* (Chrysomelidae).

Generally the femur and tibia are the longest leg segments but variations in the lengths and robustness of each segment relate to their functions. For example, walking (**gressorial**) and running (**cursorial**) insects usually have well developed femora and tibiae on all legs, whereas jumping (**saltatorial**) insects such as grasshoppers have disproportionately developed hind femora and tibiae. In aquatic beetles (Coleoptera) and bugs (Hemiptera), the tibiae and/or tarsi of one or more pairs of legs are usually modified for swimming (**na-tatorial**) with fringes of long, slender hairs. Many ground-dwelling insects, such as mole crickets (Orthoptera: Gryllotalpidae), nymphal cicadas (Hemiptera: Cicadidae) and scarab beetles (Scarabaeidae), have the tibiae of the fore legs enlarged and modified for digging (**fossorial**) (Fig. 8.2), whereas the fore legs of some predatory insects, such as mantispid lacewings (Neuroptera) and mantids (Mantodea), are specialized for seizing prey (**raptorial**) (Fig. 12.3). The tibia and basal tarsomere of each hind leg of honey bees are modified for the collection and carriage of pollen (Fig. 11.3).

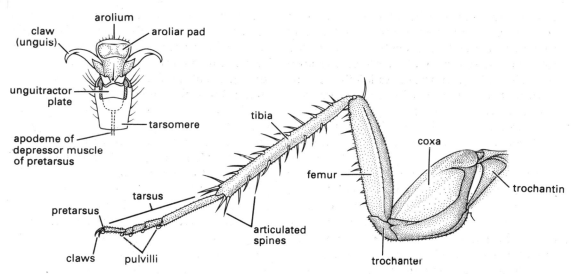

Fig. 2.19 The hind leg of a cockroach, *Periplaneta americana* (Blattodea: Blattidae), with enlargement of ventral surface of pretarsus and last tarsomere. (After Cornwell, 1968; enlargement after Snodgrass, 1935.)

These 'typical' thoracic legs are a distinctive feature of insects, whereas abdominal legs are confined to the immature stages of holometabolous insects. There have been conflicting views on whether (i) the legs on the immature thorax of the Holometabola are developmentally identical (serially homologous) to those of the abdomen, and/or (ii) the thoracic legs of the holometabolous immature stages are homologous with those of the adult. Detailed study of musculature and innervation shows similarity of development of thoracic legs throughout all stages of ametabolous (without metamorphosis, as in silverfish) and hemimetabolous (partial metamorphosis and no pupal stage) insects and in adult Holometabola, with identical innervation through the lateral nerves. Moreover, the oldest known larva (from the Upper Carboniferous) has thoracic and abdominal legs/leglets each with a pair of claws, as in the legs of nymphs and adults. Although larval legs are clearly homologous with those of adults and nymphs, the term **proleg** is used for the larval leg. Prolegs on the abdomen, especially on caterpillars, are usually lobe-like

and each bears an apical circle or band of small sclerotized hooks, or **crochets**. The thoracic prolegs may possess the same number of segments as the adult leg but the number is more often reduced, apparently through fusion. In other cases the thoracic prolegs, like those of the abdomen, are unsegmented outgrowths of the body wall, often bearing apical hooks.

2.4.2 Wings

Wings are only fully developed in the adult (or in the penultimate stage, the pre-imago, of Ephemeroptera). Typically, functional wings are flap-like cuticular projections supported by tubular, sclerotized veins. The major veins are longitudinal, running from the wing base towards the tip, and are more concentrated at the anterior margin. Additional supporting **cross-veins** are transverse struts which join the longitudinal veins, to give a more complex structure. The major veins usually contain tracheae, blood vessels and nerve fibres, with the intervening membranous areas comprising the closely appressed dorsal and ventral

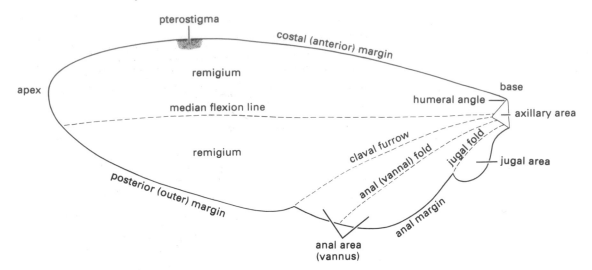

Fig. 2.20 Nomenclature for the main areas, folds and margins of a generalized insect wing.

cuticular surfaces. Generally the major veins are alternately 'convex' and 'concave' in relation to the surface plane of the wing, especially near the wing attachment; this configuration is described by plus (+) and minus (-) signs. Most veins lie in an anterior area of the wing called the **remigium** (Fig. 2.20) which, powered by the thoracic flight muscles, is responsible for most of the movements of flight. The area of wing posterior to the remigium sometimes is called the **clavus**, but more often two areas are recognized: an anterior **anal area** (or **vannus**) and a posterior **jugal area**. Wing areas are delimited and subdivided by **fold-lines**, along which the wing can be folded, and **flexion-lines** at which the wing flexes during flight. The fundamental distinction between these two types of lines is often blurred, since fold-lines may permit some flexion and *vice versa*. The **claval furrow** (a flexion-line) and the **jugal fold** (or fold-line) are nearly constant in position in different insect groups but the **median flexion-line** and the **anal** (or **vannal**) **fold** (or fold-line) form variable and unsatisfactory area boundaries. Wing folding may be very complicated; transverse folding occurs in the hind wings of Coleoptera and Dermaptera,

and the enlarged anal area of some insects may be folded like a fan.

The fore and hind wings of insects in many orders are coupled together, improving the aerodynamic efficiency of flight. The commonest coupling mechanism (seen clearly in Hymenoptera and some Trichoptera) is a row of small hooks, or **hamuli**, along the anterior margin of the hind wing that engage a fold along the posterior margin of the fore wing (**hamulate** coupling). In some other insects (e.g. Mecoptera, Lepidoptera and some Trichoptera), a jugal lobe of the fore wing overlaps the anterior hind wing (**jugate** coupling), or the margins of the fore and hind wing overlap broadly (**amplexiform** coupling), or one or more hind-wing bristles (the **frenulum**) hook under a retaining structure (the **retinaculum**) on the fore wing (**frenate** coupling). The mechanics of flight are described in section 3.1.4 and the evolution of wings is dealt with in section 7.4.

All winged insects are believed to share the same groundplan of wing venation consisting of eight veins, named from the anterior to the posterior of the wing as: **precosta** (PC), **costa** (C), **subcosta** (Sc), **radius** (R), **media** (M), **cubitus** (Cu), **anal** (A) and **jugal** (J).

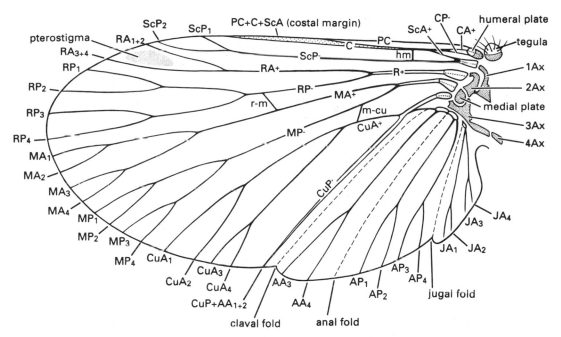

Fig. 2.21 A generalized wing of a neopteran insect (any living winged insect other than Ephemeroptera and Odonata), showing the articulation and the Kukalová-Peck nomenclatural scheme of wing venation. Notation as follows: AA, anal anterior; AP, anal posterior; Ax, axillary sclerite; C, costa ; CA, costa anterior; CP, costa posterior; CuA, cubitus anterior; CuP, cubitus posterior; hm, humeral vein; JA, jugal anterior; MA, media anterior; m-cu, cross-vein between medial and cubital areas; MP, media posterior; PC, precosta; R, radius; RA, radius anterior; r-m, cross-vein between radial and median areas; RP, radius posterior; ScA, subcosta anterior; ScP, subcosta posterior. Branches of the anterior and posterior sector of each vein are numbered, e.g. CuA_{1-4}. (After CSIRO, 1991.)

Primitively, each of these veins has an anterior convex (+) **sector** (a branch with all of its subdivisions) and a posterior concave (−) sector. In almost all extant insects, the precosta is fused with the costa and the jugal vein is rarely apparent. The wing nomenclatural system presented in Fig. 2.21 is that of Kukalová-Peck and is based on detailed comparative studies of fossil and living insects. This system can be applied to the venation of all insect orders, although as yet it has not been widely applied because the various schemes devised for each insect order have a long history of use and there is a reluctance to discard familiar systems. Thus in most textbooks, the same vein may be referred to by different names in different insect orders because the structural homologies were not recognized correctly in early studies. For example, until 1991, the venational scheme for Coleoptera labelled the radius posterior (RP) as the media (M) and the media posterior (MP) as the cubitus (Cu). Correct interpretation of venational homologies is extremely important to phylogenetic studies and the establishment of a single, universally applied scheme is desirable.

Cells are areas of the wing delimited by veins and may be **open** (extending to the wing margin) or **closed** (surrounded by veins). They are named usually according to the longitudinal veins or vein branches that they lie behind, except that certain cells are known by special names, such as the discal cell in

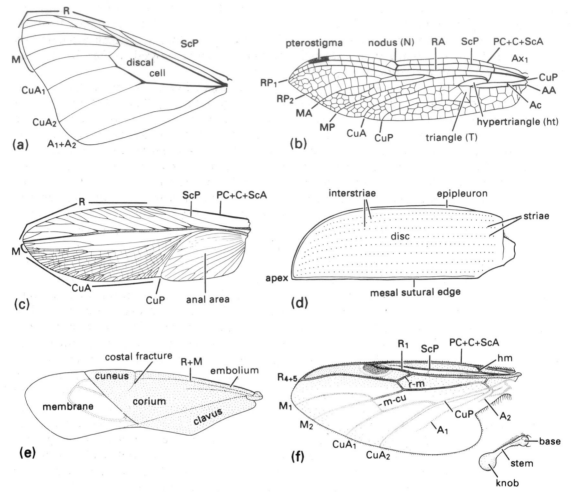

Fig. 2.22 The left wings of a range of insects showing some of the major wing modifications: (a) fore wing of a butterfly of *Danaus* (Lepidoptera: Nymphalidae); (b) fore wing of a dragonfly of *Urothemis* (Odonata: Anisoptera: Libellulidae); (c) fore wing or tegmen of a cockroach of *Periplaneta* (Blattodea: Blattidae); (d) fore wing or elytron of a beetle of *Anomala* (Coleoptera: Scarabaeidae); (e) fore wing or hemelytron of a mirid bug (Hemiptera: Heteroptera: Miridae) showing three wing areas – the membrane, corium and clavus; (f) fore wing and haltere of a fly of *Bibio* (Diptera: Bibionidae). Nomenclatural scheme of venation consistent with that depicted in Fig. 2.21; that of (b) after J.W.H. Trueman (unpublished). ((a)–(d) after Youdeowei, 1977; (f) after McAlpine, 1981.)

Lepidoptera (Fig. 2.22(a)) and the triangle in Odonata (Fig. 2.22(b)). The **pterostigma** is an opaque or pigmented spot anteriorly near the apex of the wing (Figs 2.20, 2.22(b)).

Wing venation patterns are quite consistent within groups (especially families and orders) but often differ between groups and, together with folds or pleats, provide major features used in insect classification and identification. In comparison to the basic scheme outlined above, venation may be greatly reduced by loss or fusion of veins or increased in complexity by the occurrence of numerous crossveins or substantial terminal branching. Other features that may be diagnostic of the wings of different insect groups are pigment patterns and colours, hairs and scales. Scales occur on the wings of Lepidoptera, many Trichoptera and a few psocids (Psocoptera) and flies. Hairs consist of small microtrichia, either scattered or grouped, and larger macrotrichia, typically on the veins.

Usually two pairs of functional wings lie dorsolaterally as **fore wings** on the mesothorax and as **hind wings** on the metathorax; typically the wings are membranous and transparent. However, from this basic pattern are derived many other conditions, often involving variation in the relative size, shape and degree of sclerotization of the fore and hind wings. Examples of fore-wing modification include the thickened, leathery fore wings of Blattodea, Dermaptera and Orthoptera which are called **tegmina** (Fig. 2.22(c)), the hardened fore wings of Coleoptera that form protective wing cases or **elytra** (Fig. 2.22(d)), and the **hemelytra** of heteropteran Hemiptera with the basal part thickened and the apical part membranous (Fig. 2.22(e)). Typically the heteropteran hemelytron is divided into three wing areas: the **membrane**, **corium** and **clavus**. Sometimes the corium is further divided with the **embolium** anterior to R+M, and the **cuneus** distal to a **costal fracture**. In Diptera the hind wings are modified as stabilizers (**halteres**) (Fig. 2.22(f)) and do not function as wings, whereas in male Strepsiptera the fore wings form halteres and the hind wings are used in flight (Box 12.4).

Small insects confront different aerodynamic challenges to larger insects and their wing area often is expanded. Thrips (Thysanoptera), for example, have narrow and straplike wings but a fringe of long setae to extend the wing area (Box 10.7). In termites (Isoptera) and ants (Hymenoptera: Formicidae) the winged reproductives, or alates, have large **deciduous** wings that are shed after the nuptial flight. Some insects are wingless, or **apterous**, either primitively as in silverfish (Thysanura) and bristletails (Archaeognatha), which diverged from other insect lineages prior to the origin of wings, or secondarily as

in all lice (Phthiraptera) and fleas (Siphonaptera) which evolved from winged ancestors. Secondary partial wing reduction occurs in a number of short-winged, or **brachypterous**, insects.

In all winged insects (Pterygota), a triangular **axillary area** (Fig. 2.20) at the wing base contains the movable **articular sclerites** via which the wing articulates on the thorax. These sclerites are derived, by reduction and fusion, from a band of articular sclerites in the ancestral wing. There are three different types of wing articulation among living Pterygota as a result of unique patterns of fusion and reduction of the articular sclerites. In Neoptera (all living winged insects except the Ephemeroptera and Odonata), the articular sclerites consist of the **humeral plate**, the **tegula** and usually three, rarely four, **axillary sclerites** (1Ax, 2Ax, 3Ax and 4Ax) (Fig. 2.21). The Ephemeroptera and Odonata each have a different configuration of these sclerites compared to the Neoptera (literally meaning 'new wing') and, unlike neopteran adults, odonate and ephemeropteran adults cannot fold their wings back along the abdomen. In Neoptera, the wing articulates via the articular sclerites with the **anterior** and **posterior wing processes** dorsally, and ventrally with the **pleural wing processes** and two small pleural sclerites (the basalare and subalare) (Fig. 2.18).

2.5 THE ABDOMEN

Primitively, the insect abdomen is 11-segmented although segment 1 may be reduced or incorporated into the thorax (as in many Hymenoptera) and usually the terminal segments are variously modified and/or diminished (Fig. 2.23(a)). Generally, at least the first seven abdominal segments of adults (the **pregenital** segments) are similar in structure and lack appendages. However, apterygotes (bristletails and silverfish) and many immature aquatic insects do have abdominal appendages. Apterygotes possess a pair of **styles** – rudimentary appendages that are

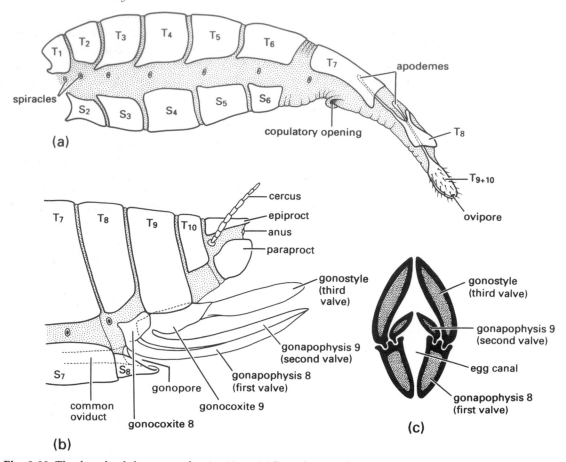

Fig. 2.23 The female abdomen and ovipositor: (a) lateral view of the abdomen of an adult tussock moth (Lepidoptera: Lymantriidae) showing the substitutional ovipositor formed from the extensible terminal segments; (b) lateral view of a generalized orthopteroid ovipositor composed of appendages of segments 8 and 9; (c) transverse section through the ovipositor of a katydid (Orthoptera: Tettigoniidae). T_1–T_{10}, terga of first to tenth segments; S_2–S_8, sterna of second to eighth segments. ((a) after Eidmann, 1929; (b) after Snodgrass, 1935; (c) after Richards and Davies, 1959.)

serially homologous with the distal part of the thoracic legs – and, mesally, one or two pairs of **protrusible** (or **exsertile**) **vesicles** on at least some abdominal segments. These vesicles are derived from the coxal and trochanteral **endites** (inner annulated lobes) of the ancestral abdominal appendages (Fig. 7.4(b)). A number of aquatic larvae and nymphs have gills laterally on some or most abdominal segments (Chapter 9). Some of these may be serially homologous with thoracic wings (e.g. the plate gills of mayfly nymphs) or with other leg derivatives. Spiracles typically are present on segments 1 to 8, but reductions in number occur frequently in association with modifications of the tracheal system (section 3.5), especially in immature insects, and with specializations of the terminal segments in adults.

2.5.1 Terminalia

The ano-genital part of the abdomen, known as the **terminalia**, generally consists of segments 8 or 9 to the abdominal apex. Segments 8 and 9 bear the genitalia; segment 10 is visible as a complete segment in many 'lower' insects but always lacks appendages; and the small segment 11 is represented by a dorsal epiproct and pair of ventral paraprocts derived from the sternum (Fig. 2.23(b)). A pair of appendages, the **cerci**, articulates laterally on segment 11; typically these are annulated and filamentous but have been variously modified (e.g. the forceps of earwigs) or reduced in different insect orders. An annulated caudal filament, the median **appendix dorsalis**, arises from the tip of the epiproct in apterygotes, most mayflies (Ephemeroptera) and a few fossil insects. A similar structure in nymphal stoneflies (Plecoptera) is of uncertain homology. These terminal abdominal segments have excretory and sensory functions in all insects, but in adults there is an additional reproductive function.

The organs specifically concerned with mating and the deposition of eggs are known collectively as the **external genitalia**, although they may be largely internal in 'higher' insects. The form of the components of the external genitalia of insects are very diverse and often have considerable taxonomic value, particularly if species appear structurally similar in many other respects. The male external genitalia have been widely used as aids in distinguishing species, whereas the female external genitalia are usually simpler and less varied. The diversity and species-specificity of genitalic structures are discussed in section 5.5.

The terminalia of adult female insects include internal structures for receiving the male copulatory organ and his spermatozoa (sections 5.4 and 5.6) and external structures used for oviposition (egg-laying; section 5.8). The external genitalia of female insects function as egg-laying tubes, or **ovipositors**, although these are absent in Isoptera, Phthiraptera, many Plecoptera and most Ephemeroptera. Ovipositors take two forms:

- The true, or **appendicular**, ovipositor formed from appendages of abdominal segments 8 and 9 (Fig. 2.23(b)).
- The **substitutional** ovipositor composed of extensible posterior abdominal segments (Fig. 2.23(a)).

Substitutional ovipositors include a variable number of the terminal segments and clearly have been derived convergently several times, even within some orders. They occur in many insects, including most Lepidoptera, Coleoptera and Diptera. In these insects, the terminalia are telescopic and can be extended as slender tubes, manipulated by muscles attached to apodemes of the modified terga (Fig. 2.23(a)) and/or sterna.

Appendicular ovipositors represent the primitive condition for female insects and are present in Archaeognatha, Thysanura, many Odonata, Orthoptera, some Hemiptera, some Thysanoptera, and Hymenoptera. In some Hymenoptera, the ovipositor is modified as a poison-injecting sting (Fig. 13.10) and the eggs are ejected at the base of the sting. In all other cases, the eggs pass down a canal in the shaft of the ovipositor (section 5.8). The shaft is composed of three pairs of **valves** (Fig. 2.23(b),(c)), supported on two pairs of **valvifers** – the coxae + trochanters, or **gonocoxites**, of segments 8 and 9 (Fig. 2.23(b)). The gonocoxites of segment 8 have a pair of trochanteral endites (inner lobe from each trochanter), or **gonapophyses**, which form the first valves, whereas the gonocoxites of segment 9 have a pair of gonapophyses (the second valves) plus a pair of **gonostyles** (the third valves) derived from the distal part of the appendages of segment 9 (and homologous with the styles of the apterygotes mentioned above). In each half of the ovipositor, the second valve slides in a tongue-and-groove fashion against the first valve (Fig. 2.23(c)), whereas the third valve generally forms a sheath for the other valves.

The external genitalia of male insects include an organ for transferring the

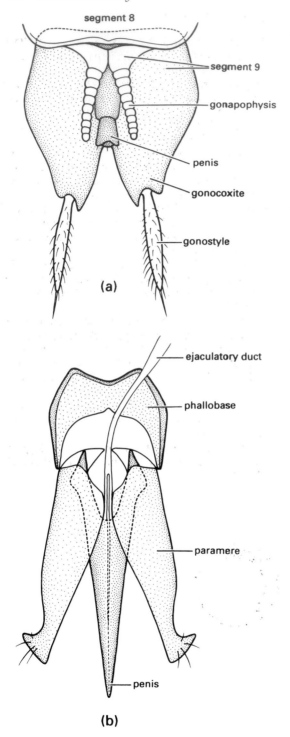

(a)

(b)

spermatozoa (either packaged in a **sperma-tophore** or free in fluid) to the female and often involve structures that grasp and hold the partner during mating. Numerous terms are applied to the various components in different insect groups and homologies are difficult to establish. Males of Archaeognatha, Thysanura and Ephemeroptera have relatively simple genitalia consisting of gonocoxites, gonostyles and sometimes gonapophyses on segment 9 (and also on segment 8 in Archaeognatha), as in the female, except with a median **penis** (**phallus**) or, if paired or bilobed, **penes**, on segment 9 (Fig. 2.24(a)). The penis or penes is believed to be derived from the fused inner lobes (endites) of either the ancestral coxae or trochanters of segment 9. In the blattoid-orthopteroid orders, the gonocoxites are reduced or absent, although gonostyles may be present (called styles), and there is a median penis with a lobe called a **phallomere** on each side of it. The fate of the gonapophyses and the origin of the phallomeres is uncertain. In the 'higher' insects – the hemipteroids and the holometabolous orders – the homologies and terminology of the male structures are even more confusing if one tries to compare the terminalia of different orders. The whole copulatory organ of higher insects generally is known as the **aedeagus** and, in addition to insemination, it may clasp and provide sensory stimulation to the female. Typically there is a median, tubular penis (although sometimes the term aedeagus is restricted to this lobe), which often has an inner tube, the **endophallus**, that is everted during insemination (Fig. 5.2(b)). The ejaculatory duct opens at the gonopore at either the tip of the

Fig. 2.24 Male external genitalia: (a) dorsal view of abdominal segment 9 (with cerci and appendix dorsalis removed) of the bristletail *Machilis variabilis* (Archaeognatha: Machilidae); (b) dorsal view of aedeagus of a click beetle (Coleoptera: Elateridae). ((a) after Snodgrass, 1957.)

penis or the endophallus. Lateral to the penis is a pair of lobes or **parameres** which may have a clasping or sensory function. Their origin is uncertain; they may be homologous with the gonocoxites and gonostyles of lower insects, with the phallomeres of blattoid-orthopteroid insects or be derived *de novo*, perhaps even from segment 10. This trilobed type of aedeagus is well exemplified in many beetles (Fig. 2.24(b)) but modifications are too numerous to describe here.

Much variation in male external genitalia correlates with mating position, which is very variable between and sometimes within orders. Mating positions include end-to-end, side-by-side, male below with his dorsum up, male on top with female dorsum up, and even venter-to-venter. In some insects, torsion of the terminal segments may take place post-metamorphosis, just prior to or during copulation and asymmetries of male clasping structures occur in many insects. Copulation and associated behaviours are discussed in more detail in Chapter 5.

FURTHER READING

Binnington, K. and Retnakaran, A. (eds) (1991) *Physiology of the Insect Epidermis*, CSIRO Publications, Melbourne.

Chapman, R.F. (1982) *The Insects. Structure and Function*, 3rd edn, Hodder and Stoughton, London, Sydney, Auckland, Toronto.

Hadley, N.F. (1986) The arthropod cuticle, *Scientific American*, **255 (1)**, 98–106.

Lawrence, J.F., Nielsen, E.S. and Mackerras, I.M. (1991) Skeletal anatomy and key to orders, in *The Insects of Australia*, 2nd edn, (CSIRO), Melbourne University Press, Carlton, pp. 3–32.

Nichols, S.W. (1989) *The Torre-Bueno Glossary of Entomology*, The New York Entomological Society in cooperation with the American Museum of Natural History, New York.

Richards, A.G. and Richards, P.A. (1979) The cuticular protuberances of insects. *International Journal of Insect Morphology and Embryology*, **8**, 143–57.

Smith, J.J.B. (1985) Feeding mechanisms, in *Comprehensive Insect Physiology, Biochemistry and Pharmacology*, Volume 4, *Regulation: Digestion, Nutrition, Excretion*, (eds G.A. Kerkut and L.I. Gilbert), Pergamon Press, Oxford, New York, Toronto, Sydney, Paris, Frankfurt, pp. 33–85.

Snodgrass, R.E. (1935) *Principles of Insect Morphology*, McGraw Hill Book Company, New York and London.

Wootton, R.J. (1992) Functional morphology of insect wings. *Annual Review of Entomology*, **37**, 113–40.

3
INTERNAL ANATOMY AND PHYSIOLOGY

Internal structures of a locust

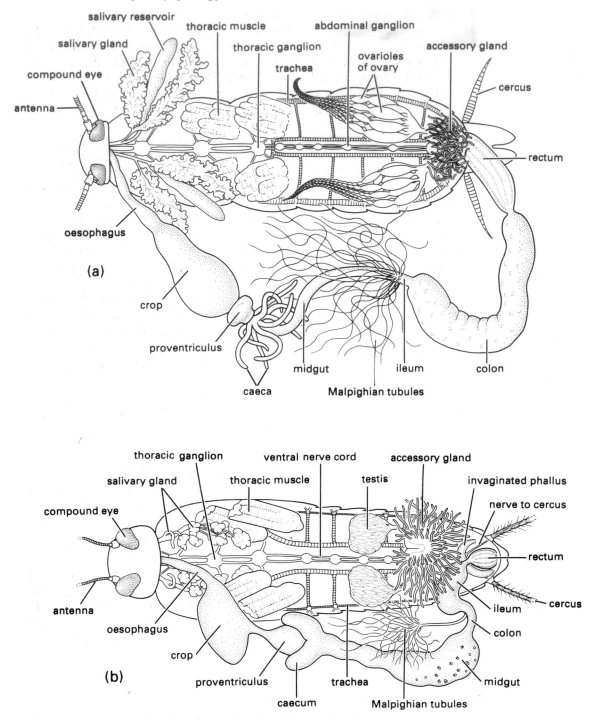

Fig. 3.1 Dissections: (a) female American cockroach, *Periplaneta americana* (Blattodea: Blattidae); (b) a male black field cricket, *Teleogryllus commodus* (Orthoptera: Gryllidae). The fat body and most of the tracheae have been removed; most details of the nervous system are not shown.

If you dissect the body of an insect, you see a complex and compact masterpiece of functional design. Figure 3.1 shows the 'insides' of two omnivorous insects, a cockroach and a cricket, which have relatively unspecialized digestive and reproductive systems. The digestive system, which includes salivary glands as well as an elongate gut, consists of three main sections. These function in storage, biochemical breakdown, absorption and excretion. Each gut section has more than one physiological role and this may be reflected in local structural modifications, such as thickening of the gut wall or diverticula (extensions) from the main lumen. The reproductive systems depicted in Fig. 3.1 exemplify the female and male organs of many insects. These may be dominated in males by very visible accessory glands, especially since the testes in many adult insects are degenerate or absent. This is because the spermatozoa are produced in the pupal or penultimate stage and stored. In gravid female insects, the body cavity may be filled with eggs at various stages of development, thereby obscuring other internal organs. Likewise, the internal structures (except the gut) of a well-fed, late-stage caterpillar may be hidden within the mass of fat body tissue.

The insect body cavity, called the **haemocoel** and filled with fluid **haemolymph**, is lined with endoderm and ectoderm. It is not a true coelom, which is defined as a mesoderm-lined cavity. Haemolymph (so called because it combines many roles of vertebrate blood (haem) and lymph) bathes all internal organs, delivers nutrients, removes metabolites and performs immune functions. Unlike vertebrate blood, haemolymph rarely has respiratory pigments and therefore has little or no role in gaseous exchange. In insects this function is performed by the **tracheal system**, a ramification of air-filled tubes (**tracheae**) which sends fine branches throughout the body. Gas entry to and exit from tracheae is controlled by sphincter-like structures called **spiracles** that open through the body wall.

Non-gaseous wastes are filtered from the haemolymph by filamentous **Malpighian tubules** (named after their discoverer) which have free ends distributed through the haemocoel. Their contents are emptied into the gut from which, after further modification, wastes are eliminated eventually via the anus.

All motor, sensory and physiological processes in insects are controlled by the nervous system in conjunction with hormones (chemical messengers). The brain and ventral nerve cord are readily visible in dissected insects but most endocrine centres, neurosecretion sites, numerous nerve fibres, muscles and other tissues are not apparent to the unaided eye.

This chapter describes insect internal structures and their functions. Topics covered are the muscles and locomotion (both walking and flight), the nervous system and co-ordination, endocrine centres and hormones, the haemolymph and its circulation, the tracheal system and gas exchange, the gut and digestion, the fat body, nutrition and micro-organisms, the excretory system and waste disposal, and lastly the reproductive organs and gametogenesis.

3.1 MUSCLES AND LOCOMOTION

Much of the success of insects can be related to their ability to sense, interpret and move around their environment (section 1.3.2). Whilst flight, which originated at least 300 million years ago, was a major innovation, terrestrial and aquatic locomotion is also well developed. Power for movement originates from muscles operating against a skeletal system, either the rigid cuticular exoskeleton or, in soft-bodied larvae, a hydrostatic skeleton.

3.1.1 Muscles

Vertebrates and many non-insect invertebrates have **striated** and **smooth** muscles but insects have only striated muscles, so called

Fig. 3.2 Muscle attachments to body wall: (a) tonofibrillae traversing the epidermis from the muscle to the cuticle; (b) a muscle attachment in an adult beetle of *Chrysobothrus femorata* (Coleoptera: Buprestidae); (c) a multicellular apodeme with a muscle attached to one of its thread-like, cuticular 'tendons'. (After Snodgrass, 1935.)

because of the overlapping of the thicker myosin and thinner actin filaments which gives a microscopic appearance of cross-banding. Each striated muscle fibre comprises many cells, with a common plasma membrane and **sarcolemma**, or outer sheath. The sarcolemma is invaginated, but not broken, where an oxygen-supplying tracheole (section 3.5; Fig. 3.10(b)) contacts the muscle fibre. Contractile **myofibrils** run the length of the fibre, arranged in sheets or cylinders. When viewed under high magnification, a myofibril comprises a thin actin filament sandwiched between a pair of thicker myosin filaments. Muscle contraction involves the sliding of filaments past each other, stimulated by nerve impulses. Innervation comes from one to three motor axons per bundle of fibres, each separately tracheated and referred to as one muscle unit, with several units grouped in a functional muscle.

There are several different muscle types. The most important division is between those that respond synchronously, with a contraction cycle once per impulse, and fibrillar muscles that contract asynchronously, with multiple contractions per impulse, for example, as in some flight muscles and the tymbal muscle of cicadas (Chapter 4).

There is no inherent difference in action between muscles of insects and vertebrates, although insects can produce prodigious muscular feats, such as the leap of a flea or the repetitive stridulation of the cicada tympanum. Insects do benefit from a reduced body size, due to the relationship between power, which is proportional to muscle cross-section and decreases with reduction in size by the square root, and the body mass, which decreases with reduction in size by the cube root. Thus the power:mass ratio increases as body size decreases.

3.1.2 Muscle attachments

In vertebrates, muscles work against an internal skeleton, but in insects the muscles must attach to the inner surface of an external skeleton. Since the musculature is of mesodermal origin and the exoskeleton is of ectodermal origin, fusion must take place. This occurs by the growth of **tonofibrillae** –

fine connecting fibrils that link the epidermal end of the muscle to the epidermal layer (Fig. 3.2(a),(b)). The tonofibrillae are discarded with the cuticle at each moult and must therefore be regrown.

At the site of tonofibrillar attachment, the inner cuticle is often strengthened through ridges or **apodemes** which, when elongated into arms, are termed **apophyses** (Fig. 3.2(c)). These muscle attachment sites, particularly the long, slender apodemes for individual muscle attachments, often include resilin to give elasticity resembling vertebrate tendons.

Some insects, such as soft-bodied larvae, have mainly thin, flexible cuticle, which lacks the rigidity to anchor muscles unless given additional strength. The body contents form a **hydrostatic skeleton**, with turgidity maintained by criss-cross body wall 'turgor' muscles that continuously contract against the incompressible fluid of the haemocoel, giving a strengthened foundation for other muscles. If the larval body wall is perforated, the fluid leaks, the haemocoel becomes compressible and the turgor muscles cause the larva to become flaccid.

3.1.3 Crawling, wriggling, jumping swimming and walking

Soft-bodied larvae with hydrostatic skeletons move by crawling. Muscular contraction in one part of the body gives equivalent extension in a relaxed part elsewhere on the body. In apodous (legless) larvae, such as dipteran 'maggots', waves of contractions and relaxation run from head to tail. Bands of adhesive hooks or tubercles successively grip and detach from the substrate to provide a forward motion. Some maggots use their mouth-hooks to grip the substrate during movement. In water, lateral waves of contraction against the hydrostatic skeleton can give a sinuous, snake-like swimming motion; anterior-to-posterior waves give an undulating motion.

Larvae with thoracic legs and abdominal prolegs, like caterpillars, develop posterior to anterior waves of turgor muscle contraction, with up to three waves visible at one time. Locomotor muscles are used in cycles of successive detachment of the thoracic legs, reaching forward and grasping the substrate. These cycles occur in concert with inflation, deflation and forward movement of the posterior prolegs.

Insects with hard exoskeletons contract and relax pairs of agonistic and antagonistic muscles attached to the cuticle. Movement by walking or running uses the six thoracic legs. Compared with crustaceans and myriapods, insects have fewer legs and they are located more ventrally and brought close together on the thorax, allowing more control and greater efficiency. Motion with six legs gives continuous contact with the ground by a tripod of fore and hind legs on one side and midleg of the opposite side, whilst the opposite legs are moved forwards (Fig. 3.3). The centre of gravity of the moving insect always lies within this tripod, giving great stability. Motion is imparted through thoracic muscles acting on the leg bases, with transmission via internal leg muscles through the leg to extend or flex the leg. Anchorage to the substrate needed to provide a lever to propel the body is through pointed claws and adhesive pads (the arolium or, in flies, pulvilli). Claws can obtain purchase on the slightest roughness in a surface, whilst the pads of some insects can adhere to perfectly smooth surfaces through the application of lubricants to the tips of numerous fine hairs and the action of close-range molecular forces between the hairs and the substrate.

Many insects jump, some prodigiously, using modified hind legs. Muscle tension is built up gradually and stored in orthopterans by distortion of the tibia and in fleas by compression of an elastic resilin pad. Release of tension is sudden, resulting in propulsion of the insect into the air.

In walking, forward motion is achieved by a rearward thrust of each leg (**retraction**) while it is in contact with the substrate, with leg

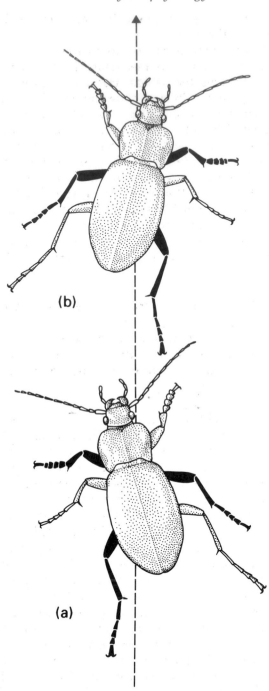

(b)

(a)

Fig. 3.3 A ground beetle (Coleoptera: Carabidae: *Carabus*) walking in the direction of the broken line. The three blackened legs are those in contact with the ground in the two positions illustrated – (a) is followed by (b). (After Wigglesworth, 1972.)

position restored (**protraction**) while out of contact with the substrate. In swimming, however, contact with the water is maintained during protraction and so it is necessary for the insect to impart more thrust to the rowing motion than to the recovery stroke if progress is to be made. This is achieved by expanding the effective leg area during retraction by extending fringes of hairs and spines. These are collapsed onto the folded leg during the recovery stroke. We have seen already how some insect larvae swim using contractions against their hydrostatic skeleton. Others, including many nymphs and the larvae of caddisflies can walk underwater and do not swim routinely, particularly in running waters.

The surface film of water is strong enough to support some specialist insects, most of which have hydrofuge (water-repelling) cuticles or hair fringes. Some, such as gerrid pond skaters, move by rowing with hair-fringed legs.

3.1.4 Flight

The development of flight allowed insects much greater mobility, which helped in food and mate location and gave much improved powers of dispersal. Importantly, flight opened up many new environments for exploitation. Plant microhabitats such as flowers and foliage are more accessible to winged insects than to those without flight.

Fully developed, functional, flying wings occur only in adult insects, although in nymphs the developing wings can be seen in later instars as wing buds. Usually two pairs of functional wings arise dorsolaterally as fore wings on the second and hind wings on the third thoracic segment. Some of the many derived variations have been described in Chapter 2.

To fly, the forces of weight (gravity) and drag (air resistance to movement) must be overcome. In gliding flight, in which the wings are held rigidly outstretched, these forces are overcome through the use of passive air movements – known as the **relative wind**. The insect achieves lift by adjusting the angle of the leading edge of the wing when orientated into the wind. As this **attack angle** increases, so lift increases until stalling occurs, that is, lift is catastrophically lost. In contrast to aircraft, nearly all of which stall at around 20°, the attack angle of insects can be raised to more than 30°, even as high as 50°, giving great manoeuvrability. Aerodynamic effects such as enhanced lift and reduced drag can come from wing scales and hairs that affect the boundary layer across the wing surface.

Most insects glide a little, and dragonflies (Odonata) and some grasshoppers (Orthoptera), notably locusts, glide extensively. However, most winged insects fly by beating their wings. Examination of wing beat is difficult because the frequency of even a large slow-flying butterfly may be five times a second (5 Hz), a bee may beat its wings at 180 Hz, and some midges emit an audible buzz with their wing beat frequency of greater than 1000 Hz. However, by running a high-speed cine film at a slow rate, the speed of an insect wing beat can be reduced until a single beat can be analysed. This reveals that a single beat comprises three interlinked movements, the first being a cycle of downward, forward motion followed by an upwards and backwards motion. Second, during the cycle each wing is rotated around its base. The third component occurs as various parts of the wing flex in response to local variations in air pressure. Unlike gliding, in which the relative wind derives from passive air movement, in true flight the relative wind is produced by the moving wings. The flying insect makes constant adjustments, so that during a wing beat, the air ahead of the insect is thrown backwards and downwards, impelling the insect upwards (lift) and forwards (thrust). In climbing, the emergent air is directed more downward, reducing thrust but increasing lift. In turning, the power of the wing on the inside of the turn is reduced by decreasing the amplitude of the beat.

Despite the elegance and intricacy of detail of insect flight, the mechanisms responsible for beating the wings are not excessively complicated. The thorax of the wing-bearing segments can be envisaged as a box with the sides (pleura) and base (sternum) rigidly fused, and the wings connected where the rigid tergum is attached to the pleura by flexible membranes. This membranous attachment and the wing hinge is composed of resilin (section 2.1) which gives crucial elasticity to the thoracic box.

Flying insects have one of two kinds of arrangements of muscles powering their flight:

- **Direct flight muscles** connected to the wings.
- An **indirect system** in which there is no muscle to wing connection, but rather muscle action deforms the thoracic box to give wing movement.

Direct flight muscles are found in the more primitive insects such as dragonflies and cockroaches. More advanced insects use indirect flight muscles; they retain direct muscles but these are used for wing orientation rather than power production.

In insects with direct flight muscles, the upward stroke is produced by contraction of direct muscles attached to the wing base inside the pivotal point (Fig. 3.4(a)). The downward stroke of the wing is produced through contraction of muscles that extend from the sternum to the wing base outside the pivotal point (Fig. 3.4(b)). In contrast, indirect flight muscles are attached to the tergum and sternum. Contraction causes the tergum, and with it the very base of the wing, to be pulled down. This movement levers the outer, main part of the wing in an upwards stroke (Fig. 3.4(c)). The down beat is powered by contraction of the second set of indirect muscles that run from front to back of the thorax, thereby

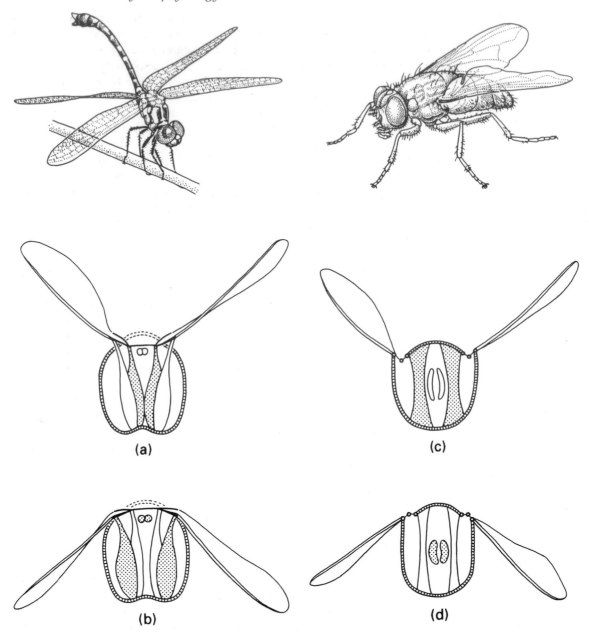

Fig. 3.4 Direct ((a),(b)) and indirect ((c),(d)) flight mechanisms. Dragonfly thorax during (a) upstroke and (b) downstroke of the wings. House-fly thorax during (c) upstroke and (d) downstroke of wings. Stippled muscles are those contracting in each illustration. (After Blaney, 1976.)

deforming the box and lifting the tergum (Fig. 3.4(d)). At each stage in the cycle, energy is conserved when the flight muscles relax, because the elasticity of the thorax restores its shape.

Primitively, the four wings may be controlled independently with small variation in timing and rate allowing alteration in direction of flight. However, excessive variation impedes controlled flight and the beat of all wings is usually harmonized (as in butterflies, bugs and bees, for example) by locking the fore and hind wings together, and also by neural control. For insects with slower wing-beat frequencies (no more than 100 Hz) such as dragonflies, one nerve impulse for each beat can be maintained, by **synchronous** muscles. However, in faster beating wings, which may attain a frequency from 100 to over 1000 Hz, one impulse per beat is impossible and **asynchronous** muscles are required. In these insects, the wing construction is such that in the wing-beat cycle only two wing positions are stable: fully up and fully down. As the wing moves from one extreme to the alternate one, it passes through an intermediate unstable position. As it passes this unstable ('click') point, thoracic elasticity snaps the wing through to the alternate stable position. Insects with this asynchronous mechanism have peculiar fibrillar flight muscles with the property that, on sudden release of muscle tension (as at the click point), the next muscle contraction is induced. Thus muscles can oscillate, contracting at a much higher frequency than the stimulatory nerve impulses, which need only be periodic to keep the insect in flight. Harmonization of the wing beat on each side is maintained through the rigidity of the thorax: as the tergum is depressed or relaxed, what happens to one wing must happen identically to the other. However, insects with indirect flight muscles retain direct muscles that are used in making fine adjustments in wing orientation during flight.

Direction and any deviations from course,
perhaps caused by air movements, are sensed by insects predominantly through their eyes and antennae. However, the true flies (Diptera) have extremely sophisticated sensory equipment with their hind wings modified as balancing organs. These halteres, which each comprise a base, stem and apical knob, beat in time but out of phase with the fore wings. The knob, which is heavier than the rest of the organ, tends to keep the halteres beating in one plane. When the fly alters direction, whether voluntarily or otherwise, the haltere is twisted. This movement is detected by the stem, which is richly endowed with sensilla, and the fly can respond accordingly.

Initiation of flight, for any reason, may involve the legs springing the insect into the air. Loss of tarsal contact with the ground causes neural firing of the direct flight muscles. In flies, flight activity originates in contraction of a mid-leg muscle, which both propels the leg downwards (and the fly upwards) and simultaneously pulls the tergum downwards to inaugurate flight. The legs are also important when landing because there is no gradual braking by running forward – all the shock is taken on the outstretched legs, endowed with pads, spines and claws for adhesion.

3.2 THE NERVOUS SYSTEM AND CO-ORDINATION

The complex nervous system of insects integrates a diverse array of external sensory and internal physiological information and generates some of the behaviours discussed in Chapter 4. In common with other animals, the basic component is the nerve cell, or **neurone**, composed of a cell body with two projections (fibres) – the **dendrite** that receives stimuli and the **axon** that transmits information, either to another neurone or to an effector organ such as a muscle. Insect neurones release a variety of chemicals at **synapses** either to stimulate or to inhibit effector

Fig. 3.5 Diagram of simple reflex mechanism of an insect. The arrows show the paths of nerve impulses along nerve fibres (axons and dendrites). (After Richards and Davies, 1959.)

neurones or muscles. In common with vertebrates, particularly important neurotransmitters include acetylcholine and catecholamines such as dopamine. Neurones (Fig. 3.5) are of at least four types:

- **Sensory neurones** receive stimuli from the insect's environment and transmit it to the central nervous system.
- **Interneurones** (or association neurones) receive information from and transmit it to other neurones.
- **Motor neurones** receive information from interneurones and transmit it to muscles.
- **Neuroendocrine cells** (section 3.3.1).

The cell bodies of interneurones and motor neurones are aggregated with the fibres interconnecting all types of nerve cells to form nerve centres called **ganglia**. Simple reflex behaviour has been well studied in insects (section 4.5) but insect behaviour can be complex, involving integration of neural information within the ganglia.

The **central nervous system** (CNS) (Fig. 3.6) is the principal division of the nervous system and consists of series of ganglia joined by paired longitudinal nerve cords called **connectives**. Primitively there is a pair of ganglia per body segment but usually the two ganglia of each thoracic and abdominal segment are fused into a single structure and the ganglia of all head segments coalesce to form two ganglionic centres – the **brain** and the **suboesophageal ganglion** (seen in Fig. 3.7). The chain of thoracic and abdominal ganglia found on the floor of the body cavity is called the **ventral nerve cord**. The brain, or the dorsal ganglionic centre of the head, is composed of three pairs of fused ganglia (from the first three head segments):

- The **protocerebrum**, associated with the eyes and thus bearing the optic lobes.
- The **deutocerebrum**, innervating the antennae.
- The **tritocerebrum**, concerned with handling the signals that arrive from the body.

Coalesced ganglia of the three mouthpart-bearing segments form the suboesophageal

Fig. 3.6 The central nervous system of various insects showing the diversity of arrangement of ganglia in the ventral nerve cord. Varying degrees of fusion of ganglia occur from the least to the most specialized: (a) three separate thoracic and eight abdominal ganglia, as in *Dictyopterus* (Coleoptera: Lycidae) and *Pulex* (Siphonaptera: Pulicidae); (b) three thoracic and six abdominal ganglia, as in *Blatta* (Blattodea: Blattidae) and *Chironomus* (Diptera: Chironomidae); (c) two thoracic and considerable abdominal fusion of ganglia, as in *Crabro* and *Eucera* (Hymenoptera: Sphecidae and Anthophorinae); (d) highly fused with one thoracic and no abdominal ganglia, as in *Musca*, *Calliphora* and *Lucilia* (Diptera: Muscidae, Calliphoridae); (e) extreme fusion with no separate suboesophageal ganglion, as in *Hydrometra* (Hemiptera: Hydrometridae) and *Rhizotrogus* (Scarabaeidae). (After Horridge, 1965.)

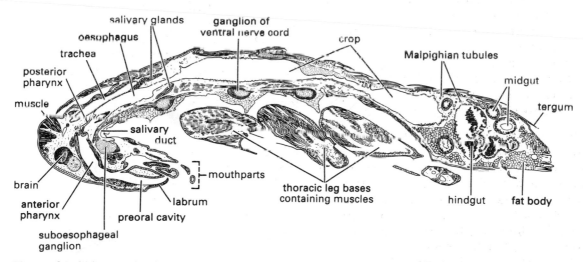

Fig. 3.7 Mediolongitudinal section of an immature cockroach of *Periplaneta americana* (Blattodea: Blattidae) showing internal organs and tissues.

BOX 3.1

MOLECULAR GENETIC TECHNIQUES AND THEIR APPLICATION TO NEUROPEPTIDE RESEARCH (after Whitten, 1989; Masler *et al.*, 1991)

Molecular biology is essentially a set of techniques for the isolation, analysis and manipulation of DNA and its RNA and protein products. Molecular genetics is concerned primarily with the nucleic acids, whereas research on the proteins and their constituent amino acids involves chemistry. Thus genetics and chemistry are integral to molecular biology. Its methods include:

- A means of cutting DNA at specific sites using restriction enzymes and of rejoining naked ends of cut fragments with ligase enzymes.
- Techniques, such as the polymerase chain reaction (PCR), that produce numerous identical copies by repeated cycles of amplification of a segment of DNA.
- Methods for rapid sequencing of nucleotides of DNA or RNA, and amino acids of proteins.
- The ability to synthesize short sequences of DNA or proteins.
- DNA–DNA affinity hybridization to compare the match of the synthesized DNA with the original sequence.
- The ability to search a genome for a specific nucleotide sequence using oligonucleotide probes, which are defined nucleic acid segments that are complementary to the sequence being sought.
- Site-directed mutation of specific *in vitro* DNA segments.
- Genetic engineering – the isolation and transfer of intact genes into other organisms, with subsequent stable transmission and gene expression.
- Cytochemical techniques to identify how, when and where genes are actually transcribed.
- Immuno and histochemical techniques to identify how, when and where a specific gene product functions.

Until relatively recently, insect peptide hormones were difficult to study because of the minute quantities produced by individual insects and their structural complexity and occasional instability. Currently, neuropeptides are the subject of an explosion of studies because of the realization that these proteins play crucial roles in most aspects of insect physiology, the availability of appropriate technologies in chemistry (e.g. gas-phase sequencing of amino acids in proteins) and genetics. Knowledge of neuropeptide amino acid sequences provides a means of using the powerful capabilities of molecular genetics. Nucleotide sequences deduced from primary protein structures allow construction of oligonucleotide probes for searching out peptide genes in other parts of the genome or, more importantly, in other organisms, especially pests. Methods such as the polymerase chain reaction (PCR) and its variants facilitate the production of probes from partial amino acid sequences and trace amounts of DNA. Genetic amplification methods, such as PCR, allow the production of large quantities of DNA and thus allow easier sequencing of genes. Of course, these uses of molecular genetic methods depend on the initial chemical characterization of the neuropeptides. Furthermore, appropriate bioassays are essential to assessment of the authenticity of any product of molecular biology.

Whitten (1989) provides an informative discussion of the uses of molecular biology in both pure and applied entomology.

ganglion with nerves emerging that innervate the mouthparts.

The **visceral** (or **sympathetic**) **nervous system** consists of three subsystems: the **stomodeal** or **stomatogastric**, which includes the frontal ganglion; the **ventral visceral**; and the **caudal visceral**. Together the nerves and ganglia of these subsystems innervate the anterior and posterior gut, several endocrine organs (corpora cardiaca and corpora allata), the reproductive organs, and the tracheal system including the spiracles.

The **peripheral nervous system** consists of all the motor neurone axons that radiate to the muscles from the ganglia of the CNS and stomodeal nervous system plus the sensory neurones of the cuticular sensory structures (the sense organs) that receive mechanical, chemical, thermal or visual stimuli from an insect's environment. Insect sensory systems are discussed in detail in Chapter 4.

3.3 THE ENDOCRINE SYSTEM AND THE FUNCTION OF HORMONES

Hormones are chemicals produced within an organism's body and transported, generally in body fluids, away from their point of synthesis to sites where they influence a remarkable variety of physiological processes even though present in extremely small quantities. Insect hormones have been studied in detail in only a handful of species but similar patterns of production and function are likely to apply to all insects. The actions and interrelationships of these chemical messengers are varied and complex but the role of hormones in the moulting process is of overwhelming importance and will be discussed more fully in this context in Chapter 6 (section 6.3). This section provides a general picture of the endocrine centres and the hormones that they export.

Historically, the implication of hormones in the processes of moulting and metamorphosis resulted from simple but elegant experiments. These utilized techniques that either removed the influence of the brain (**decapitation**), isolated the haemolymph of different parts of the body (**ligation**) or artificially connected the haemolymph of two or more insects by joining their bodies. Ligation and decapitation of insects enabled researchers to localize the sites of control of developmental and reproductive processes and to show that substances are released that affect tissues at sites distant from the point of release. In addition, critical developmental periods for the action of these controlling substances have been identified. The blood-sucking bug *Rhodnius prolixus* (Hemiptera: Reduviidae) and various moths and flies were the principal experimental insects. More refined technologies allowed microsurgical removal or transplant of various tissues, haemolymph transfusion, hormone extraction and purification, and radioactive labelling of hormone extracts. Today, molecular biological (Box 3.1) and advanced chemical analytical techniques allow hormone isolation, characterization and manipulation.

3.3.1 Endocrine centres

The hormones of the insect body are produced by neuronal, neuroglandular or glandular centres (Fig. 3.8). Hormonal production by some organs, such as the ovaries, is secondary to their main function but several tissues and organs are specialized for an endocrine role.

Neurosecretory cells (NSC), also called neuroendocrine cells, are modified neurones found throughout the nervous system (within the CNS, peripheral nervous system and the stomodeal nervous system) but they occur in major groups in the brain. These cells produce most of the known insect hormones, the notable exceptions being the production by non-neural tissues of ecdysteroids and juvenile hormones (see below). However, the synthesis and release of the latter hormones are regulated by neurohormones from NSC.

The **corpora cardiaca** are a pair of neuroglandular bodies located on either side of the aorta and behind the brain. They store and

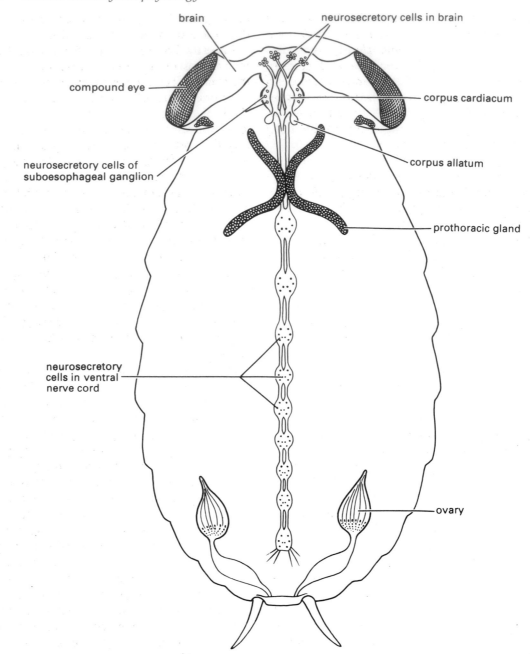

Fig. 3.8 The main endocrine centres in a generalized insect. (After Novak, 1975.)

release neurohormones, including prothoracicotropic hormone (PTTH, formerly called brain hormone or ecdysiotropin), originating from the NSC of the brain, as well as producing their own neurohormones. PTTH stimulates the secretory activity of the prothoracic glands.

The **prothoracic glands** are diffuse paired

glands generally located in the thorax or the back of the head; in cyclorrhaphous Diptera they are part of the ring gland which also contains the corpora cardiaca and corpora allata. The prothoracic glands secrete an ecdysteroid, usually ecdysone (sometimes called moulting hormone), which, after hydroxylation, elicits the moulting process of the epidermis (section 6.3).

The **corpora allata** are small, discrete, paired glandular bodies derived from the epithelium and located on either side of the foregut. In some insects, they fuse to form a single gland. Their function is to secrete juvenile hormone, which has regulatory roles in both metamorphosis and reproduction.

3.3.2 Hormones

Three hormones or hormone types are integral to the growth and reproductive functions in insects. These are the ecdysteroids, the juvenile hormones and the neurohormones.

Ecdysteroid is a general term applied to any steroid with mould-promoting activity. All ecdysteroids are derived from sterols, such as cholesterol, that insects cannot synthesize *de novo* and must obtain from their diet. Ecdysteroids occur in all insects and form a large group of compounds of which ecdysone and 20-hydroxyecdysone are the most common members. **Ecdysone** is released from the prothoracic glands into the haemolymph and is usually converted, in several peripheral tissues, to the more active hormone **20-hydroxyecdysone** (often referred to as ecdysterone or β-ecdysone in older literature), which is the most widespread and physiologically important ecdysteroid in insects. The action of ecdysteroids in eliciting moulting has been studied extensively and has the same function in different insects. Ecdysteroids also are produced by the ovary of the adult female insect and may be involved in ovarian maturation (e.g. yolk deposition) or may be packaged in the eggs to be metabolized during the formation of embryonic cuticle.

Juvenile hormones (JH) form a family of related sesquiterpenoid compounds, so that the symbol JH may denote one or a mixture of hormones, including JH-I, JH-II, JH-III and JH-0. The occurrence of insects (such as the tobacco hornworm, *Manduca sexta*) producing mixed JH adds to the complexity of unravelling the functions of the homologous JHs. These hormones have two major roles: the control of metamorphosis and regulation of reproductive development. Larval characteristics are maintained and metamorphosis is inhibited by JH; adult development requires a moult in the absence of JH (section 6.3). Thus JH controls the degree and direction of differentiation at each moult. In the adult female insect, JH stimulates the deposition of yolk in the eggs and affects accessory gland activity and pheromone production (section 5.11).

Neurohormones constitute the third and largest class of insect hormones. They are generally peptides (small proteins) – hence the alternative name **neuropeptides**. These protein messengers are the master regulators of many aspects of insect development, homeostasis, metabolism and reproduction, including the secretion of the JHs and ecdysteroids. A representative range of physiological processes reported to be affected by neurohormones in various insects is summarized in Table 3.1. The diversity and vital co-ordinating roles of these small molecules were largely unrealized until technological developments in peptide chemistry and molecular biology allowed their characterization and functional interpretation. Structural diversity among peptides of equivalent or related biological activity is a consequence of synthesis from large precursors that are cleaved and modified to form the active peptides.

Neuropeptides either reach terminal effector sites directly along nerve axons or via the haemolymph, or they indirectly exert control

Table 3.1   Examples of some important insect physiological processes mediated by neuropeptides (after Keeley and Hayes, 1987; Holman *et al.*, 1990)

Neuropeptide	*Action*
Growth and development:	
Allatostatin	Suppresses JH secretion
Allatotropin	Induces JH secretion from corpora allata
Bursicon	Activates cuticular tanning
Diapause hormone	Causes dormancy in silkworm eggs
Eclosion hormone	Controls events at eclosion
JH esterase inducing factor	Stimulates JH degradative enzyme
Prothoracicotropic hormone	Induces ecdysteroid secretion from prothoracic gland
Puparium tanning factor	Accelerates fly puparium tanning
Reproduction:	
Antigonadotropin (e.g. oostatic hormone)	Suppresses oocyte development
Egg development neurohormone (EDNH)	Activates ovaries to secrete ecdysteroids
Oviposition peptides	Stimulate egg deposition
Pheromone biosynthesis activating neuropeptide (PBAN)	Stimulates pheromone production
Metabolism and homeostasis:	
(i) Diuretic and antidiuretic peptides	
Antidiuretic hormone	Suppresses water excretion
Diuretic hormone	Enhances water excretion
Chloride-transport factor	Stimulates rectal Cl^- transport
(ii) Metabolic peptides (= AKH/RPCH family)	
Adipokinetic hormone	Releases lipid from fat body
Hyperglycaemic hormone	Releases carbohydrate from fat body
Hypoglycaemic hormone	Enhances carbohydrate uptake
Protein synthesis factors	Enhances fat body protein synthesis
(iii) Myotropic peptides	
Cardiopeptides	Increases heartbeat rate
Kinin family (e.g. leucokinins) and myosuppressins	Regulation of gut contraction
Proctolin	Increases hindgut contraction

via their action on other endocrine glands (corpora allata and prothoracic glands). Both inhibitory and stimulatory signals are involved in neurohormone regulation. The effectiveness of regulatory neuropeptides depends on stereospecific high-affinity binding sites located in the plasma membrane of the target cells.

Hormones reach their target tissues by transport (even over short distances) by the body fluid or haemolymph. Hormones are often water-soluble but some may be transported bound to proteins in the haemolymph; for example, ecdysteroid-binding proteins and JH-binding proteins are known in a number of insects. These haemolymph-binding proteins may contribute to the regulation of hormone levels by facilitating uptake by target tissues, reducing non-specific binding, or protecting from degradation or excretion.

3.4 THE CIRCULATORY SYSTEM

Haemolymph, the insect body fluid (section 3.4.1), circulates freely around the internal organs. The pattern of flow is regular between compartments and appendages; it is assisted by muscular contractions of parts of the body, especially the peristaltic contractions of a longitudinal **dorsal vessel**, part of which is sometimes called the heart. Haemolymph does not directly contact the cells because the internal organs and the epidermis are covered by a basement membrane which may regulate the exchange of materials. This open circulatory system has only a few vessels and compartments to direct haemolymph movement, in contrast to the closed network of blood-conducting vessels seen in vertebrates.

3.4.1 Haemolymph

The volume of the haemolymph may be substantial (20–40% of body weight) in soft-bodied larvae, which use the body fluid as a hydrostatic skeleton, but is less than 20% of body weight in most nymphs and adults. Haemolymph is a watery fluid containing ions, molecules and cells. It is often clear and colourless but may be variously pigmented yellow, green or blue, or rarely, in the immature stages of a few aquatic and endoparasitic flies, red due to haemoglobin. All chemical exchanges between insect tissues are mediated via the haemolymph: hormones are transported, nutrients are distributed from the gut, and wastes are removed to the excretory organs. However, insect haemolymph only rarely contains respiratory pigments and hence has a very low oxygen-carrying capacity. Local changes in haemolymph pressure are important in ventilation of the tracheal system (section 3.5.1), in thermoregulation (section 4.2.2) and at moulting to aid splitting of the old and expansion of the new cuticle. The haemolymph is also a reserve of water since its main constituent, **plasma**, is an aqueous solution of inorganic ions, lipids,

sugars (mainly trehalose), amino acids, proteins, organic acids and other compounds. High concentrations of amino acids and organic phosphates characterize insect haemolymph. The blood cells, or **haemocytes**, are of several types (mainly plasmatocytes, cystocytes and granular cells) and all are nucleate. They have four basic functions: phagocytosis – the ingestion of small particles and substances such as metabolites; encapsulation of parasites and other large foreign materials; haemolymph coagulation; and storage and distribution of nutrients.

The haemocoel contains two further types of cells. **Nephrocytes** (sometimes called pericardial cells) generally occur near the dorsal vessel and appear to function as ductless glands by sieving the haemolymph of certain substances and metabolizing them for use or excretion elsewhere. **Oenocytes** may occur in the haemocoel, fat body or epidermis and, although their functions are unclear in most insects, they appear to have a role in cuticle paraffin synthesis and, in some chironomids, they produce haemoglobins.

3.4.2 Circulation

Circulation in insects is maintained by a system of muscular pumps moving haemolymph through compartments separated by fibromuscular septa or membranes. The main pump is the pulsatile dorsal vessel. The anterior part may be called the aorta and the posterior part the heart but the two terms are inconsistently applied. The dorsal vessel is a simple tube, generally composed of one layer of myocardial cells and with segmentally arranged openings, or **ostia**. The ostia typically permit the one-way flow of haemolymph into the dorsal vessel due to valves that prevent backflow. There may be up to three pairs of thoracic ostia and nine pairs of abdominal ostia, although there is an evolutionary tendency towards reduction in number of ostia. The dorsal vessel lies in the **pericardial sinus**, a compartment above a

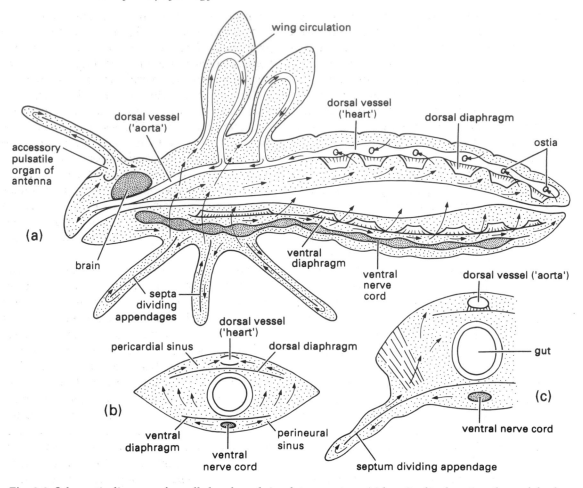

Fig. 3.9 Schematic diagram of a well-developed circulatory system: (a) longitudinal section through body; (b) transverse section of the abdomen; (c) transverse section of the thorax. Arrows indicate directions of haemolymph flow. (After Wigglesworth, 1972.)

dorsal diaphragm (a fibromuscular septum – a separating membrane) formed of connective tissue and segmental pairs of **alary muscles**. The alary muscles support the dorsal vessel but their contractions do not affect heartbeat. Haemolymph enters the pericardial sinus via segmental openings in the diaphragm and/or at the posterior border and then moves into the dorsal vessel via the ostia during a muscular relaxation phase. Waves of contraction, which normally start at the posterior end of the body, pump the haemolymph forward in the dorsal vessel and out via the aorta into the head. Next, the appendages of the head and thorax are supplied with haemolymph as it circulates posteroventrally and eventually returns to the pericardial sinus and the dorsal vessel. The direction of haemolymph circulation in the body is shown in Fig. 3.9.

Another important component of the circulation of many insects is the **ventral diaphragm** (Fig. 3.9(b)), a fibromuscular septum that lies in the floor of the body cavity associated with the ventral nerve cord. Circulation of the haemolymph is aided by active peristaltic contractions of the ventral diaphragm which direct the haemolymph backwards and laterally in the **perineural sinus** below the diaphragm. These movements are especially important in insects that use the circulation in thermoregulation (some Odonata, Diptera, Lepidoptera and Hymenoptera). Another function of the diaphragm may be to facilitate rapid exchange of chemicals between the ventral nerve cord and the haemolymph by either actively moving the haemolymph or moving the cord itself.

Haemolymph is generally circulated to appendages unidirectionally by various tubes, septa, valves and pumps (Fig. 3.9(c)). The muscular pumps are termed **accessory pulsatile organs** and occur at the base of the antennae and in the legs. Furthermore, the antennal pulsatile organs may release neurohormones that are carried to the antennal lumen to influence the sensory neurones. Wings also have a definite but variable circulation, although it may be apparent only in the young adult. Circulation may be sustained by influxes of air into the wing veins rather than by any postulated wing pulsatile organs. Pulses of air in the fine tracheal tubes of the veins are thought to push the accompanying haemolymph through the enclosed space of the veins.

The insect circulatory system displays an impressive degree of co-ordination between the activities of the dorsal vessel, fibromuscular diaphragms and accessory pumps, mediated by both the nervous and neurohormonal regulation. The physiological regulation of many body functions by the neurosecretory system occurs via neurohormones transported in the haemolymph.

3.4.3 Protection and defence by the haemolymph

Haemolymph provides various kinds of protection and defence from physical injury, the entry of disease organisms, parasites or other foreign substances, and sometimes from the actions of predators. In some insects the haemolymph contains malodorous or distasteful chemicals which are deterrent to predators (Chapter 13). Injury to the integument elicits a wound-healing process that involves haemocyte and plasma coagulation. A haemolymph clot is formed to seal the wound and reduce further haemolymph loss and bacterial entry. If disease organisms or particles enter an insect's body, immune responses are invoked. These include the cellular defence mechanisms of phagocytosis, encapsulation and nodule formation mediated by the haemocytes, as well as the actions of humoral factors such as enzymes or other proteins (e.g. lysozymes, prophenoloxidase, lectins and peptides).

The immune system of insects has little resemblance to the complex immunoglobulin-based vertebrate system. However, insects sublethally infected with bacteria can rapidly develop greatly increased resistance to subsequent infection. Haemocytes are involved in phagocytosing bacteria but, in addition, immunity proteins with antibacterial activity appear in the haemolymph after a primary infection. For example, lytic peptides called cecropins, which disrupt the cell membranes of bacteria and other pathogens, have been isolated from certain moths. Furthermore, some neuropeptides may participate in cell-mediated immune responses by exchanging signals between the neuroendocrine system and the immune system, as well as influencing the behaviour of cells involved in immune reactions. The insect immune system is much more complicated than it was once thought to be.

3.5 THE TRACHEAL SYSTEM AND GAS EXCHANGE

In common with all aerobic animals, insects must obtain oxygen from their environment and eliminate carbon dioxide respired by their cells. This is **gas exchange** – as distinguished from **respiration**, which strictly refers to

Fig. 3.10 Schematic diagram of a generalized tracheal system seen in a transverse section of the body at the level of a pair of abdominal spiracles. Enlargements show: (a) an atriate spiracle with closing valve at inner end of atrium; (b) tracheoles running to a muscle fibre. (After Snodgrass, 1935.)

oxygen-consuming, cellular metabolic processes. In almost all insects, gas exchange occurs by means of internal air-filled tracheae. These tubes branch and ramify through the body (Fig. 3.10). The finest branches contact all internal organs and tissues and are especially numerous in tissues with high oxygen requirements. Air usually enters the tracheae via spiracular openings positioned laterally on the body, primitively with one pair per post-cephalic segment. No extant insect has more than ten pairs (two thoracic

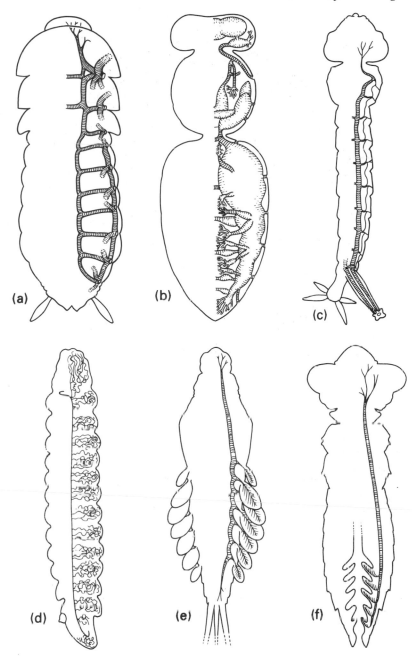

Fig. 3.11 Some basic variations in the ((a)–(c)) open and ((d)–(f)) closed tracheal systems of insects: (a) simple tracheae with valved spiracles, as in cockroaches; (b) tracheae with mechanically ventilated air-sacs, as in honey bees; (c) metapneustic system with only terminal spiracles functional, as in mosquito larvae; (d) entirely closed tracheal system with cutaneous gas exchange, as in most endoparasitic larvae; (e) closed tracheal system with abdominal tracheal gills, as in mayfly nymphs; (f) closed tracheal system with rectal tracheal gills, as in dragonfly nymphs. (After Wigglesworth, 1972; details in (a) after Richards and Davies, 1977; (b) after Snodgrass, 1956; (c) after Snodgrass, 1935; (d) after Wigglesworth, 1972.)

and eight abdominal) (Fig. 3.11(a)); most have eight or nine, and some have one (Fig. 3.11(c)), two or none (Fig. 3.11(d)–(f)). Typically, spiracles (Fig. 3.10(a)) have a chamber or **atrium** with an opening-and-closing mechanism or **valve**, either projecting externally or at the inner end of the atrium. In the latter type, a filter apparatus sometimes protects the outer opening. Each spiracle may be set in a sclerotized cuticular plate called a **peritreme**.

The tracheae are invaginations of the epidermis and thus their lining is continuous with the body cuticle. The characteristic ringed appearance of the tracheae seen in tissue sections (as in Fig. 3.7) is due to the spiral ridges or thickenings of the cuticular lining, the **taenidia**, which allow the tracheae to be flexible but resist compression (analogous to the function of the ringed hose of a vacuum cleaner). The cuticular linings of the tracheae are shed with the rest of the exoskeleton when the insect moults. Usually only the linings of the finest branches of the tracheal system, the **tracheoles**, remain largely intact through the life of an insect. Tracheoles are less than 1 μm in diameter; they end blindly and closely contact the respiring tissues (Fig. 3.10(b)), sometimes indenting into the cells that they supply.

In terrestrial and many aquatic insects the tracheae open to the exterior via the spiracles (an **open tracheal system**) (Fig. 3.11(a)–(c)). In contrast, in some aquatic and many endoparasitic larvae spiracles are absent (a **closed tracheal system**) and the tracheae divide peripherally to form a network. This covers the general body surface, allowing cutaneous gas exchange (Fig. 3.11(d)), or lies within specialized filaments or lamellae (tracheal gills; Fig. 3.11(e),(f)). Some aquatic insects with an open tracheal system carry temporary or permanent **gas gills**, such as bubbles of air (Chapter 9).

The volume of the tracheal system varies widely between perhaps 5% and 50% of the body volume, depending on species and stage of development. The more active the insect,

the more extensive is the tracheal system. In many insects, parts of tracheae are dilated or enlarged to increase the reservoir of air, and in some species the dilations form **air-sacs** (Fig. 3.11(b)), which collapse readily because the taenidia of the cuticular lining are reduced or absent. Sometimes the tracheal volume may decrease within a developmental stage as air-sacs are occluded by growing tissues. Air-sacs reach their greatest development in very active flying insects, such as bees and cyclorrhaphous Diptera. They may assist flight by increasing buoyancy, but their main function is in ventilation of the tracheal system.

3.5.1 Diffusion and ventilation

Oxygen enters the spiracle and passes through the length of the tracheae to the tracheoles and into the target cells by a combination of ventilation and diffusion along a concentration gradient, from high in the external air to low in the tissue. Whereas the net movement of oxygen molecules in the tracheal system is inwards, the net movement of carbon dioxide and (in terrestrial insects) water vapour molecules is outwards. Hence gas exchange in most terrestrial insects is a compromise between securing sufficient oxygen and reducing water loss via the spiracles. During periods of inactivity, the spiracles in many insects are kept closed most of the time, only opening periodically. In insects adapted to xeric environments, the spiracles may be small with deep atria or have a mesh of cuticular projections in the orifice.

In insects without air-sacs, such as most holometabolous larvae, diffusion appears to be the primary mechanism for the movement of gases in the tracheae and is always the sole mode of gas exchange at the tissues. The efficiency of diffusion is related to the distance of diffusion and perhaps to the diameter of the tracheae (Box 3.2). Furthermore, large or dilated tracheae may serve as an oxygen reserve when the spiracles are closed. In very

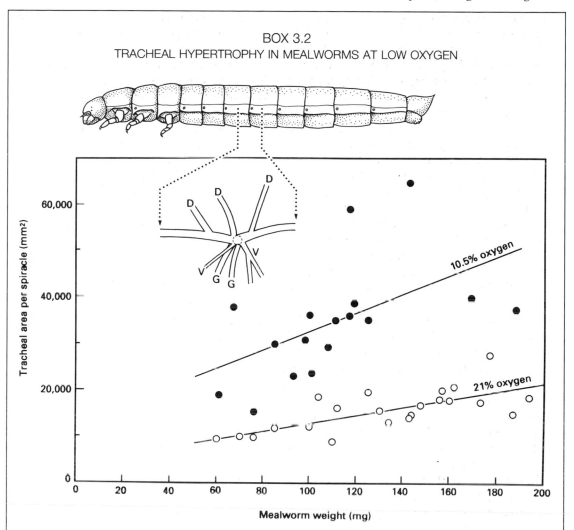

BOX 3.2
TRACHEAL HYPERTROPHY IN MEALWORMS AT LOW OXYGEN

Resistance to diffusion of gases in insect tracheal systems arises from the spiracular valves when they are partially or fully closed, the tracheae, and the cytoplasm supplied by the tracheoles at the end of the tracheae. Air-filled tracheae will have a much lower resistance per unit length than the watery cytoplasm because oxygen diffuses several orders of magnitude faster in air than in cytoplasm for the same gradient of oxygen partial pressure. Until recently, the tracheal system was believed to provide more than sufficient oxygen (at least in non-flying insects that lack air sacs), with the tracheae offering trivial resistance to the passage of oxygen. Experiments on mealworm larvae, *Tenebrio molitor* (Coleoptera: Tenebrionidae), that were reared in different levels of oxygen (all at the same total gas pressure) showed that the main tracheae supplying oxygen to the tissues in the larvae hypertrophy (increase in size) in lower oxygen levels. The dorsal (D), ventral (V) and visceral or gut (G) tracheae were affected but not the lateral longitudinal tracheae that interconnect the

---

BOX 3.2—*continued*

spiracles (the four tracheal categories are illustrated in an inset on the graph). The dorsal tracheae supply the dorsal vessel and dorsal musculature; the ventral tracheae supply the nerve cord and ventral musculature; the visceral tracheae supply the gut, fat body and gonads. The graph shows that the cross-sectional areas of the dorsal, ventral and visceral tracheae were greater when the larvae were reared in 10.5% oxygen (●) than in 21% oxygen (as in normal air)(○) (after Loudon, 1989). Each point on the graph is for a single larva and is the average of the summed areas of the dorsal, ventral and visceral tracheae for six pairs of abdominal spiracles. This hypertrophy appears to be inconsistent with the widely accepted hypothesis that tracheae contribute an insignificant resistance to net oxygen movement in insect tracheal systems. Alternatively, hypertrophy may simply increase the amount of air (and thus oxygen) that can be stored in the tracheal system, rather than reduce resistance to air flow. This might be particularly important for mealworms because they normally live in a dry environment and may minimize the opening of their spiracles. Whatever the explanation, the observations suggest that some adjustment can be made to the size of the tracheae in mealworms (and perhaps other insects) to match the requirements of the respiring tissues.

---

active insects, especially large ones, active pumping movements of the thorax or abdomen **ventilate** (pump air through) the outer parts of the tracheal system and so the diffusion pathway to the tissues is reduced. Rhythmic thoracic movements and/or dorsoventral flattening or telescoping of the abdomen expels air, via the spiracles, from extensible or some partially compressible tracheae or from air-sacs. Co-ordinated opening and closing of the spiracles usually accompanies ventilatory movements and provides the basis for the unidirectional air flow that occurs in the main tracheae of larger insects. Anterior spiracles open during inspiration and posterior ones during expiration. The presence of air-sacs, especially if large or extensive, facilitates ventilation by increasing the volume of tidal air that can be changed as a result of ventilatory movements. If the main tracheal branches are strongly ventilated, diffusion appears sufficient to oxygenate even the most actively respiring tissues, such as flight muscles. However, the design of the gas exchange system of insects places an upper limit on size because, if oxygen has to diffuse over a considerable distance, the requirements of a very large and active insect either could not be met, even with ventilatory movements, or would result in substantial loss of water through the spiracles. It is interesting that many large insects are long and thin, thereby minimizing the diffusion distance from the spiracle along the trachea to any internal organ.

3.6 THE GUT, DIGESTION AND NUTRITION

Insects of different groups consume an astonishing variety of foods, including watery xylem sap (e.g. nymphs of spittle bugs and cicadas), vertebrate blood (e.g. bed bugs and female mosquitoes), dry wood (e.g. some termites), bacteria and algae (e.g. black fly and many caddisfly larvae) and the internal tissues of other insects (e.g. endoparasitic wasp larvae). The diverse range of mouthpart types (section 2.3.1) correlates with the diets of

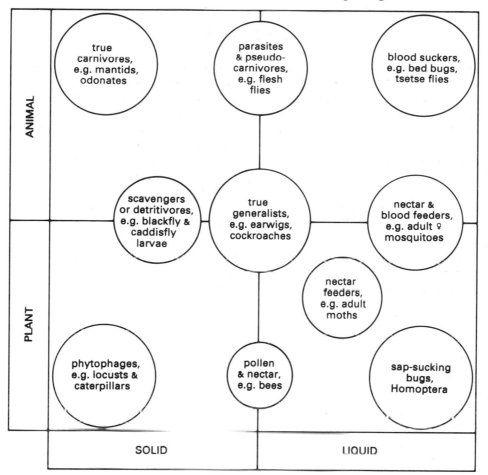

Fig. 3.12 The four major categories of insect feeding specialization. Many insects are typical of one category, but others cross two categories (or more, as in generalist cockroaches). (After Dow, 1986.)

different insects, but gut structure and function also reflect the mechanical properties and the nutrient composition of the food eaten. Four major feeding specializations can be identified, depending on whether the food is solid or liquid or of plant or animal origin (Fig. 3.12). Some insect species clearly fall into a single category, but others with generalized diets may fall between two or more of them, and most endopterygotes will occupy different categories at different life-stages (e.g. moths and butterflies switch from solid/plant as larvae to liquid/plant as adults).

Gut morphology and physiology relate to these dietary differences in the following ways. Insects that take solid food typically have a wide, straight, short gut with strong musculature and obvious protection from abrasion (especially in the midgut which has no cuticular lining). These features are most obvious in solid-feeders with rapid throughput of food, for example, plant-feeding caterpillars. In contrast, insects feeding on blood, sap or nectar usually have long, narrow, convoluted guts to allow maximal contact with the liquid food and protection from abrasion is unnecessary. The most obvious gut specialization of liquid-feeders is a

BOX 3.3
THE FILTER CHAMBER OF HOMOPTERA

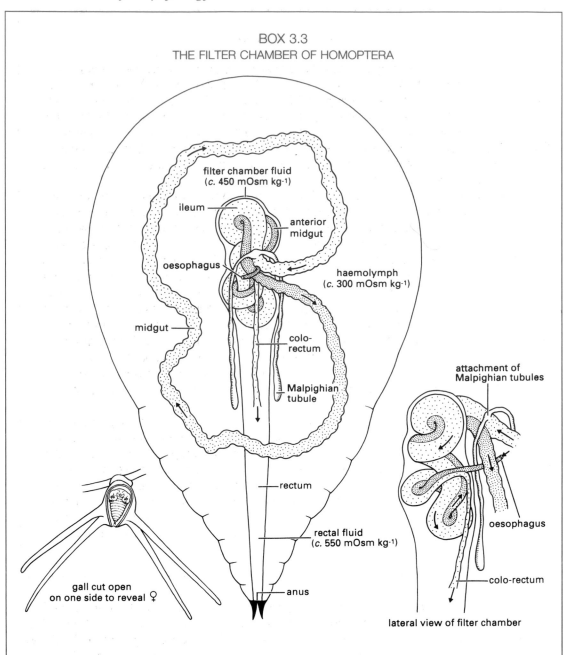

filter chamber fluid
(*c.* 450 mOsm kg⁻¹)

ileum

anterior midgut

oesophagus

haemolymph
(*c.* 300 mOsm kg⁻¹)

midgut

colo-rectum

attachment of
Malpighian tubules

Malpighian tubule

rectum

oesophagus

gall cut open
on one side to reveal ♀

rectal fluid
(*c.* 550 mOsm kg⁻¹)

colo-rectum

anus

lateral view of filter chamber

Most Homoptera have an unusual arrangement of the midgut which is related to their habit of feeding on plant fluids. An anterior and a posterior part of the gut (typically involving the midgut) are in intimate contact to allow concentration of the liquid food. This **filter chamber** enables excess water and relatively small molecules, such as simple sugars, to be passed quickly and directly from the anterior gut to the hindgut, thereby short-circuiting the main

BOX 3.3—*continued*

absorptive portion of the midgut. Thus the digestive region is not diluted by water nor congested by superabundant food molecules. Well-developed filter chambers are characteristic of cicadas and spittlebugs that feed on xylem (sap that is rich in ions, low in organic compounds and with low osmotic pressure) and leafhoppers and coccoids that feed on phloem (sap that is rich in nutrients, especially sugars, and with high osmotic pressure). The gut physiology of such sap-suckers has been rather poorly studied because accurate recording of gut fluid composition and osmotic pressure depends on the technically difficult task of taking readings from an intact gut.

Adult female coccoids of gall-forming *Apiomorpha* species (Homoptera: Eriococcidae) (Chapter 10) tap the vascular tissue of the gall wall to obtain phloem sap. Some species have a highly developed filter chamber formed from loops of the anterior midgut and anterior hindgut enclosed within the membranous rectum. Depicted here is the gut of an adult female of *A. munita* viewed from the ventral side of the body. The thread-like sucking mouthparts (Fig. 10.4(c)) in series with the cibarial pump connect to a short oesophagus, which can be seen here in the main drawing and in the enlarged lateral view of the filter chamber. The oesophagus terminates at the anterior midgut, which coils upon itself as three loops of the filter chamber. It emerges ventrally and forms a large midgut loop lying free in the haemolymph. Absorption of nutrients occurs in this free loop. The Malpighian tubules enter the gut at the commencement of the ileum, before the ileum enters the filter chamber where it is closely apposed to the much narrower anterior midgut. Within the irregular spiral of the filter chamber, the fluids in the two tubes move in opposite directions (as indicated by the arrows).

The filter chamber of these coccoids apparently transports sugar (perhaps by active pumps) and water (passively) from the anterior midgut to the ileum and then via the narrow colo-rectum to the rectum, from which it is eliminated as honeydew. In *A. munita* the honeydew is mostly sugar (accounting for 80% of the total osmotic pressure of about 550 mOsm/kg). Remarkably, the osmotic pressure of the haemolymph (about 300 mOsm/kg) is much lower than that within the filter chamber (about 450 mOsm/kg) and rectum. Maintenance of this large osmotic difference may be facilitated by the impermeability of the rectal wall. (Osmolarity values are from the unpublished data of P.D. Cooper and A.T. Marshall).

---

mechanism for removing excess water to concentrate nutrient substances prior to digestion, as seen in homopterans (Box 3.3). From a nutritional viewpoint, most plant-feeding insects need to process large amounts of food because nutrient levels in leaves and stems are often low. The gut is usually short and without storage areas since food is available continuously. By comparison, a diet of animal tissue is nutrient rich and, at least for predators, well balanced. However, the food may be available only intermittently (such as when a predator captures prey or a blood meal is obtained) and the gut normally has large storage capacity.

3.6.1 Structure of the gut

There are three main regions to the insect gut (or alimentary canal), with sphincters (valves)

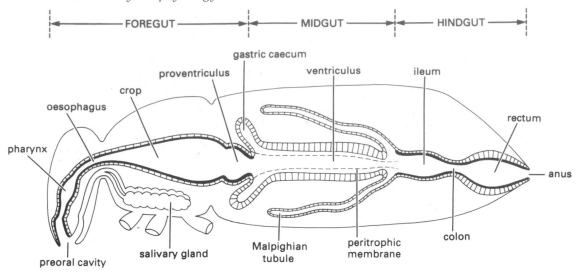

Fig. 3.13 Generalized insect alimentary canal showing division into three regions. The cuticular lining of the foregut and hindgut are indicated by thicker black lines. (After Dow, 1986.)

controlling food/fluid movement between regions (Fig. 3.13). The **foregut (stomodeum)** is concerned with ingestion, storage, grinding and transport of food to the next region, the **midgut (mesenteron)**. Here digestive enzymes are produced and secreted and absorption of the products of digestion occur. The material remaining in the gut lumen together with urine from the Malpighian tubules then enter the **hindgut (proctodeum)**, where absorption of water, salts and other valuable molecules occurs prior to elimination of the faeces through the anus. The gut epithelium is one cell layer thick throughout the length of the canal and rests on a basement membrane surrounded by a variably developed muscle layer. Both the foregut and hindgut have a cuticular lining, whereas the midgut does not.

Each region of the gut displays several local specializations which are variously developed in different insects, depending on diet. Typically the foregut is subdivided into a **pharynx**, an **oesophagus** and a **crop** (food storage area); and in insects that ingest solid food there is often a grinding organ, the **proventriculus** (or gizzard). The proventriculus is especially well developed in orthopteroid insects, such as cockroaches, crickets and termites, in which the epithelium is folded longitudinally to form ridges on which the cuticle is armed with spines or teeth. At the anterior end of the foregut, the mouth opens into a preoral cavity bounded by the bases of the mouthparts and often divided into an upper area or **cibarium** and a lower part or **salivarium** (Fig. 3.14a). The paired **labial** or **salivary glands** vary in size and arrangement from simple elongated tubes to complex branched or lobed structures. Complicated glands occur in many Hemiptera that produce two types of saliva (section 3.6.2). In Lepidoptera, the labial glands produce silk, whereas mandibular glands secrete the saliva. Several types of secretory cell may occur in the salivary glands of one insect. The secretions from these cells are transported along cuticular ducts and emptied into the ventral part of the preoral cavity. In insects that store meals in their foregut, the crop may take up the greater portion of the food and is often capable of extreme distension, with a posterior sphincter controlling food retention. The crop may be an enlargement of part of the tubular gut (Fig. 3.7) or a lateral diverticulum.

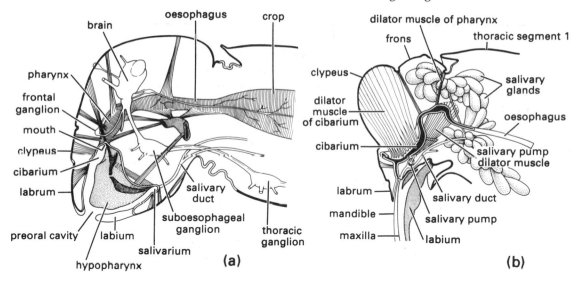

Fig. 3.14 Preoral and anterior foregut morphology in (a) a generalized orthopteroid insect and (b) a xylem feeding cicada. Musculature of the mouthparts and the (a) pharyngeal or (b) cibarial pump are indicated but not fully labelled. Contraction of the respective dilator muscles causes dilation of the pharynx or cibarium and fluid is drawn into the pump chamber. Relaxation of these muscles results in elastic return of the pharynx or cibarial walls and expels food upwards into the oesophagus. (After Snodgrass, 1935.)

In generalized insects there are two main areas of the midgut: the tubular **ventriculus** and blindly ending lateral diverticula called **caeca**. Most cells of the midgut are structurally similar, being columnar with microvilli (finger-like protrusions) covering the inner surface. The distinction between the almost indiscernible foregut epithelium and the thickened epithelium of the midgut is usually visible in histological sections (Fig. 3.15). The midgut epithelium of most insects is separated from the food by a thin sheath called the **peritrophic membrane**. It consists of a network of chitin fibrils in a protein–carbohydrate matrix and is either delaminated from the whole midgut or produced by cells at the anterior of the midgut. Hemiptera are exceptional in lacking a peritrophic membrane.

Typically, the beginning of the hindgut is defined by the entry point of the Malpighian tubules, often into a distinct **pylorus** forming a muscular pyloric sphincter, followed by the ileum, colon and rectum. The main functions of the hindgut are the absorption of water, salts and other useful substances from the faeces and urine; a detailed discussion of structure and function is presented in section 3.7.1.

3.6.2 Saliva and food ingestion

Salivary secretions dilute the ingested food and adjust its pH and ionic content. The saliva often contains digestive enzymes and, in blood-feeding insects, anticoagulants may be present. In insects with extra-intestinal digestion, such as predatory Hemiptera, digestive enzymes are exported into the food and the resulting liquid is ingested. Most Hemiptera produce an alkaline watery saliva that is a vehicle for enzymes (either digestive or lytic) and a proteinaceous solidifying saliva that forms either a complete sheath around the mouthparts (stylets) as they pierce and penetrate the food or just a securing flange at the

Fig. 3.15 Longitudinal section through the anterior body of a caterpillar of the cabbage white butterfly, *Pieris rapae* (Lepidoptera: Pieridae). Note the thickened epidermal layer lining the midgut.

point of entry (section 10.2.3; Fig. 10.4(c)). Stylet-sheath feeding is characteristic of phloem- and xylem-feeding Homoptera, such as aphids, scale insects (coccoids) and spittle-bugs, which leave visible tracks formed of exuded solidifying saliva in the plant tissue on which they have fed. The sheath may function to guide the stylets, prevent loss of fluid food from the feeding puncture or seal wounds to reduce defensive reaction by the plant. By comparison a macerate-and-flush strategy is typical of Heteroptera, such as mirids and coreids. These insects disrupt the tissues of plants or other insects by thrusting of the stylets or by addition of salivary enzymes. The macerated or partly digested food is 'flushed out' with saliva and then ingested by sucking.

In fluid-feeding insects there are prominent dilator muscles which attach to the walls of the pharynx or the preoral cavity to form a pump (Fig. 3.14(b)), although some sort of pharyn-geal pump is present in most other insects

(Fig. 3.14(a)) to allow drinking and also air intake to facilitate cuticle expansion during a moult.

3.6.3 Digestion of food

Most digestion occurs in the midgut, where the epithelial cells produce and secrete digestive enzymes and also absorb the resultant food breakdown products. Insect food consists principally of polymers of carbohydrates and proteins. Digestion involves breaking these large molecules into small monomers via the action of a series of enzymes. The midgut pH is usually 6–7.5, although very alkaline values (pH 9–12) occur in many plant-feeding insects that extract hemicellu-loses from plant cell walls and very low pH values occur in many Diptera. High pH may prevent or reduce the binding of dietary tannins to food proteins, thereby increasing the digestibility of ingested plants, although

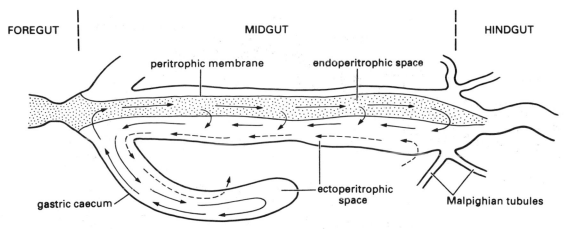

Fig. 3.16 Generalized scheme of the endo–ectoperitrophic circulation of digestive enzymes in the midgut. (After Terra and Ferreira, 1981.)

the significance of this mechanism *in vivo* is unclear. In some insects, gut lumenal surfactants (detergents) may have an important role in preventing the formation of tannin–protein complexes, particularly in insects with near neutral gut pH.

In most insects, the midgut epithelium is separated from the food bolus by the peritrophic membrane, which constitutes a very efficient high-flux sieve. It is perforated by pores which allow passage of small molecules while restricting large molecules, bacteria and food particles from directly accessing the midgut cells. In some insects, all or most midgut digestion occurs inside the peritrophic membrane in the **endoperitrophic space**. In others only initial digestion occurs there and smaller food molecules then diffuse out into the **ectoperitrophic space**, where further digestion takes place (Fig. 3.16). A final phase of digestion usually occurs on the surface of the midgut microvilli where certain enzymes are either trapped in a mucopolysaccharide coating or bound to the cell membrane. Thus the peritrophic membrane forms a permeability barrier and helps to compartmentalize the phases of digestion, in addition to mechanical protection of the midgut cells, which was once believed to be its principal function. Fluid

containing partially digested food molecules and digestive enzymes is thought to circulate through the midgut in a posterior direction in the endoperitrophic space and forwards in the ectoperitrophic space, as indicated in Fig. 3.16. This endo–ectoperitrophic circulation may facilitate digestion by moving food molecules to sites of final digestion and absorption or by conserving digestive enzymes that are removed from the food bolus before it passes to the hindgut.

3.6.4 The fat body

In many insects, especially the larvae of holometabolous groups, fat body tissue is a conspicuous component of the internal anatomy (Figs 3.7 and 3.15). Typically, it forms a white or yellow tissue formed of loose sheets or ribbons of cells lying in the haemocoel. The structure of this organ is ill-defined and taxonomically variable, but often caterpillars and other larvae have a peripheral layer of fat body beneath the cuticle and a central layer around the gut. The fat body is an organ of multiple metabolic functions, including the metabolism of carbohydrates, lipids and nitrogenous compounds, the storage of glycogen, fat and protein, the synthesis and regulation of blood sugar, and the synthesis of major haemolymph proteins (such as haemoglobins, vitellogenins for yolk formation, and storage proteins). Fat body cells can switch their activities in response to nutritional and hormonal signals to supply the requirements of insect growth, metamorphosis and reproduction. For example, specific storage proteins are synthesized by the fat body during the last larval instar of holometabolous insects and accumulate in the haemolymph to be used during metamorphosis as a source of amino acids for the synthesis of proteins during pupation. **Calliphorin**, a haemolymph storage protein synthesized in the fat body of larval blow flies (Diptera: Calliphoridae: *Calliphora*) may comprise about 75% of the haemolymph protein of a late-instar maggot or about 7 mg; the amount of calliphorin falls to around 3 mg at the time of pupariation and to 0.03 mg after emergence of the adult fly. The production and deposition of proteins specifically for amino acid storage is a feature that insects share with seed plants but not with vertebrates. Humans, for example, excrete any dietary amino acids that are in excess of immediate needs.

The principal cell type found in the fat body is the **trophocyte**, which is responsible for most of the above metabolic and storage functions. Visible differences in the extent of the fat body in different individuals of the same insect species reflect the amount of material stored in the trophocytes; little body fat indicates either active tissue construction or starvation. Two other cell types – urocytes and mycetocytes – may occur in the fat body of some insect groups. **Urocytes** temporarily store spherules of urates, including uric acid, one of the nitrogenous wastes of insects. Studies of cockroaches show that urocytes carry out recycling of urate nitrogen, perhaps with assistance of mycetocyte bacteria, rather than being permanent stores of excreted waste uric acid (storage excretion). **Mycetocytes** contain symbiotic micro-organisms and are scattered through the fat body of cockroaches or contained within special organs (section 3.6.5), sometimes surrounded by fat body. These bacteria-like symbionts appear to have various functions in insect nutrition.

3.6.5 Nutrition and micro-organisms

Broadly defined, nutrition concerns the nature and processing of foods needed to meet the requirements for growth and development. It involves feeding behaviour and digestion, which are dealt with in Chapter 2 and above in section 3.6. Insects often have unusual or restricted diets. Sometimes, even though only one or a few foods are eaten, the diet provides a complete range of the chemicals essential to metabolism. In these cases, monophagy is a behavioural specialization without nutritional limitations. In other cases, a restricted diet may necessitate the utilization of micro-organisms in digesting or supplementing the directly available nutrients. In particular, sterols (required for moulting hormone) and carotenoids (used in visual pigments) cannot be synthesized by insects and must come from the diet or micro-organisms.

Insects may harbour extra- or intracellular micro-organisms (referred to as **symbionts** because they are dependent on their insect hosts) which are known or assumed to contribute to the nutrition of their hosts by functioning in sterol, vitamin, carbohydrate or amino acid synthesis, or metabolism. Symbiotic micro-organisms may be bacteria or bacteroids, yeasts or other unicellular fungi, or protozoa. Studies on their function are hampered by difficulties in removing them (for example, with antibiotics) without harming the host insect. However, the diets of their hosts provide some clues as to the functions of these micro-organisms. Insect hosts include many sap-sucking homopterans (such as aphids, psyllids, whiteflies, scale insects, leafhoppers and cicadas) and sap- and blood-sucking heteropterans (Hemiptera), lice (Phthiraptera), some wood-feeding insects (such as termites and some longicorn beetles and weevils), many seed- or grain-feeding insects (certain beetles) and some omnivorous insects (such as cockroaches, some termites and some ants). Predatory insects never seem to contain such symbionts. This suggests that micro-organisms are required by insects with suboptimal nutrition, that is, those lacking essential chemicals or containing substances (e.g. lignin and cellulose) indigestible by the insect unaided.

Transmission from one individual insect to another is by one of two main methods, depending upon where the symbionts are located within the insect. The first is oral uptake by the offspring, which is appropriate for insects with gut symbionts. Micro-organisms may be acquired from the anus or the excreta of other individuals or eaten at a specific time, as in some bugs in which the newly-hatched young eat the contents of special symbiont-containing capsules deposited with the eggs. Extracellular symbionts may be free in the gut lumen or housed in diverticula or pockets of the midgut or hindgut. For example, the hindguts of some termites contain bacteria or flagellate protozoa that digest the cellulose in wood. Intracellular symbionts may be housed in the gut epithelium, as in lygaeid bugs and some weevils.

However, most insects with intracellular micro-organisms house them in mycetocytes (symbiont-containing cells) in the body cavity, usually associated with the fat body or the gonads, and often in special aggregations of mycetocytes forming an organ called a **mycetome**. In such insects, the symbionts are transferred to the ovary and then to the eggs or embryos prior to oviposition or parturition – a process referred to as **transovarial transmission**. This method of transmission, found in many Homoptera and cockroaches, indicates a very close association or coevolution of the insects and their micro-organisms.

Some insects that maintain fungi essential to their diet cultivate them externally as a means of converting woody substances to an assimilable form. Examples are the fungus gardens of some ants (Formicidae) and termites (Termitidae) (section 8.5) and the fungi transmitted by certain timber pests, namely, wood wasps (Hymenoptera: Siricidae) and ambrosia beetles (Coleoptera: Scolytidae).

3.7 THE EXCRETORY SYSTEM AND WASTE DISPOSAL

Excretion – the removal from the body of waste products of metabolism, especially nitrogenous compounds – is essential. It differs from defecation in that excretory wastes have been metabolized in cells of the body rather than simply passing directly from the mouth to the anus (sometimes essentially unchanged chemically). Of course, insect faeces, either in liquid form or packaged in pellets and known as **frass**, contains both undigested food and metabolic excretions. Aquatic insects excrete dilute wastes from their anus directly into water, and so their faecal material is flushed away. In comparison, terrestrial insects generally must conserve water. This requires efficient waste disposal in a concentrated or even dry form but simultaneously avoiding the potentially toxic effects of nitrogen. Furthermore, both terrestrial and aquatic insects must conserve ions, such as sodium (Na^+), potassium (K^+) and chloride (Cl^-), that may be limiting in their food or, in aquatic insects, lost into the water by diffusion. Production of insect urine or frass is therefore a result of two intimately related processes: excretion and **osmoregulation** – the maintenance of a favourable body fluid composition (osmotic and ionic homeostasis). The system responsible for excretion and osmoregulation is referred to loosely as the excretory system and its activities are performed largely by the Malpighian tubules and hindgut as outlined below. However, in freshwater insects, haemolymph composition must be regulated in response to constant loss of salts (as ions) to the surrounding water and ionic regulation involves both

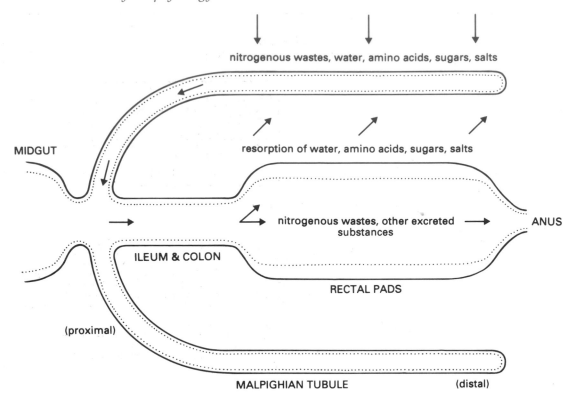

Fig. 3.17 Schematic diagram of a generalized excretory system showing the path of elimination of wastes. (After Daly *et al.*, 1978.)

the typical excretory system and special cells, called **chloride cells**, which are usually associated with the hindgut. Chloride cells are capable of absorbing inorganic ions from very dilute solutions and are best studied in larval dragonflies and damselflies.

3.7.1 The Malpighian tubules and rectum

The main organs of excretion and osmoregulation in insects are the Malpighian tubules acting in concert with the rectum or ileum. Malpighian tubules are outgrowths of the alimentary canal and consist of long thin tubules (Fig. 3.1) formed of a single layer of cells surrounding a blind-ending lumen. They range in number from as few as two in most scale insects (coccoids) to more than 200 in large locusts. Generally they are free, waving

around in the haemolymph where they filter out solutes. The vignette for this chapter shows the gut of *Locusta* but with only a few of the many Malpighian tubules depicted. Only aphids lack Malpighian tubules. Similar structures are believed to have arisen convergently in different arthropod groups, such as myriapods and arachnids, in response to the physiological stresses of life on dry land. Traditionally, insect Malpighian tubules are considered to belong to the hindgut and be ectodermal in origin. Their position marks the junction of the midgut and the cuticle-lined hindgut.

The anterior hindgut is called the **ileum**, the generally narrower middle portion is the **colon** and the expanded posterior section is the **rectum** (Fig. 3.13). In many terrestrial insects the rectum is the only site of water and

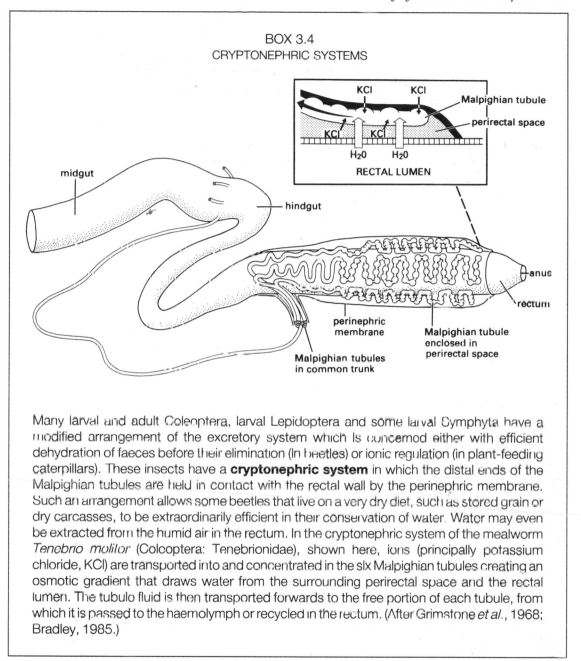

BOX 3.4
CRYPTONEPHRIC SYSTEMS

Many larval and adult Coleoptera, larval Lepidoptera and some larval Symphyta have a modified arrangement of the excretory system which is concerned either with efficient dehydration of faeces before their elimination (in beetles) or ionic regulation (in plant-feeding caterpillars). These insects have a **cryptonephric system** in which the distal ends of the Malpighian tubules are held in contact with the rectal wall by the perinephric membrane. Such an arrangement allows some beetles that live on a very dry diet, such as stored grain or dry carcasses, to be extraordinarily efficient in their conservation of water. Water may even be extracted from the humid air in the rectum. In the cryptonephric system of the mealworm *Tenebrio molitor* (Coleoptera: Tenebrionidae), shown here, ions (principally potassium chloride, KCl) are transported into and concentrated in the six Malpighian tubules creating an osmotic gradient that draws water from the surrounding perirectal space and the rectal lumen. The tubule fluid is then transported forwards to the free portion of each tubule, from which it is passed to the haemolymph or recycled in the rectum. (After Grimstone *et al.*, 1968; Bradley, 1985.)

solute resorption from the excreta, but in other insects, for example the desert locust *Schistocerca gregaria* (Orthoptera: Acrididae), the ileum makes some contribution to osmoregulation. In a few insects, such as the cockroach *Periplaneta americana* (Blattodea: Blattidae), even the colon may be a potential site of some nutrient absorption. The resorptive role of the rectum (and sometimes the anterior hindgut) is indicated by its anatomy.

In most insects, specific parts of the rectal epithelium are thickened to form **rectal pads** or papillae composed of aggregations of columnar cells; typically there are six pads arranged longitudinally but there may be fewer pads or many papillate ones.

The general picture of insect excretory processes outlined here is applicable to most freshwater species and to the adults of many terrestrial species. The Malpighian tubules produce a filtrate (the primary urine) which is isosmotic but ionically dissimilar to the haemolymph; and then the hindgut, especially the rectum, selectively reabsorbs water and certain solutes but eliminates others (Fig. 3.17).

Details of Malpighian tubule and rectal structure and of filtration and absorption mechanisms differ between taxa, in relation to both taxonomic position and dietary composition (Box 3.4 gives an example of one type of specialization – cryptonephric systems) but the excretory system of the desert locust *S. gregaria* (Fig. 3.18) exemplifies the general structure and principles of insect excretion. The Malpighian tubules of the locust produce an isosmotic filtrate of the haemolymph which is high in K^+ and low in Na^+, and in which Cl^- is the major anion. The active transport of ions, especially K^+, into the tubule lumen generates an osmotic pressure gradient so that water follows passively (Fig. 3.18(a)). Sugars and most amino acids are also passively filtered from the haemolymph (probably via junctions between the tubule cells), whereas the amino acid proline (later used as an energy source by the rectal cells) and non-metabolizable and toxic organic compounds are actively transported into the tubule lumen. Sugars, such as sucrose and trehalose, are resorbed from the lumen and returned to the haemolymph. The continuous secretory activity of each Malpighian tubule leads to a flow of primary urine from its lumen towards and into the gut. In the rectum, the urine is modified by removal of solutes and water to maintain fluid and ionic homeostasis of the

locust's body (Fig. 3.18(b)). Specialized cells in the rectal pads carry out active recovery of Cl^- under hormonal stimulation. This pumping of Cl^- generates electrical and osmotic gradients that lead to some reabsorption of other ions, water, amino acids and acetate.

3.7.2 Nitrogen excretion

Many predatory, blood-feeding and even plant-feeding insects ingest nitrogen, particularly certain amino acids, far in excess of requirements. Most insects excrete nitrogenous metabolic wastes at some or all stages of their life although some nitrogen is stored in the fat body or as proteins in the haemolymph in some insects. Many aquatic insects and some flesh-feeding flies excrete large amounts of ammonia, whereas in terrestrial insects wastes generally consist of **uric acid** or certain of its salts (urates), often in combination with urea, pteridines, certain amino acids and/or relatives of uric acid, such as hypoxanthine, allantoin and allantoic acid. Amongst these waste compounds, ammonia is relatively toxic and must usually be excreted as a dilute solution, or else rapidly volatilized from the cuticle or faeces (as in cockroaches). Urea is less toxic but more soluble, requiring much water for its elimination. Uric acid and urates require less water for synthesis than either ammonia or urea (Fig. 3.19), are non-toxic and, being insoluble in water (at least in acidic conditions), can be excreted essentially dry. Waste dilution can be achieved easily by aquatic insects, but water conservation is essential for terrestrial insects and uric acid excretion (**uricotelism**) is highly advantageous.

Deposition of urates in specific cells of the fat body (section 3.6.4) has been viewed as 'excretion' by storage of uric acid but may be actually a metabolic store for recycling by the insect, perhaps with the assistance of symbiotic micro-organisms, as in cockroaches that house bacteria in their fat body. These cockroaches, including *P. americana*, do not excrete

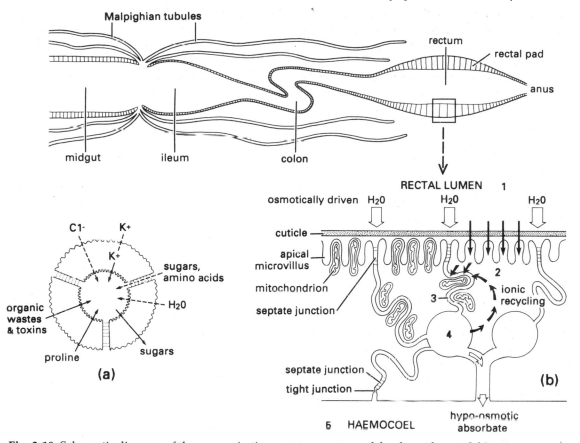

Fig. 3.18 Schematic diagram of the organs in the excretory system of the desert locust *Schistocerca gregaria* (Orthoptera: Acrididae). Few of the >100 Malpighian tubules are drawn. (a) Transverse section of one Malpighian tubule showing probable transport of ions, water and other substances between the surrounding haemolymph and the tubule lumen; active processes are indicated by solid arrows and passive processes by dashed arrows. (b) Diagram illustrating the movements of solutes and water in the rectal pad cells during fluid resorption from the rectal lumen. Pathways of water movement are represented by open arrows and solute movements by black arrows. Ions are actively transported from the rectal lumen (compartment 1) to the adjacent cell cytoplasm (compartment 2) and then to the intercellular spaces (compartment 3). Mitochondria are positioned to provide the energy for this active ion transport. Fluid in the spaces is hyperosmotic (higher ion concentration) to the rectal lumen and draws water by osmosis from the lumen via the septate junctions between the cells. Water thus moves from compartment 1 to 3 to 4 and finally to 5, the haemolymph in the haemocoel. (After Bradley, 1985.)

uric acid in the faeces even if fed a high-nitrogen diet but produce large quantities of internally-stored urates.

By-products of feeding and metabolism need not be excreted as waste – for example, the anti-feedant defensive compounds of plants may be sequestered directly or may form the biochemical base for synthesis of chemicals used in communication (Chapter 4) including warning and defence. White-pigmented uric acid derivatives colour the epidermis of some insects and provide the white in the wing scales of certain butterflies (Lepidoptera: Pieridae).

uric acid   ammonia   urea

N : H = 1.00   N : H = 0.33   N : H = 0.50

Fig. 3.19 Molecules of the three common nitrogenous excretory products. The high N:H ratio of uric acid relative to both ammonia and urea means that less water is used for uric acid synthesis (since hydrogen atoms are derived ultimately from water).

3.8 REPRODUCTIVE ORGANS

The reproductive organs of insects exhibit a incredible variety of forms, but there is a basic design and function to each component so that even the most aberrant reproductive system can be understood in terms of a generalized plan. Individual components of the reproductive system can vary in shape (e.g. of gonads and accessory glands), position (e.g. of the attachment of accessory glands) and number (e.g. of ovarian or testicular tubes, or sperm storage organs) between different insect groups, and sometimes even between different species in a genus. Knowledge of the homology of the components assists in interpreting structure and function in different insects. Generalized male and female systems are depicted in Fig. 3.20 and a comparison of the corresponding reproductive structures of male and female insects is provided in Table 3.2. Many other aspects of reproduction, including copulation and regulation of physiological processes, are discussed at length in Chapter 5.

3.8.1 The female system

The main functions of the female reproductive system are egg production, including the provision of a protective coating in many insects, and the storage of the male's spermatozoa until the eggs are ready to be fertilized. Transport of the spermatozoa to the female's storage organ and their subsequent controlled release requires movement of the spermatozoa, which in some species is known to be mediated by muscular contractions of parts of the female reproductive tract.

The basic components of the female system (Fig. 3.20(a)) are paired **ovaries**, which empty their mature **oocytes** (eggs) via the **calyces** (sing. **calyx**) into the **lateral oviducts** which unite to form the **common** (or **median**) **oviduct**. The **gonopore** (opening) of the common oviduct is usually concealed in an inflection of the body wall that typically forms a cavity, the **genital chamber**. This chamber serves as a copulatory pouch during mating and thus is often known as the **bursa copulatrix**. Its external opening is the **vulva**. In many insects the vulva is narrow and the genital chamber becomes an enclosed pouch or tube, referred to as the **vagina**. Two sorts of ectodermal glands open into the genital chamber. The first is the **spermatheca** which stores spermatozoa until they are needed for egg fertilization. Typically the spermatheca is

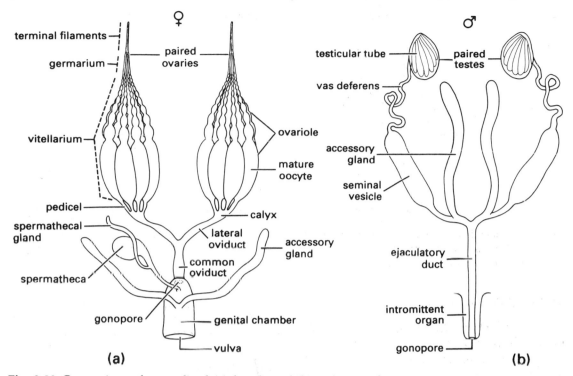

Fig. 3.20 Comparison of generalized (a) female and (b) male reproductive systems. (After Snodgrass, 1935.)

Table 3.2 The corresponding female and male reproductive organs of insects

Female reproductive organs	Male reproductive organs
Paired ovaries composed of ovarioles (ovarian tubes)	Paired testes composed of follicles (testicular tubes)
Paired oviducts (ducts leading from ovaries)	Paired vasa deferentia (ducts leading from testes)
Egg calyces (if present, reception of eggs)	Seminal vesicles (sperm storage)
Common (median) oviduct and vagina	Median ejaculatory duct
Accessory glands (ectodermal origin: colleterial or cement glands)	Accessory glands (2 types): (a) ectodermal origin (b) mesodermal origin
Bursa copulatrix (copulatory pouch) and spermatheca (sperm storage)	No equivalent
Ovipositor (if present)	Genitalia (if present): aedeagus and associated structures

single and generally sac-like with a slender duct, and often has a diverticulum that forms a tubular **spermathecal gland**. The gland or glandular cells within the storage part of the spermatheca provide nourishment to the contained spermatozoa. The second type of ectodermal gland, known collectively as accessory glands, opens more posteriorly in the genital chamber and has a variety of functions depending on the species.

Each ovary is composed of a cluster of egg or ovarian tubes, the **ovarioles**, each consisting of a terminal filament, a **germarium** (in which mitosis gives rise to primary oocytes), a **vitellarium** (in which oocytes grow by deposition of yolk in a process known as vitellogenesis – section 5.11.1) and a **pedicel** (or stalk). An ovariole contains a series of developing oocytes each surrounded by a layer of follicle cells forming an epithelium (the oocyte with its epithelium is termed a follicle); the youngest oocytes occur near the apical germarium and the most mature near the pedicel. Recognition of different types of ovariole is based on the manner in which the oocytes are nourished. A **panoistic ovariole** lacks specialized nutritive cells so that it contains only a string of follicles, with the oocytes obtaining nutrients from the haemolymph via the follicular epithelium. Ovarioles of the other two types contain trophocytes (nurse cells) that contribute to the nutrition of the developing oocytes. In a **telotrophic** (or **acrotrophic**) **ovariole** the trophocytes are confined to the germarium and remain connected to the oocytes by cytoplasmic strands as the oocytes move down the ovariole. In a **polytrophic ovariole** a number of trophocytes are connected to each oocyte and move down the ovariole with it, providing nutrients until depleted; thus individual oocytes alternate with groups of successively smaller trophocytes. Different suborders or orders of insects usually have only one of these three ovariole types.

Accessory glands of the female reproductive tract are often referred to as **colleterial** or **cement glands** because in most insect orders their secretions surround and protect the eggs or cement them to the substrate (Chapter 5). In other insects the accessory glands may function as **poison glands** (as in many Hymenoptera) or as **'milk' glands** in the few insects (e.g. tsetse flies, *Glossina* species) that exhibit adenotrophic viviparity (section 5.9). Accessory glands of a variety of forms and functions appear to have been derived independently in females of different orders and sometimes may even be non-homologous within an order, as in Coleoptera.

3.8.2 The male system

The main functions of the male reproductive system are the production and storage of spermatozoa and their transport in a viable state to the reproductive tract of the female. Morphologically, the male tract consists of paired **testes** each containing a series of testicular tubes or follicles (in which spermatozoa are produced) which open separately into the mesodermally derived sperm duct or **vas deferens** which usually expands posteriorly to form a sperm storage organ or **seminal vesicle** (Fig. 3.20(b)). Typically, tubular paired **accessory glands** are formed as diverticula of the vasa deferentia, but sometimes the vasa deferentia themselves are glandular and fulfil the functions of accessory glands. The paired vasa deferentia unite where they lead into the ectodermally derived **ejaculatory duct** – the tube that transports the semen or the sperm package to the gonopore. In a few insects, particularly certain flies, the accessory glands consist of an enlarged glandular part of the ejaculatory duct.

Thus the accessory glands of male insects can be classified into two types according to their mesodermal or ectodermal derivation. Almost all are mesodermal in origin and the few apparently ectodermal ones have been poorly studied. Furthermore, the mesodermal structures of the male tract frequently differ morphologically from the basic paired sacs or tubes described above. For example, in male cockroaches and many other orthopteroids the seminal vesicles and the numerous accessory gland tubules (Fig. 3.1) are clustered into a single median structure called the **mushroom body**. Secretions of the male accessory glands form the spermatophore (the package that surrounds the spermatozoa of many insects), contribute to the seminal fluid which nourishes the spermatozoa during transport to the female, are involved in activation (induction of motility) of the spermatozoa, and may alter female behaviour (induce non-receptivity to further males and/or stimulate oviposition; Chapter 5).

FURTHER READING

Buchner, P. (1965) *Endosymbiosis of Animals with Plant Microorganisms,* Wiley Interscience, New York.

Chapman, R.F. (1991) General anatomy and function, in *The Insects of Australia,* (CSIRO), Melbourne University Press, Carlton, pp. 33–67.

Davey, K.G. (1985) The male reproductive tract, The female reproductive tract, in *Comprehensive Insect Physiology, Biochemistry, and Pharmacology, Volume 1, Embryogenesis and Reproduction,* (eds G.A. Kerkut and L.I. Gilbert), Pergamon Press, Oxford, New York, Toronto, Sydney, Paris, Frankfurt, pp. 1–14, 15–36.

Dow, J.A.T. (1986) Insect midgut function. *Advances in Insect Physiology,* **19,** 187–328.

Holman, G.M., Nachman, R.J. and Wright, M.S. (1990) Insect neuropeptides. *Annual Review of Entomology,* **35,** 201–17.

Kerkut, G.A. and Gilbert, L.I. (eds) (1985) *Comprehensive Insect Physiology, Biochemistry, and Pharmacology; Volume 3, Integument, Respiration and Circulation; Volume 4, Regulation: Digestion, Nutrition, Excretion; Volumes 7 and 8, Endocrinology I and II,* Pergamon Press, Oxford, New York, Toronto, Sydney, Paris, Frankfurt.

Menn, J.J., Kelly, T.J. and Masler, E.P. (eds) (1991) *Insect Neuropeptides: Chemistry, Biology, and Action,* American Chemical Society, Washington.

Raabe, M. (1989) *Recent Developments in Insect Neurohormones,* Plenum Press, New York and London.

Terra, W.R. (1990) Evolution of digestive systems of insects. *Annual Review of Entomology,* **35,** 181–200.

4

SENSORY SYSTEMS AND BEHAVIOUR

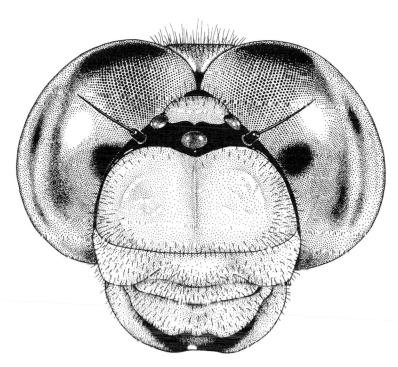

Head of a dragonfly showing enormous compound eyes

We introduced this book in the first chapter by suggesting that insects were successful because they could sense and interpret their surroundings and discriminate on a fine scale. Insects can identify and respond selectively to cues from a heterogeneous environment. They can differentiate between hosts, both plant and animal, and distinguish between many microclimatic factors, such as variations in humidity, temperature and air flow.

Sensory complexity facilitates both simple and complex behaviours of insects. For example, controlled flight is possible only with the ability to sense the aerial environment and make appropriate responses. Since much insect activity is nocturnal, orientation and navigation cannot rely solely on the conventional visual cues, and odours and sounds play a major role in communication in many night-active species. It is important to realize that the range of sensory information used by insects differs from that of humans. We rely heavily on visual information and, although many insects have well developed vision, most insects make greater use of olfaction and hearing than do humans.

The insect is isolated from its external surroundings by the relatively inflexible, insensitive and impermeable barrier of the cuticle. The answer to the enigma of how this armoured insect can perceive its immediate environment lies in frequent and abundant cuticular modifications that detect external stimuli. Sensory organs (**sensilla**) protrude from the cuticle, or sometimes lie within or beneath it. Specialized cells detect stimuli which may be categorized as mechanical, thermal, chemical and visual. Other cells (the neurones) transmit messages to the central nervous system, where they are integrated. The nervous system instigates and controls appropriate behaviours, such as posture, movement, feeding and behaviours associated with mating and oviposition.

This chapter surveys sensory systems and presents selected behaviours that are elicited or modified by environmental stimuli. The

means of detection and, where relevant, the production of these stimuli are treated in the following sequence: touch, position, sound, temperature, chemicals (with particular emphasis on communication chemicals called pheromones) and light. The chapter concludes with a section on insect behaviour in relation to the preceding discussion on stimuli.

4.1 MECHANICAL STIMULI

The stimuli grouped under this heading are those associated with distortion caused by mechanical movement due to the environment itself, the insect in relation to the environment, or to internal forces derived from the muscles. The mechanical stimuli sensed include touch, body stretching and stress, position, pressure, gravity and vibrations, including pressure changes of the air and substrate involved in sound transmission and hearing.

4.1.1 Tactile mechanoreception

The bodies of insects are clothed with cuticular projections. These are called microtrichia if many arise from one cell, or hairs, bristles, setae or macrotrichia if they are of multicellular origin. Most flexible projections arise from an innervated socket. These are sensilla, termed **trichoid sensilla** (literally hair-like little sense organs), and develop from epidermal cells that switch from cuticle production. Three cells are involved (Fig. 4.1):

- The **trichogen cell** grows the conical hair.
- The **tormogen cell** grows the socket.
- The nerve cell, or **sensory neurone**, grows a **dendrite** into the hair and an **axon** that winds inwards to link with other axons to form a nerve connected to the central nervous system.

The fully developed trichoid sensilla fulfil tactile functions. As touch sensilla they respond to the movement of the hair by firing

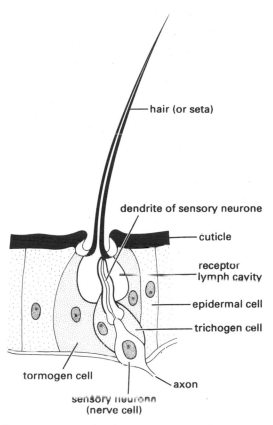

hair (or seta)

dendrite of sensory neurone

cuticle

receptor lymph cavity

epidermal cell

trichogen cell

tormogen cell

axon

sensory neurone (nerve cell)

Fig. 4.1 Longitudinal section of a trichoid sensillum showing the arrangement of the three associated cells. (After Chapman, 1991.)

impulses from the dendrite at a frequency related to the extent of the deflection. Touch sensilla are only stimulated during actual movement of the hair. The sensitivity of each hair varies, with some being so sensitive that they respond to vibrations of air particles caused by noise, as in some caterpillars that respond to the vibrations caused by the audible approach of parasitic wasps.

4.1.2 Position mechanoreception (proprioceptors)

Insects require continuous knowledge of the relative position of their body parts such as limbs or head, and need to detect how the orientation of the body relates to gravity. This

information is conveyed by **proprioceptors** (self-perception receptors), of which three types are described here. One type of trichoid sensillum gives a continuous sensory output at a frequency that varies with the position of the hair. Sensilla often form a bed of grouped small hairs, a **hair plate**, at joints or at the neck, in contact with the cuticle of an adjacent body part (Fig. 4.2(a)). The degree of flexion of the joint gives a variable stimulus to the sensilla, thereby allowing monitoring of the relative positions of different parts of the body.

The second type, stretch receptors, are internal proprioceptors associated with muscles such as those of the abdominal and gut walls. Alteration of the length of the muscle fibre is detected by multiple-inserted neurone endings, producing variation in the rate of firing of the nerve cell. Stretch receptors monitor body functions such as abdominal or gut distension, or ventilation rate. The third type are stress detectors on the cuticle via stress receptors called **campaniform sensilla**. Each sensillum comprises a central cap or peg surrounded by a raised circle of cuticle and with a single neurone per sensillum (Fig. 4.2(b)). These sensilla are located on joints, such as those of legs and wings, and other places liable to distortion. Locations include the haltere (the knob-like modified hind wing of Diptera), at the base of which there are dorsal and ventral groups of campaniform sensilla that respond to distortions created during flight.

4.1.3 Sound reception

Sound is a pressure fluctuation transmitted in a wave form via movement of the air or the substrate, including water. Sound and hearing are terms often applied to the quite limited range of frequencies of airborne vibration that humans perceive with their ears, usually in adults from 20 to 20 000 Hz (1 hertz (Hz) is a frequency of one cycle per second). Such a definition of sound is restrictive, particularly

Fig. 4.2 Proprioceptors: (a) sensilla of a hair plate located at a joint, showing how the hairs are stimulated by contacting adjacent cuticle; (b) campaniform sensillum on the haltere of a fly. ((a) After Chapman, 1982; (b) after Snodgrass, 1935; McIver, 1985.)

since many insects receive vibrations ranging from as low as 1–2 Hz to ultrasound frequencies perhaps as high as 100 kHz. Specialized emission and reception across this range of frequencies of vibration are considered here. Insects receive these frequencies using a variety of organs, none of which resemble mammalian ears.

An important role of insect sound is in intraspecific acoustic communication. For example, much orthopteran courtship is acoustic, with males producing species-specific sounds ('songs') that the predominantly non-singing females detect and upon which they base their choice of mate. A second, and perhaps primary, function of hearing is to detect predators. Some orthopterans, moths and lacewings can hear and evade insectivorous bats that use ultrasound in hunting. Each species of insect probably detects sound within a relatively narrow range of frequencies related to function. Thus bat-avoiding moths detect the ultrasound frequencies in the 30–100 kHz range produced by insectivorous bats, and crickets and katydids (also

called bushcrickets) respond to sounds produced by their own species during courtship (including frequencies audible to the human ear).

(a) Non-tympanal vibration reception

A simple form of sound reception occurs in species that have very sensitive, elongate trichoid sensilla that respond to vibrations produced by a close source of sound. For example, thoracic hairs of caterpillars of the noctuid moth *Barathra brassicae* are about 0.5 mm long and respond optimally to vibrations of 150 Hz. This system is effective in air only for locally produced sounds. For low frequency sounds in a more viscous medium such as water, longer distance transmission is possible. Currently, however, few aquatic insects are known to communicate through underwater sounds. Notable examples are the 'drumming' sounds that some aquatic larvae produce to assert territory, and the noises produced by diving hemipterans such as

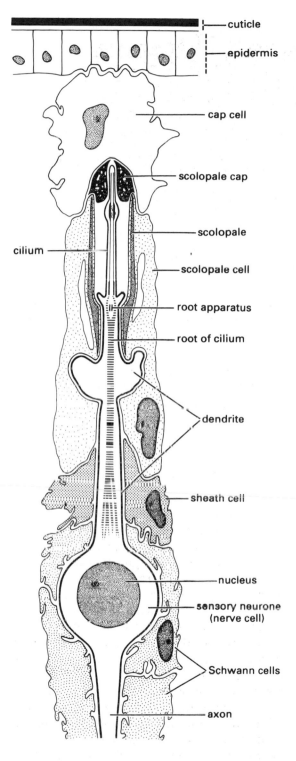

corixids and nepids. Water-striders (Hemiptera: Gerridae) that live on the aquatic surface film send pulsed waves across the water surface to communicate in courtship and aggression. Moreover, they can detect the vibrations produced by the struggles of prey that fall onto the water surface.

The specialized sensory organs that receive vibrations are subcuticular mechanoreceptors called **chordotonal organs**. An organ consists of one to many **scolopidia**, each of which consists of three linearly arranged cells: a subtympanal **cap cell** placed on top of a sheath cell (**scolopale cell**) which envelops the end of a nerve cell dendrite (Fig. 4.3). All adult insects and many larvae have a particular chordotonal organ, **Johnston's organ** lying within the pedicel, the second antennal segment. The primary function is to sense movements of the antennal flagellum relative to the rest of the body, as in detection of flight speed by air movement. Additionally, it functions in hearing in some insects. In male mosquitoes (Culicidae) and midges (Chironomidae), many scolopidia are contained in the swollen pedicel. These scolopidia are attached at one end to the pedicel wall and at the other, sensory, end to the base of the third antennal segment. This strongly modified Johnston's organ is the male receptor for the female wing tone (see below), as shown when amputation of the terminal flagellum or arista of the antenna renders males unreceptive to the sound of the female.

Detection of substrate vibration involves the **subgenual organ**, a chordotonal organ located in the proximal tibia of each leg. Subgenual organs are found in most insects, except the Coleoptera and Diptera. The organ consists of a semi-circle of many sensory cells lying in the haemocoel, connected at one end to the inner cuticle of the tibia and at the other to the trachea. There are subgenual organs

Fig. 4.3 Longitudinal section of a scolopidium, the basic unit of a chordotonal organ. (After Gray, 1960.)

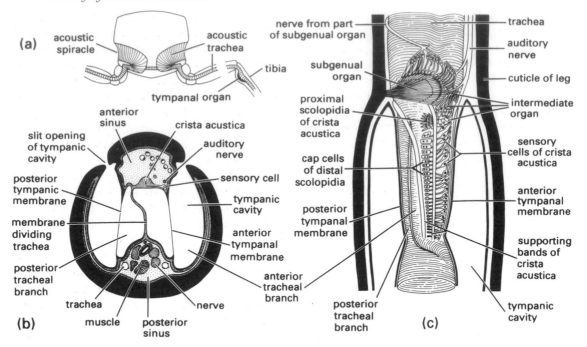

Fig. 4.4 Tympanal organs of a katydid, *Decticus* (Orthoptera: Tettigoniidae): (a) transverse section through the fore legs and prothorax to show the acoustic spiracles and tracheae; (b) transverse section through the base of the fore tibia; (c) longitudinal break-away view of the fore tibia. (After Schwabe, 1906, in Michelsen and Larsen, 1985.)

within all legs: the organs of each pair of legs may respond specifically to substrate-borne sounds of differing frequencies. Vibration reception may involve either direct transfer of low frequency substrate vibrations to the legs, or there may be more complex amplification and transfer. Airborne vibrations can be detected if they cause vibration of the substrate and hence of the legs.

(b) Tympanal reception

The most elaborate sound reception system in insects involves a specific receptor structure, the **tympanum**. This is a membrane that responds to sounds produced at some distance, transmitted by airborne vibration. Tympanal membranes are linked to chordotonal organs and may be associated with complex morphological structures, such as

modifications of the trachea, that enhance sound reception. Tympanal organs are located on:

- The ventral thorax between the metathoracic legs of mantids.
- The metathorax of many noctuid moths.
- The prothoracic legs of many orthopterans.
- The abdomen of other orthopterans, cicadas and some moths.
- The wings of certain moths and lacewings.

The location of these organs in different positions and the occurrence in distantly related insect groups indicates that tympanal hearing evolved several times in insects. Neuroanatomical studies suggest that all insect tympanal organs evolved from proprioceptors and the wide distribution of proprioceptors throughout the insect cuticle must account for the variety of positions of tympanal organs.

Tympanal sound reception is particularly well developed in orthopterans, notably in the crickets and katydids. In most of these ensiferan Orthoptera the tympanal organs are on the tibia of each fore leg (Figs 4.4, 8.2(a)). Behind the paired tympanal membranes lies an acoustic trachea that runs from a prothoracic spiracle down each leg to the tympanal organ (Fig. 4.4(a)).

Crickets and katydids have similar hearing systems. The system in crickets appears to be less specialized because their acoustic tracheae remain connected to the ventilatory spiracles of the prothorax. In contrast, the acoustic tracheae of katydids form a system completely isolated from the ventilatory tracheae, opening via a separate pair of acoustic spiracles. In many katydids, the base of the tibia has two separated longitudinal slits, each of which leads into a tympanic chamber (Fig. 4.4(b)). The acoustic trachea, which lies centrally in the leg, is divided in half at this point by a membrane, so that one half is in close connection with the anterior and the other half with the posterior tympanal membrane. The primary route of sound to the tympanal organ is usually from the acoustic spiracle and along the acoustic trachea to the tibia. The change in cross-sectional area from the enlargement of the trachea behind each spiracle (sometimes called a tracheal vesicle) to the tympanal organ in the tibia approximates the function of a horn and amplifies the sound. Some researchers believe that the slits of the tympanic chambers serve as sound guides allowing directional hearing and that these chambers may resonate to particular sound frequencies, but this view is disputed. The slits do allow the entry of sound, although their exact function is uncertain.

Whatever the major route of sound entry to the tympanal organs, air- and substrate-borne acoustic signals cause the tympanal membranes to vibrate. This is sensed by three chordotonal organs: the **subgenual organ**, the **intermediate organ** and the **crista acustica** (Fig. 4.4(c)). The subgenual organs, which have a form and function like those of non-orthopteroid insects, are present on all legs but the crista acustica and intermediate organs are found only on the fore legs in conjunction with the tympana. This implies that the tibial hearing organ is a serial homologue of the proprioceptor units of the mid and hind legs.

The crista acustica consists of a row of up to 60 scolopidial cells attached to the acoustic trachea and is the main sensory organ for airborne sound in the 5–50 kHz range. The intermediate organ, which consists of 10–20 scolopidial cells, is posterior to the subgenual organ and virtually continuous with the crista acustica. The role of the intermediate organ is uncertain but it may respond to airborne sound of frequencies of 2–14 kHz. Each of the three chordotonal organs is innervated separately, but the neuronal connections between the three imply that signals from the different receptors are integrated.

Observation and experiments show that hearing insects such as crickets and katydids can identify the direction of a point source of sound. Exactly how they do so is controversial and may vary between taxa. Localization of sound directionality clearly depends upon detection of differences in the sound received by the tympanal membranes within one leg and/or between legs. Sound reception varies with the orientation of the body relative to the sound source, allowing some precision in locating the source. High frequencies are often used for location at a distance. Close-up orientation is provided by lower frequency vibrations, often in association with visual or olfactory cues.

4.1.4 Sound production

The commonest method of sound production by insects is by **stridulation** in which one specialized body part, the **scraper**, is rubbed against another, the **file**. The file is a series of

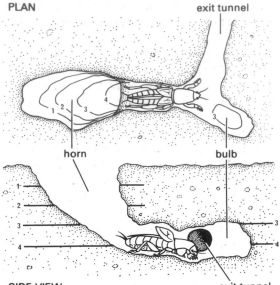

PLAN

exit tunnel

horn

bulb

SIDE VIEW

exit tunnel

Fig. 4.5 The singing burrow of a mole cricket, *Scapteriscus acletus* (Orthoptera: Gryllotalpidae), in which the singing male sits with his head in the bulb and tegmina raised across the throat of the horn. (After Bennet-Clark, 1989.)

teeth, ridges or pegs, which vibrate through contact with a ridged or plectrum-like scraper. The file itself makes little noise, and amplification must be provided in order to generate airborne sound. The horn-shaped burrow of the mole cricket is an excellent sound enhancer (Fig. 4.5), and insects produce many modifications of the body, particularly of wings and internal air-sacs of the tracheal system, to produce amplification and resonance.

Sound production by stridulation occurs in some species of many orders of insects, but the Orthoptera show most elaboration and diversity. All stridulating orthopterans enhance their sounds using the tegmina (the modified fore wings). The file of katydids and crickets is formed from a basal vein of one or both tegmina that rasps against a scraper on the other wing. Grasshoppers and locusts (Acrididae) rasp a file on the fore femora against a similar scraper on the tegmen.

Many insects lack the body size, power or sophistication to produce high-frequency airborne sounds, but they can produce and transmit low-frequency sound by vibration of the substrate (such as wood, soil or a host plant), which is a denser medium. Substrate vibrations are also a by-product of airborne sound production as in acoustic signalling insects, such as some katydids, whose whole body vibrates whilst producing audible airborne stridulatory sounds. Body vibrations, which are transferred through the legs to the substrate (plant or ground), are of low frequencies of 1–5000 Hz. Substrate vibrations can be detected by the female and appear to be used in closer range localization of the calling male, in contrast to the airborne signal used at greater distance.

A second means of sound production involves alternate muscular distortion and relaxation of a specialized area of elastic cuticle, the **tymbal**, to give individual clicks or variably modulated pulses of sound. Tymbal sound production is most audible to the human ear from cicadas but many other hemipterans and some moths produce sounds from a tymbal. In the cicadas, only the males have these paired tymbals, which are located dorsolaterally, one on each side, on the first abdominal segment. The tymbal membrane is supported by a variable number of ribs. A strong tymbal muscle distorts the membrane and ribs to produce a sound; on relaxation, the elastic tymbal returns to rest. To produce sounds of high frequency, the tymbal muscle contracts asynchronously, with many contractions per nerve impulse (section 3.1.1). A group of chordonotal sensilla is present and there is also a smaller tensor muscle that controls the shape of the tymbal, thereby allowing alteration of the acoustic property. The noise of one or more clicks is emitted as the tymbal distorts and further sounds may be produced during the elastic return on relaxation. The first abdominal segment contains air-sacs – modified tracheae – that are tuned to resonate at or close to the natural frequency of tymbal vibration.

Cicada calls are species-specific, generally in the range of 4–7 kHz and usually of high intensity. They may travel over 1 km even in thick forest. Sound is received by tympanic membranes that are present in both sexes and lie ventral to the position of the male tymbal on the first abdominal segment. In other sound-producing hemipterans, both sexes may possess tymbals and, because they lack abdominal air-sacs, the sound is very damped compared to that of cicadas. The sounds produced by *Niloparvata lugens* (the brown planthopper) and probably other non-cicadan hemipterans, are transmitted by vibration of the substrate.

Recently moths have been shown to hear the ultrasound produced by predatory bats and they themselves can produce sound using metepisternal tymbals. Arctiid moths produce high-frequency sounds that cause bats to veer away from attack. It is speculated that moth clicks might jam the bat sonar systems, leading a bat to believe that it is approaching an object too large to be food (but see page 361).

The humming or buzzing sound characteristic of swarming mosquitoes, gnats and midges is a flight tone produced by the frequency of wing beat. This tone, which can be virtually species-specific, differs between the sexes: the male produces a higher tone than the female. The tone also varies with age and ambient temperature for both sexes. Male insects that form nuptial (mating) swarms recognize the swarm site by species-specific environmental markers rather than audible cues (section 5.1) but they are insensitive to the wing tone of males of their species. Neither can the male detect the wing tone of immature females; the Johnson's organ in his antenna responds only to the wing tone of physiologically receptive females.

4.2 THERMAL STIMULI

4.2.1 Thermoreception

Insects evidently detect variation in temperature, as seen by their behaviour (section 4.2.2), yet it is not clear where the receptors are located nor how they function. In the cockroach *Periplaneta americana*, temperature receptors are present on the ariolum and pulvilli of the tarsi, and thermoreceptors have been found on the legs of the few other insects investigated. The temperature preferences of insects with amputated antennae are different to those of intact insects, indicating the presence of antennal receptors. There must be central temperature sensors to detect internal temperature, but the only experimental evidence is from a large moth in which tharacic neural ganglia were found to have a role in instigating temperature-dependent flight muscle activity.

4.2.2 Thermoregulation

Insects are poikilothermic, in that they lack a means to maintain a constant temperature independent of fluctuations in the ambient (surrounding) conditions. Inactive insects closely track external temperature but many can substantially regulate their temperatures upwards or downwards, even if only for a short time. Insects can vary their temperature by behavioural and physiological mechanisms, the former use external heat, the latter use internally generated heat sources.

(a) Behavioural thermoregulation (ectothermy)

The extent to which radiant energy (either solar or substrate) influences body temperature is related to the aspect that a diurnal insect adopts. Basking, by which many insects maximize heat uptake, involves both posture and orientation relative to the source of heat. The wing position and orientation may enhance heat absorption or, alternatively, provide shading from excessive solar radiation. Cooling may include shade-seeking behaviour, such as seeking cooler environmental microhabitats or altered orientation on plants. Many desert insects avoid temperature extremes by burrowing. Some insects living in exposed places may avoid excessive heating by 'stilting': they raise themselves on extended legs to elevate most of the body out of the narrow boundary layer close to the ground. Conduction of heat from the substrate is reduced, and convection is enhanced in the cooler moving air above the boundary layer.

There is a complex (and disputed) relationship between temperature regulation and insect colour and surface sculpturing. Amongst some desert beetles (Tenebrionidae), black species become active earlier in the day at lower ambient temperatures than do pale ones, which in turn can remain active longer during hotter times. The application of white paint to black tenebrionid beetles results in substantial body temperature changes: black beetles warm up more rapidly at a given ambient temperature and overheat more quickly compared with white ones, which have greater reflectivity to heat. These physiological differences correlate with the observed differences in thermal ecology between dark and pale species. Further evidence of the role of colour comes from a bicoloured cicada (Hemiptera: *Cacama valvata*) in which basking involves directing the dark dorsal surface towards the sun, in contrast to cooling, when only the pale ventral surface is exposed.

For aquatic insects in which body temperature must follow water temperature, there is little or no ability to regulate body temperature beyond seeking microclimatic differences within a water body.

(b) Physiological thermoregulation (endothermy)

Some insects can be endothermic because the thoracic flight muscles have a very high metabolic rate and produce much heat. The thorax can be maintained at a relatively constant high temperature during flight. Temperature regulation may involve clothing the thorax with insulating scales or hairs but insulation must be balanced with the need to dissipate any excess heat generated during flight. Some butterflies and locusts alternate flight (which produces heat) with gliding (which allows cooling) but many insects must fly continuously and cannot glide. Bees and many moths prevent thoracic overheating in flight by increasing the heart rate and circulating haemolymph from the thorax to the poorly insulated abdomen. In some bumble-bees (*Bombus*) and carpenter bees (*Xylocopa*) a counter-current system that normally prevents heat loss is by-passed during flight to augment abdominal heat loss.

The insects that produce elevated temperatures during flight often require a warm thorax before they can take off. When ambient temperatures are low, these insects use the flight muscles to generate heat prior to switching them for use in flight. There are different mechanisms according to whether the flight muscles are synchronous or asynchronous (section 3.1.4). Insects with synchronous flight muscles warm up by contracting antagonistic muscle pairs synchronously or synergistic muscles alternately. This activity generally produces some wing vibration (as seen, for example, in odonates). Insects with asynchronous flight muscles warm up by operating the flight muscles whilst the wings are uncoupled, or the thoracic box is held rigid by accessory muscles to prevent wing movement. Usually no wing movement is seen, though ventilatory pumping movements of the abdomen may be visible. When the thorax is warm but the insect is sedentary (for example, whilst feeding), many insects maintain temperature by shivering, which may be prolonged. In contrast, foraging honey bees may cool off and must then warm up before take-off.

4.3 CHEMICAL STIMULI

In comparison with vertebrates, insects show a more profound use of chemicals in communication, particularly with other individuals of their own species. Chemicals are produced by insects for many purposes. Their perception in the external environment is through specific chemoreceptors.

4.3.1 Chemoreception

The chemical senses may be divided into **taste**, for detection of aqueous chemicals, and **smell**, for airborne ones – but the distinction is relative. Alternative terms are contact (taste, gustatory) and distant (smell, olfactory) chemoreception. For aquatic insects, all chemicals sensed are in aqueous solution, and strictly all chemoreception should be termed 'taste'. However, if an aquatic insect has a chemoreceptor that is structurally and functionally equivalent to one in a terrestrial insect that is olfactory, then the aquatic insect is said to 'smell' the chemical.

Chemosensors trap chemical molecules, which are transferred to a site for recognition, where they specifically depolarize a membrane and stimulate a nerve impulse. Effective trapping involves localization of the chemoreceptors. Thus many contact (taste) receptors occur on the mouthparts, such as the labella of higher Diptera (Box 14.4) where salt and sugar receptors occur, and on the ovipositor, to assist with identification of suitable oviposition sites. The antennae, which are often forward-directed and prominent, encounter sensory stimuli first and are endowed with many distant chemoreceptors, some contact chemoreceptors and many mechanoreceptors. The legs, particularly the tarsi which are in contact with the substrate, also have many chemoreceptors. In butterflies, stimulation of the tarsi by sugar solutions evokes an automatic extension of the proboscis. In blow flies, a complex sequence of stereotyped feeding behaviours is induced when a tarsal chemoreceptor is stimulated with sucrose. The proboscis starts to extend and, following sucrose stimulation of the chemoreceptors on the labellum, further proboscis extension occurs and the labellar lobes open. With more sugar stimulus, the source is sucked until

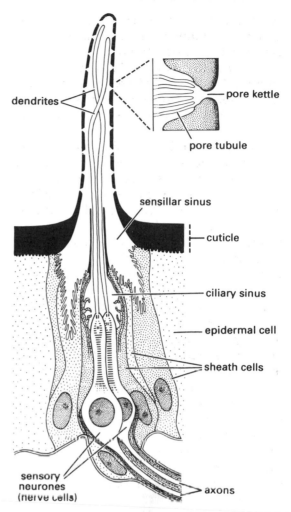

Fig. 4.6 Longitudinal section through a multiporous olfactory sensillum. (After Birch and Haynes, 1982; Zacharuk, 1985.)

stimulation of the mouthparts ceases. When this happens, a quite predictable pattern of search for further food follows.

Insect chemoreceptors are sensilla with one or more pores (holes). Two classes of sensilla can be defined based on their ultrastructure: **uniporous**, with one pore, and **multiporous**, with several to many pores on the apex of the sensilla. Uniporous sensilla range in appearance from hairs to pegs, plates or simply pores

BOX 4.1
THE ELECTROANTENNOGRAM

Electrophysiology is the study of the electrical properties of biological material, such as all types of nerve cells, including the peripheral sensory receptors of insects. Insect antennae bear a large number of sensilla and are the major site of olfaction in most insects. Electrical recordings can be made from either individual sensilla on the antenna (single cell recordings)

BOX 4.1—*continued*

or the whole antenna (electroantennogram), as explained by Rumbo, 1989. The electroantennogram (EAG) technique measures the total response of insect antennal receptor cells to particular stimuli. Recordings can be made using the antenna either excised, or attached to an isolated head or to the whole insect. In the illustrated example, the effects of a particular biologically active compound (a pheromone), blown across the isolated antenna of a male moth, are being assessed. The recording electrode, connected to the apex of the antenna, detects the electrical response which is amplified and depicted as a trace on a microcomputer or an oscilloscope, as in the EAG set-up illustrated in the upper drawing. Antennal receptors are very sensitive and specifically perceive particular odours, such as the sex pheromone of potential conspecific partners or volatile chemicals released by the insect's host. Different compounds usually elicit different EAG responses from the same antenna, as depicted in the two traces on the lower right.

This elegant and simple technique has been used extensively in pheromone identification studies as a quick method of bioassaying compounds for activity. For example, the antennal responses of a male moth to the natural sex pheromone obtained from conspecific female moths are compared with responses to synthetic pheromone components or mixtures. Clean air is blown continuously over the antenna at a constant rate and the samples to be tested are introduced into the air stream and the EAG response is observed. The same samples can be passed through a gas chromatograph (GC), which can be interfaced with a mass spectrometer to determine molecular structure of the compounds being tested. Thus the biological response from the antenna can be related directly to the chemical separation (seen as peaks in the GC trace), as illustrated here in the graph on the lower left (after Struble and Arn, 1984).

In addition to lepidopteran species, EAG data have been collected for cockroaches, beetles, flies, bees and other insects to measure antennal responses to a range of volatile chemicals affecting host attraction, mating, oviposition and other behaviours. EAG information is of greatest utility when interpreted in conjunction with behavioural studies.

in a cuticular depression, but all have relatively thick walls and a simple permeable pore which may be apical or central. The hair or peg contains a chamber, which is in basal contact with a dendritic chamber lying beneath the cuticle. The outer chamber is often seen to contain (and extrude) a viscous liquid, presumed to assist in the entrapment of chemicals and in their transfer to the dendrites. It is assumed that these uniporous chemoreceptors predominantly detect chemicals by contact, although there is evidence for some olfactory function.

The major olfactory role comes from multiporous sensilla, which are hair-like or peg-like setae, with many round pores or slits in the thin walls, leading into a chamber known as the **pore kettle**. This is richly endowed with pore tubules that run inwards to meet multi branched dendrites (Fig. 4.6).

Investigations of the electrophysiology of chemoreception are hindered by the small size of sensilla and the apparently complex relationship between the reception of the chemical, the production of electrical potential at the receptor membrane, and impulses generated in the neurone. Some of these difficulties are alleviated by the use of an electroantennogram (Box 4.1) to study the summed response of all chemoreceptors in the antenna, particularly in relation to pheromones.

Very sensitive chemoreceptors are involved in chemical communications between conspecific insects. Great sensitivity is achieved by spreading very many receptors over as great an area as possible, whilst allowing the maximum volume of air to flow across the receptors. Thus the antennae of many male moths are large and frequently the surface area is enlarged by pectinations that form a sieve-like basket (Fig. 4.7). Each antenna of the male silkworm moth (Saturniidae: *Bombyx mori*) has about 17 000 sensilla that respond to the specific sex-signalling chemicals produced by the female (sex pheromones; section 4.3.2). Since each sensillum has up to 3000 pores, each of 10–15 nm in diameter, there are some 45 million pores per moth. Some calculations concerning the silkworm moth suggest that just a few molecules of pheromone could stimulate a nerve impulse above the background rate and behavioural change may be elicited by less than a hundred molecules of sex pheromone.

4.3.2 Semiochemicals: pheromones

Many insect behaviours rely on the sense of smell. Chemical odours, termed **semiochemicals**, are especially important in communication. Semiochemicals are used in both interspecific and intraspecific communication. The latter is particularly highly developed in insects, and involves the use of chemicals called **pheromones**. When they were first recognized in the 1950s, pheromones were defined as:

> substances that are secreted to the outside by an individual and received by a second individual of the same species in which they release a specific reaction, for example, a definite behaviour or developmental process.

This definition remains valid today, despite the discovery of a hidden complexity of pheromone cocktails.

Pheromones are predominantly volatile but sometimes they are liquid contact chemicals. All pheromones are produced by exocrine glands (those that secrete to the outside of the body) derived from epidermal cells. The scent organs may be located almost anywhere on the body: for example, sexual scent glands on female Lepidoptera lie in eversible sacs or pouches between the eighth and ninth abdominal segments; the organs are mandibular in the female honey bee but are on the swollen hind tibiae of female aphids and within the midgut and genitalia in cockroaches.

Classification of pheromones by chemical structure reveals that many naturally occurring compounds (such as host odours) and pre-existing metabolites (like cuticular waxes) have been co-opted by insects to serve in the biochemical synthesis of a wide variety of compounds that function in communication. Chemical classification, whilst of interest, is of less value to many entomologists than the behaviours that the chemicals elicit. Very many insect behaviours are governed by chemicals; nevertheless we can distinguish pheromones that **release** specific behaviours from those that **prime** long-term, irreversible physiological changes. Thus the stereotyped sexual behaviour of a male moth is released by the female-emitted sex pheromone, whereas the crowding pheromone of locusts will prime maturation of gregarious phase individuals (section 6.7.5). Here, further classification of pheromones is based on five categories of behaviour associated with sex, aggregation, spacing, trail forming and alarm.

(a) Sex pheromones

Male and female conspecific insects often communicate with chemical **sex pheromones**. Mate location and courtship may involve chemicals in two stages, with **sex attraction pheromones** acting at a distance, followed by

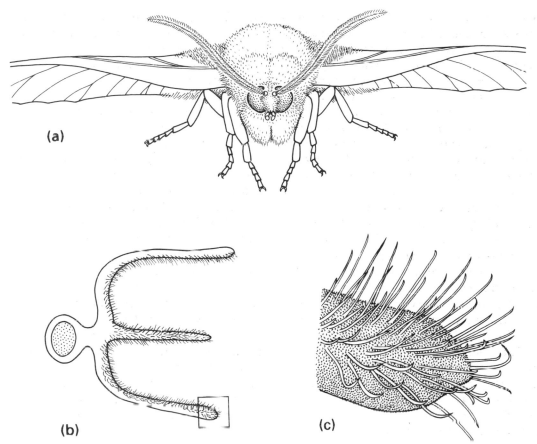

Fig. 4.7 The antennae of a male moth of *Trictena atripalpis* (Lepidoptera: Hepialidae): (a) anterior view of head showing tripectinate antennae of this species, (b) cross-section through the antenna showing the three branches; (c) enlargement of tip of outer branch of one pectination showing olfactory sensilla.

close-up **courtship pheromones** employed prior to mating. The sex pheromones involved in attraction often differ from those used in courtship. Production and release of sex attractant pheromones tend to be restricted to one sex, mostly the female, although there are lepidopterans and scorpionflies in which males are the releasers of distance attractants that lure females. The producer releases volatile pheromones that stimulate characteristic behaviour in those members of the opposite sex within range of the odorous plume. An aroused recipient raises the antennae, orientates towards the source and walks or flies upwind, often in a zig-zag track until

the source is reached (Fig. 4.8). Each successive action appears to depend upon an increase in concentration of this airborne pheromone. As the insect approaches the source, cues such as sound and vision may be involved in close-up courtship behaviour.

Courtship (section 5.2), which involves co-ordination of the two sexes, may require close-up chemical stimulation of the partner with courtship pheromones. This pheromone may be simply a high concentration of the attractant pheromone, but 'aphrodisiac' chemicals exist, as seen in the queen butterfly (Nymphalidae: Danainae: *Danaus gilippus*). The males of this species, as with several other

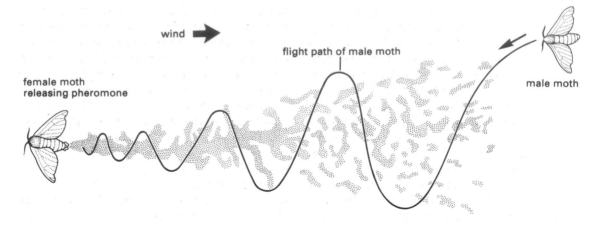

Fig. 4.8 Location of pheromone-emitting female by male moth tacking upwind. The pheromone trail forms a somewhat discontinuous plume because of turbulence, intermittent release and other factors. (After Haynes and Birch, 1985.)

Fig. 4.9 A pair of queen butterflies, *Danaus gilippus* (Lepidoptera: Nymphalidae: Danainae), showing aerial 'hairpencilling' by the male. The male (above) has splayed hairpencils (at his abdominal apex) and is applying pheromone to the female (below). (After Brower *et al.*, 1965.)

lepidopterans, have extrusible abdominal hairpencils (brushes) which produce a pheromone that is dusted directly onto the antennae of the female, whilst both are in flight (Fig. 4.9). The effect of this pheromone is to placate the natural escape reaction of the female, who alights, folds her wings and allows copulation. In *D. gilippus*, this male courtship pheromone, a pyrrolixidine alkaloid called danaidone, is essential to successful courtship. However, the butterfly cannot synthesize it without acquiring the chemical precursor by feeding on selected plants as an adult. In the arctiid moth, *Creatonotus gangis*, the precursor of the male courtship pheromone likewise cannot be synthesized, but it is sequestered by the larva in the form of a toxic alkaloid from the host plant. The larva uses the chemical in its defence and at metamorphosis the toxins are transferred to the adult. Both sexes use them as defensive compounds, with the male additionally converting them to his pheromone. This he emits from inflatable abdominal tubes, called coremata, whose development is regulated by the alkaloid pheromone precursor.

Sex pheromones were thought to be unique, species-specific chemicals. Actually, they are often chemical blends. The same chemical (for example, a particular 14-carbon chain alcohol) may be present in a range of related and unrelated species but it occurs in a blend of different proportions with several other chemicals. An individual component may elicit only one part of the sex attraction behaviour, or a partial or complete mixture may be required. Often the blend produces a greater response than any individual component, a **synergism** which is widespread in insects that produce pheromone mixtures. Chemical structural similarity of pheromones may indicate systematic relationship amongst the producers. However, obvious anomalies arise when identical or very similar pheromones are synthesized from chemicals derived from identical diets by unrelated insects.

Even if individual components are shared by many species, the mixture of pheromones is very often species-specific. It is evident that pheromones, and the stereotyped behaviours that they evoke, are highly significant in maintenance of reproductive isolation between species. The species-specificity of sex pheromones avoids cross-species mating before males and females come into contact.

(b) Aggregation pheromones

The release of an **aggregation pheromone** causes conspecific insects of both sexes to crowd around the source of the pheromone. Aggregation may lead to increased likelihood of mating but, in contrast to many sex pheromones, both sexes may produce and respond to aggregation pheromones. The potential benefits provided by the response include security from predation, maximum utilization of a scarce food resource, overcoming of host resistance and cohesion of social insects, as well as the chance to mate.

Aggregation pheromones are known in six insect orders, including cockroaches, but their presence and mode of action have been studied in most detail in Coleoptera, particularly in economically damaging species such as stored-grain beetles (from several families) and timber and bark beetles (Scolytidae). A well studied example of a complex suite of aggregation pheromones is provided by the Californian western pine beetle, *Dendroctonus brevicomis* (Scolytidae), which attacks ponderosa pine (*Pinus ponderosa*). On arrival at a new tree, colonizing females release the pheromone *exo*-brevicomin together with myrcene, a terpene originating from the damaged pine tree. Both sexes of western pine beetle are attracted by this mixture, and newly arrived males then add to the chemical mix by releasing another pheromone, frontalin. The cumulative lure of frontalin, *exo*-brevicomin and myrcene is synergistic, that is, greater than any one of these chemicals alone. The aggregation of many pine beetles overwhelms the tree's defensive secretion of resins.

(c) Spacing pheromones

There is a limit to the number of western pine beetles (*D. brevicomis*) that attack a single tree. Cessation is assisted by reduction in the attractant aggregation pheromones, but deterrent chemicals are also produced. After the beetles mate on the tree, both sexes produce 'anti-aggregation' pheromones called verbenone and *trans* verbenone. These deter further beetles from landing close by and this encourages spacing out of new colonists. When the resource is saturated, further arrivals are repelled.

Such semiochemicals, called **spacing, epideictic or dispersion pheromones**, may effect appropriate spacing on food resources, as with some phytophagous insects. Several species of tephritid flies lay eggs singly in fruit where the solitary larva is to develop. Spacing occurs because the ovipositing female deposits an oviposition-deterrent pheromone on the fruit on which she has laid an egg, thereby deterring subsequent oviposition. Social insects, which by definition are aggregated, utilize pheromones to regulate many aspects of their behaviour, including the spacing between colonies. Spacer pheromones of colony-specific odours may be used to ensure an even dispersal of colonies of conspecifics, as in African weaver ants (Formicidae: *Oecophylla longinoda*).

(d) Trail-marking pheromones

Many social insects use pheromones to mark their trails, particularly to food and the nest. **Trail-marking pheromones** are volatile and short-lived chemicals that evaporate within days unless reinforced (perhaps as a response to a food resource that is longer-lived than usual). Trail pheromones in ants are commonly metabolic waste products excreted by the poison gland. These need not be species-specific – several species share some common chemicals. Dufour's gland secretions of some ant species may be more species-specific chemical mixtures associated with marking of territory and pioneering trails. Ant trails appear to be non-polar, that is the direction to nest or food resource cannot be determined by the trail odour.

In contrast to trails laid on the ground, an airborne trail – an odour plume – becomes directional due to increasing concentration of the odour towards the source. An insect may rely upon angling its flight path relative to the direction of the wind that brings the odour, resulting in a zig-zag upwind flight towards the source. Each directional shift is produced where the odour diminishes at the edge of the plume (Fig. 4.8).

(e) Alarm pheromones

Nearly two centuries ago it was shown that honey-bee (*Apis mellifera*) workers were alarmed by a freshly extracted sting. In the intervening years many aggregating insects have been found to produce chemical releasers of alarm behaviour – **alarm pheromones** – that are characteristic of most social

insects (termites and eusocial hymenopterans). In addition, alarm pheromones occur in several hemipterans, including subsocial treehoppers (Membracidae), aphids (Aphididae) and some other true bugs. Alarm pheromones are volatile, non-persistent compounds that are readily dispersed throughout the aggregation. Alarm is provoked by the presence of a predator, or in many social insects by a threat to the nest. The behaviour elicited may be rapid dispersal, such as in hemipterans that drop from the host plant, or escape from an unwinnable conflict with a large predator, as in poorly defended ants living in small colonies. The alarm behaviour of many eusocial insects is most familiar to humans when nest disturbance induces many ants, bees or wasps to an aggressive defence. Alarm pheromones attract aggressive workers and these recruits attack the cause of the disturbance by biting, stinging or firing repellent chemicals. Emission of more alarm pheromone mobilizes further defenders. Alarm pheromone may be daubed over an intruder to aid in directing the attack.

Alarm pheromones may have evolved from chemicals used as general anti-predator devices (allomones, see below), utilizing glands co-opted from many different parts of the body to produce the substances. For example, hymenopterans commonly produce alarm pheromones from mandibular glands but also from poison glands, metapleural glands, the sting shaft and even the anal area. All these glands may also be production sites for defensive chemicals.

4.3.3 Semiochemicals: kairomones, allomones and synomones

Communication chemicals (semiochemicals) may function between individuals of the same species, in which case they are called pheromones, or between different species, in which case they are **allelochemicals**. Interspecific semiochemicals may be grouped according to the benefits they provide to the producer and receiver. Those that benefit the receiver but disadvantage the producer are **kairomones**. **Allomones** confer advantage on the producer by modifying the behaviour of the receiver although having a neutral effect on the receiver. **Synomones** are chemicals that benefit the producer and also benefit the receiver. This terminology has to be made in the context of the specific behaviour induced in the recipient, as seen in the examples discussed below. A particular chemical can act as an intraspecific pheromone and may also fulfil all three categories of interspecific communication, depending on circumstances.

(a) Kairomones

Myrcene, the terpene produced by a ponderosa pine when it is damaged by the western pine beetle, acts as a synergist with aggregation pheromones to lure more beetles. Thus myrcene and other terpenes produced by damaged conifers can be kairomones (chemicals that disadvantage the producer), in this case by luring damaging timber beetles. A kairomone need not be a product of insect attack: elm bark beetles (Scolytidae: *Scolytus* spp.) respond to alpha cubebene, a product of the Dutch elm disease fungus *Ceratocystis ulmi* that indicates a weakened or dead elm tree (*Ulmus*). Elm beetles will inoculate previously healthy elm trees with the fungus, but pheromone-induced beetle aggregations form only when the kairomone (fungal alpha-cubenene) indicates suitability for colonization. Host-plant detection by phytophagous insects also involves reception of plant chemicals that therefore act as kairomones.

Insects produce many communication chemicals, with clear benefits. However, these semiochemicals may also act as kairomones when other insects recognize them. In 'hijacking' the chemical messenger for their own use, specialist parasitoids (Chapter 12) use chemicals emitted by the host, or plants attacked by the host, to locate a suitable host for the development of its offspring.

A spectacular example of kairomones occurs in bolas spiders, which do not build a web but whirl a single thread terminating in a sticky globule towards moths (like gauchos using a bolas to hobble cattle). The spiders lure male moths to within reach of the bolas using synthetic lures of sex-attractant pheromone cocktails. The proportions of the components vary according to the abundance of particular moth species available as prey. Similar principles using deceitful signalling are applied by humans to control pest insects by pheromone lures (Chapter 15).

(b) Allomones

Allomones are chemicals that benefit the producer but have neutral effects on the recipient. For example, defensive or repellent chemicals are allomones that advertise distastefulness and protect the producer from lethal experiment by prospective predators. The effect on a potential predator is considered to be neutral since it is warned from wasting energy in seeking a distasteful meal.

The worldwide beetle family Lycidae has many distasteful and warning-coloured (aposematic) members, including species of *Metriorrhynchus* that are protected by odorous alkylpyrazine allomones. In Australia, several distantly related beetle families include many mimics that are modelled visually on *Metriorrhynchus*. Some mimics are remarkably convergent in colour and distasteful chemicals and possess nearly identical alkylpyrazines. Others share the allomones but differ in distasteful chemicals, whereas some have the warning chemical but appear to lack distastefulness. Other insect mimicry complexes involve allomones. Mimicry and insect defences in general are considered further in Chapter 13.

(c) Synomones

The terpenes produced by damaged pines are kairomones for pest beetles, but if identical chemicals are used by beneficial parasitoids to locate and attack the bark beetles, the terpenes are acting as synomones (benefiting both the producer and the receiver). Thus alpha-pinene and myrcene produced by damaged pines, are kairomones for species of *Dendroctonus* but synomones for the pteromalid hymenopterans that parasitize these timber beetles. In like manner, alpha-cubebene produced by Dutch elm fungus is a synomone for the braconid hymenopteran parasitoids of elm bark beetles (for which it is a kairomone).

An insect parasitoid may respond to host plant odour directly, like the phytophage it seeks to parasitize, but this means of searching does not guarantee that the phytophage host is actually present. There is a greater chance of success for the parasitoid if it can identify and respond to the specific plant chemical defences that the phytophage provokes. If a damaged host plant produced a repellent odour, then the chemical would act as:

- An allomone that deters non-specialist phytophages.
- A kairomone that attracts a specialist phytophage.
- A synomone that lures the parasitoid of the phytophage.

Of course phytophagous, parasitic and predatory insects do not rely solely on odours to locate potential hosts or prey and visual discrimination is implicated in resource location.

4.4 INSECT VISION

With the exception of a few blind subterranean and endoparasitic species, most insects can see to some extent and many possess highly developed visual systems. The basic components needed for vision are a lens to condense the visual field onto an organ comprising cells containing light-sensitive chemicals and a nervous system complex enough to interpret and act upon visual information. Compound eyes are the most obvious and familiar visual organs of insects but there are three other means by which an insect may perceive light: dermal detection, stemmata and ocelli. The dragonfly head depicted in the vignette for this chapter is dominated by its huge compound eyes with the three ocelli and paired antennae in the centre.

4.4.1 Dermal detection

In insects able to detect light through their body surface, there are no specific sense organs but there are receptors on the general surface. Evidence for this general responsivity to light comes from the persistence of photic responses after covering all visual organs of, for example, cockroaches and lepidopteran larvae. Some blind cave insects, with no recognizable visual organs, can detect and avoid light, as do decapitated cockroaches. In most cases a connection between the sensitive cells and the central nervous system has yet to be discovered. However, within the brain itself, aphids have light sensitive cells which detect changes in day length – an environmental cue that controls the mode of reproduction (i.e. either sexual or parthenogenetic). The setting of the biological clock (Box 4.2) relies upon the ability to detect photoperiod.

4.4.2 Stemmata

The only visual organs of larval holometabolous insects are **stemmata** (Fig. 4.10(a)). These organs are on the head and vary from a single pigmented spot on each side to six or more larger stemmata, each with numerous nerve cells. In the simplest stemma, a cuticular lens overlies a crystalline lens secreted by three cells. The image formed by the lens falls

BOX 4.2
BIOLOGICAL CLOCKS

Locality 1, Day 1, Afternoon

Locality 2, Day 2, Morning

Seasonal changes in environmental conditions allow insects to adjust their life histories to optimize the use of suitable conditions and minimize the impact of unsuitable ones (e.g. through diapause; section 6.8). Similar physical fluctuations on a daily scale encourage a diurnal (daily) cycle of activity and quiescence. **Nocturnal** insects are active at night, **diurnal** ones by day, and **crepuscular** insect activity occurs at dusk and dawn when light intensities are transitional. The external physical environment, such as light/dark or temperature, controls some daily activity patterns, called **exogenous rhythms**. However, many other periodic activities are internally driven **endogenous rhythms** that have a clock-like or calendar-like frequency irrespective of external conditions. Endogenous periodicity is frequently about 24 hours (**circadian**), but lunar and tidal periodicities govern the emergence of adult aquatic midges from large lakes and marine intertidal zones respectively. This unlearned, once-in-a-lifetime rhythm that allows synchronization of eclosion, independent of the environment, demonstrates the innate ability of insects to measure passing time.

Experimentation is required to discriminate between exogenous and endogenous rhythms. This involves observing what happens to rhythmic behaviour when external environmental cues are either altered, removed or made invariate. Such experiments show that inception (setting) of endogenous rhythms is found to be day length, with the clock then free-running, without daily reinforcement by the light/dark cycle, often for a considerable period. Thus, if nocturnal cockroaches that become active at dusk are kept at constant temperature in constant light or dark, they will maintain the dusk commencement of their activities at a circadian rhythm of 23–25 hours. Rhythmic activities of other insects may require an occasional clock-setting (such as darkness) to prevent the circadian rhythm

---

BOX 4.2—*continued*

drifting, either through adaptation to an exogenous rhythm or into arrhythmy.

Biological clocks allow solar orientation – the use of the sun's elevation above the horizon as a compass – provided that there is a means of assessing (and compensating for) the passage of time. Some ants and honey bees use a 'light-compass', finding direction from the sun's elevation and using the biological clock to compensate for the sun's movement across the sky. Evidence came from an elegant experiment with honey bees trained to forage in the late afternoon at a feeding table (F) placed 180 m NW of their hive, as depicted in the left figure (after Lindauer, 1960). Overnight the hive was moved to a new location to prevent use of familiar landmarks in foraging, and a selection of four feeding tables (F_{1-4}) was provided at 180 m NW, SW, SE and NE from the hive. In the morning, despite the sun being at a very different angle to that during the previous afternoon's training, 15 of the 19 bees were able to locate the NW table (as depicted in the figure on the right). The honey bee 'dance-language' that communicates direction and distance of food to other workers (Chapter 11) depends upon the capacity to calculate direction from the sun.

The circadian pacemaker (oscillator) that controls the rhythm is located in the brain; it is not an external photoperiod receptor. Experimental evidence shows that the pacemaker of some silkworm activities lies in the cerebral lobes of the brain, but in cockroaches and crickets it may lie in the optic lobes. In all other insects investigated an oscillator site appears predominantly within the brain, often in the suboesophageal ganglion, but perhaps also elsewhere in the body. There is increasing evidence of more than one oscillator.

---

on the **rhabdom**(s) – the central photoreceptive structure formed from part of several adjacent **retinula** (or nerve) **cells** – consisting of closely packed microvilli containing visual pigment. Vision comes from the sum of all stemmata, each of which points to a different part of the visual field. Since the stemmata cover an incomplete field, the insect sees only a few points of light according to the number of stemmata. Movement of the head allows scanning of an increased field and estimation of shape and even distance and size, but vision is poor: the lens acts more to concentrate light than to form sharp images. In some larvae, such as those of sawflies, there is only a single stemma on each side of the head but it is a proper image-forming eye with a two-layered lens that can form a relatively clear image on an extended retina composed of many rhabdoms.

4.4.3 Ocelli

Many adult insects and larvae or nymphs (of hemimetabolous insects) have dorsal **ocelli** in addition to compound eyes. Typically three small organs lie in a triangle on top of the head. They appear to enhance light detection by the compound eyes and to register cyclical changes in light intensity that correlate with diurnal behavioural rhythms.

The cuticle covering an ocellus may be curved as a lens; it is transparent and it overlies transparent epidermal cells, so that light passes to grouped retinal nerve cells below. Each cell in a group has an inner specialized area containing visual pigment, the **rhabdomere**, which contributes to the rhabdom of a group of cells (Fig. 4.10(b)). Light falling on the rhabdom stimulates the nerve. Individual groups of cells

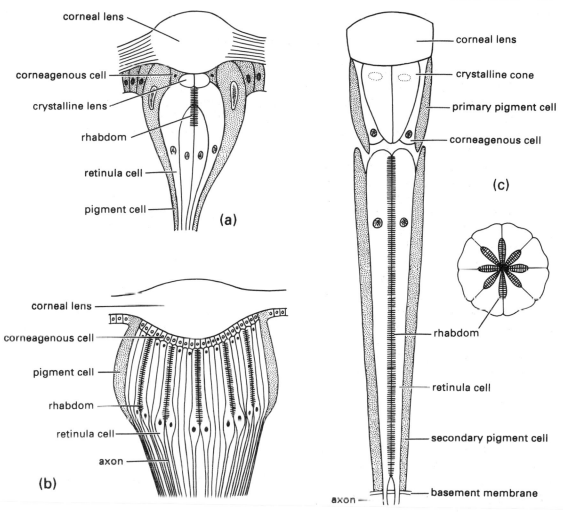

Fig. 4.10 Longitudinal sections through the rhabdoms in simple and compound eyes: (a) a simple stemma of a lepidopteran larva; (b) dorsal ocellus of a adult bug; (c) an ommatidium from a compound eye, with enlargement showing a transverse section. ((a) After Snodgrass, 1935; (b) after Link, 1909; (c) after CSIRO, 1970.)

or the complete light sensitive organ are surrounded by pigment cells or by a reflective layer.

Even if there is a cuticular lens, a sharp image does not fall on the layer of pigmented cells (the retina). Thus the ocelli do not allow vision, but they are very sensitive to low light and to subtle changes in intensity. Detection of these changes is enhanced by a nervous system in which the many primary light-sensitive nerve cells converge to fewer second

order cells connected to the brain. In the ocellus of the dragonfly *Sympetrum*, some 675 nerve cells are reduced in the ocellar nerve to one large axon, two medium-sized ones and a few small ones.

4.4.4 Compound eyes

The culmination of the development of invertebrate visual organs is the **compound eye**. Virtually all adult insects and nymphs have large, prominent compound eyes that give them a broad visual field.

The compound eye is based on repetition of many individual units called **ommatidia**. Each ommatidium resembles a simple stemma (Fig. 4.10(c)): it has a cuticular lens overlying a conical crystalline lens or cone above eight (or maybe six to ten) elongate nerve (retinula) cells (seen in transverse section in Fig. 4.10(c)). The inner part or rhabdomere of each cell contains the visual pigment(s) and contributes to the central rhabdom. The nerve cells are clustered around the longitudinal axis of each ommatidium and are surrounded by a ring of light-absorbing pigmented cells, which isolates an ommatidium from its neighbours.

It appears that each ommatidium can form an image from its field of view but, as with the stemmata, there are too few cells to record this detailed image and only overall light intensity is registered. Light falls on an ommatidium from an angle of up to 20°, but this angle has to be restricted somehow, otherwise adjacent ommatidia would receive overlapping images which, when combined, would depict no more than a uniform light intensity. The actual image formed by the compound eye is of a series of apposed points of light of different intensities, termed an **apposition image**.

An apposition eye does not function well in low light, and nocturnal insects have a somewhat different visual system: the **superposition eye**. In this, the ommatidia function in bright light as in the apposition eye, with screening pigment isolating each unit. In subdued light, the screening pigment moves towards the outer surface of the eye, allowing light to pass between adjacent ommatidia. Images formed in this way are spread and partially overlap on one another and the retinula cells respond to variation in light intensity across the image. Because these superposition images are not exactly congruent, they are less sharp than those produced by an apposition eye. However, less light is lost by absorption by the screening pigment. Further conservation occurs through reflection of light from the back of the eye through the ommatidia for a second time, utilizing a mirror-like **tapetum** of small tracheae at the base of the retinula cells as a reflector.

In comparison with a vertebrate eye, the insect compound eye shows a much reduced resolving power. Detection of pattern, shape and contrast are more significant than fine detail. However, the possession of very many ommatidia allows good perception of movement, particularly when the head is moved relative to the field of view – for example, as in flight. Compound eyes allow accurate perception of distance through binocular vision, an important attribute of visual predators and highly mobile insects such as odonate larvae and adults.

Throughout the insects there are many species that possess colour vision, particularly amongst flower visitors such as bees and butterflies. Variations in the insect visual pigments allow detection of different wavelengths of light (colours). Insects see poorly at the red end of the spectrum but well in the ultraviolet, and many insect-frequented plants provide distinctive alluring patterns visible only in ultraviolet.

Light emanating from the sky is polarized, that is with greater vibration in some planes than others, relative to the position of the sun. Some insects can detect the plane of polarization of light and utilize this in navigation. The major sensor appears to be microvilli (orientated visual pigments) in each rhabdomere that act as polarizing filters. Ocelli and stemmata may also detect the polarized plane.

4.4.5 Light production

The most spectacular visual displays of insects involve light production, or **bioluminescence**. Some insects co-opt symbiotic luminescent bacteria or fungi, but self-luminescence is found in a few Collembola, one hemipteran (the fulgorid lantern bug), a few dipteran fungus gnats and a diverse group of several families of coleopterans. The latter are members of the Phengodidae, Drilidae, and notably the Elateridae and Lampyridae, commonly given colloquial names including fireflies, glow-worms and lightning bugs. Any or all stages and sexes in the life history may glow, using one to many luminescent organs that may be located almost anywhere on the body. Light emitted may be white, yellow, red or green.

The light-emitting mechanism studied in the lampyrid firefly *Photinus pyralis* may be typical of luminescent Coleoptera. The enzyme **luciferase** oxidizes a substrate, **luciferin**, in the presence of an energy source of adenosine triphosphate (ATP) and oxygen, to produce oxy-luciferin, carbon dioxide and light. Variation in ATP release controls the rate of flashing, and differences in pH may allow variation in the frequency (colour) of light emitted.

The principal role of light emission is in courtship signalling. This involves species-specific variation in the duration, number and rate of flashes in a pattern and the frequency of repetition of the pattern. Generally a mobile male advertises his presence by instigating the signalling with one or more flashes and a sedentary female indicates her location with a flash in response. As with all communication systems, there is scope for abuse, for example, that involving prey-luring by a carnivorous female lampyrid of *Photurus* (section 12.1.2).

Bioluminescence is involved in both prey-luring and mate-finding in Australian and New Zealand cave-dwelling *Arachnocampa* fungus gnats (Diptera: Mycetophilidae). All developmental stages use a reflector to concentrate light that they produce from modified Malpighian tubules. In the dark zone of a cave, the larval light lures prey, particularly small flies, onto a sticky thread suspended by the larva from the cave ceiling. The flying adult males locate the luminescent female while she is still in the pharate state and wait for the opportunity to mate upon her emergence.

4.5 INSECT BEHAVIOUR

Many of the insect behaviours mentioned in this chapter appear very complex, but behaviourists attempt to reduce them to simpler components. Thus, individual **reflexes** (simple responses to simple stimuli) can be identified, such as the flight response when the legs lose contact with the ground, and cessation of flight when contact is regained. Some extremely rapid reflex actions, such as the feeding lunge of odonate nymphs or the 'escape reactions' of many insects, depend upon a reflex involving **giant axons** that conduct impulses rapidly from sense organs to the muscles. The integration of multiple reflexes associated with movement of the insect may be divisible into:

- **Kinesis** (plural **kineses**), in which unorientated action varies according to stimulus intensity.
- **Taxis** (plural **taxes**), in which movement is directly towards or away from the stimulus

Kineses include **akinesis**, unstimulated lack of movement, **orthokinesis**, in which speed depends upon stimulus intensity and **klinokinesis**, which is a 'random walk' with course changes (turns) being made when unfavourable stimuli are perceived and with the frequency of turns depending on the intensity of the stimulus. Increased exposure to unfavourable stimuli leads to increased tolerance (**acclimation**), so that random walking and acclimation will lead the insect to a favourable environment. The male response to the plume of sex attractant (Fig. 4.8) is an example of klinokinesis to a chemical stimulus. Ortho- and klinokineses are effective responses to diffuse stimuli, such as temperature or humidity, but different and more efficient responses are seen when an insect is confronted by less diffuse, gradient or point-source stimuli.

Kineses and taxes can be defined with respect to the type of stimulus. Appropriate prefixes include **anemo-** for air currents, **astro-** for solar, lunar or astral (including polarized light), **chemo-** for taste and odour, **geo-** for gravity, **hygro-** for moisture, **phono-** for sound, **photo-** for light, **rheo-** for water current, and **thermo-** for temperature. Orientation and movement may be positive or negative with respect to the stimulus source so that, for example, resistance to gravity is termed negative geotaxis, attraction to light is positive phototaxis, and repulsion from moisture is negative hygrotaxis.

In **klinotaxic** behaviour, an insect moves relative to a gradient (or cline) of stimulus intensity, such as a light source or a sound emission. To do this the strength of the stimulus must be compared on each side of the body, either by moving the receptors from side to side (as in head waving of ants when they follow an odour trail), or by paired receptors that detect differences in stimulus intensity between the two sides of the body, as the tympanal organs detect the direction of the sound source by differences in intensity between the two organs. Orientation with respect to a constant angle of light is termed **menotaxis** and includes the 'light-compass' referred to in Box 4.2. Visual fixation of an object, such as prey, is termed **telotaxis**.

Often the relationship between stimulus and behavioural response is complex, since a **threshold** intensity may be required before an action ensues. A particular stimulatory intensity is termed a **releaser** for a particular behaviour. Furthermore, complex behaviour elicited by a single stimulus may comprise several sequential steps, each of which may have a higher threshold, requiring an increased stimulus. As described in section 4.3.2(a), a male moth responds to a low-level sex pheromone stimulus by raising the antennae; at higher levels he orientates towards the source; and at an even higher threshold flight is initiated. Increasing concentration encourages continued flight and a second,

higher threshold may be required before courtship ensues. In other behaviours, several different stimuli are involved, such as for courtship through to mating. This sequence can be seen as a long **chain reaction** of stimulus, action, new stimulus, next action and so on, with each successive behavioural stage depending upon the occurrence of an appropriate new stimulus. An inappropriate stimulus during a chain reaction (such as the presentation of food while courting) is not likely to elicit the usual response.

Most insect behaviours are considered to be **innate** – that is, they are programmed genetically to arise stereotypically upon first exposure to the appropriate stimulus. However, many behaviours are environmentally and physiologically modified: for example, virgins and mated females respond in very different ways to identical stimuli and immature insects often respond to different stimuli than their adults. Furthermore, there is experimental evidence that learning can modify innate behaviour. By experimental teaching (using training and reward), bees and ants can learn to run a maze and butterflies can be induced to alter their favourite flower colour. However, study of natural behaviour (ethology) is more relevant to understanding the role that insects' behavioural plasticity, including the ability to modify behaviour through learning, plays in their evolutionary success. In pioneering ethological studies, Niko Tinbergen showed that digger wasps (Sphecidae: *Philanthus triangulum*) learn the location of their chosen nest site by making a short flight to memorize elements of the local terrain. Adjustment of prominent landscape features around the nest misleads the homing wasp. However, since wasps identify landmark relationships rather than individual features, the confusion may be only temporary. Closely related *Bembix* digger wasps learn nest locations through more distant and subtle markers, including the appearance of the horizon, and are not tricked by investigator ethologists moving local small scale landmarks.

FURTHER READING

Atkins, M.D. (1980) *Introduction to Insect Behavior*, Macmillan Publishing Co., Inc., New York, Collier Macmillan Publishers, London.

Birch, M.C. and Haynes, K.F. (1982) *Insect Pheromones*, Studies in Biology no. 147, Edward Arnold, London.

Birch, M. and Hefetz, A. (1986) The mating of moths. *New Scientist*, **112**, 48–51.

Chapman, R.F. (1982) *The Insects. Structure and Function*, 3rd edn, Hodder and Stoughton, London, Sydney, Auckland, Toronto.

Fuller, J.H. and Yack, J.E. (1993) The evolutionary biology of insect hearing. *Trends in Ecology and Evolution*, **8**, 248–252.

Kerkut, G.A. and Gilbert, L.I. (eds) (1985) *Comprehensive Insect Physiology, Biochemistry, and Pharmacology, Volume 6, Nervous System: Sensory; Volume 9, Behaviour*, Pergamon Press, Oxford, New York, Toronto, Sydney, Paris, Frankfurt.

Lewis, T. (ed.) (1984) *Insect Communication*, Academic Press, London, Orlando, San Diego, New York, Toronto, Montreal, Sydney, Tokyo.

Wood, D.L. (1982) The role of pheromones, kairomones and allomones in host selection and colonization by bark beetles. *Annual Review of Entomology*, **27**, 411–446.

5

REPRODUCTION

Male–male combat (part of sexual selection)

After behaviour associated with survival, much of an insect's life is concerned with reproduction. Most insects are sexual and mature males and females must be present at the same time and place for reproduction to take place. Since insects are generally short-lived, their life history, behaviour and reproductive condition must be synchronized. This requires finely tuned and complex physiological responses to the external environment. Furthermore, reproduction depends also on monitoring of internal physiological stimuli, and the neuroendocrine system plays a key regulatory role. We are still far from understanding the control and regulation of insect reproduction, particularly compared to our knowledge of vertebrate reproduction. For example, although mating and egg production in many flies are controlled by a series of hormonal and behavioural changes, little is known of the underlying mechanisms.

These complex regulatory systems are highly successful. For example, look at the rapidity with which pest insect outbreaks occur. A combination of short generation time, high fecundity and population synchronization to environmental cues allows many insect populations to react extremely rapidly under appropriate environmental conditions, such as a crop monoculture or release from a controlling predator. In these situations, temporary or obligatory loss of males (**parthenogenesis**) has proved to be another effective means by which some insects rapidly exploit temporarily (or seasonally) abundant resources.

This chapter examines the different mechanisms associated with courtship and mating, avoidance of interspecies mating, ensuring paternity, and determination of sex of offspring. A later section examines the avoidance of sex and shows some extreme cases in which the adult insect has been dispersed with altogether. These observations are related to theories concerning sexual selection, including those relating to why insects have such remarkable diversity of genitalic structures.

The concluding summary of the physiological control of reproduction emphasizes the extreme complexity and sophistication of mating and oviposition in insects.

5.1 BRINGING THE SEXES TOGETHER

Insects are often at their most conspicuous when synchronizing the time and place for mating. The flashing lights of fireflies, the singing of crickets and cacophony of cicadas are spectacular examples. However, there is a wealth of less ostentatious behaviour, of equal significance in getting the sexes together and signalling readiness to mate to other members of the species. All signals are species-specific, serving to attract members of the opposite sex of the same species, but abuse of these communication systems can take place, as when females of one predatory species of firefly lure males of another species to their death by emulating the flashing signal of that species.

Swarming is a characteristic and perhaps fundamental behaviour of insects, since it occurs amongst some primitive insects such as mayflies and odonates and also in many higher insects such as flies and butterflies. Swarming sites are identified by visual markers and are species-specific. Swarms are predominantly of the male sex only, though female-only swarms do occur. Swarms are most evident when many individuals are involved, such as when midge swarms are so dense that they have been mistaken for smoke from burning buildings, but small swarms may be more significant in evolution. A single male insect holding station over a spot is a swarm of one – he awaits the arrival of a receptive female that has responded equally to visual cues that identify the site. The precision of swarm sites allows more effective mate-finding than searching, particularly when individuals are rare, or dispersed and at low density. The formation of a swarm allows insects of differing genotypes to meet and outbreed. This is of particular importance in

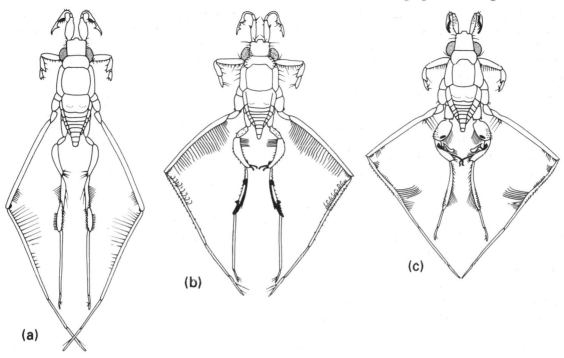

Fig. 5.1 Males of three species of the water-strider genus *Rheumatobates*, showing species-specific antennal and leg modifications (mostly flexible setae). These non-genitalic male structures are specialized for contact with the female during mating, when the male rides on her back. Females of all species have a similar body form. (a) *R. trulliger*; (b) *R. rileyi*; (c) *R. bergrothi*. (After Hungerford, 1954.)

circumstances where larval development sites are patchy and locally dispersed; if adults did not disperse, inbreeding would occur.

The meeting site need not be visually determined. For species whose larval development medium is discrete, such as rotting fruit, animal dung or a specific host plant or vertebrate host, where better for the sexes to meet and court? The olfactory receptors by which the female dung fly finds a fresh pile of dung for egg laying (the larval development site) can be employed by both sexes to facilitate meeting.

Another odoriferous communication involves one or both sexes producing and emitting a **pheromone**, which is a chemical or mixture of chemicals perceptible to another member of the species (section 4.3.2). Substances emitted with the intention of altering the sexual behaviour of the recipient are termed **sex pheromones**. Generally these are produced by the female and announce her presence and sexual availability to conspecific males. Recipient males that detect the odour plume become aroused and orientate from downwind towards the source. More and more insects investigated are found to have species-specific sex pheromones, the diversity and specificity of which are important in maintenance of species reproductive isolation.

When the sexes are in proximity, mating in some species takes place with little further ado. For example, when a conspecific female arrives at a swarm of male flies, a nearby male, recognizing her by the particular sound of her wingbeat frequency, immediately copulates with her. However, more elaborate and specialized close-range behaviours, termed courtship, are commonplace.

5.2 COURTSHIP

Although the long-range attraction mechanisms discussed above reduce the number of species present at a prospective mating site, there will remain an excess of potential partners. Further discrimination between species and conspecific individuals usually takes place. Courtship is the close-range behaviour that induces sexual receptivity prior to mating and acts as a mechanism to reduce cross-species mating.

Courtship may include visual displays, predominantly by males, including movements of adorned parts of the body, such as antennae, eyestalks and 'picture' wings, and ritualized movements ('dancing'). Tactile stimulation such as rubbing and stroking often occurs later in courtship, often immediately prior to mating, and may continue during copulation. Parts of the anatomy used in tactile stimulation include the antennae, palps, head horns, external genitalia and legs (Fig. 5.1).

Insects such as crickets, which use long-range calling, may have different calls for use in close-range courtship. Others, such as fruit flies (*Drosophila*), have no long-distance call and only sing (by wing vibration) in close-up courtship. In some predatory insects, including empidid flies and mecopterans, the male courts a prospective mate by offering a prey item as a nuptial gift (Box 5.1).

If the sequence of display proceeds correctly, courtship grades into mating. Sometimes the complete sequence need not be completed before copulation commences. On other occasions courtship must be prolonged and repeated. It may be unsuccessful if one sex fails to respond or makes inappropriate responses. Generally members of different species differ in some elements of their courtships and interspecies matings do not occur. The great specificity and complexity of

insect courtship behaviours can be interpreted in terms of mate location, synchronization and species recognition, and viewed as having evolved as a pre-mating isolating mechanism. Important as this view is, there is equally compelling evidence that courtship is an extension of a wider phenomenon of competitive communication and involves sexual selection.

5.3 SEXUAL SELECTION

Many insects are sexually dimorphic, usually with the male adorned with secondary sexual characteristics, some of which have been noted above in relation to courtship display. Darwin's knowledge of sexual dimorphism in insects such as staghorn beetles, song in orthopterans and cicadas, and wing colour in butterflies and odonates helped him to recognize the operation of sexual selection – the elaboration of features associated with sexual competition rather than directly with survival. For example, certain male flies have antlers (which may include eyestalks) which are used in male–male combat in competition for access to the oviposition site visited by females (as shown for the tephritid *Phytalmia mouldsi* in the vignette for this chapter). Insects are highly amenable to studies of sexual selection because of short generation time, facility of manipulation in the laboratory and relative ease of observation in the field.

Observations on insects show that courtship in many mating systems can be viewed as intraspecific competition for mates, with certain male behaviours inducing female response in ways that can increase the male's success. Since females differ in their responsiveness to male stimuli, females can be said to choose between males. Thus, courtship can be seen to be competitive. Female choice might involve no more than selection of the winners of male–male interactions, or may be as subtle as to discriminate between the sperm of different males. In this interpretation all elements of communication associated with

BOX 5.1
COURTSHIP AND MATING IN MECOPTERA

Sexual behaviour has been well studied in hangingflies (Bittacidae) of the North American *Hylobittacus* (*Bittacus*) *apicalis* and *Bittacus* species and the Australian *Harpobittacus* species, and in the Mexican *Panorpa* scorpionflies (Panorpidae). Adult males hunt for arthropod prey, such as caterpillars, bugs, flies and katydids. These same food items may be presented to a female as a nuptial offering to be consumed during copulation. Females are attracted by a sex pheromone emitted from one or more eversible vesicles or pouches near the end of the male's abdomen as he hangs in the foliage using prehensile fore tarsi.

Courting and mating in Mecoptera is exemplified by the sexual interactions in *Harpobittacus australis* (Bittacidae). The female closely approaches the 'calling' male; he then ends pheromone emission by retracting the abdominal vesicles. Usually the female probes the prey briefly, presumably testing its quality, while the male touches or rubs her abdomen and seeks her genitalia with his own. If the female rejects the nuptial gift, she refuses to copulate. However, if the prey is suitable, the genitalia of the pair couple and the male temporarily withdraws the prey with his hind legs. The female lowers herself until she hangs head downwards, suspended by her terminalia. The male then surrenders the nuptial offering (in the illustration, a caterpillar) to the female and she feeds as copulation proceeds. At this stage the male frequently supports the female

BOX 5.1—*continued*

by holding either her legs or the prey on which she is feeding. The derivation of the common name hangingflies is obvious!

Detailed field observations and manipulative experiments have demonstrated female choice of male partners in species of Bittacidae. Both sexes mate several times per day with different partners. Females discriminate against males that provide small or unsuitable prey either by rejection or by copulating only for a short time, insufficient to pass the complete ejaculate. Given an acceptable nuptial gift, the duration of copulation correlates with the size of the offering. Each copulation in field populations of *H. australis* lasts from one to a maximum of about 17 minutes for prey 3–14 mm long. In the larger *Hy. apicalis*, copulations involving prey of the size of house flies and above (19–50 mm^2) last 20–29 minutes, resulting in maximal sperm transfer, increased oviposition and the induction of a **refractory period** (female non-receptivity to other males) of several hours. Copulations lasting less than 20 minutes reduce or eliminate male fertilization success.

gaining fertilization of the female, from long-distance sexual calling through to insemination, are seen as competitive courtship between males. By this reasoning, members of a species avoid hybrid matings because of a specific-mate recognition system which evolved under the direction of female choice, rather than as a mechanism to promote species cohesion.

5.4 COPULATION

The evolution of male external genitalia made it possible for insects to transfer sperm directly from male to female during copulation. This freed all but the most primitive insects from reliance on indirect methods, such as the male depositing a **spermatophore** (sperm packet) for the female to pick up from the substrate, as in Collembola, Diplura and apterygote insects. In pterygote insects, copulation (sometimes referred to as mating) involves the physical apposition of male and female genitalia and insemination – the transfer of sperm via the insertion of part of the male's aedeagus, the penis, into the reproductive tract of the female. In males of many species the extrusion of the aedeagus during copulation is a two-stage process: first, the whole aedeagus

is extended from the abdomen, and then the intromittent organ is everted or extended, producing a long structure (sometimes called the flagellum) capable of depositing semen deep within the female's reproductive tract (Fig. 5.2). In many insects the male terminalia have specially modified claspers, which lock with specific parts of the female terminalia to maintain the connection of their genitalia during sperm transfer.

This mechanistic definition of copulation ignores the sensory stimulation that is a vital part of the copulatory act in insects, as in other animals. In over a third of all insect species surveyed, the male indulges in copulatory courtship – behaviour that appears to stimulate the female during mating. The male may stroke, tap or bite the body or legs of the female, wave antennae, produce sounds or thrust or vibrate parts of his genitalia.

Sperm is received by the female insect in a copulatory pouch (genital chamber, vagina or bursa copulatrix) or directly into a spermatheca or its duct (as in *Oncopeltus*; Fig. 5.2). A spermatophore is the means of sperm transfer in most orders of insects; only some Heteroptera, Coleoptera, Diptera and Hymenoptera deposit unpackaged sperm. Sperm transfer requires lubrication, obtained from the

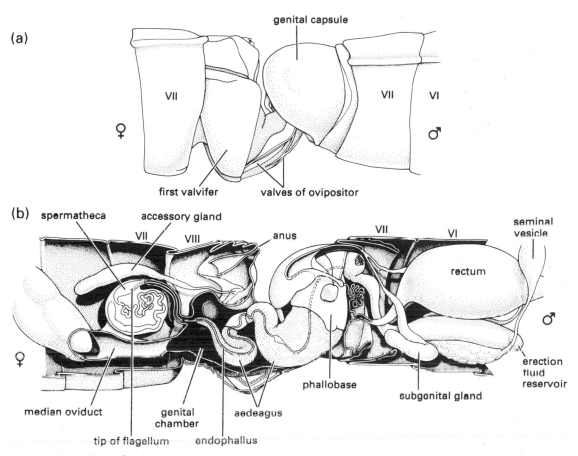

Fig. 5.2 Posterior ends of a pair of copulating milkweed bugs, *Oncopeltus fasciatus* (Hemiptera: Lygaeidae). Mating commences with the pair facing in the same direction, then the male rotates his eighth abdominal segment (90°) and genital capsule (180°), erects the aedeagus and gains entry to the female's genital chamber, before he swings around to face in the opposite direction. The bugs may copulate for several hours, during which they walk around with the female leading and the male walking backwards. (a) Lateral view of the terminal segments, showing the valves of the female's ovipositor in the male genital chamber; (b) longitudinal section showing internal structures of reproductive system, with the tip of the male's aedeagus in the female's spermatheca. (After Bonhag and Wick, 1953.)

seminal fluids, and, in insects that use a spermatophore, packaging of sperm. Secretions of the male accessory glands serve both of these functions as well as sometimes facilitating the final maturation of sperm, supplying energy for sperm maintenance, regulating female physiology and, in a few species, providing nourishment to the female (Box 5.2). The male accessory secretions may elicit one or two major responses in the female – induction of oviposition (egg-laying) and/or repression of sexual receptivity – by entering the female haemolymph and acting on her nervous or endocrine system.

BOX 5.2
NUPTIAL FEEDING

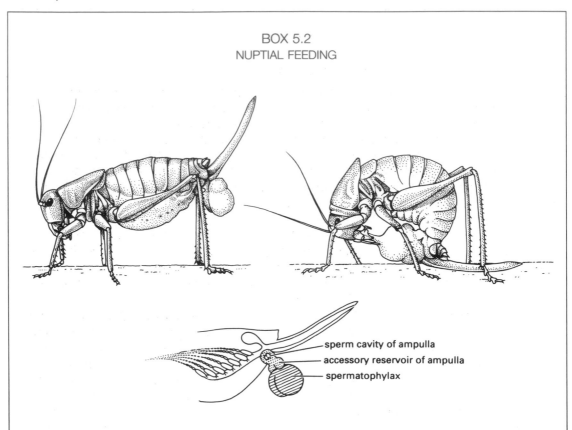

- sperm cavity of ampulla
- accessory reservoir of ampulla
- spermatophylax

Feeding of the female by the male before, during or after copulation has evolved independently in several different insect groups. From a female perspective, feeding takes three forms:

- They may receive nourishment from food collected, captured or regurgitated by the male (Box 5.1).
- They may obtain nourishment from a glandular product of the male.
- They may cannibalize males during or after copulation.

Nuptial feeding may represent parental investment (provided that the male can be sure of paternity) since it may increase the number or survival of the male's offspring indirectly via nutritional benefits to the female. Alternatively, courtship feeding may increase the male's fertilization success by preventing the female from interfering with sperm transfer. These two hypotheses concerning the function of nuptial feeding are not necessarily mutually exclusive; their explanatory value appears to vary between insect groups and may depend, at least partly, on the nutritional status of the female at the time of mating. Studies of mating in Mecoptera, Orthoptera and Mantodea exemplify the three nuptial feeding types seen in insects and continuing research on these groups addresses the relative importance of the two competing hypotheses that seek to explain the selective advantage of such feeding.

BOX 5.2—*continued*

Mating in katydids (Orthoptera: Tettigoniidae)

During copulation the males of many species of katydids transfer elaborate spermatophores which are attached externally to the female's genitalia. Each spermatophore consists of a large, proteinaceous, sperm-free portion, the **spermatophylax**, which is eaten by the female after mating, and a sperm ampulla, eaten after the spermatophylax has been consumed and the sperm has been transferred to the female. The illustration on the upper left shows a recently mated female Mormon cricket, *Anabrus simplex*, with a spermatophore attached to her gonopore; in the illustration on the upper right, the female is consuming the spermatophylax of the spermatophore (after Gwynne, 1981). The schematic illustration underneath depicts the posterior of a female Mormon cricket showing the two parts of the spermatophore: the spermatophylax (cross-hatched) and the sperm ampulla (stippled) (after Gwynne, 1990). During consumption of the spermatophylax, sperm is transferred from the ampulla along with substances that 'turn off' female receptivity to further males. Insemination also stimulates oviposition by the female, thereby increasing the probability that the male supplying the spermatophore will fertilize the eggs.

There are two hypotheses for the adaptive significance of this form of nuptial feeding. The spermatophylax may serve as a sperm protection device by preventing the ampulla from being removed until after the complete ejaculate has been transferred. Alternatively, the spermatophylax may be a form of parental investment in which nutrients from the male increase the number or size of the eggs sired by that male. Of course the spermatophylax may serve both purposes and indeed there is evidence to support each hypothesis. Experimental alteration of the size of the spermatophylax has demonstrated that females take longer to eat larger ones but in some katydid species the spermatophylax is larger than necessary to allow complete insemination and, in this case, the nutritional bonus to the female benefits the male's offspring. The function of the spermatophylax apparently varies between species and sometimes even intraspecifically depending on the quality of the female's diet, especially protein availability.

Cannibalistic mating in mantids (Mantodea)

The sex life of mantids is the subject of some controversy, partly as a consequence of behavioural observations made under unnatural conditions in the laboratory. For example, there are many reports of the male being eaten by the generally larger female before, during or after mating. Males decapitated by females are even known to copulate more vigorously because of the loss of the suboesophageal ganglion that normally inhibits copulatory movements. Sexual cannibalism has been attributed to food deprivation in confinement but female mantids of at least some species indeed may eat their partners in the wild.

Courtship displays may be complex or absent, depending on species. Typically, the male approaches the female cautiously, arresting movement if she turns her head towards him, and then he leaps onto her back from beyond her strike reach. Once mounted, he crouches to elude his partner's grasp. Copulation usually lasts at least half an hour and may continue for several hours, during which sperm is transferred from the male to the female in a

---

BOX 5.2—*continued*

spermatophore. After mating, the male retreats hastily. If the male were in no danger of becoming the female's meal, his distinctive behaviour in the presence of the female would be inexplicable. Furthermore, suggestions of gains in reproductive fitness of the male via indirect nutritional benefits to his offspring are negated by the obvious unwillingness of the male to participate in the ultimate nuptial sacrifice – his own life.

Whereas there is no evidence yet for an increase in male reproductive success as a result of sexual cannibalism, females that obtain an extra meal by eating their mate may gain a selective advantage, especially if food is limiting. This hypothesis is supported by experiments with captive females of the Asian mantid *Hierodula membranacea* that were fed different quantities of food. The frequency of sexual cannibalism was higher for females of poorer nutritional condition and, among the females on the poorest diet, those which ate their mates produced significantly larger oothecae (egg packages) and hence more offspring. The cannibalized males would be making a parental investment only if their sperm fertilized the eggs that they have nourished. The crucial data on sperm competition in mantids are not available and so currently the advantages of this form of nuptial feeding are attributed entirely to the female.

5.5 DIVERSITY IN GENITALIC MORPHOLOGY

The components of the terminalia of insects are very diverse in structure and frequently exhibit species-specific morphology (Fig. 5.3), even in otherwise similar species. Variations in external features of the male genitalia often allow differentiation of species, whereas external structures in the female are usually simpler and less varied. Conversely, the internal genitalia of female insects often show greater diagnostic variability than do internal structures of the males. In general, external genitalia of both sexes are much more sclerotized than the internal genitalia, even though parts of the reproductive tract are lined with cuticle. Increasingly, characteristics of insect soft tissues are recognized as allowing species delineation and providing evidence of phylogenetic relationships.

Observations that genitalia are frequently complex and species-specific in form, sometimes appearing to correspond tightly between the sexes, led to formulation of the **'lock-and-key'** hypothesis as an explanation of this phenomenon. Species-specific male genitalia (the 'keys') were believed to fit only the conspecific female genitalia (the 'locks'), thus preventing interspecific mating or fertilization. For example, in some katydids interspecific copulations are unsuccessful in transmitting spermatophores because the specific structure of the male claspers (modified cerci) fails to fit the subgenital plate of the 'wrong' female. The lock-and-key hypothesis was first postulated in 1844 and has been the subject of controversy ever since. In many insects, mechanical exclusion of 'incorrect' male genitalia by the female is seen as unlikely for several reasons:

- The morphological correlation between conspecific male and female parts may be poor.
- Interspecific, intergeneric and even interfamilial hybrids can be induced.
- Amputation experiments have demonstrated that male insects do not need all parts of the genitalia to inseminate conspecific females successfully.

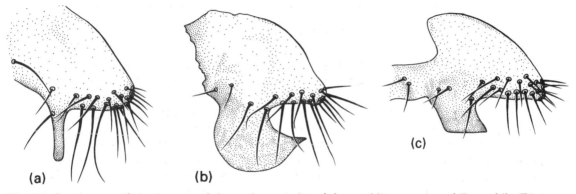

Fig. 5.3 Species-specificity in part of the male genitalia of three sibling species of *Drosophila* (Diptera: Drosophilidae). The epandrial processes of tergite nine in: (a) *D. mauritiana*; (b) *D. simulans*; (c) *D. melanogaster*. (After Coyne, 1983.)

Some support for the lock-and-key hypothesis comes from studies of certain noctuid moths in which structural correspondence in the internal genitalia of the male and female is thought to indicate their function as a post-copulatory but prezygotic isolating mechanism. Laboratory experiments involving interspecific matings support a lock-and-key function for the internal structures of other noctuid moths. Interspecific copulation can occur but, unless the fit of the male's vesica (the flexible tube everted from the aedeagus during insemination) into the female's bursa (genital pouch) is precise, the sperm may be discharged from the spermatophore to the cavity of the bursa, instead of into the duct that leads to the spermatheca, resulting in fertilization failure. In conspecific pairings, the spermatophore is positioned so that its opening lies opposite that of the duct (Fig. 5.4).

Mechanical reproductive isolation is not the only available explanation of species-specific genital morphology. Four other hypotheses have been advanced: pleiotropy, genitalic recognition, female choice, and male–male competition. The first two of these are further attempts to account for reproductive isolation of different species, whereas the last two are concerned with sexual selection, a topic that is addressed in more detail in sections 5.3 and 5.7.

The pleiotropy hypothesis explains genitalic differences between species as chance effects of genes that primarily code for other vital characteristics of the organism. This idea fails to explain why genitalia should be more affected than other parts of the body. Nor can pleiotropy explain genital morphology in groups (such as the Odonata) in which organs other than the primary male genitalia have an intromittent function (like those on the anterior abdomen in the case of odonates). Such secondary genitalia consistently become subjected to the postulated pleiotropic effects whereas the primary genitalia do not, a result inexplicable by the pleiotropy hypothesis.

The hypothesis of genitalic recognition concerns reproductive isolation of species via female sensory discrimination between different males based upon genitalic structures, both internal and external. The female thus responds only to the appropriate genital stimulation of a conspecific male and never that of any male of another species.

In contrast, the female choice hypothesis involves female sexual discrimination amongst conspecific males based on qualities that can vary intraspecifically and for which the female shows preference. This idea has nothing to do with the origin of reproductive isolation, although female choice may lead to reproductive isolation or speciation as a by-product. The female choice hypothesis

Fig. 5.4 Spermatophores lying within the bursae of the female reproductive tracts of moth species from four different genera (Lepidoptera: Noctuidae). The sperm leave via the narrow end of each spermatophore, which has been deposited so that its opening lies opposite the 'seminal duct' leading to the spermatheca (not drawn). The bursa on the far right contains two spermatophores, indicating that the female has remated. (After Williams, 1941; Eberhard, 1985.)

predicts diverse genitalic morphology in taxa with promiscuous females but uniform genitalia in strictly monogamous taxa. This prediction seems to be fulfilled in some insects. For example, in the neotropical *Heliconius* butterflies, species in which females remate are more likely to have species-specific male genitalia than species in which females mate only once. The greatest reduction in external genitalia (to near absence) occurs in termites, which, according to prediction, form monogamous pairs.

Another possibility is that species-specific elaborations of male genitalia may result from interactions between conspecific males vying for inseminations. Selection may act on male genitalic clasping structures to prevent usurpation of the female during copulation or act on the intromittent organ itself to produce structures that can remove or displace the sperm of other males (section 5.7). However, although sperm displacement has been documented in a few insects, this phenomenon is unlikely to

be a general explanation of male genitalic diversity because the penis of most male insects cannot reach the sperm storage organ(s) of the female.

Functional generalizations about the species-specific morphology of insect genitalia are controversial because different explanations no doubt apply in different groups. For example, male–male competition (via sperm removal and displacement; Box 5.3) may be important in accounting for the shape of odonate penises but appears irrelevant as an explanation in noctuid moths. Female choice and male–male competition may have little selective effect on genitalic structures of insect species in which the female mates with only one male (as in termites). In such species, sexual selection may affect features that determine which male is chosen as a partner but not how the male's genitalia are shaped. Furthermore, both mechanical and sensory locks-and-keys will be unnecessary if isolating mechanisms, such as courtship behaviour or seasonal or ecological differences, are well developed. So we might predict morphological constancy (or a high level of similarity, if we allow for a bit of pleiotropy) in genitalic structures among species in a genus that has species-specific pre-copulatory displays involving non-genital structures followed by a single insemination of each female.

5.6 SPERM STORAGE, FERTILIZATION AND SEX DETERMINATION

Many female insects store the sperm that they receive from one or more males in their sperm storage organ or spermatheca. Females of most insect orders have a single spermatheca but some flies are notable in having more, often two or three. Sometimes sperm remain viable in the spermatheca for a considerable time, even three or more years in the case of honey bees. During storage, the viability of sperm is maintained by secretions from the female's spermathecal gland.

Eggs are fertilized as they pass down the median oviduct and vagina. The sperm enter the egg via one or more **micropyles**, which are narrow canals that pass through the eggshell. The micropyle or micropylar area is oriented towards the opening of the spermatheca during egg passage, facilitating sperm entry. In many insects, the release of sperm from the spermatheca appears to be precisely controlled in timing and number. In queen honey bees as few as 20 sperm per egg may be released, suggesting extraordinary economy of use.

The fertilized eggs of most insects give rise to both males and females, with the sex dependent upon specific determining mechanisms, which are predominantly genetic. Most insects are diploid, that is with a set of chromosomes from each parent. The most common mechanism is for sex of the offspring to be determined by the inheritance of sex (X-) chromosomes (heterochromosomes), which are differentiated from the remaining autosomes. Individuals are thus allocated to sex according to the presence of one (X0) or two (XX) sex chromosomes, but whilst XX is usually female and X0 male, this allocation varies within and between taxonomic groups. Mechanisms involving multiple sex chromosomes also occur and there are related observations of complex fusions between sex chromosomes and autosomes. Matters are complicated further by the frequent inability to recognize sex chromosomes, particularly since sex is determined by single genes in certain flies such as some mosquitoes and midges. Additional complications with the determination of sex arise with the interaction of both the internal and external environment on the genome (epigenetic factors). Furthermore, great variation is seen in sex ratios at birth; although the ratio is often one male to one female, there are many deviations ranging from 100% of one sex to 100% of the other.

One of the more frequent of the 'aberrant' sex determining systems is haplodiploidy (male haploidy) in which the male sex is determined by having only one set of chromosomes. This may arise through his development from an unfertilized egg (containing half of the female chromosome complement following meiosis) or from a fertilized egg in which the paternal set of chromosomes is inactive and eliminated (as in many male scale insects).

In honey bees, females (queens and workers) develop from fertilized eggs whereas males (the drones) come from unfertilized eggs – an example of arrhenotoky (section 5.10). The haplodiploid method of sex determination allows the female to control the sex of her offspring through her ability to store sperm and control fertilization of eggs. All the evidence points to a precise control of sperm release from storage but very little is known about this process in most insects. It seems that the presence of an egg in the genital chamber stimulates contractions of the spermathecal walls, leading to sperm release.

5.7 SPERM COMPETITION

Multiple matings are common in many insect species. The occurrence of remating under natural conditions can be determined by observing the mating behaviour of individual females or by dissection to establish the amount of ejaculate or the number of spermatophores present in the female's sperm storage organs. Some of the best documentation of remating comes from studies of many Lepidoptera, in which part of each spermatophore persists in the bursa copulatrix of the female throughout her life. These data show that remating occurs, to some extent, in almost all species of Lepidoptera for which adequate field data are available.

The combination of internal fertilization, sperm storage, multiple mating by females and the overlap within a female of ejaculates from different males leads to a phenomenon known as **sperm competition**. This occurs within the reproductive tract of the female at the time of oviposition when sperm from two or more males compete to fertilize the eggs. Both physiological and behavioural mechanisms determine the outcome of sperm competition. Thus events inside the female's reproductive tract, combined with various attributes of mating behaviour, determine which sperm will succeed in reaching the eggs. It is important to realize that male reproductive fitness is measured in terms of the number of eggs fertilized or offspring fathered and not simply the number of copulations achieved, although these measures are sometimes correlated. Often there may be a trade-off between the number of copulations that a male can secure and the number of eggs that he will fertilize at each mating. A high copulation frequency is generally associated with low time or energy investment per copulation but also with low certainty of paternity. At the other extreme, males that exhibit substantial parental investment, such as feeding their mates (see Boxes

Fig. 5.5 A copulating pair of stink or shield bugs of the genus *Poecilometis* (Hemiptera: Pentatomidae). Many heteropteran bugs engage in prolonged copulation, which prevents other males from inseminating the female until either she becomes non-receptive to further males or she lays the eggs fertilized by the 'guarding' male.

5.1 and 5.2 on nuptial feeding) and other adaptations that more directly increase certainty of paternity, will inseminate fewer females over a given period.

There are two main types of sexually selected adaptations in males that increase certainty of paternity. The first strategy

BOX 5.3
SPERM PRECEDENCE

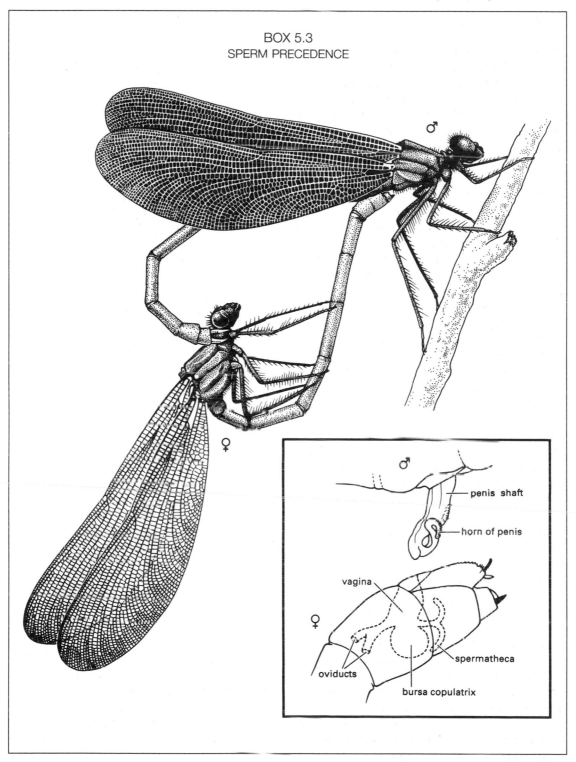

BOX 5.3—*continued*

The penis or aedeagus of a male insect may be modified to facilitate placement of his own sperm in a strategic position within the spermatheca of the female or even to remove a rival's sperm. Sperm displacement of the former type, called **stratification**, involves pushing previously deposited sperm to the back of spermatheca in systems in which a 'last in, first out' principle operates (i.e. the most recently deposited sperm is the first to be used when the eggs are fertilized). Last-male sperm precedence occurs in the majority of insect species that have been examined; in other species either first-male precedence or no precedence is found. In some dragonflies, males appear to use inflatable lobes on the penis to reposition rival sperm. Such sperm packing enables the copulating male to place his sperm closest to the oviduct. However, stratification of sperm from separate inseminations may occur in the absence of any deliberate repositioning, by virtue of the tubular design of the storage organs.

A second strategy of sperm displacent is **removal**, which can be achieved either by direct scooping out of existing sperm prior to depositing an ejaculate or, indirectly, by flushing out of a previous ejaculate with a subsequent one. An unusually long penis that could reach well into the spermathecal duct may facilitate flushing of a rival's sperm from the spermatheca. A number of structural and behavioural attributes of male insects can be interpreted as devices to facilitate this form of sperm precedence, but some of the best known examples come from odonates.

Copulation in Odonata involves the female placing the tip of her abdomen against the underside of the male's anterior abdomen, where his sperm is stored in a reservoir of his secondary genitalia. In some dragonflies and most damselflies, such as the pair of copulating *Calopteryx* damselflies (Calopterygidae) illustrated here in the wheel position (after Zanetti, 1975), the male spends the greater proportion of the copulation time physically removing the sperm of other males from the female's sperm storage organs (spermathecae and bursa copulatrix) and only at the last minute introducing his own. In these species, the male's penis is structurally complex, sometimes with an extensible head used as a scraper and a flange to trap the sperm plus lateral horns or hook-like distal appendages with recurved spines to remove rival sperm (inset to figure; after Waage, 1986). A male's ejaculate may be lost if another male mates with the female before she oviposits. Thus it is not surprising that male odonates guard their mates, which explains why they are so frequently seen as pairs flying in tandem.

involves mechanisms by which males can ensure that females use their sperm preferentially. Such **sperm precedence** is usually achieved by displacing the ejaculate of males that have mated previously with the female (Box 5.3). The second strategy is to reduce the effectiveness or occurrence of subsequent inseminations by other males. Various mechanisms appear to achieve this result, including mating plugs, prolonged copulation (Fig. 5.5), guarding of females and improved structures for gripping the female during copulation to prevent 'take-over' by other males. A significant selective advantage would accrue to any male that could both achieve sperm precedence and prevent other males from successfully inseminating the female until his sperm had fertilized at least some of her eggs.

The factors that determine the outcome of sperm competition are not totally under male control. Female choice is a complicating influence, as shown in the above discussions on sexual selection and on morphology of genitalic structures. Female choice of sexual partners may be two-fold. Firstly, there is good evidence that the females of many species choose among potential mating partners. For example, females of many mecopteran species mate selectively with males that provide food of a certain minimum size and quality (Box 5.1). In a few moth and katydid species, females have been shown to prefer larger males as mating partners. Secondly, after copulation, the female might discriminate between partners as to which sperm will be used. One idea is that variation in the stimuli of the male genitalia induce the female to use one male's sperm in preference to that of another, based upon an 'internal courtship'.

5.8 OVIPARITY (EGG-LAYING)

The vast majority of female insects are **oviparous**, that is, they lay eggs. Generally, ovulation – expulsion of eggs from the ovary into the oviducts – is followed rapidly by fertilization and then oviposition. Ovulation is controlled by hormones released from the brain, whereas oviposition appears to be under both hormonal and neural control. Oviposition, the process of the egg passing

Fig. 5.6 Oviposition by a South African ladybird beetle, *Chilomenes lunulata* (Coleoptera: Coccinellidae). The eggs adhere to the leaf surface because of a sticky secretion applied to each egg. (After Blaney, 1976.)

from the external genital opening or vulva to the outside of the female (Fig. 5.6), is often associated with behaviours such as digging or probing into an egg-laying site, but often the eggs are simply dropped to the ground or into water. Usually the eggs are deposited on or near the food required by the offspring upon hatching.

An insect egg within the female's ovary is complete when an oocyte becomes covered with an outer protective coating, the eggshell, formed of the **vitelline membrane** and the **chorion**. The chorion may be composed of any or all of the following layers: wax layer, innermost chorion, endochorion and exochorion (Fig. 5.7). The eggshell is produced by the ovarian follicle cells (Chapter 3) and the surface sculpturing of the chorion usually reflects the outline of these cells. Typically, the eggs are yolk-rich and thus large relative to the size of the adult insect; egg cells range in length from 0.2 mm to about 20 mm. Embryonic development within the egg begins after egg activation and is described in Chapter 6.

The eggshell has a number of important functions. Its design allows selective entry of the sperm at the time of fertilization (section 5.6). Its elasticity facilitates oviposition, especially for species in which the eggs are compressed during passage down a narrow egg-laying tube, as described below. Its structure and composition afford the embryo protection from deleterious conditions such as unfavourable humidity and temperature and microbial infection, while also allowing the exchange of oxygen and carbon dioxide between the inside and outside of the egg.

The differences seen in composition and complexity of layering of the eggshell in different insect groups generally are correlated with the environmental conditions encountered at the site of oviposition. In parasitic wasps the eggshell is usually thin and relatively homogeneous, allowing flexibility during passage down the narrow ovipositor, but, because the embryo develops

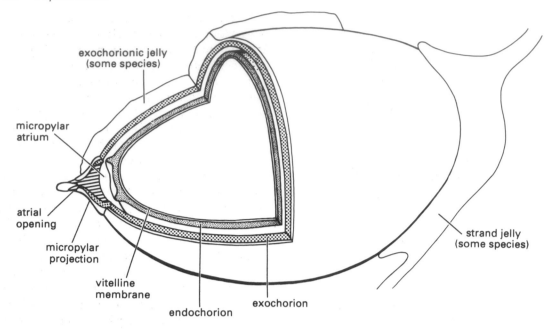

Fig. 5.7 The generalized structure of a libelluloid dragonfly egg (Odonata: Corduliidae, Libellulidae). Libelluloid dragonflies oviposit into freshwater but always exophytically (outside of plant tissues). The endochorionic and exochorionic layers of the eggshell are separated by a distinct gap in some species. A gelatinous matrix may be present on the exochorion or as connecting strands between eggs. (After Trueman, 1991.)

within host tissues where desiccation is not a hazard, the wax layer of the eggshell is absent. In contrast, many insects lay their eggs in dry places and here the problem of avoiding water loss while obtaining oxygen is often acute because of the high surface area to volume ratio of most eggs. The majority of terrestrial eggs have a hydrofuge waxy chorion that contains a meshwork holding a layer of gas in contact with the outside atmosphere via narrow holes or aeropyles.

The females of many insects (e.g. Thysanura, many Odonata, Orthoptera, some Hemiptera, some Thysanoptera, and Hymenoptera) have appendages of the eighth and ninth abdominal segments modified to form an egg-laying organ or ovipositor (section 2.5.1). In other insects (e.g. many Lepidoptera, Coleoptera and Diptera) it is the posterior segments rather than appendages of

the female's abdomen that function as an ovipositor (a 'substitutional' ovipositor). Often these segments can be protracted into a telescopic tube in which the opening of the egg passage is close to the distal end. The ovipositor or the modified end of the abdomen enables the insect to insert its eggs into particular sites, such as crevices, soil, plant tissues or, in the case of many parasitic species, into an arthropod host. Other insects, such as Isoptera, Phthiraptera and many Plecoptera, lack an egg-laying organ and eggs are simply deposited on a surface.

In certain Hymenoptera (some wasps, bees and ants) the ovipositor has lost its egg-laying function and is used as a poison-injecting sting. The stinging Hymenoptera eject the eggs from the opening of the genital chamber at the base of the modified ovipositor. However, in most wasps the eggs pass down the

These variations are exemplified by the scales found in the conspicuous ovipositors of crickets and katydids (Orthoptera: Gryllidae and Tettigoniidae). The ovipositor of the field cricket *Teleogryllus commodus* (Fig. 5.9) possesses overlapping plate-like scales and scattered, short sensilla along the length of the egg canal. These sensilla may provide information on the position of the egg as it moves down the canal, whereas a group of larger sensilla at the apex of each dorsal valve presumably signals that the egg has been expelled. In addition, in *T. commodus* and some other insects, there are scales on the outer surface of the ovipositor tip which are oriented in the opposite direction to those on the inner surface. These are thought to assist with penetration of the substrate and holding the ovipositor in position during egg-laying.

In addition to the eggshell, many eggs are provided with a proteinaceous secretion or cement which coats and fastens them to a substrate, such as a vertebrate hair in the case of sucking lice or a plant surface in the case of

Fig. 5.8 A female of the parasitic wasp *Megarhyssa nortoni* (Hymenoptera: Ichneumonidae) probing a pine log with her very long ovipositor in search of a larva of the sirex wood wasp, *Sirex noctilio* (Hymenoptera: Siricidae). If a larva is located, she stings and paralyses it before laying an egg on it.

canal of the ovipositor shaft, even if the shaft is very narrow (Fig. 5.8). In some parasitic wasps with very slender ovipositors the eggs are extremely compressed and stretched as they move through the narrow canal of the shaft.

The valves of an insect ovipositor usually are held together by interlocking tongue-and-groove joints, which prevent lateral movement but allow the valves to slide back and forth on one another. Such movement, and sometimes also the presence of serrations on the tip of the ovipositor, are responsible for the piercing action of the ovipositor into an egg-laying site. Movements of eggs down the ovipositor tube are possible due to many posteriorly directed 'scales' (microsculpturing) located on the inside surface of the valves. Ovipositor scales vary in shape, from plate-like to spine-like, and in arrangement among insect groups; they are seen best in a scanning electron microscope.

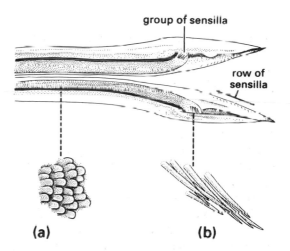

(a)   **(b)**

Fig. 5.9 Tip of the ovipositor of a female of the black field cricket, *Teleogryllus commodus* (Orthoptera: Gryllidae), split open to reveal the inside surface of the two halves of the ovipositor. Enlargements show: (a) posteriorly directed ovipositor scales; (b) distal group of sensilla. (After Austin and Browning, 1981.)

many beetles (Fig. 5.6). Such secretions are produced by colleterial glands, accessory glands of the female reproductive tract. In other insects, groups of thin-shelled eggs are enclosed in an **ootheca**, which protects the developing embryos from desiccation. The colleterial glands produce the tanned, purse-like ootheca of cockroaches (Box 8.6) and the frothy ootheca of mantids, whereas the foamy ootheca that surrounds locust and other orthopteran eggs in the soil is formed from the accessory glands in conjunction with other parts of the reproductive tract.

5.9 OVOVIVIPARITY AND VIVIPARITY

Most insects are oviparous, with laying involved in initiation of egg development. However, some species are viviparous, with initiation of egg development taking place within the mother. The life cycle is shortened by retention of eggs and even developing young within the mother. Four main types of viviparity are observed in different insect groups, and many of the specializations are prevalent in various higher dipterans.

Ovoviviparity, in which fertilized eggs containing yolk are incubated inside the reproductive tract of the female, occurs in Thysanoptera, some cockroaches (Blattidae), a few beetles and some flies (Muscidae, Calliphoridae and Tachinidae).

Pseudoplacental viviparity occurs when a yolk-deficient egg develops in the genital tract of the female. The mother provides a special placenta-like tissue, through which nutrients are transferred to developing embryos. There is no oral feeding and larvae are laid upon hatching. This form of viviparity occurs in aphids (Hemiptera), some earwigs (Dermaptera), psocids (Psocoptera) and polyctenid bugs (Hemiptera).

Haemocoelous viviparity involves embryos developing free in the female's haemolymph, with nutrients taken up by osmosis. This form of internal parasitism occurs only in Strepsiptera, in which the larvae exit through a

brood canal (Box 12.4), and in some gall midges (Diptera: Cecidomyiidae) where the larvae may consume the mother (as in paedogenetic development, below).

Adenotrophic viviparity occurs when a poorly developed larva hatches and feeds orally from accessory ('milk') gland secretions within the 'uterus' of the mother's reproductive system. The full-grown larva is deposited and pupariates immediately. The Dipteran 'pupiparan' families, namely the Glossinidae (tsetse flies), Hippoboscidae (louse or wallaby flies, keds) and Nycteribiidae and Streblidae (bat flies), demonstrate adenotrophic viviparity.

5.10 ATYPICAL MODES OF REPRODUCTION

Sexual reproduction with separate male and female individuals (**gonochorism**) is the usual mode of reproduction in insects. However, other modes are not uncommon. Various types of asexual reproduction occur in many insect groups; development from unfertilized eggs is a widespread phenomenon, whereas the production of multiple embryos from a single egg is rare. Some species exhibit alternating sexual and asexual reproduction, depending on season or food availability. A few species possess both male and female reproductive systems in one individual (**hermaphroditism**) but self-fertilization has been established for species in just one genus.

5.10.1 Parthenogenesis, paedogenesis and neoteny

Representatives of virtually every insect order have dispensed with mating, with females producing viable eggs even though unfertilized. In others, notably the Hymenoptera, mating occurs but the sperm need not be used in fertilizing all the eggs. Development from unfertilized eggs is called **parthenogenesis**, which in some species may be **obligatory**, but in many others is **facultative**. **Sporadic** parthenogenesis occurs, but by its very nature it is ill-studied. The female may produce parthenogenetically only female eggs (**thelytokous parthenogenesis**), only male eggs (**arrhenotokous parthenogenesis**) or eggs of both sexes (**amphitokous** or **deuterotokous parthenogenesis**. The largest insect group showing arrhenotoky is the Hymenoptera. Within the Homoptera, aphids display thelytoky and some scale insects and whiteflies are arrhenotokous. Certain Diptera and a few Coleoptera are thelytokous and Thysanoptera display all three types of parthenogenesis. Facultative parthenogenesis, and variation in sex of egg produced, may be a response to fluctuations in environmental conditions, as occurs in aphids that vary sex of offspring and mix parthenogenetic and sexual cycles according to season.

Some insects cut short their life cycles by loss of the adult or even adult and pupal stages. In this precocious stage, reproduction is almost exclusively by parthenogenesis. **Larval paedogenesis**, the production of young by the larval insect, has arisen several times in the gall midges (Diptera, Cecidomyiidae) and once in the Coleoptera. In some gall midges, in an extreme case of haemocoelous viviparity, the precocially developed eggs hatch internally and the larvae consume the body of the mother-larva. In the well-studied gall midge *Heteropeza pygmaea*, eggs (fertilized or unfertilized) develop into female larvae, which may either metamorphose to female adults or become paedogenetic and produce more larvae. These larvae, in turn, may be males, females or a mixture of both sexes. Female larvae may become adult females or repeat the larval paedogenetic cycle, whereas male larvae must develop to adulthood.

In **pupal paedogenesis**, which sporadically occurs in gall midges, embryos are formed in the haemocoel of a paedogenetic mother-pupa, termed a hemipupa since it differs morphologically from the 'normal' pupa. This production of live young in pupal paedogenetic insects also destroys the mother-pupa from within, either by larval perforation

of the cuticle or by the eating of the mother by the offspring. Paedogenesis appears to have evolved to allow maximum use of locally abundant but ephemeral larval habitats, such as a mushroom fruiting body. When the oviposition site is detected by a gravid female, eggs are deposited, and the larval population builds up rapidly through paedogenetic development. Adults are developed only in response to conditions adverse to larvae, such as food depletion and overcrowding. Adults may be female only, or males may occur in some species under specific conditions.

In what was thought to be a similar phenomenon, a species of aquatic chironomid midge produces viable eggs in the pupal stage, and the flying adult appears to be dispensed with. However, in this case eggs are laid by the pharate female and retained within the pupal skin. Sporadic or facultative female emergence takes place in some populations but males remain unknown.

In true paedogenetic taxa there are no reproductive adaptations beyond precocious egg development. In contrast, in **neoteny** a non-terminal instar develops reproductive features of the adult, including ability to locate a mate, copulate and deposit eggs (or larvae) in a conventional manner. For example, the scale insects (Hemiptera: Coccoidea) appear to have neotenous females. Whereas a moult to the winged adult male follows the final immature instar, development of the reproductive female involves omission of one or more instars relative to the male. In appearance the female is a sedentary larviform instar, resembling a larger version of the previous (second or third) instar in all respects but for the presence of developing eggs. Neoteny also occurs in all members of the order Strepsiptera; in these insects female development ceases at the puparium stage. In some other insects (marine midges, for example) the adult appears larva-like, but this is evidently not due to neoteny because complete metamorphic development is retained, including a pupal instar. Their larviform appearance therefore results from suppression of adult features, rather than the paedogenetic acquisition of reproductive ability in the larval stage.

5.10.2 Hermaphroditism

Several of the species of *Icerya* (Hemiptera: Margarodidae) that have been studied cytologically are gynomonoecious hermaphrodites since they are female-like but possess an ovotestis (a gonad that is part testis, part ovary). In these species, occasional males arise from unfertilized eggs and are apparently functional, while self-fertilization is assured by production of male gametes prior to female gametes (protandry of the hermaphrodite). Without doubt, hermaphroditism greatly assists the spread of the pestiferous cottony-cushion scale, *I. purchasi*, since single nymphs of this and other hermaphroditic *Icerya* species can initiate new infestations if dispersed or accidently transported to new plants. Furthermore, all iceryine margarodids are arrhenotokous with unfertilized eggs developing into males and fertilized eggs into females.

5.10.3 Polyembryony

This form of asexual reproduction involves the production of two or more embryos from one egg by subdivision. It is largely restricted to parasitic insects, in particular one species of strepsipteran and genera in several wasp families. In these parasitic wasps, the number of larvae produced from a single egg varies in different genera but is influenced by the size of the host, with from fewer than ten to several thousand individuals arising from one small, yolkless egg. Nutrition for a large number of developing embryos obviously cannot be supplied by the original egg and is acquired from the host's haemolymph through a specialized enveloping membrane called the **trophamnion**.

5.11 PHYSIOLOGICAL CONTROL OF REPRODUCTION

The initiation and termination of some reproductive events often depend on environmental factors, such as temperature, humidity, photoperiod, availability of food or a suitable egg-laying site. In addition, the effects of external influences may be modified by internal factors like nutritional condition and the state of maturation of the oocytes. Copulation may trigger oocyte development, oviposition and inhibition of sexual receptivity in the female via enzymes or peptides transferred to the female reproductive tract in male accessory gland secretions (Box 5.4). Fertilization following mating normally triggers embryogenesis via egg activation (Chapter 6). Regulation of reproduction is complex and involves sensory receptors, neuronal transmission and integration of messages in the brain, as well as chemical messengers (hormones) transported in the haemolymph or via the nerve axons to target tissues or to other endocrine glands. Certain parts of the nervous system, particularly neurosecretory cells in the brain, produce neurohormones or neuropeptides (proteinaceous messengers) and also control the synthesis of two groups of insect hormones – the ecdysteroids and the juvenile hormones (JH). There are more detailed discussions of the regulation and functions of all of these hormones in Chapters 3 and 6. Neuropeptides, steroid hormones and JH all play essential roles in the regulation of reproduction, as summarized in Fig. 5.10.

Juvenile hormones and/or ecdysteroids are essential to most stages of reproduction, with JH mostly triggering the functioning of organs such as the ovary or accessory glands, whereas ecdysteroids mainly influence morphogenesis. Neuropeptides play various roles at different stages of reproduction since they regulate endocrine function (via the corpora allata and prothoracic glands) and also directly influence reproductive events, especially ovulation and oviposition or larviposition.

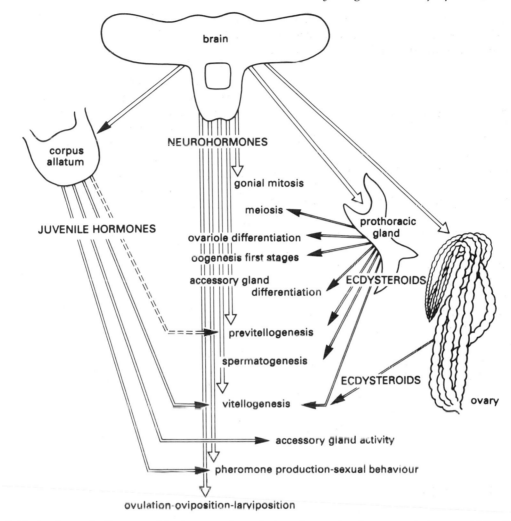

Fig. 5.10 A schematic diagram of the hormonal regulation of reproductive events in insects. (After Raabe, 1986.)

Control of reproduction via neuropeptides is an expanding area of research, made possible by the recent advent of new technologies, especially in biochemistry and molecular biology. To date, most studies have concentrated on the Diptera (especially *Drosophila*, mosquitoes and house flies), the Lepidoptera (especially the tobacco hornworm, *Manduca sexta*) and cockroaches.

5.11.1 Vitellogenesis and its regulation

In the ovary, both nurse cells (or trophocytes) and ovarian follicle cells are associated with the oocytes (Chapter 3). These cells pass nutrients to the growing oocytes. The relatively slow period of oocyte growth is followed by a rapid period of yolk deposition or **vitellogenesis**, which mostly occurs in the

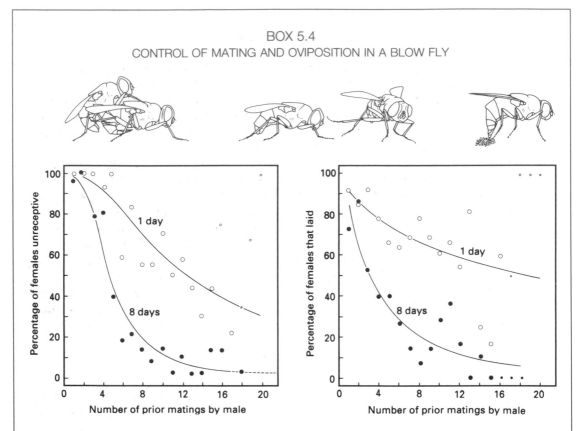

BOX 5.4
CONTROL OF MATING AND OVIPOSITION IN A BLOW FLY

The sheep blow fly, *Lucilia cuprina* (Diptera: Calliphoridae), costs the Australian sheep industry many millions of dollars annually due to losses caused by myiases or 'strikes'. This pestiferous fly is believed to have been introduced to Australia from Africa in the late nineteenth century. The reproductive behaviour of *L. cuprina* has been studied in some detail because of its relevance to a control programme for this pest. Ovarian development and reproductive behaviour of the adult female are highly stereotyped and readily manipulated via precise feeding of protein. Most females are anautogenous, that is, they require protein to develop their eggs; they usually mate after a protein meal and before their oocytes have reached early vitellogenesis. After their first mating females reject further mating attempts for several days. The 'switch-off' is activated by a protein produced in the accessory glands of the male and transferred to the female during mating. Mating also stimulates oviposition; virgin females rarely lay eggs, whereas mated females readily do so. The eggs of each fly are laid in a single mass of a few hundred (illustration on top right) and then a new ovarian cycle commences with another batch of synchronously developing oocytes. Females may lay one to four egg masses before remating.

Unreceptive females respond to male mating attempts by curling their abdomen under their body (illustrations on top left), by kicking at the males (illustration at top centre) or by actively avoiding them. Receptivity gradually returns to previously mated females, in contrast to their gradually diminishing tendency to lay. If remated, such non-laying females resume

BOX 5.4—*continued*

laying. Neither the size of the female's sperm store nor the mechanical stimulation of copulation can explain these changes in female behaviour. Experimentally, it has been demonstrated that the female mating refractory period and readiness to lay are related to the amount of male accessory gland substance deposited in the female's bursa copulatrix during copulation. If a male repeatedly mates during one day (a multiply-mated male), less gland material is transferred at each successive copulation. Thus if one male is mated, during one day, to a succession of females which are later tested at intervals for their receptivity and readiness to lay, then the proportion of females either unreceptive or laying is inversely related to the number of females with which the male had previously mated. The graph on the left shows the percentage of females unreceptive to further mating when tested one day (O) or eight days (●) after having mated with multiply-mated males. The percentage unreceptive values are based on one to 29 tests of different females. The graph on the right shows the percentage of females that laid eggs during six hours of access to oviposition substrate presented one day (O) or eight days (●) after mating with multiply-mated males. The percentage laid values are based on tests of 1–15 females. These two plots represent data from different groups of 30 males; sample sizes of fewer than five female flies are represented by smaller symbols. (After Bartell *et al.*, 1969; Barton Browne *et al.*, 1990; Smith *et al.*, 1990.)

terminal oocyte of each ovariole and leads to the production of fully developed eggs. Vitellogenesis involves the production (mostly by the fat body) of specific female lipoglycoproteins called vitellogenins, followed by their passage into the oocyte. Once inside the oocyte these proteins are called vitellins and their chemical structure may differ slightly from that of vitellogenins. Lipid bodies – mostly triglycerides from the follicle cells, nurse cells or fat body – are also deposited in the growing oocyte.

Vitellogenesis has been a favoured area of insect hormone research because it is amenable to experimental manipulation with artificially supplied hormones and analysis is facilitated by the large amounts of vitellogenins produced during egg growth. The regulation of vitellogenesis varies among insect taxa with juvenile hormone from the corpora allata, ecdysteroids from the prothoracic glands or the ovary, and brain neurohormones (neuropeptides such as egg

development neurohormone) considered to induce or stimulate vitellogenin synthesis to varying degrees in different insect species (Fig. 5.10).

Inhibition of egg development in ovarian follicles in the previtellogenic stage is mediated by antigonadotropins. This inhibition ensures that only some of the oocytes undergo vitellogenesis in each ovarian cycle. The antigonadotropins responsible for this suppression are peptides termed **oostatic hormones**. In most of the insects studied, oostatic hormones are produced by the ovary or neurosecretory tissue associated with the ovary and, depending on species, may either:

- inhibit the release or synthesis of egg development neurohormone (EDNH); or
- affect ovarian development by inhibiting proteolytic enzyme synthesis and blood digestion in the midgut, as in mosquitoes; or
- inhibit the action of JH on vitellogenic

follicle cells, thus preventing the ovary from accumulating vitellogenin from the haemolymph, as in the blood-sucking bug *Rhodnius prolixus*.

Until recently it was firmly believed that JH controlled vitellogenesis in most insects. Then, in certain insects, the importance of ecdysteroids was discovered. Now we are becoming increasingly aware of the part played by neuropeptides, a group of proteins for which reproductive regulation is but one of an array of functions in the insect body.

FURTHER READING

Alcock, J. (1979) Selective mate choice by females of *Harpobittacus australis* (Mecoptera: Bittacidae). *Psyche*, **86**, 213–17.

Austin, A.D. and Browning, T.O. (1981) A mechanism for movement of eggs along insect ovipositors. *International Journal of Insect Morphology and Embryology*, **10**, 93–108.

Bailey, W.J. and Ridsdill-Smith, J. (eds) (1991) *Reproductive Behaviour of Insects*, Chapman & Hall, London.

Birkhead, T.R. and Hunter, F.M. (1990) Mechanisms of sperm competition. *Trends in Ecology and Evolution*, **5**, 48–52.

Birkhead, T. and Lawrence, S. (1988) Life and loves of a sexual cannibal. *New Scientist*, **118(1617)**, 63–66.

Eberhard, W.G. (1985) *Sexual Selection and Animal Genitalia*, Harvard University Press, Cambridge, Massachusetts, and London, England.

Eberhard, W.G. (1990) Animal genitalia and female choice. *American Scientist*, **78**, 134–41.

Gwynne, D.T. (1988) Courtship feeding and the fitness of female katydids (Orthoptera: Tettigoniidae). *Evolution*, **42**, 545–55.

Kerkut, G.A. and Gilbert, L.I. (eds) (1985) *Comprehensive Insect Physiology, Biochemistry, and Pharmacology, Volume 1, Embryogenesis and Reproduction*, Pergamon Press, Oxford, New York, Toronto, Sydney, Paris, Frankfurt.

Menn, J.J., Kelly, T.J. and Masler, E.P. (eds) (1991) *Insect Neuropeptides: Chemistry, Biology, and Action*. American Chemical Society, Washington, D.C.

Mikkola, K. (1992) Evidence for lock-and-key mechanisms in the internal genitalia of the *Apamea* moths (Lepidoptera, Noctuidae). *Systematic Entomology*, **17**, 145–53.

Parker, G.A. (1970) Sperm competition and its evolutionary consequences in insects. *Biological Reviews*, **45**, 5, 525–67.

Raabe, M. (1986) Insect reproduction: regulation of successive steps. *Advances in Insect Physiology*, **19**, 29–154.

Shapiro, A.M. and Porter, A.H. (1989) The lock-and-key hypothesis: evolutionary and biosystematic interpretation of insect genitalia. *Annual Review of Entomology*, **34**, 231–45.

Thornhill, R. and Alcock, J. (1983) *The Evolution of Insect Mating Systems*, Harvard University Press, Cambridge, Massachusetts.

Wedell, N. and Arak, A. (1989) The wartbiter spermatophore and its effect on female reproductive output (Orthoptera: Tettigoniidae, *Decticus verrucivorus*). *Behavioral Ecology and Sociobiology*, **24**, 117–25.

West-Eberhard, M.J. (1984) Sexual selection, competitive communication and species-specific signals in insects, in *Insect Communication*, (ed. T. Lewis), Academic Press, London, Orlando, San Diego, New York, Toronto, Montreal, Sydney, Tokyo, pp. 283–324.

6
INSECT DEVELOPMENT AND LIFE HISTORIES

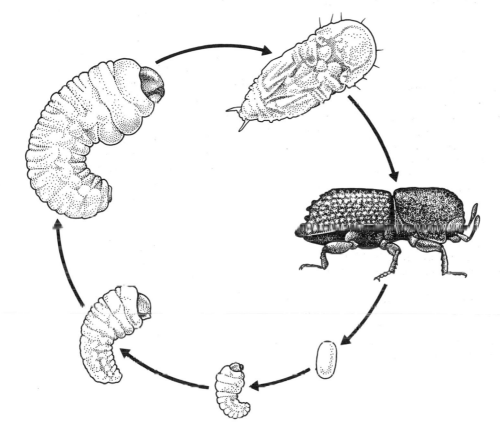

Life cycle of a bark beetle (Ips)

This chapter discusses the ontogeny – the pattern of growth from egg to adult – and life histories of insects. It deals with the various growth phases from the egg, through immature development to the emergence of the adult. It considers the significance of different kinds of metamorphosis and suggests that complete metamorphosis reduces competition between conspecific juveniles and adults, by providing a clear differentiation between immature and adult stages. Amongst the different aspects of life histories covered are voltinism, age determination, allometry, and genetic and environmental effects on development, including resting stages and the co-existence of different forms within a single species. The influence of environmental factors (temperature, photoperiod, humidity, toxins and biotic interactions) upon life-history traits is vital to any applied entomological research. Likewise, knowledge of the process and hormonal regulation of moulting is fundamental to insect control.

Insect life-history characteristics are very diverse and the heterogeneity of strategies seen in many higher taxa implies that these traits are highly adaptive, with little phylogenetic constraint. For example, diverse environmental factors trigger termination of egg dormancy in different species of *Aedes* although the species in this genus are phylogenetically close.

The chapter concludes by considering how data on insect growth and development can be modelled and used in a predictive way to answer questions in applied entomology.

6.1 GROWTH

Insect growth is discontinuous, at least for the sclerotized cuticular parts of the body, because the rigid cuticle limits expansion. Size increase is by **moulting** – periodical formation of new cuticle of greater surface area and shedding of the old cuticle. Thus for sclerite-bearing body segments and appendages,

increases in body dimensions are confined to the post-moult period immediately after moulting before the cuticle stiffens and hardens (Chapter 2). Hence, the sclerotized head capsule of a beetle or moth larva increases in dimensions in a saltatory manner (in major increments) during development, whereas the membranous nature of body cuticle allows the larval body to grow more or less continuously.

Studies concerning insect development involve two components of growth. The first, the **moult increment**, is the increment in size occurring between one **instar** (growth stage, or the form of the insect between two successive moults) and the next. Generally increase in size is measured as the increase in a single dimension (length or width) of some sclerotized body part, rather than a weight increment that may be misleading due to variability in food or water intake. The second component of growth is the **intermoult period** or interval, better known as the **stadium** or instar duration, which is defined as the time between two successive moults, or more precisely between successive ecdyses (Fig. 6.1 and section 6.3). The magnitude of both moult increments and intermoult periods may be affected by food supply, temperature, larval density and physical damage such as loss of appendages (section 6.7), and may differ between the sexes of a species.

In collembolans, diplurans and apterygote insects, growth is **indeterminate** – the animals continue to moult until they die. There is no definitive terminal moult in such animals but they do not continue to increase in size throughout their adult life. In the vast majority of insects, growth is **determinate** since there is a distinctive instar that marks the cessation of growth and moulting. All insects with determinate growth become reproductively mature in this final instar, called the adult or imaginal instar. This reproductively mature individual is called an adult or **imago**. In most insect orders it is fully winged, although secondary wing loss has occurred

Fig. 6.1 Schematic drawing of the life cycle of a non-biting midge (Diptera: Chironomidae, *Chironomus*) showing the various events and stages of insect development.

independently in the adults of a number of groups, such as lice, fleas and certain parasitic flies, and in the adult females of all scale insects (Hemiptera: Coccoidea). In just one order of insects, the Ephemeroptera or mayflies, there is a **subimaginal instar** immediately preceding the final or imaginal instar. The **subimago**, although capable of flight, is rarely reproductive; in the few mayfly groups in which the female mates as a subimago she dies without moulting to an imago, so that the subimaginal instar is actually the final growth stage.

In some pterygote taxa the total number of pre-adult growth stages or instars may vary within a species depending on environmental conditions, such as developmental temperature, diet and larval density. In many other species, the total number of instars (although not necessarily final adult size) is genetically determined and constant regardless of environmental conditions.

6.2 LIFE-HISTORY PATTERNS AND PHASES

Growth is an important component of an individual's **ontogeny**, the developmental history of that organism from egg to adult.

Equally significant are the changes, both subtle and dramatic, that take place in body form as most insects moult and grow larger. Morphological changes during ontogeny affect both external structures and internal organs, but only the external changes are apparent at each moult. Changes in form during ontogeny allow us to recognize three broad patterns of development, based on the degree of external alteration that occurs in the post-embryonic phases of development.

In the primitive developmental pattern, **ametaboly**, the insect hatches from the egg in a form essentially resembling a miniature adult but lacking genitalia. This pattern is retained by the primitively wingless orders, the silverfish (Thysanura) and bristletails (Archaeognatha) (Box 8.3), in which adults continue to moult after they become sexually mature. In contrast, all pterygote insects undergo a more or less marked change in form, a **metamorphosis**, between the immature phase of development and the winged or secondarily apterous adult or imaginal phase. These insects can be subdivided according to two broad patterns of development, **hemimetaboly** (partial or incomplete metamorphosis;

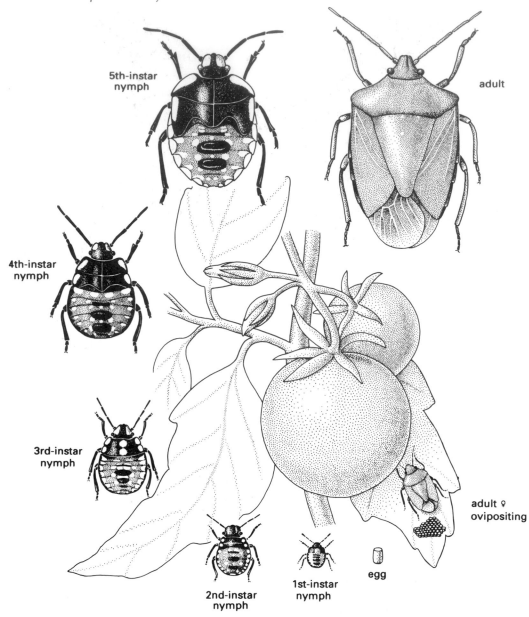

Fig. 6.2 The life cycle of a hemimetabolous insect, the southern green stink bug or green vegetable bug, *Nezara viridula* (Hemiptera: Pentatomidae), showing the eggs, nymphs of the five instars and the adult bug on a tomato plant. This cosmopolitan and polyphagous bug is an important world pest of food and fibre crops. (After Hely *et al.*, 1982.)

Fig. 6.2) and **holometaboly** (complete metamorphosis; Fig. 6.3 and the vignette for this chapter, which shows the life cycle of the scolytid beetle *Ips grandicollis*). Essentially these patterns are distinguished by quantitative rather than qualitative differences in the morphogenesis that occurs during metamorphosis. In all but the youngest immature

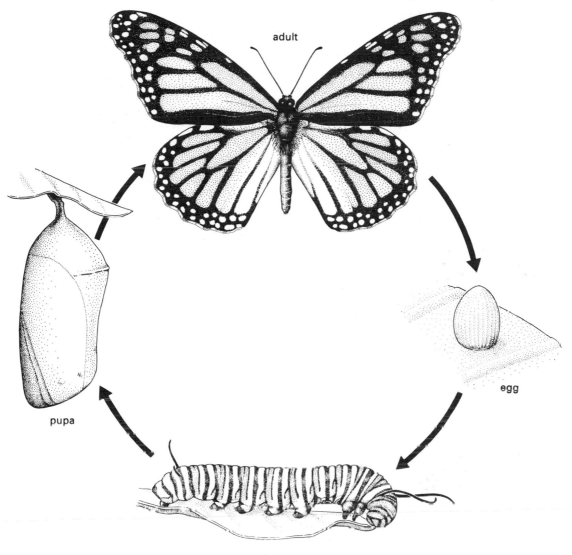

Fig. 6.3 Life cycle of a holometabolous insect, the monarch or wanderer butterfly, *Danaus plexippus* (Lepidoptera: Nymphalidae), showing the egg, a mature larva, the pupa or chrysalis, and the adult butterfly. There are five caterpillar (larval) instars.

instars of hemimetabolous insects, the developing wings are visible in external sheaths on the dorsal surface of the nymphs or larvae and the term **exopterygote** has been applied to this type of wing growth. In older literature, the insect orders with hemimetabolous and exopterygote development were grouped as the 'Hemimetabola' or 'Exopterygota' – terms that are recognized to be inappropriate now because they apply to a grade of organization rather than to a phylogenetic unit (Chapter 7). In contrast, the pterygote orders that display holometabolous development share the unique evolutionary innovation of a resting

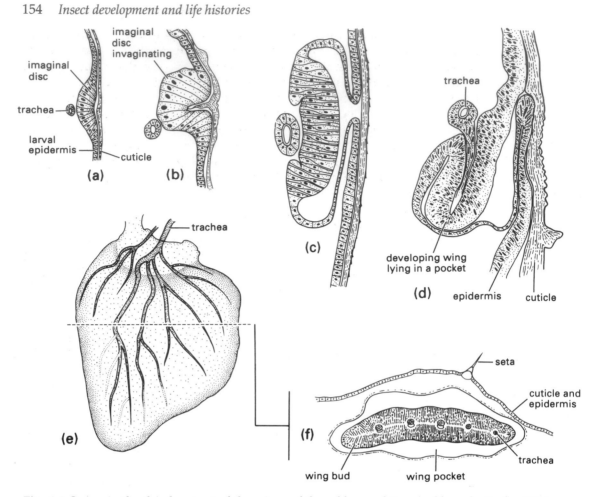

Fig. 6.4 Stages in the development of the wings of the cabbage white or cabbage butterfly, *Pieris rapae* (Lepidoptera: Pieridae). A wing imaginal disc in (a) first-instar larva, (b) second-instar larva, (c) third-instar larva and (d) fourth-instar larva; (e) the wing bud as it appears if dissected out of the wing pocket or (f) cut in cross-section in a fifth-instar larva. ((a)–(e) After Mercer, 1900.)

stage or **pupal instar** in which the major structural differences between the larval and adult phases are bridged. Structures unique to adults, particularly wings and genitalia, are present internally in the larvae as groups of cells called **imaginal discs** or **buds** (Fig. 6.4), which are scarcely visible until the pupal instar. Such wing development is called **endopterygote** because the wings develop in invaginated pockets of the integument and are everted only at the larval–pupal moult.

The orders that share this pattern of development represent a valid phylogenetic unit called the **Endopterygota** or **Holometabola**.

6.2.1 Phases in insect ontogeny

(a) Embryonic phase

The egg stage begins as soon as the mature egg is deposited by the female. For practical reasons, the age of an egg is estimated from the time of its deposition even though the egg existed before oviposition. The beginning of the egg stage, however, need not mark the commencement of an individual insect's ontogeny, which actually begins when embryonic development within the egg is triggered by **activation**. This trigger usually results from fertilization in sexually reproducing insects but in parthenogenetic species appears to be induced by various events at oviposition, including the entry of oxygen to the egg or mechanical distortion.

Following activation of the insect egg cell, the zygote nucleus subdivides by mitotic division to produce many daughter nuclei. These nuclei and their surrounding cytoplasm, called cleavage **energids**, migrate to the egg periphery where the membrane infolds leading to cellularization of the superficial layer to form a blastoderm, which is one cell thick. This distinctive superficial cleavage during early embryogenesis in insects is the result of the large amount of yolk in the egg. The blastoderm usually gives rise to all the cells of the larval body, whereas the central yolky part of the egg provides the nutrition for the developing embryo and will be used up by the time of **eclosion**, or emergence from the egg. Regional differentiation of the blastoderm leads to the formation of the **germ anlage** or germ disc (Fig. 6.5(a)) which is the first sign of the developing embryo, whereas the remainder of the blastoderm becomes a thin membrane, the **serosa** or embryonic cover. Next, the germ anlage develops an infolding in a process called **gastrulation** (Fig. 6.5(b)) and sinks into the yolk, forming a two-layered embryo containing the amniotic cavity (Fig. 6.5(c)). After gastrulation, the germ anlage becomes the **germ band** which externally is characterized by segmental organization (commencing in Fig. 6.5(e) with the formation of the procephalon). The germ band essentially forms the ventral regions of the future body which progressively differentiates with the head, body segments and appendages becoming increasingly well

defined (Figs 6.5(e)–(g)). Near the end of embryogenesis (Fig. 6.5(h),(i)), the edges of the germ band grow over the remaining yolk and fuse mid-dorsally to form the lateral and dorsal parts of the insect – a process called **dorsal closure**. Functional specialization of cells and tissues occurs during the latter period of embryonic development, so that by the time of hatching (Fig. 6.5(j)) the embryo is a tiny larva crammed into an eggshell.

(b) Larval or nymphal phase

Eclosion from the egg marks the beginning of the first stadium when the young insect is said to be in the first instar (Fig. 6.1). This stage ends at the first ecdysis when the old cuticle is shed to reveal the insect in its second instar. A third and often subsequent instars generally follow. Thus the development of the immature insect is characterized by repeated moults separated by periods of feeding, with hemimetabolous insects generally undergoing more moults to reach adulthood than holometabolous insects.

All immature holometabolous insects are called **larvae**. Immature individuals of aquatic hemimetabolous insects (Odonata, Ephemeroptera and Plecoptera) are also frequently referred to as larvae (or naiads), even though they possess external wing pads in at least the later instars. On the other hand, immature terrestrial insects with hemimetabolous development are called **nymphs**. Thus there are cockroach, grasshopper, mantid and bug nymphs, but dragonfly, damselfly, mayfly and stonefly larvae. Larvae of every instar look quite different from the final adult form, whereas nymphs more closely approach the adult appearance at each successive moult. Furthermore, the diets and lifestyles of larvae are very different to those of their adults, whereas nymphs often eat the same food and live in the same place as the adults of their species. Competition is thus rare between larvae and their adults but likely to be prevalent between nymphs and their adults.

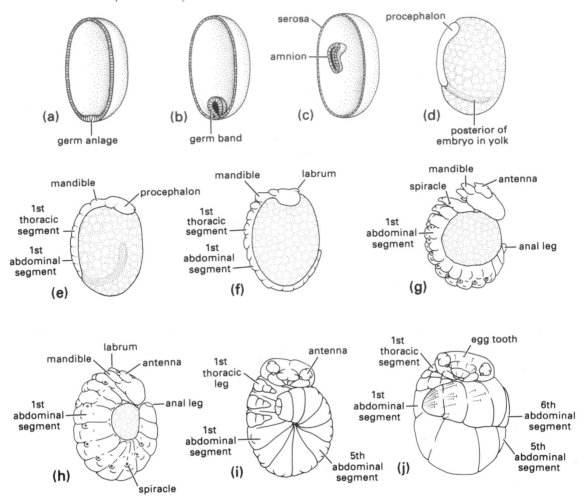

Fig. 6.5 Embryonic development of the scorpionfly *Panorpodes paradoxa* (Mecoptera: Panorpodidae): (a)–(c) schematic drawings of egg halves from which yolk has been removed to show position of embryo; (d)–(j) gross morphology of developing embryos at various ages. Age from oviposition: (a) 32 hours; (b) 2 days; (c) 7 days; (d) 12 days; (e) 16 days; (f) 19 days; (g) 23 days; (h) 25 days; (i) 25–26 days; (j) fully grown at 32 days. (After Suzuki, 1985.)

The great variety of endopterygote larvae can be classified into a few functional rather than phylogenetic types. Often the same larval type occurs convergently in unrelated orders. The three commonest forms are the polypod, oligopod and apod larvae. Lepidopteran caterpillars (Fig. 6.6(a),(b)) are the commonest example of **polypod larvae**, characterized by cylindrical bodies with short thoracic legs and abdominal prolegs (pseudopods). Mecoptera, symphytan Hymenoptera (Fig. 6.6 (c)) and some Trichoptera also have polypod larvae. Such larvae are rather inactive and are mostly phytophagous. **Oligopod larvae** (Fig. 6.6(d)–(f)) lack abdominal prolegs but have functional thoracic legs and frequently prognathous mouthparts. Many are active predators but others are slow-moving

POLYPOD LARVAE   OLIGOPOD LARVAE   APOD LARVAE

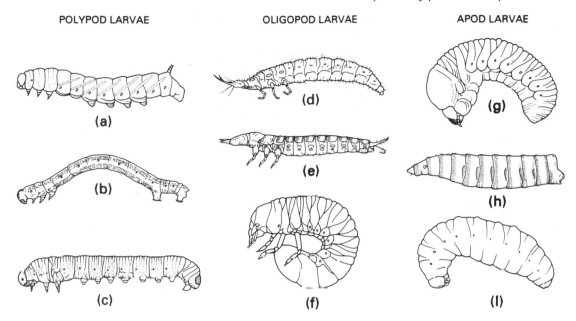

Fig. 6.6 Examples of larval types. Polypod larvae: (a) Lepidoptera: Sphingidae; (b) Lepidoptera: Geometridae; (c) Hymenoptera: Diprionidae. Oligopod larvae: (d) Neuroptera: Osmylidae; (e) Coleoptera: Carabidae; (f) Coleoptera: Scarabaeidae. Apod larvae: (g) Coleoptera: Scolytidae; (h) Diptera: Calliphoridae; (i) Hymenoptera: Vespidae. ((a),(e),(f),(g) After Chu, 1949; (b),(c) after Borror *et al.*, 1989; (h) after Ferrar, 1987; (i) after CSIRO, 1970.)

detritivores living in soil or are phytophages. This larval type occurs in at least some members of most orders of insects but not in the Lepidoptera, Mecoptera, Diptera, Siphonaptera or Strepsiptera. **Apod larvae** (Fig. 6.6(g)–(i)) lack true legs and are usually worm- or maggot-like, living in soil, mud, dung, decaying plant or animal matter, or within the bodies of other organisms as parasitoids (Chapter 12). The Siphonaptera, aculeate Hymenoptera, nematoceran Diptera and many Coleoptera typically have apod larvae with a well developed head, whereas in the maggots of higher Diptera the mouth hooks may be the only obvious evidence of the cephalic region. The grub-like apod larvae of some parasitic and gall-forming wasps and flies are greatly reduced in external structure and are difficult to identify to order level even by a specialist entomologist. Furthermore, the early-instar larvae of some parasitic wasps resemble a naked embryo but change into typical apod larvae in later instars.

A major change in form during the larval phase, such as different larval types in different instars, is called **larval heteromorphosis** (or **hypermetamorphosis**). In some beetles and in the Strepsiptera this involves an active first-instar larva or **triungulin**, followed by several grub-like, inactive, sometimes legless, later-instar larvae. This developmental phenomenon occurs most commonly in parasitic insects in which a mobile first instar is necessary for host location and entry. Larval heteromorphosis and diverse larval types are typical of many parasitic wasps, as mentioned above.

(c) Metamorphosis

All pterygote insects undergo varying degrees of transformation from the immature to the adult phase of their life history. Some exopterygotes, such as cockroaches, show only slight morphological changes during post-embryonic development, whereas the body is completely reconstructed at metamorphosis in many endopterygotes. Only the Endopterygota have a metamorphosis involving a pupal stadium, during which larval structures are destroyed and adult structures elaborated. Alterations in body shape, which are the essence of metamorphosis, are brought about by differential growth of various body parts. Organs that will function in the adult but that were suppressed in the larva grow at a faster rate than the body average. The accelerated growth of wing pads is the most obvious example but legs, genitalia, gonads and other internal organs may increase in size and complexity to a considerable extent.

The onset of metamorphosis generally is associated with the attainment of a certain body size, which is thought to programme the brain for metamorphosis, resulting in altered hormone levels. Metamorphosis in most beetles, however, shows considerable independence from the influence of the brain, especially during the pupal instar. In most insects, a reduction in the amount of circulating juvenile hormone (due to reduction of corpora allata activity) is essential to the initiation of metamorphosis. (The physiological events are described in section 6.3.)

The moult into the pupal instar is called **pupation** or the larval–pupal moult. Many insects survive conditions unfavourable for development in the 'resting', non-feeding pupal stage, but often what appears to be a pupa is actually a fully developed adult within the pupal cuticle, referred to as a **pharate** (cloaked) adult. Typically, a protective cell or cocoon surrounds first the pupa and then, prior to emergence, the pharate adult; only certain Coleoptera, Diptera, Lepidoptera and Hymenoptera have unprotected pupae.

Several pupal types (Fig. 6.7) are recognized and these appear to have arisen convergently in different orders. Most pupae are **exarate** –

their appendages (e.g. legs, wings, mouthparts and antennae) are not closely appressed to the body; the remainder of pupae are **obtect** – their appendages are cemented to the body and the cuticle is often heavily sclerotized (as in almost all Lepidoptera). Exarate pupae can have articulated mandibles (**decticous**), which the pharate adult uses to cut through the cocoon, or the mandibles can be non-articulated (**adecticous**), in which case the adult usually first sheds the pupal cuticle and then uses its mandibles and legs to escape the cocoon or cell. In some cyclorrhaphous Diptera (the Schizophora) the adecticous exarate pupa is enclosed in a **puparium** – the sclerotized cuticle of the last larval instar. Escape from the puparium is facilitated by eversion of a membranous sac on the head of the emerging adult, the **ptilinum**. Insects with obtect pupae may lack a cocoon, as in coccinellid beetles and most nematocerous and orthorrhaphous Diptera. If one is present, as in most Lepidoptera, either the pupa forces itself through the cocoon using backwardly directed abdominal spines and a projection on the head or the adult emerges from the pupal cuticle before escaping the cocoon, sometimes with the aid of a fluid that dissolves the silk.

(d) Imaginal or adult phase

As discussed above, all except the mayflies have an adult phase with a single instar. This imaginal stage has a reproductive role and is often the dispersive stage in insects with relatively sedentary larvae. After the imago emerges from the cuticle of the previous instar (eclosion), it may be reproductively competent almost immediately or there may be a period of maturation in readiness for sperm transfer or oviposition. Depending on species and food availability, there are from one to several reproductive cycles in the adult stadium. The adults of certain species, such as some mayflies, midges and male scale insects, are very short-lived. These insects have reduced or no mouthparts and fly for only a

EXARATE
DECTICOUS

EXARATE ADECTICOUS

OBTECT ADECTICOUS

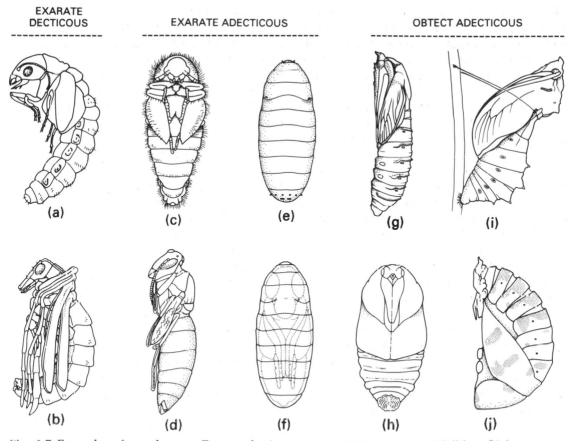

Fig. 6.7 Examples of pupal types. Exarate decticous pupae: (a) Megaloptera: Sialidae; (b) Mecoptera: Bittacidae. Exarate adecticous pupae: (c) Coleoptera: Dermestidae; (d) Hymenoptera: Vespidae, (e),(f) Diptera: Calliphoridae, puparium and pupa within. Obtect adecticous pupae: (g) Lepidoptera: Cossidae; (h) Lepidoptera: Saturniidae; (i) Lepidoptera: Papilionidae, chrysalis; (j) Coleoptera: Coccinellidae. ((a) After Evans, 1978; (b),(c),(e),(g) after CSIRO, 1970; (d) after Chu, 1949, (h) after Common, 1990; (i) after Common and Waterhouse, 1972; (j) after Palmer, 1914.)

few hours or at the most a day or two: they simply mate and die. Most adult insects live at least a few weeks, often a few months and sometimes for several years; termite reproductives and queen ants and bees are particularly long-lived.

Adult life begins at eclosion from the pupal cuticle. Metamorphosis, however, may have been complete for some hours or days previously and the pharate adult may have rested in the pupal cuticle until the appropriate environmental trigger for emergence.

Changes in temperature or light and perhaps chemical signals may synchronize adult emergence in most species. The hormonal control of emergence has been studied most comprehensively in Lepidoptera, especially by Truman and his colleagues in the tobacco hornworm, *Manduca sexta* (Lepidoptera: Sphingidae). At least four hormones are involved in successful eclosion. A few days prior to eclosion the ecdysteroid level declines, initiating a series of physiological and behavioural events in preparation for ecdysis,

Fig. 6.8 The pupal–imaginal moult of a male dragonfly of *Aeshna cyanea* (Odonata: Aeshnidae). The final-instar larva climbs out of the water prior to the shedding of its cuticle. The old cuticle splits mid-dorsally; the teneral adult frees itself, swallows air and must wait many hours for its wings to expand and dry. (After Blaney, 1976.)

including the release of eclosion hormone. This neuropeptide increases cuticle extensibility to allow withdrawal from the old cuticle, triggers ecdysis behaviour (such as seeking a site suitable for ecdysis), activates adult behaviours and permits the release of further neurohormones – **bursicon** and **cardiopeptides** – that are involved in wing expansion after ecdysis. The cardiopeptides stimulate the heart, facilitating movement of haemolymph into the thorax and thus into the wings. Bursicon induces a brief increase in cuticle plasticity to permit wing expansion, followed by sclerotization of the cuticle in its expanded form.

The newly emerged or **teneral** adult has soft cuticle which permits expansion of the body surface by swallowing air, by taking air into the tracheal sacs and by locally increasing haemolymph pressure by muscular activity. The wings normally hang down (Fig. 6.8), which aids their inflation. Pigment deposition in the cuticle and epidermal cells occurs just before or after emergence and is either linked to, or followed by, sclerotization of the body cuticle under the influence of the neurohormone bursicon.

Following emergence from the pupal cuticle, many holometabolous insects void a faecal fluid called the **meconium**. This represents the metabolic wastes that have accumulated during the pupal stadium. Sometimes the teneral adult retains the meconium in the rectum until sclerotization is complete, thus aiding increase in body size.

Reproduction (Chapter 5) is the main function of adult life and the length of the imaginal stadium, at least in the female, is related to the duration of egg production. Senescence correlates with termination of reproduction and death may be predetermined in the ontogeny of an insect. Females may die after egg deposition and males may die after mating. An extended post-reproductive life is important in distasteful, aposematic insects to allow predators to learn the distastefulness of the prey at a developmental period when prey individuals are expendable (Chapter 13).

6.3 THE PROCESS AND CONTROL OF MOULTING

For practical reasons an instar is defined from ecdysis to ecdysis (Fig. 6.1), but morphologically and physiologically a new instar comes into existence at the time of **apolysis** when the epidermis separates from the cuticle of the previous stage. Apolysis is difficult to detect in most insects but knowledge of its occurrence may be important because many insects spend a substantial period in the pharate state (cloaked within the cuticle of the previous instar) awaiting conditions favourable for emergence as the next stage. Insects often survive adverse conditions as pharate pupae or pharate adults because in this state the double cuticular layer restricts water loss during a developmental period during which metabolism is reduced and requirements for gaseous exchange are minimal.

Moulting is a complex process involving hormonal, behavioural, epidermal and cuticular changes that lead up to the shedding of the old cuticle. The epidermal cells are actively involved in moulting: they are responsible for partial breakdown of the old cuticle and formation of the new cuticle. The moult commences with the retraction of the epidermal cells from the inner surface of the old cuticle, usually in an anteroposterior direction. This separation is not total because muscles and sensory nerves retain their connection with the old cuticle. Apolysis is either correlated with or followed by mitotic division of the epidermal cells leading to increases in the volume and surface area of the epidermis. The subcuticular or **apolysial space**, formed after apolysis, becomes filled with the secreted but inactive moulting fluid. The chitinolytic and proteolytic enzymes of the moulting fluid are not activated until the protective outer layer of the new cuticle has been laid down by the epidermal cells. Then the inner part of the old cuticle (the endocuticle) is lysed and presumably resorbed, while the new pharate cuticle continues to be deposited as an undifferentiated procuticle.

Ecdysis commences with the remnants of the old cuticle splitting along the dorsal midline due to increase in haemolymph pressure. The cast cuticle (**exuviae**) consists of the indigestible protein, lipid and chitin of the old epicuticle and exocuticle. Once free of the constraints of this previous 'skin', the newly ecdysed insect expands the new cuticle by swallowing air or water and/or by increasing haemolymph pressure in different body parts to smooth out the wrinkled and folded epicuticle and stretch the procuticle. After cuticular expansion, some or much of the body surface may become sclerotized by the chemical stiffening and darkening of the procuticle to form exocuticle (Chapter 2). However, in larval insects most of the body cuticle remains membranous and exocuticle is confined to the head capsule. Following ecdysis, more proteins and chitin are secreted from the epidermal cells thus adding to the inner part of the procuticle, the endocuticle, which may continue to be deposited well into the intermoult period. Sometimes the endocuticle is partially sclerotized during the stadium and frequently the outer surface of the cuticle is covered in wax secretions. Finally the stadium draws to an end and apolysis is initiated once again.

The above events are effected by hormones acting on the epidermal cells to control the cuticular changes and also on the nervous system to co-ordinate the behaviours associated with ecdysis. Hormonal regulation of moulting has been studied most thoroughly at metamorphosis when endocrine influences on moulting *per se* are difficult to separate from those involved in the control of morphological change. The classical view of the hormonal regulation of moulting and metamorphosis is presented schematically in Fig. 6.9; the endocrine centres and their hormones are described in more detail in Chapter 3.

Moulting and metamorphosis are controlled by three major types of hormones:

- Neuropeptides including **prothoracicotropic hormone** (PTTH) and **eclosion hormone** (EH).
- **Ecdysteroids**.
- **Juvenile hormone** (JH), which may occur in several different forms even in the same insect.

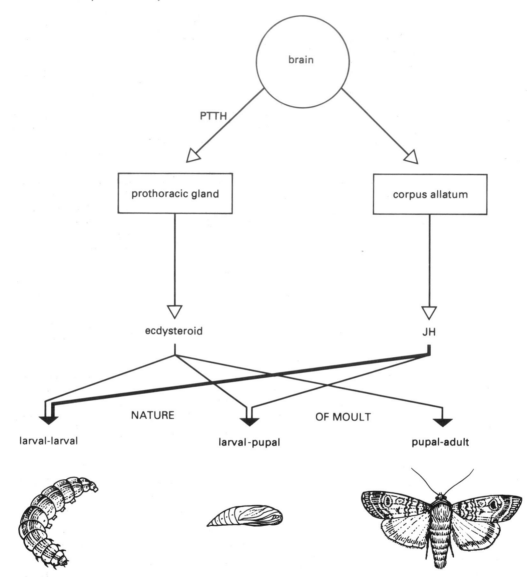

Fig. 6.9 Schematic diagram of the classical view of endocrine control of the epidermal processes that occur in moulting and metamorphosis in an endopterygote insect. This scheme simplifies the complexity of ecdysteroid and JH secretion and does not indicate the influence of neuropeptides such as eclosion hormone. JH, juvenile hormone; PTTH, prothoracicotropic hormone. (After Richards, 1981.)

Neurosecretory cells in the brain secrete PTTH which passes down nerve axons to the corpora cardiaca, a pair of neuroglandular bodies that store and later release PTTH into the haemolymph. The PTTH stimulates ecdysteroid synthesis and secretion by the prothoracic or moulting glands. Ecdysteroid release then initiates the changes in the epidermal cells that lead to the production of new cuticle. The characteristics of the moult are regulated by JH from the corpora allata: JH inhibits the expression of adult features so

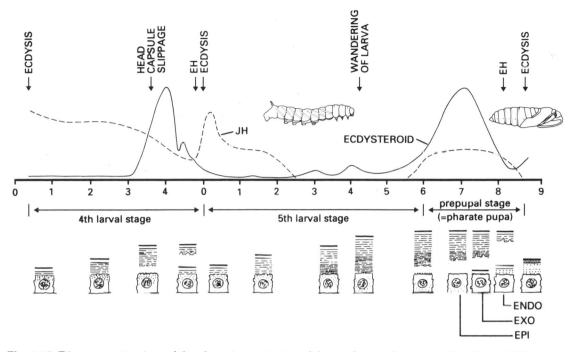

Fig. 6.10 Diagrammatic view of the changing activities of the epidermis during the fourth and fifth larval instars and prepupal (= pharate pupal) development in the tobacco hornworm, *Manduca sexta* (Lepidoptera: Sphingidae) in relation to the hormonal environment. The dots in the epidermal cells represent granules of the blue pigment insecticyanin; HCS, head capsule slippage; JH, juvenile hormone; EH, eclosion hormone; EPI, EXO, ENDO, deposition of pupal epicuticle, exocuticle and endocuticle, respectively. The numbers on the x-axis represent days. (After Riddiford, 1991.)

that a high haemolymph level (titre) of JH is associated with a larval–larval moult, a lower titre with a larval–pupal moult and absence at the pupal–adult moult.

Ecdysis is mediated by eclosion hormone (EH), which appears to be important at every moult in the life history of perhaps all insects. This neuropeptide acts on a steroid-primed central nervous system to evoke the co-ordinated motor activities associated with escape from the old cuticle. Eclosion hormone derives its name from the pupal–adult ecdysis or eclosion, for which its importance was first discovered and before its wider role was realized. Indeed, the association of EH with ecdysis behaviours appears to represent an evolutionarily ancient relationship. In many insects, another neuropeptide, bursicon,

controls sclerotization of the exocuticle and post-moult deposition of endocuticle.

The relationship between the hormonal environment and the epidermal activities that control moulting and cuticular deposition in a lepidopteran, the tobacco hornworm *Manduca sexta*, are presented in Fig. 6.10. Only now are we beginning to understand how hormones regulate moulting and metamorphosis at the cellular and molecular levels. However, detailed studies on the tobacco hornworm clearly show the correlation between the ecdysteroid and JH titres and the cuticular changes that occur in the last two larval instars and in prepupal development. During the moult at the end of the fourth larval instar, the epidermis responds to the surge of ecdysteroid by halting synthesis of endocuticle and the blue pigment, insecticyanin. A new epicuticle is synthesized, much of the old cuticle is digested and resumption of endocuticle and insecticyanin production occurs by the time of ecdysis. In the final larval instar the JH declines to undetectable levels, allowing small rises in ecdysteroid that first stimulate the epidermis to produce a stiffer cuticle with thinner lamellae and then elicit wandering in the larva. When ecdysteroid initiates the next moult, the epidermal cells produce pupal cuticle due to the activation of many new genes. It seems that the decline in ecdysteroid level towards the end of each moult is essential to and may be the physiological trigger for ecdysis to occur by rendering the tissues sensitive to EH and permitting its release into the haemolymph. (Section 6.2.1(d) discusses further the actions of eclosion hormone.) Apolysis at the end of the fifth larval instar marks the beginning of a pre-pupal period when the developing pupa is pharate within the larval cuticle. Differentiated exocuticle and endocuticle appear at this larval–pupal moult. During larval life, the epidermal cells covering most of the body do not produce exocuticle and so the caterpillar's cuticle is soft and flexible, allowing considerable growth within an instar as a result of feeding.

6.4 VOLTINISM

Insects are short-lived creatures, whose lives can be measured by their **voltinism** – the numbers of generations per year. Generation times in excess of one year are found amongst some inhabitants of the polar extremes, where suitable conditions for development may exist for only a few weeks in each year. Furthermore, some large insects that rely upon nutritionally poor diets develop slowly over many years. For example, periodic cicadas feeding on sap in roots may take 13 or 17 years to mature, and beetles that develop within dead wood have been known to emerge after more than 20 years' development. Most insects, however, take a year or less to develop, with either one generation per year (**univoltine** insects), or two (**bivoltine**) or more than two (**multivoltine** or polyvoltine).

Univoltinism is common amongst temperate insects, particularly those that use seasonally restricted resources – for example, insects whose aquatic immature stages rely on spring algal bloom or phytophagous insects using short-lived plants. Multivoltine insects tend to be small and fast-developing, using resources that are more evenly available throughout the year. Bivoltine insects include those that develop slowly on evenly spread resources, and those that track a bimodally distributed factor, such as spring and autumn temperature. Some species have fixed voltinism patterns, whilst others may vary with geography, particularly insects with broad latitudinal or elevation ranges.

Most insects do not develop throughout the year but arrest their development during unfavourable times by quiescence or diapause (section 6.8). Many univoltine and some bivoltine species diapause as adults, awaiting suitable conditions before commencing the next generation. For some of these univoltine insects, many social insects and others that take longer than a year to develop, adult longevity may extend to several years. In contrast, the adult life of small multivoltine insects may be as little as a few hours at low tide for marine midges such as *Clunio* (Diptera: Chironomidae), or a single evening for many Ephemeroptera.

6.5 POLYMORPHISM AND POLYPHENISM

The existence of several generations per year is often associated with morphological change between generations. Similar variation may occur contemporaneously within a population, such as the simultaneous existence of winged and flightless forms ('**morphs**'). Sexual differences between males and females, and the existence of strong differentiation in social insects such as ants and bees, are further obvious examples of the phenomenon. The term **polymorphism** encompasses all such discontinuities which occur in the same life-history phase at a frequency greater than might be expected from repeated mutation alone. It is defined as the simultaneous or recurrent occurrence of distinct morphological differences, reflecting and often including physiological, behavioural or ecological differences among conspecific individuals.

6.5.1 Genetic polymorphism

The distinction between the sexes is an example of a particular polymorphism, namely sexual dimorphism, which is almost totally genetically determined in insects. Environmental factors may affect sexual expression, as in castes of some social insects or in feminization of genetically male insects by mermithid nematode infections. Apart from the dimorphism of the sexes, different genotypes may co-occur within a single species, maintained by natural selection at specific frequencies that vary from place to place and time to time throughout the range. For example, adults of some gerrid bugs are fully winged and migratory, whilst other coexisting individuals of the same species are brachypterous and cannot fly. Intermediates are at a selective disadvantage and the two genetically determined morphs coexist in a balanced polymorphism.

Some of the most complex genetically-based polymorphisms have been discovered in butterflies that mimic chemically protected butterflies of another species (the model) for purposes of defence from predators (Chapter 13). Some butterfly species may mimic more than one model, and the accuracy of the several distinct mimicry patterns in these species is maintained because inappropriate intermediates are not recognized by predators as being distasteful and get eaten. Mimetic polymorphism is predominantly restricted to the females; the males are generally monomorphic and non-mimetic. The basis for switching between the different mimetic morphs is simple Mendelian genetics involving either relatively few genes or supergenes.

It is a common observation that, within species with a wide range of latitudinal distributions, there exist different life-history strategies according to location. For example, populations living at high latitudes (nearer the poles) or high elevation may be univoltine, with a long dormant period, whilst populations nearer the equator or lower in elevation may be multivoltine and develop continuously without dormancy. Dormancy is induced by the environment (section 6.8) but the ability of the insect to recognize and respond to these cues is programmed genetically. At least some geographic variation seen in life-histories is the result of genetic polymorphism.

6.5.2 Environmental polymorphism or polyphenism

A phenotypic difference between generations that lacks a genetic basis and is entirely environmentally determined is often termed **polyphenism**. An example is the temperate to tropical Old World pierid butterfly *Eurema hecabe*, which shows a seasonal change in wing colour between summer and autumn morphs. Photoperiod induces morph change: a dark-winged summer morph is induced by a long day of more than 13 hours, whereas a short day of fewer than 12 hours induces the paler-winged autumn morph, particularly at temperatures of under 20 °C, with temperature affecting males more than females.

Amongst the most complex polyphenisms are those seen in the aphids. Within parthenogenetic lineages (i.e. in which there is absolute genetic identity) the females may show up to eight distinct phenotypes, in addition to polymorphisms in sexual forms. These female aphids may vary in morphology, physiology, fecundity, offspring timing and size, development time, longevity, and host-plant choice and utilization. Environmental cues responsible for alternative morphs are similar to those that govern diapause in many insects (section 6.8), including photoperiod, temperature, and maternal effects such as elapsed time (rather than number of generations) since the winged founding mother. Overcrowding triggers many aphid species to produce a winged dispersive phase. Crowding also is responsible for one of the most dramatic examples of polyphenism, the phase transformation from the solitary young locusts (hoppers) to the gregarious phase (section 6.7.5). Studies on the physiological mechanisms that link environmental cues to these phenotype changes have implicated juvenile hormone in many aphid morph shifts.

If aphids show the greatest number of polyphenisms, the social insects come a close second and undoubtedly have a greater degree of morphological differentiation between morphs, termed **castes**. This is discussed in more detail in Chapter 11; suffice to say that maintenance of the phenotypic differences between castes as different as queens, workers and soldiers includes physiological mechanisms such as pheromones transferred with food, olfactory and tactile stimuli, and endocrine control including juvenile hormone and ecdysone. Superimposed on these polymorphisms are the dimorphic differences between the sexes, which impose some limits on variation.

6.6 AGE-GRADING

Identification of the growth stages or ages of insects in a population can be very important in ecological or applied entomology. Information on the proportion of a population in different developmental stages and the proportion of the adult population at reproductive maturity can be used to construct time-specific life-tables or budgets to determine factors that cause and regulate fluctuations in population size and dispersal rate, and to monitor fecundity and mortality factors in the population. Such data are integral to predictions of pest outbreaks as a result of climate and to the construction of models of population response to the introduction of a control programme.

Many different techniques have been proposed for estimating either the growth stage or the age of insects. Some provide an estimate of chronological (calendar) age within a stadium, whereas most estimate either instar number or relative age within a stadium, in which case the term **age-grading** is used in place of age determination.

6.6.1 Age-grading of immature insects

For many population studies it is important to know the number of larval or nymphal instars in a species and to be able to recognize the instar to which any immature individual belongs. Generally such information is available or its acquisition is feasible for species with a constant and relatively small number of immature instars, especially those with a life span of a few months or less. However, it is logistically difficult to obtain such data for species with either many or a variable number of instars, or with overlapping generations. The latter situation may occur in species with many asynchronous generations per year or in species with a life-cycle of longer than one year. In some species there are readily discernible qualitative (e.g. colour) or meristic (e.g. antennal segment number) differences between consecutive immature instars. More frequently, the only obvious difference between successive larval or nymphal instars is the increase in size that occurs after each moult (the moult increment). Thus it should be possible to determine the actual number of instars in the life history of a species from a frequency histogram of measurements of a sclerotized body part (Fig. 6.11).

Entomologists have sought to quantify this size progression for a range of insects. One of

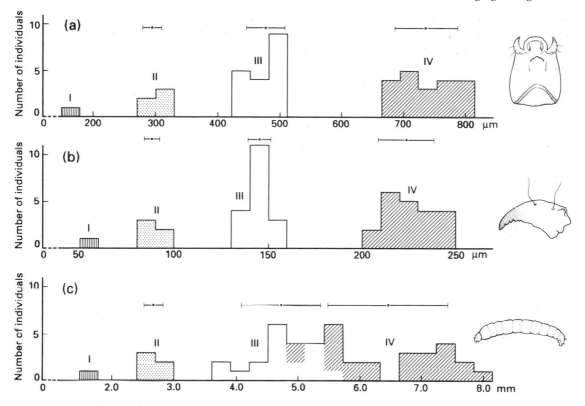

Fig. 6.11 Growth and development in a marine midge, *Telmatogeton* (Diptera: Chironomidae), showing increases in: (a) head capsule length, (b) mandible length and (c) body length between the four larval instars (I–IV). The dots and horizontal lines above each histogram represent the means and standard deviations of measurements for each instar. Note that the lengths of the sclerotized head and mandible fall into discrete size classes representing each instar, whereas body length is an unreliable indicator of instar number especially for separating the third and fourth-instar larvae.

the earliest attempts was that of H.G. Dyar, who in 1890 established a 'rule' from observations on the caterpillars of 28 species of Lepidoptera. Dyar's measurements showed that the width of the head capsule increased in a regular linear progression in successive instars by a ratio (range 1.3–1.7) that was constant for a given species. **Dyar's rule** states that: *postmoult size/premoult size (or moult increment) = constant*. Thus if logarithms of measurements of some sclerotized body part in different instars are plotted against the instar number, a straight line should result; any deviation from a straight line indicates a missing instar. In practice, however, there are

many departures from Dyar's rule since the progression factor is not always constant, especially in field populations subject to variable conditions of food and temperature during growth.

A related empirical 'law' of growth is Przibram's rule, which states that: *an insect's weight is doubled during each instar and at each moult all linear dimensions are increased by a ratio of 1.26*. The growth of most insects shows no general agreement with this rule, which assumes that the dimensions of a part of the insect body should increase at each moult by the same ratio as the body as a whole. In reality, growth in most insects is allometric, that is, the parts grow at rates peculiar to themselves, and often very different to the growth rate of the body as a whole.

6.6.2 Age-grading of adult insects

The age of an adult insect is not determined easily. However, adult age is of great significance, particularly in the insect vectors of disease. For instance, it is crucial to epidemiology that the age (longevity) of an adult female mosquito be known, since this relates to the number of blood meals taken and therefore the number of opportunities for pathogen transmission. Most techniques for assessing the age of adult insects estimate relative (not chronological) age and hence age-grading is the appropriate term.

Three general categories of age assessment have been proposed, relating to age-related changes in physiology and morphology of the reproductive system; in somatic structures; and in external wear and tear. The latter approach has proved unreliable and only two of the many methods involving the two former have wide applicability.

In the first method, age is graded according to reproductive physiology in a technique applicable only to females. Examination of an ovary of a **parous** insect (one that has laid at least one egg) shows that after each egg is laid (or even resorbed) evidence remains in the form of a **follicular relic** that denotes an irreversible change in the epithelium. Associated with each egg is a follicle, whose deposition, together with contraction of the previously distended membrane, leaves one follicular relic per egg. The actual shape and form of the follicular relic varies between species, but one or more residual dilatations of the lumen, with or without pigment or granules, is common in the Diptera. Females that have no follicular relic have not developed an egg and are termed **nulliparous**.

The number of follicular relics can give a comparative measure of the physiological age of a female insect, for example allowing discrimination of parous from nulliparous individuals, and often allowing further segregation within parous individuals according to the number of ovipositions. The chronological age can be calculated if the time between successive ovipositions (the **ovarian cycle**) is known. However, if there is one ovarian cycle per blood meal, as in many medically significant flies, it is the physiological age (number of cycles) that is of greater significance than the precise chronological age.

The second generally applicable method of age determination has a more direct relationship with chronology and most of the somatic features that allow age estimation are present in both sexes. Estimates of age can be made from measures of cuticle growth, fluorescent pigments, fat body size, cuticle hardness and, in females only, colour or patterning of the abdomen. Cuticular growth estimates of age rely upon there being a daily rhythm of deposition of the endocuticle. In exopterygotes, cuticular layers are more reliable, whilst in endopterygotes the apodemes (internal skeletal projections upon which muscles attach) are more dependable. The daily layers are most distinctive when the temperature for cuticle formation is not attained for part of each day. This use of growth rings is confounded by development temperatures too cold for deposition, or too high for the daily cycle of deposition and cessation. A further drawback to the technique is that deposition ceases after a certain age is attained, perhaps only 10–15 days after eclosion. Physiological age can be determined by measuring the pigments that accumulate in the aging cells of many animals, including insects. These pigments exhibit fluorescence and can be studied by fluorescence microscopy. Lipofuscin from post-mitotic cells in most body tissues and pteridine eye pigments have been measured in this way, especially in flies.

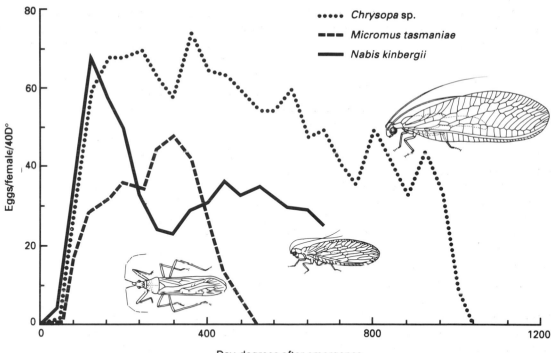

Fig. 6.12 Age-specific oviposition rates of three predators of cotton pests, *Chrysopa* sp. (Neuroptera: Chrysopidae), *Micromus tasmaniae* (Neuroptera: Hemerobiidae) and *Nabis kinbergii* (Hemiptera: Nabidae), based on physiological time above respective development thresholds of 10.5°, –2.9° and 11.3°. (After Samson and Blood, 1979.)

6.7 ENVIRONMENTAL EFFECTS ON DEVELOPMENT

The rate at which, or the manner in which, an insect develops or grows may depend upon a number of factors, including the type and amount of food, the amount of moisture (for terrestrial species) and heat (measured as temperature), or the presence of environmental signals (e.g. photoperiod), mutagens and toxins, or other organisms, either predators or competitors. Two or more of these factors may interact to complicate interpretation of growth characteristics and patterns.

6.7.1 Temperature

Most insects are poikilothermic – body temperature varies more or less directly with environmental temperature – and so heat is the force driving the rate of growth and development when food is not limiting. An increase in temperature, within a favourable range, will speed up the metabolism of an insect and consequently increase its rate of development. Each species and each stage in the life history may develop at its own rate in relation to temperature. Thus, **physiological time**, a measure of the amount of heat required over time for an insect to complete development or a stage of development, is more meaningful as a measure of development time than age in calendar time. Knowledge of temperature–development relationships and the use of physiological time allows comparison of the life cycles or fecundity of pest species in the same system (Fig. 6.12), and prediction of the larval feeding

BOX 6.1
CALCULATION OF DAY-DEGREES

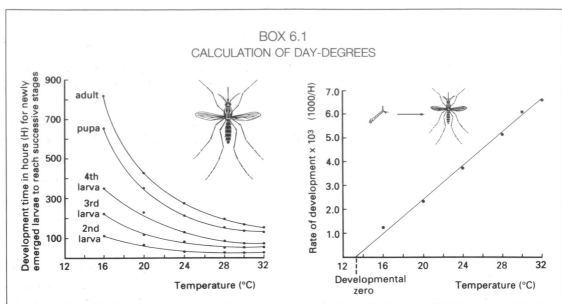

An outline of a simple method to estimate day-degrees (after Daly *et al.*, 1978) is exemplified by data on the relationship between temperature and development in the yellow fever mosquito, *Aedes aegypti* (Diptera: Culicidae) (after Bar-Zeev, 1958).

1. In the laboratory, establish the average time required for each stage to develop at different constant temperatures. The graph on left shows the time in hours (H) for newly hatched larvae of *Ae. aegypti* to reach successive stages of development when incubated at various temperatures.

2. Plot the reciprocal of development time (1/H), the development rate, against temperature to obtain a sigmoid curve with the middle part of the curve approximately linear. The graph on the right shows the linear part of this relationship for the total development of *Ae. aegypti* from the newly hatched larva to the adult stage. A straight line would not be obtained if extreme development temperatures (e.g. higher than 32 °C or lower than 16 °C) had been included.

3. Fit a linear regression line to the points and calculate the slope of this line. The slope represents the amount in hours by which development rates are increased for each one degree of increased temperature. Hence the reciprocal of the slope gives the number of hour-degrees, above threshold, required to complete development.

4. To estimate the developmental threshold, the regression line is projected to the x-axis (abscissa) to give the developmental zero, which in the case of *Ae. aegypti* is 13.3 °C. This zero value may differ slightly from the actual developmental threshold determined experimentally, probably because at low (or high) temperatures the temperature–development relationship is rarely linear. For *Ae. aegypti*, the developmental threshold actually lies between 9 and 10 °C.

5. The equation of the regression line is $1/H = k(T° - T^t)$, where H = development period, T° = temperature, T^t = development threshold temperature, and k = slope of line.

Thus the physiological time for development is $H(T° - T^t) = 1/k$ hour-degrees; or $H(T° - T^t)/24 = 1/k = K$ day-degrees, where K = thermal constant or K-value.

BOX 6.1—*continued*

By inserting the values of H, T° and T^t for the data from *Ae. aegypti* in the equation given above, the value of K can be calculated for each of the experimental temperatures from 14–36 °C:

Temp.(°C)	14	16	20	24	28	30	32	34	36
K	1008	2211	2834	2921	2866	2755	2861	3415	3882

Thus the K value for *Ac. acgypti* is approximately independent of temperature, except at extremes (14 and 34–36 °C), and averages about 2740 hour-degrees or 114 day-degrees between 16 and 32 °C.

---

periods, generation length and time of adult emergence under variable temperature conditions that exist in the field. Such predictions are especially important for pest insects, since control measures must be timed carefully to be effective.

Physiological time is the cumulative product of total development time (in hours or days) multiplied by the temperature (in degrees) above the **developmental** (or **growth**) **threshold** or the temperature below which no development occurs. Thus physiological time is commonly expressed as **day-degrees** (D°) or **hour-degrees** (h°). Normally, physiological time is estimated for a species by rearing a number of individuals of the life-history stage of interest under different constant temperatures in several identical growth cabinets. The developmental threshold is estimated by the linear regression x-axis method (Box 6.1) although more accurate threshold estimates can be obtained by more time-consuming methods.

In practice, the application of laboratory-estimated physiological time to natural populations may be complicated by several factors. Under fluctuating temperatures, especially if extremes are experienced by the insects, growth may be retarded or accelerated compared with the same number of day-degrees under constant temperatures. Furthermore,

the temperatures actually experienced by the insects, in their often sheltered microhabitats on plants or in soil or litter, may be several degrees different to the temperatures recorded at a meteorological station even just a few metres away. Insects may select microhabitats that ameliorate cold night conditions or reduce or increase daytime heat. Thus predictions of insect life-cycle events based on extrapolation from laboratory to field temperature records may be inaccurate. For these reasons, the laboratory estimates of physiological time should be corroborated by calculating the hour degrees or day degrees required for development under more natural conditions, but using the laboratory-estimated developmental threshold, as follows:

- Place newly laid eggs or newly hatched larvae in their appropriate field habitat and record temperature each hour (or calculate a daily average – a less accurate method).
- Estimate the time for completion of each instar by discarding all temperature readings below the developmental threshold of the instar and subtracting the developmental threshold from all other readings to determine the effective temperature for each hour (or simply subtract the developmental threshold temperature from the

daily average temperature). Sum the degrees of effective temperature for each hour from the beginning to the end of the stadium. This procedure is called thermal summation.

- Compare the field-estimated number of hour-degrees (or day-degrees) for each instar with that predicted from the laboratory data. If there are discrepancies, then microhabitat and/or fluctuating temperatures may be influencing insect development or the developmental zero read from the graph may be a poor estimate of the developmental threshold.

Another problem with laboratory estimation of physiological time is that insect populations maintained for lengthy periods under laboratory conditions frequently undergo acclimation to constant conditions or even genetic change in response to the altered environment or as a result of population reductions that produce genetic 'bottlenecks'. Therefore, insects maintained in rearing cages may exhibit different temperature–development relationships from individuals of the same species in wild populations.

For all of the above reasons any formula or model that purports to predict insect response to environmental conditions must be tested carefully for its fit with natural population responses.

6.7.2 Photoperiod

Many insects – perhaps most – do not develop continuously all year round but avoid some seasonally adverse conditions by a resting period (section 6.8). Summer dormancy (aestivation) and winter dormancy (hibernation) are two examples of avoidance of seasonal extremes. The most predictable environmental indicator of changing seasons is **photoperiod** – the length of the daily light phase or, more simply, daylength. Near the equator, although sunrise to sunset of the longest day may be only a few minutes longer

than on the shortest day, if the period of twilight is included then total daylength shows more marked seasonal change. The photoperiod response is to duration rather than intensity but there is a critical threshold intensity of light below which the insect does not respond; this threshold is often as dim as twilight but rarely as low as bright moonlight. Many insects appear to measure the duration of the light phase of the 24 hours. Some have been shown experimentally to measure the duration of dark. Others recognize long days by light falling within the 'dark' half of the day.

Most insects can be described as 'long-day' species, with growth and reproduction in summer, and with dormancy commencing with decreasing daylength. Others show the reverse pattern, with 'short-day' (often autumn and spring) activity and summer aestivation. In some species the life-history stage in which photoperiod is assessed is in advance of the stage that reacts, as is the case when the photoperiodic response of the maternal generation of silkworms affects the eggs of the next generation.

The ability of insects to recognize seasonal photoperiod and other environmental cues requires some means of measuring time between the cue and the subsequent onset or cessation of diapause. This is achieved through a 'biological clock' (Box 4.2), which may be driven by internal (endogenous) or external (exogenous) daily cycles called **circadian rhythms**. Interactions between the short time periodicity of circadian rhythms and longer term of seasonal photoperiod recognition are complex and diverse, and have probably evolved many times within insects.

6.7.3 Humidity

The high ratio of surface area to volume in insects means that loss of body water is a serious hazard in a terrestrial environment, especially a dry one. Low moisture content of the air can affect the physiology and thus the development, longevity and oviposition of many insects. Air holds more water vapour at high than at low temperatures.

The relative humidity (RH) at a particular temperature is the ratio of actual water vapour present to that necessary for saturation of the air at that temperature. At low relative humidities, development may be retarded, for example in many stored products pests, but at high relative humidities or in saturated air (100% RH) insects or their eggs may drown or be infected more readily by pathogens. The fact that stadia may be greatly lengthened by unfavourable humidity has serious implications for estimates of development times, whether calendar or physiological time is used. The complicating effects of low, and sometimes even high, air moisture levels should be taken into account when gathering such data.

6.7.4 Mutagens and toxins

Stressful conditions induced by toxic or mutagenic chemicals may affect insect growth and form to varying degrees, ranging from death at one extreme to slight phenotypic modifications at the other end of the spectrum. Some life-history stages may be more sensitive to mutagens or toxins than others and sometimes the phenotypic effects may not be easily measured by crude estimates of stress, such as percentage survival. One sensitive and efficient measure of the amount of genetic or environmental stress experienced by insects during development is the incidence of **fluctuating asymmetry**, or the quantitative differences between the left and right sides of each individual in a sample of the population. Insects are usually bilaterally symmetrical if grown under ideal conditions, so that the left and right halves of their bodies are mirror images (except for obvious differences in structures such as the genitalia of some male insects). If grown under stressful conditions, however, the degree of asymmetry tends to increase.

The measurement of fluctuating asymmetry has many potential uses in theoretical and economic entomology and in assessment of environmental quality. For example, it can be used as an indicator of developmental stability to determine the effect on non-target organisms of exposure to insecticides or vermicides, such as avermectins. Bush flies (*Musca vetustissima*) breeding in the dung of cattle treated with Avermectin B_1 have been shown to be significantly more asymmetric for two morphometric wing characters than flies breeding in the dung of untreated cattle. As a measure of environmental quality, fluctuating asymmetry can be used, for example, to assess water quality by comparing the amount of asymmetry in aquatic insects reared in polluted and clean water. In industrially polluted waters, particular bloodworms (larvae of chironomid midges) may survive but often exhibit gross developmental abnormalities. However, at lower levels of pollutants, more subtle effects can best be measured by examining deviations from symmetry compared with clean-water controls. In addition, measures of developmental effects on non-target insects are frequently used to assess the specificity of biocides prior to marketing.

6.7.5 Biotic effects

In most insect orders, adult size has a strong genetic component and growth is strongly determinate. In many Lepidoptera, for example, final adult size is relatively constant within a species; reduction in food quality or availability usually delays caterpillar growth rather than reducing final adult size, although there are exceptions. In contrast, in flies that have limited or ephemeral larval resources, such as a dung pat or temporary pool, cessation of larval growth would result in death as the habitat shrinks. Thus larval crowding or limitation of food supply tend to shorten development time and reduce final adult size. In some mosquitoes and midges, success in short-lived pool habitats is attained by a small proportion of the larval population developing with extreme rapidity relative to their slower siblings. In paedogenetic gall midges (Chapter 5), crowding with reduced food supply terminates larva-only reproductive cycles and induces production of adults, allowing dispersal to more favourable habitats.

Food quality appears important in all these cases but there may be related effects – for example, due to crowding. Clearly it can be difficult to segregate food effects from other potentially limiting factors. In the California red scale, *Aonidiella aurantii* (Hemiptera: Diaspididae), development and reproduction on orange trees is fastest on fruit, intermediate on twigs and slowest on leaves. Whilst these differences may reflect differing nutritional status, a microclimatic explanation cannot be excluded: fruit, because of relatively greater volume, may retain heat longer than stems and leaves and such slight temperature differences might affect the development of the insects.

The effects of crowding on development are well understood in some insects, as in locusts that have **solitary** and **gregarious** phases (Fig. 6.13), which differ in morphometrics, colour and behaviour. At low densities locusts develop into the solitary phase, with a characteristic uniformly coloured 'hopper' (nymph), and large-sized adult with large hind femora. At high densities, induced in nature by high survivorship of eggs and young nymphs under favourable climatic conditions, changes occur and a dark-striped nymph develops to a smaller locust with shorter hind femora. The most conspicuous difference is the behaviour, with gregarious phase individuals aggregating, often into enormous and mobile swarms, whilst solitary phase individuals shun each other's company and have low vagility. The behavioural shift is entirely induced by crowding, as can be shown by splitting a single locust egg pod into

SOLITARY

GREGARIOUS

Fig. 6.13 Solitary and gregarious females of the migratory locust, *Locusta migratoria* (Orthoptera: Acrididae). The solitaria adults have a pronounced pronotal crest and the femora are larger relative to the body and wing than in the gregaria adults. Intermediate morphologies occur in the transiens (transient stage) during the transformation from solitaria to gregaria or the reverse.

two: rearing the offspring at low densities induces solitary locusts, whereas their siblings reared under crowded conditions develop into gregarious locusts. The response to high population density results from the integration of several cues, including the sight, touch and sometimes the odour (pheromone) of conspecifics, which lead to endocrine or neuroendocrine changes involved in the phase transformation.

6.8 DIAPAUSE

The developmental progression from egg to adult is often interrupted by a period of dormancy. This occurs particularly in temperate areas when environmental conditions are unsuitable, such as seasonal extremes of high or low temperatures, or drought. Dormancy may be through either quiescence or diapause. **Quiescence** is a halted or slowed development as a direct response to unfavourable conditions, with development resuming immediately favourable conditions return. In contrast, **diapause** involves arrested development combined with adaptive physiological changes, with development recommencing not necessarily on return of suitable conditions but only following particular physiological stimuli. The distinction between quiescence and diapause may not always be clear-cut.

If diapause occurs at a fixed time regardless of environmental conditions, it is termed **obligatory**. Univoltine insects (those with one generation per year) often have obligatory diapause to extend an essentially short life-cycle to one full year. Diapause that is optional is termed **facultative**, and this occurs widely in insects, including many bi- or multivoltine insects in which diapause occurs only in the one generation that must pass the unfavourable conditions. Diapause can last from days to months, and can occur in any life-history stage from egg to adult. The diapausing stage predominantly is fixed within any species and can vary between close relatives. Egg and pupal diapause are common, probably because these stages are relatively closed systems, with only gases being exchanged during embryogenesis and metamorphosis respectively, allowing better survival during environmental stress.

Much research on diapause has been carried out in Japan in relation to silk production from cultured silkworms (*Bombyx mori*). Optimal silk production, which is by an egg-diapausing generation, conflicts with the requirement for continuous production, which comes from non-diapausing eggs. The complex mechanisms that promote and break diapause in this species are now well understood but may not have generality. Several different mechanisms may be at play in different, even closely related, insects and much is still to be discovered.

Major environmental cues that induce or terminate diapause are photoperiod, temperature, food quality, moisture, pH, and chemicals including oxygen, urea and plant secondary compounds. Identification of the contribution of each may be difficult, as for example in species of the mosquito genus *Aedes* that lay diapausing eggs into seasonally dry pools or containers. Flooding of the oviposition site at any time may terminate embryonic diapause in some *Aedes* species. In others, many successive inundations may be required to break diapause, the cues apparently including chemical changes such as lowering of pH by microbial decomposition of pond detritus. Furthermore, one environmental cue may enhance or override a previous one. For example, if an appropriate diapause-terminating cue of inundation occurs while the photoperiod or temperature is 'wrong', then diapause may not break, or only a small proportion of eggs may hatch.

Photoperiod is undoubtedly of major significance in diapause because changes in daylength are effective predictors of future seasonal environmental conditions. Thus photoperiod increases as summer heat approaches and diminishes towards winter cold. Insects are able to detect daylength changes (photoperiodic stimuli), sometimes with extreme accuracy, through brain photoreceptors rather than compound eyes or ocelli. The insect brain also stores the 'programming' for diapause, such that transplant of a diapausing moth's pupal brain into a non-diapausing pupa will induce diapause in the recipient. The reciprocal operation causes resumption of development in a diapausing recipient. This programming may long precede the diapause and even span a generation, such that maternal conditions can govern the diapause in the developing stages of her offspring.

Many studies have shown diapause to be under endocrine control, but substantial variation in mechanisms for the regulation of diapause reflects the multiple independent evolution of this phenomenon. Generally in diapausing larvae, the production of ecdysone moulting hormone from the prothoracic gland ceases, and juvenile hormone plays a role in termination of diapause. Increased levels of ecdysone products of the

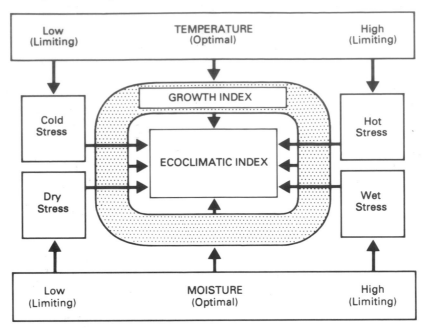

Fig. 6.14 Flow diagram depicting the derivation of the 'Ecoclimatic Index' (EI) as the product of population growth index and four stress indices. The EI value describes the climatic favourability of a given location for a given species. Comparison of EI values allows different locations to be assessed for their relative suitability to a particular species. (After Sutherst and Maywald, 1985.)

prothoracic glands appear essential for the termination of pupal diapause. Juvenile hormone is important in diapause regulation in adult insects but, as with the immature stages, may not be the only regulator. In larvae, pupae and adults of *Bombyx mori*, complex antagonistic interactions occur between a **diapause hormone**, originating from paired neurosecretory cells in the suboesophageal ganglion, and juvenile hormone from the corpora allata. The adult female produces diapause eggs when the ovariole is under the influence of diapause hormone, whereas non-diapause eggs are produced in the absence of this hormone and in the presence of juvenile hormone.

6.9 CLIMATIC MODELLING OF INSECT DISTRIBUTIONS

Section 6.7 described how various environmental factors affect insect development.

Here we briefly examine the application of such information to predictive modelling of insect abundance and distribution. The abundance of any poikilothermic species is determined largely by interactions between proximate ecological factors including climate, habitat, food availability, and the presence or absence of predators and competitors. The distribution of an insect species is the result not only of these ecological factors but also of a historical component. Ecology determines whether a species can continue to live in an area; history determines whether it does, or ever had the chance to live there. This difference is related to timing; given enough time, an ecological factor will be considered an historical factor. In the context of present-day studies of where pest insects occur and what the limits of their spread might be, history may account for the original or native distribution of a pest but knowledge of ecology will allow prediction of potential or future

BOX 6.2
CLIMATIC MODELLING FOR FRUIT FLIES

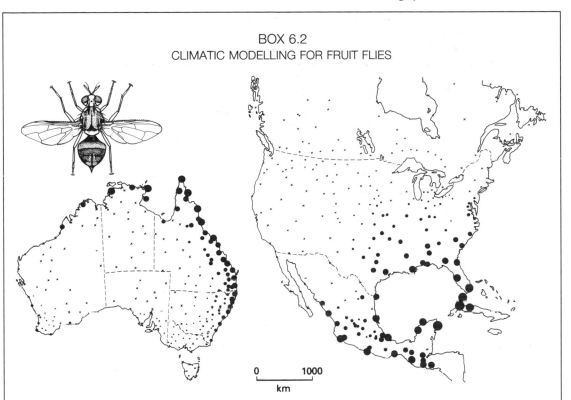

The Queensland fruit fly, *Bactrocera tryoni*, is a pest of most commercial fruits. The females oviposit into the fruit, which is quickly destroyed by larval feeding followed by rotting. Even if damage in an orchard is insignificant, any infestation is serious because of restrictions on interstate and overseas marketing of fruit infested by fruit fly.

CLIMEX has been used by Sutherst and Maywald to describe the response of *B. tryoni* to Australia's climate. The growth and stress indices of CLIMEX were estimated by inference from maps of the geographical distribution and from estimates of the relative abundance of this fly in different parts of its range in Australia. The map of Australia depicts the ecoclimatic indices describing the favourableness of each site for permanent colonization by *B. tryoni*. The area of each circle is proportional to its ecoclimatic index (EI). Crosses indicate that the fly could not permanently colonize the site.

The potential survival of *B. tryoni* as an immigrant pest in North America can be predicted using CLIMEX by climate-matching with the fly's native range. Accidental transport of this fly could lead to its establishment at the point of entry or it might be taken to other areas with climates more favourable to its persistence. Should *B. tryoni* become established in north America, the eastern seaboard from New York to Florida and west to Kansas, Oklahoma and Texas in the United States and much of Mexico are most at risk. Canada and most of the central and western United States are unlikely to support permanent colonization. Thus only certain regions of the continent are at high risk of infestation by *B. tryoni* and appropriate vigilance should be maintained by quarantine authorities in those places. (After Sutherst and Maywald, 1991.)

distributions under changed environmental conditions (e.g. Greenhouse Effect) or as a result of accidental human dispersal. Thus ecological knowledge of insect pests and their natural enemies, especially information on climatic influences on their development, is vital for the prediction of pest outbreaks and for successful pest management.

There is a wide variety of computer-based models pertaining to the population biology of economic insects, especially those affecting major crop systems in western countries. One example of a climatic model of arthropod distribution and abundance is the computer-based system called CLIMEX (developed by Sutherst and Maywald) which enables the prediction of an insect's potential relative abundance and distribution around the world, using biological data and the recorded geographical distribution. An annual 'eco-climatic index' (EI), describing the climatic favourability of a given location for permanent colonization of an insect species, is derived from a climatic database combined with estimates of the response of the organisms to temperature, moisture and daylength. The ecoclimatic index (Fig. 6.14) is calculated as follows. First, a population growth index (GI) is determined from weekly values averaged over a year to obtain a measure of the potential for population increase of the species. The GI is estimated from data on the seasonal incidence and relative abundance in different parts of the species' range. Second, the GI is reduced by incorporation of four stress indices, which are measures of the deleterious effects of cold, heat, dry and wet. Commonly, the existing geographical distribution and seasonal incidence of a pest species are known but biological data pertaining to climatic effects on development are scanty. Fortunately, the limiting effects of climate on a species can usually be estimated reliably from observations on the geographical distribution. The climatic tolerances of the species are inferred from the climate of the sites where the species is known to occur and are described by the stress indices of the CLIMEX model. The values of the stress indices are progressively adjusted until the CLIMEX predictions agree with the observed distribution of the species. Of course, other information on the climatic tolerances of the species should be used if possible because the above procedure relies on the assumption that climate limits the present distribution, which might be an oversimplification.

This type of climatic modelling based on world data has been carried out for tick species and for insects such as the Russian wheat aphid, the Colorado potato beetle, screwworm flies, biting flies of *Haematobia* species and fruit flies (Box 6.2). The output of such models has great utility in applied entomology, namely in epidemiology, quarantine, management of insect pests, and entomological control of weeds and animal pests (including other insects).

FURTHER READING

Binnington, K. and Retnakaran, A. (eds) (1991) *Physiology of the Insect Epidermis*, CSIRO Publications, Melbourne.

Daly, H.V. (1985) Insect morphometrics. *Annual Review of Entomology*, 30, 415–38.

Kerkut, G.A. and Gilbert, L.I. (eds) (1985) *Comprehensive Insect Physiology, Biochemistry, and Pharmacology, Volumes 7 and 8, Endocrinology I and II*, Pergamon Press, Oxford, New York, Toronto, Sydney, Paris, Frankfurt.

Moran, N.A. (1992) The evolution of aphid life cycles. *Annual Review of Entomology*, 37, 321–48.

Pender, M.P. (1991) Locust phase polymorphism and its endocrine relations. *Advances in Insect Physiology*, 23, 1–79.

Richards, G. (1981) Insect hormones in development. *Biological Reviews of the Cambridge Philosophical Society*, 56, 501–549.

Sander, K., Gutzeit, H.O. and Jäckle, H. (1985) Insect embryogenesis: morphology, physiology, genetical and molecular aspects, in *Comprehensive Insect Physiology, Biochemistry, and Pharmacology, Volume 1, Embryogenesis and Reproduction*, (eds G.A. Kerkut and L.I. Gilbert), Pergamon Press, Oxford, New York, Toronto, Sydney, Paris, Frankfurt, pp. 319–85.

Sehnal, F. (1985) Growth and life cycles, in *Comprehensive Insect Physiology, Biochemistry, and Pharmacology, Volume 2, Postembryonic Development*, (eds G.A. Kerkut and L.I. Gilbert), Pergamon Press, Oxford, New York, Toronto, Sydney, Paris, Frankfurt, pp. 1–86.

Sutherst, R.W. and Maywald, G.F. (1985) A computerised system for matching climates in ecology. *Agriculture, Ecosystems and Environment,* **13,** 281–99.

Švácha, P. (1992) What are and are not imaginal discs: reevaluation of some basic concepts (Insecta, Holometabola). *Developmental Biology,* **154,** 101–17.

Tyndale-Biscoe, M. (1984) Age-grading methods in adult insects: a review. *Bulletin of Entomological Research,* **74,** 341–77.

Worner, S.P. (1988) Ecoclimatic assessment of potential establishment of exotic pests. *Journal of Economic Entomology,* **81,** 9731–83.

7

INSECT SYSTEMATICS –
PHYLOGENY AND EVOLUTION

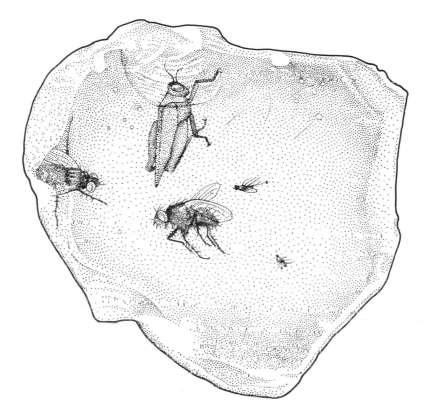

Fossils: insects preserved in amber

There is a common misconception that the classification of the living world is virtually complete; after all, we can use any of a variety of books to look up the identity and classification of any common organism. However, if we carefully compared several texts we would be bound to find some differences between treatments, say as to the taxonomic status of a geographic race of bird, or of the family placement of a species of flowering plant. This does not mean that scientists have a perverse desire to confuse things, but reflects ongoing differences in opinions concerning the most appropriate classification, as well as the continuing acquisition of additional knowledge. A further problem arises in entomology because there are so many insects that the basic **taxonomic** work of recognizing, naming and classification is far from complete.

A second popular misconception concerns the pedestrian nature of the work of scientists engaged in **systematics** – the study of the kinds and diversity of organisms and their interrelationships. Certainly the taxonomic component involves many time-consuming activities, including exhaustive literature and specimen searches, curation of collections, measurements of features from specimens, and sorting of perhaps thousands of individuals into species and perhaps hundreds of species into higher groupings. These essential tasks require considerable skill and are fundamental to the wider science of systematics – investigation of the origin, diversification and distribution, both historic and current, of the organisms studied.

Taxonomy provides the data base for systematics. The collection of these data and their interpretation, though once seen as a matter of personal taste, is the subject of a challenging debate in which entomological systematists feature as prominent participants.

In this chapter the different methods of interpretation are reviewed, followed by a brief summary of fossil and contemporary evidence concerning the evolution of the insects. Some ideas about the reasons for the success of the insects are presented. The chapter concludes with a current classification based on the postulated evolutionary relationships within the Hexapoda, of which the Insecta comprise the largest of three classes.

7.1 PHYLOGENETICS

The unravelling of evolutionary history, **phylogenetics**, is a controversial area of biology, not least for the insects. Although the various groups (**taxa**), especially the orders, are quite well defined, the phylogenetic relationships among insect taxa are a matter of much conjecture, even at the level of orders. For example, the beetles (Coleoptera) form a discrete group, easily recognized by having their forewings modified as elytra, but they have no obvious close relative. Stoneflies (Plecoptera) and mayflies (Ephemeroptera) resemble each other quite closely, but this is a superficial resemblance and is misleading as an indication of relationship. The stoneflies are more closely related to the orthopteroids (cockroaches, termites, mantids, earwigs, grasshoppers, crickets and their allies) than to mayflies. Resemblance need not indicate evolutionary relationships because, although similarity may derive from relatedness, equally it can arise through **homoplasy**, that is convergently or through parallel evolution of structures by either chance or selection for similar functions. Only similarity due to common ancestry (**homology**) can provide information regarding phylogeny.

Until recently, morphology provided most data upon which insect phylogeny was reconstructed. Some of the ambiguity and lack of clarity regarding insect relationships was blamed on inherent deficiencies in the phylogenetic informativeness of such characteristics. This led to the increased use of molecular data in attempts to resolve some of the unanswered questions, particularly pertaining to higher relationships between insects. At the time of writing, it is increasingly

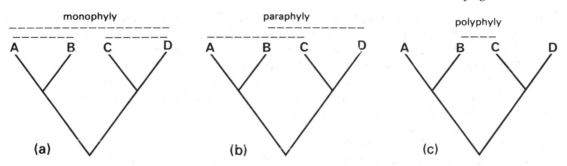

Fig. 7.1 A cladogram showing the relationships of four species, A, B, C and D, and examples of (a) the three monophyletic groups, (b) two of the four possible (ABC, ABD, ACD, BCD) paraphyletic groups, and (c) one of the four possible (AC, AD, BC and BD) polyphyletic groups that could be recognized based on this cladogram.

evident that both types of data suffer equally from homoplasy and molecules are not providing the anticipated clarity any more than the traditional morphology. The most justifiable approach is a holistic one, using data from as many sources as possible and retaining an awareness that not all similarities are equally informative in revealing phylogenetic patterns.

7.1.1 Phylogenetic methods

The various methods that attempt to recover the pattern produced by phylogenetic history rely on observations on living and fossil organisms. As a simplification, three differing methods can be identified. The **phenetic** method (phenetics) relies on estimates of overall similarity, usually derived from morphology but sometimes from behaviour and other traits, and increasingly from molecular evidence. Many of those who apply phenetics claim that evolution is unknowable and the technique identifies only patterns of resemblance; however, other scientists believe that the phenetic pattern revealed is as good an estimate of phylogeny as can be obtained. Alternative methods to phenetics are based on the premise that the pattern produced by evolutionary processes can be uncovered and, furthermore, ought to be reflected in the classification. Overall similarity, the criterion

of phenetics, may not recover this pattern of evolution and phenetic classifications are therefore artificial.

The **cladistic** method (cladistics) seeks patterns of special similarity based only on shared, evolutionarily novel features. Shared ancestral features do not indicate relationship. Furthermore, features that are unique to a particular group but unknown outside the group do not indicate inter-group relationships. Construction of a **cladogram** (Fig. 7.1), the tree-like diagram portraying the phylogenetic branching pattern, is fundamental to cladistics. From this tree, **monophyletic** groups, or **clades**, their relationships to each other, and a classification, can be inferred directly. Sister groups are taxa which are each other's closest relatives. A monophyletic group contains a hypothetical ancestor and **all** of its descendants. Further groupings can be identified from Fig. 7.1: **paraphyletic** groups lack one clade from amongst the descendants of a common ancestor, and often are created by the recognition (and removal) of a derived subgroup; **polyphyletic** groups fail to include two or more unrelated clades from amongst the descendants of a common ancestor. Thus when we recognize the monophyletic Pterygota (winged or secondarily apterous insects), the remainder of the Insecta – the apterygotes – are rendered paraphyletic. If we were to recognize a group of flying insects with wings restricted to the mesothorax (dipterans, male coccoids and a few ephemeropterans), this would be a polyphyletic grouping. Paraphyletic groups should be avoided if possible because their only defining features are primitive ones shared with other indirect relatives. Thus the absence of wings in the paraphyletic apterygotes is a primitive feature shared by many other invertebrates. The mixed ancestry of polyphyletic groups means that they are biologically uninformative and such artificial taxa should never be included in any classification.

Evolutionary systematics also uses estimates of derived similarity but, in contrast to cladistics, estimates of the amount of evolutionary change are included with the branching pattern in order to produce a classification. Thus, an evolutionary approach emphasizes distinctness, granting higher taxonomic status to taxa separated by 'gaps'. These gaps may be created by accelerated morphological innovation in a lineage, and/or by extinction of intermediate linking forms. Thus, the highly specialized ants frequently are or were given superfamily rank (the Formicoidea) within the Hymenoptera because ants have many unique features that make them look very different from their nearest relatives. Phylogenetic study indicates that ants should be placed as a family, the Formicidae, in the superfamily Vespoidea.

Current classifications of insects mix all three practices, with most orders being based on groups (taxa) with distinctive morphology. These are not necessarily monophyletic – some are almost certainly paraphyletic – but it is unlikely that any higher-level groups are polyphyletic. In many cases, the present groupings coincide with the earliest colloquial observations on insects, for example the term 'beetles' for Coleoptera. However, in other cases, such old colloquial names cover disparate modern groupings, as with the term 'flies', now seen to encompass unrelated orders from mayflies (Ephemeroptera) to true flies (Diptera). Refinements continue, as

classification is found to be out of step with our developing understanding of the phylogeny. Current classifications reflect a historical legacy, variably reconciled with phylogeny.

7.1.2 Taxonomy and classification

Difficulties with attaining a comprehensive, coherent classification of the insects arise from the obscuring of phylogeny through complex evolutionary diversifications. These include radiations associated with phytophagy and parasitism, conflicting evidence from immature and adult insects (Chapter 6) but, above all, variety from the sheer number of species.

Those scientists who study the taxonomy of insects – that is, describe, name and classify them – face a daunting task. Whereas virtually all the world's vertebrates are described, with their distributions verified and ecologies studied, perhaps only 5–20% of the estimated number of insect species have been formally described, let alone studied biologically. The disproportionate allocation of taxonomic resources is exemplified by Q.D. Wheeler's report for the United States of seven described mammal species per mammal taxonomist in contrast to 425 described insects per insect taxonomist. These ratios, which probably have worldwide application, become even more alarming if one includes estimates of undescribed species in the calculations. Whereas there are very few unnamed mammals, estimates of global insect diversity include many millions of undescribed species. The burden of descriptive taxonomy is an impediment to insect systematics.

7.2 THE ANTIQUITY OF INSECTS

7.2.1 The insect fossil record

The earliest hexapod, *Rhyniella praecursor*, known from the Lower Devonian (380 million years before present (mybp)) of Scotland, is a

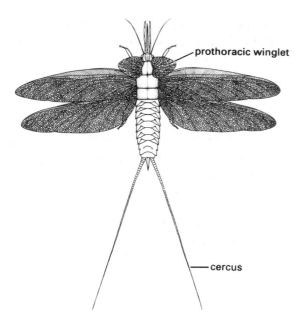

prothoracic winglet

cercus

Fig. 7.2 Reconstruction of *Stenodictya lobata* (Palaeodictyoptera: Dictyoneuridae). (After Kukalová, 1970.)

collembolan. The earliest ectognathous insects are fragmentary archaeognathans found in both contemporaneous and slightly younger Devonian deposits of north-eastern North America. *Eopterum devonicum*, from the Russian Upper Devonian, was once claimed to be the first true winged insect but is now widely regarded as a crustacean tail-fan. Evidence for early Insecta or their relatives is unlikely to be found since, even if they existed prior to the Devonian, the appropriate fresh water fossil deposits are scarce.

In the Carboniferous there was extensive radiation, evidenced by substantial Upper Carboniferous fossils. Lower Carboniferous fossils are unknown, again due to lack of freshwater deposits. By some 300 million years ago, there were several now-extinct groups usually given ordinal rank, including the Palaeodictyoptera (Fig. 7.2), Meganisoptera (= Protodonata), Megasecoptera and the Diaphanopterodea. The Palaeodictyoptera showed variation in size (with wing spans

up to 56 cm), diversity (over 70 genera in 21 families are known) and in morphology, notably in mouthparts and wing articulation and venation. This diversified order became extinct at the end of the Permian. The only extant orders clearly represented by Carboniferous fossils are Ephemeroptera, Blattodea and Orthoptera (Fig. 7.3). Putative hemipterans are best referable to the blattoid–orthopteroid complex. The beak-like, piercing mouthparts and expanded clypeus of some Carboniferous insects indicate an early origin of feeding on plants, although it was not until the Permian (270–225 mybp) that gymnosperms (conifers and allies) became abundant in the previously fern-dominated flora. Concurrently, there was a dramatic increase in ordinal diversity, with some 30 orders known from the Permian, some to become extinct, but major extant ones, excepting Hymenoptera and Lepidoptera, were present. The evolution of the plant-sucking Homoptera may have been associated with the newly available plants with thin cortex and subcorticular phloem. Other Permian insects included those that ate pollen, another resource that was previously in restricted supply.

The orders of modern insects, excepting Hymenoptera and Lepidoptera, appear to have been well established by the beginning of the Triassic (225 mybp). Many earlier lineages had disappeared by the late Permian in what may have been a major Permian extinction. Fossil evidence indicates that the numerical dominance of insects coincides with the angiosperm (flowering plant) diversification in the Cretaceous (135–65 mybp).

By the end of the Cretaceous (65 mybp), the fossil insect fauna looks quite modern, with many fossils able to be allocated to modern genera. In the Cretaceous and at later times throughout the Tertiary period (65–1.6 mybp), excellent arthropod specimens have been preserved in amber – a resinous plant secretion that trapped insects and hardened into a clear preservative (shown in the vignette for this chapter). Cretaceous amber occurs in

	PALAEOZOIC						MESOZOIC			CAENOZOIC						
											Tertiary					Quaternary
	Cambrian	Ordovician	Silurian	Devonian	Carboniferous	Permian	Triassic	Jurassic	Cretaceous	Palaeocene	Eocene	Oligocene	Miocene	Pliocene	Pleistocene	Recent
Approximate age in 10^6 years:	500	440	400	360		285	245	210	145	65	55	37	25	5	1.6	0.01

VASCULAR LAND PLANTS
　FERNS
　GYMNOSPERMS
　(conifers, cycads, etc.)
　ANGIOSPERMS
　(flowering plants)
BRYOPHYTES
　(mosses, liverworts, etc.)

HEXAPODA
　COLLEMBOLA
　DIPLURA
　INSECTA
　　"APTERYGOTA"
　　　ARCHAEOGNATHA
　　　MONURA†
　　　THYSANURA
　　PTERYGOTA
　　"PALAEOPTERA"
　　　DIAPHANOPTERODEA†
　　　PALAEODICTYOPTERA†
　　　MEGASECOPTERA†
　　　PERMOTHEMISTIDA†
　　　EPHEMEROPTERA
　　　MEGANISOPTERA†
　　　ODONATA
　　NEOPTERA
　　　ORTHOPTERA
　　　PHASMATODEA
　　　GRYLLOBLATTODEA
　　　DERMAPTERA
　　　ISOPTERA
　　　BLATTODEA
　　　MANTODEA
　　　PLECOPTERA
　　　EMBIOPTERA
　　　FOSSIL HEMIPTEROIDS†
　　　THYSANOPTERA
　　　HEMIPTERA
　　　PSOCOPTERA
　　ENDOPTERYGOTA
　　　MIOMOPTERA†
　　　COLEOPTERA
　　　STREPSIPTERA
　　　NEUROPTERA
　　　MEGALOPTERA
　　　RAPHIDIOPTERA
　　　HYMENOPTERA
　　　TRICHOPTERA
　　　LEPIDOPTERA
　　　MECOPTERA
　　　SIPHONAPTERA
　　　DIPTERA

Fig. 7.3 The geological history of insects in relation to plant evolution. Taxa that contain only fossils are indicated by the symbol †. The fossil record of Protura, Phthiraptera and Zoraptera only dates from the Recent epoch, presumably because of the small size and/or rarity of these organisms. Miomoptera represent the earliest fossil endopterygotes. (Insect records after Kukalová-Peck, 1991.)

Siberia and the Lebanon, with Dominican Republic amber dated to the Oligocene epoch (37–25 mybp). An abundant and well-studied amber derives from an Eocene/Oligocene epoch deposit, preserved and now partially exposed beneath the northern European Baltic Sea and, to a lesser extent, the southern North Sea. Periodic storms bring to the shore many pieces of amber, some with insect inclusions. The excellence of preservation of whole insects in amber contrasts favourably with compression fossils that may comprise little more than crumpled wings. Many early fossil records of extant higher taxa derive from these well preserved amber specimens, but the inherent sampling bias must be recognized.

Many species and higher taxa (groups above species level), especially genera and families, are much older than was previously thought. At species level, all northern temperate, subarctic and arctic zone fossil insects dating from the last million years or so appear to be morphologically identical to existing species. The vast majority of these fossils are beetles (particularly their elytra) but there is no reason to believe that the situation is any different for other insects. It seems that Pleistocene climatic fluctuations (glacial and interglacial cycles) caused range changes of taxa, via movements and extinctions of individuals, but resulted in the isolation of few, if any, new species, as morphologically defined. The implication is that if species of insect are typically greater than a million years old, then insect genera and other higher taxa are likely to be of immense age.

In using fossils for phylogenetic reconstruction there are several temptations that ought to be resisted:

- The inference that the oldest fossil of a group represents the phylogenetically earliest taxon.
- The inference that all character states in the fossil are in the primitive condition.
- The inference that any fossil represents an actual ancestral form of later taxa.

Palaeontology has much to bring to studies on insect evolution but phylogenetic inferences are not the exclusive province of the palaeontologist. Phylogenies can be reconstructed from the examination of characters of extant material alone, with fossils allowing the dating of the origin of diagnostic derived characters, or, optimally, all data (fossil and extant) ought to be reconciled into a single estimate of phylogeny.

7.2.2 Living insects: the evidence for antiquity

Evidence from biogeography confirms the antiquity of a number of insect lineages. The disjunct distribution, specific ecological requirements and restricted vagility of insects in a number of genera suggest that their constituent species were derived from ancestors that existed prior to the continental movements of the Jurassic and Cretaceous (commencing some 155 million years ago). For example, the occurrence of species of the chironomid midge genus *Archaeochlus* only in southern Africa and west-central Australia, suggests the taxon was fragmented by separation of the continental masses during the break-up of the supercontinent Gondwana, giving a minimum age for *Archaeochlus* of 120 million years. This dating is substantiated by Cretaceous amber fossil specimens, only slightly younger than the age of the continental break-up. Similarly, the aphid subtribe Melaphidina, all species of which have complex life cycles involving gall formation on sumac (*Rhus*) species, have a distribution disjunct between Asia and North America. This biogeographic evidence provides an estimated minimum age for this aphid–sumac association of 48 million years, again substantiated by fossil material. This is the oldest known continuous association between an insect and a plant lineage. For both examples, and in an increasing number of other cases, the possibility of recent dispersal can be eliminated.

The demonstrated morphological conservatism and great antiquity of many insect species needs to be reconciled with the obvious species and genetic diversity discussed in Chapter 1. The occurrence of similar species assemblages in the Pleistocene as are seen today (albeit at different geographic locations) suggests considerable physiological, ecological and morphological constancy of species. In comparative terms, insects display slower rates of evolution than is apparent in many larger animals such as mammals. For example, *Homo sapiens* is thought to be merely 100 000 years old and, if we classify humans as morphologically highly derived chimpanzees, then the genus containing humans and the two chimpanzee species is only about 5 million years old. Perhaps the difference lies in mammals having had a recent radiation and already having undergone major extinctions, including significant losses in the Pleistocene. In contrast, insects have had an early and several subsequent radiations, each followed by relative stasis and high persistence of lineages.

7.3 INSECT RADIATIONS

An estimated half of all insect species chew, suck, gall or mine the living tissues of higher plants (**phytophagy**), yet only nine of the 29 extant insect orders are primarily phytophagous. This imbalance suggests that when a barrier to phytophagy (e.g. plant defences) is breached, an asymmetry in species number occurs, with the phytophagous lineage being much more speciose than the lineage of its closest relative (the sister group) of different feeding mode. For example, the tremendous diversification of the almost universally phytophagous Lepidoptera can be compared with that of its sister group, the relatively species-poor, non-phytophagous Trichoptera. Likewise, the enormous phytophagous beetle group Phytophaga (Chrysomeloidea plus Curculionoidea) is overwhelmingly more diverse than the entire Cucujoidea, the whole or part of which forms the sister group to the Phytophaga. Clearly the diversification of insects and flowering plants are intimately related, a phenomenon explored further in Chapter 10. By analogy, the diversification of phytophagous insects should be accompanied by the diversification of their insect parasites or parasitoids, as discussed in Chapter 12. Such parallel species diversifications obviously require that the phytophage or parasite be able to seek out and recognize its host(s); indeed, the high level of host-specificity observed for insects is possible only because of their highly developed sensory and neuro-motor systems.

An asymmetry, similar to that seen when phytophagy is compared with non-phytophagy, is evident when flightedness is contrasted to aptery. The monophyletic Pterygota (winged or secondarily apterous insects) are vastly more speciose than either their immediate sister group, the Thysanura (silverfish), or the totality of primitively wingless apterygotes. The conclusion is unavoidable: the possession of flight correlates with a radiation under any definition of the term. Secondary aptery occurs in some Neoptera – amongst many isolated species, some genera and in all members of two small orders, the Phthiraptera (parasitic lice) and Siphonaptera (fleas). These two orders radiated as ectoparasites, but even these are less diverse than their flighted sister groups, respectively part or all of the Psocoptera and all of the diverse and speciose Diptera.

Flight allows insects the increased mobility necessary to use patchy food resources and habitats and to evade non-winged predators. These abilities may enhance species survival by reducing the threats of extinction, but wings also allow insects to reach novel habitats by dispersal across a barrier or by expansion of their range. Thus vagile pterygotes may be more prone to species formation by the two modes of geographic (**allopatric**) speciation: small isolated populations formed by the vagaries of chance dispersal by winged adults may be the progenitors of new species, or the continuous range of widely distributed species may become fragmented into isolates by **vicariance** (range division) events such as vegetation fragmentation or geological changes. New species arise as the genotypes of isolated populations diverge from those of parental populations. Such isolation may be phenological (temporal or behavioural), as well as spatial or geographic, since **sympatric** speciation as a result of host transfers or changes in breeding times is better documented for insects than for any other organisms.

The Endopterygota contains the numerically largest orders, namely the Diptera, Lepidoptera, Hymenoptera and Coleoptera (section 1.3; Fig. 1.3). One possible explanation for their success is the occurrence of metamorphosis, allowing the possibility of the larval and adult stage to differ or overlap in phenology, depending upon timing of suitable conditions. Alternative food resources or habitats may be used by a sedentary larva and vagile adult, enhancing species survival by avoidance of intra-specific competition. Furthermore, deleterious conditions for some life-history stages, such as extreme temperatures, low water levels or shortage of food, may be tolerated by a less or non-susceptible life-history stage – for example, a diapausing larva, non-feeding pupa or migratory adult.

No single factor explains the astonishing success of the insects. An early origin and an elevated rate of taxon formation at least during the Cretaceous (Chapter 10), combined with high species persistence through time, leaves us with the great number of species we see today.

7.4 EVOLUTION OF WINGS

Much of the success of insects can be attributed to the wings, found in the numerically dominant pterygotes. Since we cannot observe the origins of flight, and since fossils (though relatively abundant) have not greatly assisted in interpretation, any hypotheses of the origins of flight must be speculative. Several ideas have been promoted, and the area remains controversial.

A long-standing hypothesis attributes the origin of the wings to postulated lobes, derivations from the thoracic terga called **paranota**. These lobes were not articulated and tracheation, innervation, venation and musculature would have been of secondary derivation. The paranotal lobe hypothesis has been displaced in favour of one inferring wing origination from serially repeated, pre-existing, mobile structures of the pleuron. These were most probably the outer appendage (**exite**) and inner appendage (**endite**) of a basal leg segment, the **epicoxa** (Fig. 7.4(a)). Each 'proto-wing' or winglet was formed by the fusion of an exite and endite lobe of the respective ancestral leg. Exites and endites would have had tracheation and articulation. Fossil evidence indicates that articulated winglets were present on all body segments and were likely to have been best developed on the thorax (Fig. 7.4(b)).

The exite–endite theory can be reconciled with another recurring view: that wings were derivations from tracheal gills in an ancestral aquatic 'protopterygote'. Thus winglets of the primitive leg segments are seen as homologous with abdominal gills of aquatic mayfly nymphs, and to be serially homologous with thoracic wings. In function, winglets are postulated to have been used in aquatic juveniles in gas exchange or ventilation or even to assist in swimming, with the terrestrial adult co-opting them for an aerodynamic function.

Hypotheses concerning early wings make a common assumption that winglets originally had a non-flight function, since small winglets could have little or no use in flapping flight. Suggestions for pre-adaptive functions have included any (or all!) of the following: protection of the legs, covers for the spiracles, thermoregulation, sexual display, aid in

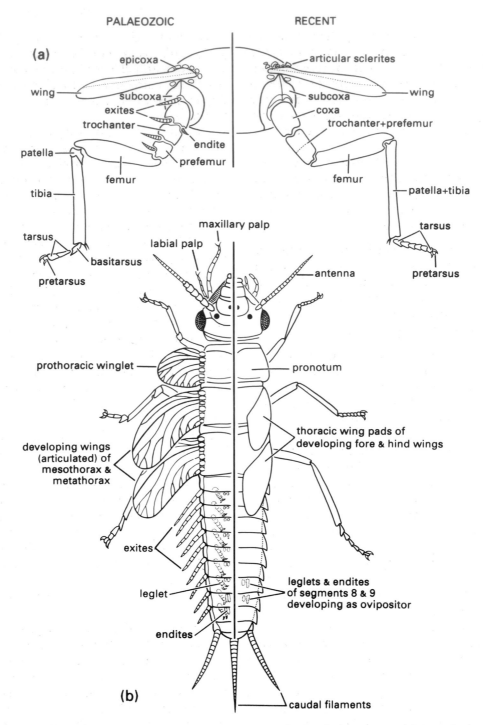

Fig. 7.4 Appendages of hypothetical primitive Palaeozoic (left of each diagram) and Recent (right of each diagram) pterygotes (winged insects): (a) thoracic segment of adult showing generalized condition of appendages; (b) dorsal view of nymphal morphology. (Modified from Kukalová-Peck, 1991, to incorporate ideas of J.W.H. Trueman (unpublished).)

concealment by breaking up the outline, and predator avoidance by extension of escape jump by gliding. Aerodynamic function came only after enlargement. However, aquatic nymphal gills may even have been large enough to give some immediate significant aerial advantage if retained in a terrestrial adult.

The manner in which flight evolved is also highly speculative and contentious but, however winglets originated, they came to assume some aerodynamic function. Three routes to flight have been argued:

- Via floating, in which small insects were assisted in passive dispersal by convection.
- Via paragliding, in which winglets assisted in stable gliding or parachuting from trees and tall vegetation.
- Via running-jump to flying.

The first two models require a fixed, non-articulated wing or an articulated wing held rigidly extended. Articulated winglets and flapping flight can only be incorporated into the run/jump-start scenario of developing flight.

The 'floating' route to flight suffers from the flaw that wings actually hinder passive dispersal, and selection would tend to favour diminution in body size and reduction in the wings with commensurate increase in features such as long hairs. Concerning the third 'running-jump' route, no insect could attain the velocity necessary for flight originating from the ground. Only the scenario of a powered leap to allow limited gliding or flight is at all plausible.

Aerodynamic theory has been applied to the problem, and models have been constructed for wind-tunnel testing. Although a size-constrained and fixed-wing model lacks realism, the following evidence has been produced. An unwinged insect model 1 cm long lacks control in a glide, and even quite small winglets give an immediate advantage by allowing some retarding of velocity. The possession of caudal filaments and/or paired

cerci would give greater gliding stability, particularly when associated with the reduction and eventual loss of posterior abdominal winglets. Additional control over gliding/flight would come with increases in body and winglet size.

Section 7.5.4 shows that there is a basic structural division of the pterygotes into 'Palaeoptera' with movable, non-folding wings and the Neoptera, with complex wing articulation that allows folding of the wings over the body. Some authors have suggested that the two wing-base types are so different that flight must have originated at least twice. However, it can be demonstrated that there is a basic venational pattern common to all pterygotes irrespective of the articulation, implying monophyly (a single origin) of wings but not necessarily of flight. The primitive pterygote wing base apparently involved many articulated sclerites: such a system is seen in fossil palaeopterans and, in a variably modified form, in extant neopterans. In the Ephemeroptera and Odonata these basal sclerites have undergone extensive fusion that prevents the wing from flexing backwards. However, the nature of these fusions, and others that have occurred within the neopteran lineages, indicate that many different pathways have been used, and fusion *per se* does not indicate monophyly of the Palaeoptera. The likelihood that the primitive winglet had complex articulation provides a major criticism of the aerodynamic model above and the supposition of fixed wings renders suspect the conclusions of these unrealistic models.

The traditional proposal for the origin of venation involves tracheated supporting or strengthening ridges on the exite/endite. Alternatively, the veins arose along the courses of haemolymph canals that supplied the winglets, in a manner seen in the gills of some aquatic insects. The basic venational pattern consists of eight veins each arising from a basal blood sinus, named from anterior to posterior: precosta, costa, subcosta, radius, media, cubitus, anal and jugal. Each vein (perhaps excepting the media) branched basally into anterior concave and posterior convex components, with additional dichotomous branching away from the base, and a polygonal pattern of cells. Evolution of the insect wing frequently involved reduction in the number of cells, development of bracing struts (cross-veins), selected increase in division of some veins and reduction in complexity or complete loss of others, and changes in the muscles used for powered flight and in the phases of wing beat (section 3.1.4). Alteration in function has taken place, with protection of one pair of wings by tegmina or elytra in some groups. Increased manoeuvrability has been gained in some other groups by coupling the fore and hind wings as a single unit, and in Diptera by reduction of the metathoracic wings to halteres that function like gyroscopes.

7.5 CLASSIFICATION AND SELECTED DIAGNOSTIC FEATURES OF EXTANT HEXAPODS

The superclass Hexapoda (Protura + Collembola + Diplura + Insecta) consists of all six-legged arthropods. The relationships between these taxa is unclear, although the cladogram of Fig. 7.5 and the classification presented in the sections that follow reflect one current view. Until recently the Collembola, Protura and Diplura were grouped as the 'Entognatha', an assemblage based largely on the similar morphology of the mouthparts that are enclosed in folds of the head, but this condition is now believed to have evolved convergently in these groups. In comparison, the Insecta have exposed mouthparts, or ectognathy. The position of the Protura, in particular, is far from established, since they share some features with myriapods (centipedes, millipedes and their allies). Traditionally, the myriapods have been regarded as the closest relatives of hexapods, but there is some convincing evidence for a sister-group

relationship between the Hexapoda and the Crustacea based on morphological (especially compound eye and nervous system) and molecular data. Such apparent homologies suggest a more recent shared ancestry for hexapods and crustaceans than for hexapods and myriapods. The phylogenetic history of these groups is far from resolved and only further data and more rigorous interpretation of character homologies will provide a clearer picture of hexapod evolution.

The diagnostic features of the Hexapoda include: unique **tagmosis** (specialization of successive body segments that more or less unite to form sections or **tagmata**, namely the head, thorax and abdomen); head composed of pregnathal region (often considered to be 3 segments) and 3 gnathal segments bearing mandibles, maxillae and labium; eyes variously developed, sometimes absent, but primitively with two primary pigment cells; thorax of 3 segments each bearing one pair of legs; each thoracic leg with maximum of 6 segments in extant forms but primitively 11-segmented with up to five exites, a coxal endite and two terminal claws; abdomen primitively with 11 segments plus telson; abdominal limbs, if present, smaller and weaker than those of thorax, primitively on all segments except 10 and with endites and exites.

7.5.1 Class and order Protura (proturans)

Small (<2 mm long), delicate, elongate, mostly unpigmented; lacking eyes and antennae; mouthparts entognathous, consisting of slender mandibles and maxillae, slightly protruding from pleural fold cavity, maxillary and labial palps present; thorax poorly differentiated; 5-segmented legs; 12-segmented abdomen with gonopore between segments 11 and 12, terminal anus; cerci absent. Respiration predominantly cutaneous. Larval development **anamorphic** (segments are added posteriorly during development). Relationships controversial, perhaps more closely related to

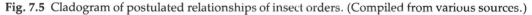

Fig. 7.5 Cladogram of postulated relationships of insect orders. (Compiled from various sources.)

myriapods (centipedes, millipedes and allies) than to any other hexapods. Sometimes grouped with Collembola to form the Ellipura (Parainsecta) but probably unjustified phylogenetically.

7.5.2 Class and order Collembola (springtails)

Minute to small (<10 mm long); soft bodied; often with rudimentary eyes or ocelli; antennae 4–6-segmented; mouthparts entognathous, consisting predominantly of elongate maxillae and mandibles enclosed by pleural folds of head, maxillary and labial palps absent; 4-segmented legs; 6-segmented abdomen with sucker-like ventral tube, retaining hook and furcula (forked jumping organ) on segments 1, 3 and 4, respectively, with gonopore on segment 5, anus on segment 6; cerci absent. Respiration usually cutaneous. Larval development **epimorphic** (segment number constant through development).

7.5.3 Class and order Diplura (diplurans)

Small to medium-sized (2–5 mm, exceptionally to 50 mm long), mostly unpigmented; lacking eyes; long, moniliform (like string of beads) antennae; mouthparts entognathous, with tips of well-developed mandibles and maxillae protruding from pleural fold cavity, maxillary and labial palps reduced; thorax poorly differentiated; 5-segmented legs; 10-segmented abdomen, some segments with small styles and protrusible vesicles; with gonopore between segments 8 and 9, terminal anus; cerci filiform to forceps-like. Tracheal system relatively well developed. Larval development epimorphic.

7.5.4 Class Insecta (true insects)

Minute to large (0.2–300 mm long), very variable in appearance; ocelli primarily present, compound eyes present; ectognathous, with mouthparts exposed, maxillary and labial palps primarily well developed; thorax variably developed in immature stages, distinct in adults with degree of development dependent on presence of wings; more than 5-segmented leg; primitively 11-segmented abdomen, gonopore nearly always on segment 8 in female, 9 in male; cerci primitively present. Respiration predominantly tracheal with pleural spiracles on thorax and abdomen, but variably reduced or absent (e.g. in many immature stages). Larval/nymphal development epimorphic.

Informal grouping 'Apterygota' (Archaeognatha and Thysanura)

Primitively wingless; thoracic segments subequal, unfused; pleura poorly developed; 11-segmented abdomen, some segments with styles and eversible vesicles; spiracles sometimes without muscular closing apparatus; pair of filiform cerci plus a long, multisegmented medial terminal filament (appendix dorsalis) medio-dorsal on tergum of segment 11 (from epiproct); mating indirect, by transfer of spermatophore without genital contact; ametabolous development with ecdysis repeated throughout life. The Archaeognatha (bristletails) and Thysanura (silverfish) formerly were united as a single taxon (based on shared primitive features) but recent interpretations of morphological data place the Thysanura as the sister group of the Pterygota in the subclass Dicondylia, with the Archaeognatha and the fossil Monura as more basal groups (Fig. 7.5).

Infraclass Pterygota

Primarily winged or secondarily apterous; adult thoracic segments usually large with meso- and metathorax variably united to form pterothorax; pleura well developed; 8–11 abdominal segments, without styles or vesicular appendages; spiracles primarily with muscular closing apparatus; median terminal filament absent except in Ephemeroptera;

mating by copulation; metamorphosis hemi-
to holometabolous, with no adult ecdysis,
except for ephemeropteran subimago (*q.v.*).

Informal grouping 'Palaeoptera'

Wings unable to be folded against body at rest
because articulation is via **axillary plates** that
are fused with veins; extant orders typically
have **triadic** veins (paired main veins with
intercalated longitudinal veins of opposite
convexity/concavity to the adjacent main
veins) and a network of cross-veins.

This wing venation and articulation, substantiated by palaeontological studies on
similar features, has suggested that the Odonata and Ephemeroptera form a monophyletic
group, the Palaeoptera, sister to all remaining
extant primarily winged orders, the Neoptera.
However, reassessment of the traditional
morphology of extant basal lineages and
increasing molecular evidence argue against
monophyly of the Palaeoptera. These data
imply that the Ephemeroptera is sister to
Odonata + Neoptera, with Odonata alone as
the sister group to the Neoptera, a view that
gives a higher classification of the Pterygota
into three Divisions:

Division (and order) Ephemeroptera

Diagnosis as for 'Palaeoptera', with a few
unique features, including the non-
functional, strongly reduced adult mouthparts, one axillary plate in wing articulation,
and modified male fore legs for grasping the
female during copulatory flight. The retention
of a subimago is unique. The nymphs (larvae)
(Box 9.1) are aquatic and may have a diagnostic mandible articulation.

Division (and order) Odonata

Diagnosis as for 'Palaeoptera', with many
unique features, including two axillary plates
(humeral and posterior axillary) in wing
articulation and many features associated

with specialized copulatory behaviour (Box
5.3) including possession of secondary copulatory apparatus on ventral segments 2–3 of
the male and the formation of a tandem wheel
during copulation. The immature stages (Box
9.2) are aquatic, possessing a highly modified
prehensile labium.

Division Neoptera

Wings capable of folding back against abdomen when at rest, with articulation through
separate movable sclerites in the wing base;
wing venation with fewer (or lacking completely) triadic veins and mostly lacking
anastomosing cross-veins; larvae in all habitats, including aquatic.

The phylogeny (and hence classification) of
the neopteran orders is still the subject of
debate, mainly concerning (a) the placement
of many extinct orders described only from
fossils of variably adequate preservation, (b)
the position of the Plecoptera, (c) the composition of the orthopteroid and blattoid
groups of orders, and (d) the relationships of
the highly derived Siphonaptera and Strepsiptera.

However, the following summary (see also
cladogram, Fig. 7.5) reflects a consensus
amongst current interpretations. In the absence of a robust phylogeny, ranking of the
component groups is more subjective than
usual and 'informal' ranks abound.

Informal Orthopteroid assemblage
The Dictyoptera has long been recognized as a
robust grouping comprising the Blattodea
(cockroaches), Isoptera (termites) and Mantodea (mantids). Although each order is
distinctive, features of the head skeleton
(perforated tentorium), mouthparts (para-
glossal musculature), digestive system
(toothed proventriculus) and female genitalia
(shortened ovipositor above large subgenital
plate) demonstrate monophyly.

No other monophyletic groupings are as
well supported as the Dictyoptera, but a sister

group relationship between the highly derived Dermaptera (earwigs) and Dictyoptera can be supported. The Orthoptera (crickets, katydids, grasshoppers, locusts, etc.) and Phasmatodea (stick insects or phasmids) may be sister groups, or the phasmids may be derived from within the Orthoptera. The Grylloblattodea (now apterous, but with winged fossils) may be linked to the Orthoptera/Phasmatodea or to the Dictyoptera/ Dermaptera according to which of the relatively few informative characters are stressed.

Informal Plecopteroid assemblage

There has been (and remains) great uncertainty concerning the phylogenetic position of the Plecoptera (stoneflies). Molecular evidence now indicates that similarities to the very specialized Embioptera (webspinners) and Zoraptera may reflect common ancestry. This grouping has been argued to be the sister to all the other Neoptera, and the Plecoptera certainly possess many primitive features. However, whether the plecopteroid assemblage takes this position, or is sister to the orthopteroid assemblage alone, is unresolved.

The remaining Neoptera can be divided readily into two monophyletic groups, given the rank of subdivision, namely the Paraneoptera and Endopterygota (Holometabola).

Subdivision Paraneoptera (= Hemipteroid orders)

This group is defined by derived mouthpart features, including the maxillary lacinia being slender, elongate and separated from the stipes and the post-clypeus being swollen to contain an enlarged cibarium. Further support comes from the reduction in tarsomere number to three or less.

Within the Paraneoptera, the Psocodea comprising Phthiraptera (parasitic lice) and Psocoptera (book lice) is monophyletic. However, although the Phthiraptera are monophyletic, there is strong evidence that the parasitic lice arose from within the Psocoptera, rendering that group paraphyletic. The second group (without formal name), comprises the Thysanoptera (thrips) and Hemiptera (bugs) and forms the sister group of the Psocodea.

Subdivision Endopterygota

Insects with immature (larval) instars very different from adults, lacking true ocelli in larvae; wings and genitalia developing in imaginal discs evaginated at penultimate moult (section 6.2); inactive non-feeding pupa preceding an often active pharate adult. Unique derived features less evident in adults than in immature stages.

Two or three groups are proposed amongst the endopterygotes, of which the strongest is a sister group relationship termed Amphiesmenoptera between the Trichoptera (caddisflies) and Lepidoptera (butterflies and moths). There is a plausible scenario of an ancestral amphiesmenopteran taxon having a larva living in damp soil amongst liverworts and mosses followed by radiation into water (Trichoptera) or into terrestrial phytophagy (Lepidoptera).

A second, postulated relationship is between Diptera (true flies), Siphonaptera (fleas) and Mecoptera (scorpionflies and hangingflies), formally named as Antliophora. Debate remains concerning which group is the closest relative of the Siphonaptera, with the weight of current evidence favouring the Diptera. The phylogenetic position of the Nannochoristidae, a southern hemisphere mecopteran taxon currently treated as a subfamily, has a significant bearing on internal relationships within the Antliophora.

A third grouping is the Neuropterida (or neuropteroid orders), comprising the Neuroptera, Megaloptera and Raphidioptera, orders that have sometimes been included in a single order despite the shortage of unambiguous unifying features.

The position of two endopterygote major

orders, Coleoptera and Hymenoptera, remains to be considered: several positions have been proposed for the Coleoptera. Current evidence derived from female genitalia and ambivalent evidence from eye structure supports a sister group relationship to the Neuropterida. The Hymenoptera may be the sister group of the Antliophora plus Amphiesmenoptera but the presence of many highly derived features of adults, and reductions in immature stages, limits morphological justification for this position.

The presence of many highly derived features makes phylogenetic placement of the parasitic Strepsiptera difficult. An often cited placement as sister group to the Coleoptera may be correct but further evidence, perhaps from molecular sequences, is required to confirm or refute this position.

FURTHER READING

Carpenter, F.M. (1992) *Treatise on Invertebrate Paleontology, Part R, Arthropoda 4, Volumes 3 & 4: Superclass Hexapoda*. The Geological Society of America, Inc., and the University of Kansas, Boulder, Colorado, and Lawrence, Kansas. (This book is a valuable compendium of fossil data but many ideas, especially regarding taxonomic status and classification, are in conflict with accepted interpretations.)

Coope, R. (1991) The study of the 'nearly-fossil' *Antenna*, **15**, 158–63.

Cranston, P.S., Gullan, P.J. and Taylor, R.W. (1991) Principles and practice of systematics, in *The Insects of Australia*, 2nd edn, (CSIRO), Melbourne University Press, Carlton, pp. 109–24.

Forey, P.L., Humphries, C.J., Kitching, I.J., Scotland, R.W., Siebert, D.J. and Williams, D.M. (1992) *Cladistics. A Practical Course in Systematics*, Clarendon Press, Oxford.

Gupta, A.P. (ed.) (1979) *Arthropod Phylogeny*, Van Nostrand Reinhold Company, New York, Cincinnati, Atlanta, Dallas, San Francisco, London, Toronto, Melbourne.

Kristensen, N.P. (1981) Phylogeny of insect orders. *Annual Review of Entomology*, **26**, 135–57.

Kristensen, N.P. (1991) Phylogeny of extant hexapods, in *The Insects of Australia*, 2nd edn, (CSIRO), Melbourne University Press, Carlton, pp. 125–40.

Kukalová-Peck, J. (1983) Origin of the insect wing and wing articulation from the arthropodan leg. *Canadian Journal of Zoology*, **61**, 1618–69.

Kukalová-Peck, J. (1987) New Carboniferous Diplura, Monura, and Thysanura, the hexapod ground plan, and the role of thoracic side lobes in the origin of wings (Insecta). *Canadian Journal of Zoology*, **65**, 2327–45.

Kukalová-Peck, J. (1991) Fossil history and the evolution of hexapod structures, in *The Insects of Australia*, 2nd edn, (CSIRO), Melbourne University Press, Carlton, pp. 141–79.

McClintock Turbeville, J., Pfeifer, D.M., Field, K.G. and Raff, R.A. (1991) The phylogenetic status of arthropods, as inferred from 18s rRNA sequences. *Molecular Biology and Evolution*, **8**, 669–86.

Mitter, C., Farrell, B. and Wiegmann, B. (1988) The phylogenetic study of adaptive zones: has phytophagy promoted insect diversification? *The American Naturalist*, **132**, 107–28.

Moran, N.A. (1989) A 48-million-year-old aphid-host plant association and complex life cycle: biogeographic evidence. *Science*, **245**, 173–5.

Poinar, G.O., Jr (1993) Insects in amber. *Annual Review of Entomology*, **46**, 145–59.

Trueman, J.W.H. (1990) Evolution of insect wings: a limb-exite-plus-endite model. *Canadian Journal of Zoology*, **68**, 1333–5.

Wheeler, W.C., Cartwright, P. and Hayashi, C.Y. (1993) Arthropod phylogeny: a combined approach. *Cladistics*, **9**, 1–39.

Wootton, R.J. (1986) The origin of insect flight: where are we now? *Antenna*, **10**, 82–6.

8
INSECTS OF SOIL, LITTER, CARRION AND DUNG

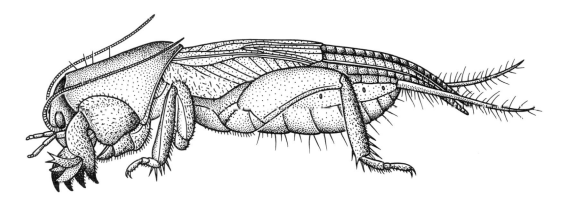

A mole cricket

A profile of a typical soil shows an upper layer of recently derived vegetational material, termed **litter**, overlying more decayed material intergrading with **humus**-enriched organic soils. These organic materials lie above the mineralized soil layers which vary with local geology and climate, such as rainfall and temperature. For subterranean organisms, particle size and soil moisture are important in influencing their micro-distributions.

An integral part of the soil system is the decompositional habitat, comprising decaying wood, leaf litter, carrion and dung. The processes of decay of vegetation and animal matter and return of nutrients to the soil involve many organisms, notably fungi; their hyphae and fruiting bodies provide a medium exploited by many insects. All decompositional substrates have associated faunas that include insects and other hexapods.

This chapter considers the taxonomic range and ecology of soil and decompositional faunas in relation to the differing macrohabitats of soil and decaying vegetation and humus, dead and decaying wood, dung and carrion. The importance of insect–fungal interactions is stressed, and two intimate associations are examined. There is a description of a specialized subterranean habitat – caves – followed by a discussion of some uses of terrestrial hexapods in environmental monitoring. The chapter concludes with five taxonomic boxes that deal with: non-insect hexapods (Collembola, Protura and Diplura); primitively wingless bristletails and silverfish (Archaeognatha and Thysanura); three small hemimetabolous orders (Grylloblattodea, Embioptera and Zoraptera); earwigs (Dermaptera); and cockroaches (Blattodea).

8.1 INSECTS AND LITTER AND SOIL

Litter is fallen vegetative debris, comprising materials such as leaves, twigs, wood, fruit and flowers in various states of decay. The processes that lead to the incorporation of recently fallen vegetation into the humus layer of the soil involve degradation by micro-organisms, such as bacteria, protozoa and fungi. The actions of nematodes, earthworms and terrestrial arthropods, including crustaceans, mites and a range of hexapods (Fig. 8.1), mechanically break down large particles and deposit finer particles as faeces. Acari (mites), termites (Isoptera), ants (Formicidae) and many beetles (Coleoptera) are important arthropods of litter and humus-rich soils. The immature stages of many insects, including beetles, flies (Diptera) and moths (Lepidoptera), may be abundant in litter and soils. In Australian forests and woodlands, the eucalypt leaf litter is consumed by the larvae of many oecophorid moths and some chrysomelid leaf beetles. The soil fauna also includes many of the non-insectan hexapods (Collembola, Protura and Diplura) and the primitively wingless insects, the Archaeognatha and Thysanura. Many Blattodea, Orthoptera and Dermaptera occur only in terrestrial litter – a habitat to which several of the minor orders of insects, the Zoraptera, Embioptera and Grylloblattodea are restricted. Soils that are permanently or regularly waterlogged, such as marshes and riparian (stream marginal) habitats, intergrade into the fully aquatic habitats described in Chapter 9 and have many faunal similarities.

In a profile of soil the transition from the upper, recently fallen litter through the lower well-decomposed litter to the humus-rich soil below may be gradual. Certain arthropods may be confined to a particular layer or depth and show a distinct behaviour and morphology appropriate to the depth in the soil. For example, amongst the Collembola, *Onychurus* lives in deep soil layers and has reduced appendages, is blind and white, and lacks a furcula, the characteristic collembolan springing organ. At intermediate soil depths, *Hypogastrura* has simple eyes, and short appendages with the furcula shorter than half the body length. In contrast, Collembola such as *Orchesella* that live amongst the superficial

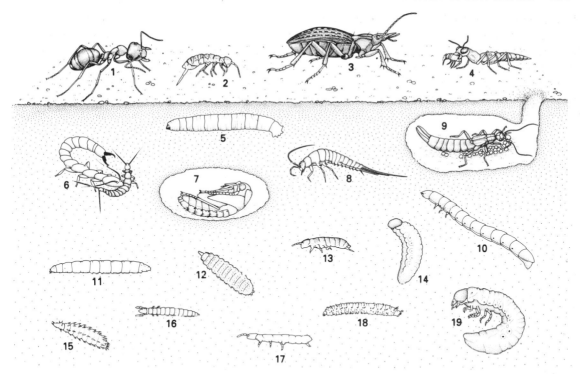

Fig. 8.1 Diagrammatic view of a soil profile showing some typical litter and soil insects and other hexapods. Note that organisms living on the soil surface and in litter have longer legs than those found deeper in the ground. Organisms occurring deep in the soil are usually legless or have reduced legs; they are unpigmented and often blind. The organisms depicted are: (1) worker of a wood ant (Hymenoptera: Formicidae); (2) springtail (Collembola: Isotomidae); (3) ground beetle (Coleoptera: Carabidae); (4) rove beetle (Coleoptera: Staphylinidae) eating a springtail; (5) larva of a crane fly (Diptera: Tipulidae); (6) japygid dipluran (Diplura: Japygidae) attacking a smaller campodeid dipluran; (7) pupa of a ground beetle (Coleoptera: Carabidae); (8) bristletail (Archaeognatha: Machilidae); (9) female earwig (Dermaptera: Labiduridae) tending her eggs; (10) wireworm, larva of a tenebrionid beetle (Coleoptera: Tenebrionidae); (11) larva of a robber fly (Diptera: Asilidae); (12) larva of a soldier fly (Diptera: Stratiomyiidae); (13) springtail (Collembola: Isotomidae); (14) larva of a weevil (Coleoptera: Curculionidae); (15) larva of a muscid fly (Diptera: Muscidae); (16) proturan (Protura: Sinentomidae); (17) springtail (Collembola: Isotomidae); (18) larva of a march fly (Diptera: Bibionidae); (19) larva of a scarab beetle (Coleoptera: Scarabaeidae). (Individual organisms after various sources, especially Eisenbeis and Wichard, 1987.)

leaf litter have larger eyes, longer appendages and an elongate furcula, more than half as long as the body.

A suite of morphological variations can be seen in soil insects. Larvae often have well-developed legs to permit active movement through the soil, and pupae frequently have spinose transverse bands that assist movement to the soil surface for eclosion. Many adult soil-dwelling insects have reduced eyes and their wings are either protected by hardened fore wings, reduced (brachypterous), lost altogether (apterous) or, as in the reproductives of ants and termites, shed after the dispersal flight (deciduous, or caducous). Flightlessness (either primary or secondary) in ground-dwelling organisms may be countered by development of jumping as a means of evading predation: the collembolan furcula is a spring mechanism and the alticine

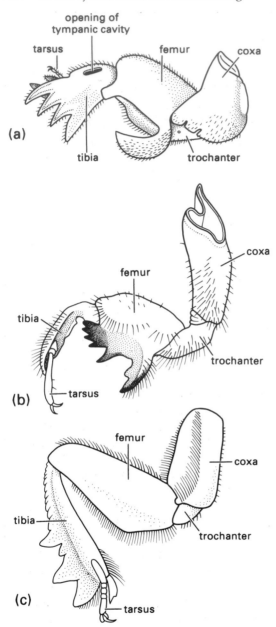

Fig. 8.2 Fossorial fore legs of: (a) a mole cricket of *Gryllotalpa* (Orthoptera: Gryllotalpidae); (b) a nymphal periodical cicada of *Magicicada* (Hemiptera: Cicadidae); (c) a scarab beetle of *Canthon* (Coleoptera: Scarabaeidae). ((a) After Frost, 1959; (b) after Snodgrass, 1967; (c) after Richards and Davies, 1977.)

Coleoptera ('flea-beetles') and terrestrial Orthoptera can jump to safety. However, jumping is of little value in subterranean organisms. In these insects, the fore legs may be developed as **fossorial** limbs, modified for digging (Fig. 8.2), as seen in some groups that construct tunnels, such as mole crickets (as depicted in the vignette of this chapter), immature cicadas and many beetles.

The distribution of subterranean insects shows seasonal changes. The constant temperatures at greater soil depths are attractive in winter as a means of avoiding low temperatures above ground. The level of water in the soil is important in governing both vertical and horizontal distributions. Frequently, larvae of subterranean insects that live in moist soils will seek drier sites for pupation, perhaps to reduce the risks of fungal disease during the immobile pupal stage. The subterranean nests of ants are usually located in drier areas, or the nest entrance is elevated above the soil surface to prevent flooding during rain, or the whole nest may be elevated to avoid excess ground moisture. Location and design of the nests of ants and termites is very important to the regulation of both humidity and temperature because, unlike social wasps and bees, they cannot ventilate their nests by fanning, although they can migrate within nests or, in some species, between them. The passive regulation of the internal nest environment is exemplified by termites of *Amitermes* (Fig. 11.8) and *Macrotermes* (Fig. 11.9) which maintain an internal environment suitable for the growth of particular fungi that serve as food (section 11.2.4).

Many soil-dwelling hexapods derive their nutrition from ingesting large volumes of soil containing dead and decaying vegetable and animal debris and associated microorganisms. These bulk-feeders, known as **saprophages** or **detritivores**, include hexapods such as some Collembola, beetle larvae and certain termites (Isoptera: Termitinae, including *Termes* and relatives) that have not been demonstrated to possess symbiotic gut

BOX 8.1
GROUND PEARLS

1ST-INSTAR NYMPH   **2ND-INSTAR NYMPH (CYST STAGE)**   **ADULT FEMALE**

In some parts of Africa, the encysted nymphs ('ground pearls') of certain subterranean scale insects are sometimes made into necklaces by the local people. These nymphal insects have few cuticular features, except for their spiracles and sucking mouthparts. They secrete a transparent or opaque, glassy or pearly covering that encloses them, forming spherical to ovoid 'cysts' of greatest dimension 1–8 mm, depending on species. Ground pearls belong to several genera of Margarodinae (Hemiptera: Margarodidae) including *Eumargarodes*, *Margarodes*, *Neomargarodes*, *Porphyrophora* and *Promargarodes*. They occur worldwide in soils among the roots of grasses, especially sugarcane, and grape vines (*Vitis vinifera*). They may be abundant and their nymphal feeding can cause loss of plant vigour and death; in lawns, feeding results in brown patches of dead grass. In South Africa they are serious vineyard pests; in Australia different species reduce sugarcane yield; whereas one species is a grass pest in the south-eastern United States.

Plant damage is caused mostly by the female insects because many species are parthenogenetic, or at least males have never been found, and when males are present they are smaller than the females. There are three female instars (as illustrated here for *Margarodes* (= *Sphaeraspis*) *capensis*, after De Klerk *et al.*, 1982): the first-instar nymph disperses in the soil seeking a feeding site on roots, where it moults to the second instar or cyst stage; the adult female emerges from the cyst from spring to autumn (depending on species) and, in species with males, comes to the soil surface where mating occurs. The female then buries back into the soil, digging with its large fossorial fore legs. The fore leg coxa is broad, the femur is massive and the tarsus is fused with the strongly sclerotized claw. In parthenogenetic species, females may never leave the soil. Adult females have no mouthparts and do not feed; in the soil, they secrete a waxy mass of white filaments – an ovisac – which surrounds their several hundred eggs.

BOX 8.1—*continued*

Although ground pearls can feed via their thread-like stylets, which protrude from the cyst, second-instar nymphs of most species are capable of prolonged dormancy (up to 17 years has been reported for one species) and can be kept dry in the laboratory for one to several years and still be capable of 'hatching' as adults. This long life and ability to rest dormant in the soil together with resistance to desiccation, means that they are difficult to eradicate from infested fields and even crop rotations do not eliminate them effectively. Furthermore, the protection afforded by the cyst wall and the subterranean existence makes insecticidal control largely inappropriate. Many of these curious pestiferous insects are probably African and South American in origin and, prior to quarantine restrictions, may have been transported within and between countries as cysts in soil or on rootstocks.

---

protozoa yet appear to be able to digest cellulose from the humus layers of the soil. Copious excreta is produced, and these organisms clearly play a significant role in structuring soils.

For arthropods that consume humic soils, the sub-soil parts of plants (the roots) will be encountered frequently. The fine parts of roots often have particular associations with fungal mycorrhizae and rhizobacteria, forming a zone called the **rhizosphere**. Bacterial and fungal densities are an order of magnitude higher in soil close to the rhizosphere compared with soil distant from roots, and microarthropod densities are correspondingly higher close to the rhizosphere. The selective grazing of Collembola, for example, can curtail growth of fungi that are pathogenic to plants and their movements aid in the transport of beneficial fungi and bacteria to the rhizosphere. Furthermore, interactions between microarthropods and fungi in the rhizosphere and elsewhere may aid in mineralization of nitrogen and phosphates, making these elements available to plants, but further experimental evidence is required to quantify these beneficial roles.

It is unlikely that roots of plants are selectively avoided by soil-feeding insects. Where there are high densities of fly larvae

that eat soil in pastures, such as Tipulidae ('leatherjackets') and Bibionidae ('March flies'), roots are damaged by their activities. Within the soil, many subterranean beetle larvae such as wireworms (Elateridae), false wireworms (Tenebrionidae), weevil grubs (Curculionidae) and scarab grubs (Scarabaeidae) and a few moth larvae attack the roots of plants and may cause economic damage in managed pastures. Sap-sucking insects on the roots of plants include aphids, mealybugs and scale insects (Box 8.1), which can cause plant injury and death.

8.2 INSECTS AND DECAYING WOOD

Tree deaths may involve insects that play a role in the transmission of pathogenic fungi between trees. Thus wood wasps of the genera *Sirex* and *Urocercus* (Hymenoptera: Siricidae) carry *Amylostereum* fungal spores in invaginated intersegmental sacs connected to the ovipositor. During oviposition spores and mucus are injected into the sapwood of trees, notably *Pinus* species, causing mycelial infection. The infestation causes locally drier conditions around the xylem, which allows optimal development of larval *Sirex*. In Australia and New Zealand the fungal disease can cause death of fire-damaged trees or those stressed by drought conditions. The role of *Scolytus* beetles (Coleoptera: Scolytidae) in the spread of Dutch Elm Disease is discussed in section 4.3.3(a) and there are several other insect-borne fungal diseases that are transmitted to live trees and may result in tree mortality. The continued decay of these trees, and those that die of natural causes, often involves further interactions between insects and fungi.

The ambrosia beetles (Platypodidae and some Scolytidae) are involved in a notable association between beetles, fungus and dead wood. Adult beetles excavate tunnels, predominantly in dead wood though some attack live wood. Mines are made in the phloem, wood, twigs or woody fruits, which the beetles infect with wood-inhabiting ectosymbiotic 'ambrosia' fungi. The fungi, which come from a wide taxonomic range, break down wood and make it more nutritious for the beetles. Both larvae and adults feed on the conditioned wood and directly on the extremely nutritious fungi. The cultivation of fungi on nutritive-poor, cellulose-rich woody substrates is particularly well developed in leafcutter ants and certain termites (section 8.5).

Fallen, rotten timber provides a valuable resource for a wide variety of detritivorous insects if they can overcome the problems of living on a substrate rich in cellulose and deficient in vitamins and sterol. Termites are able to live entirely on this diet either through the use of gut symbionts (section 11.2.4), or by external cellulose digestion, or perhaps through the possession of cellulase enzymes in their digestive systems. Certain cockroaches (Blattidae) likewise utilize gut symbionts to aid the digestion of cellulose from rotting wood or from fallen bark, leaves and twigs. Other **xylophagous** (wood-eating) strategies of insects include very long life cycles with slow development and probably also the use of xylophagous micro-organisms and fungi as food.

8.3 INSECTS AND DUNG

The excreta or dung produced by vertebrates may be a rich source of nutrients. In the grasslands and rangelands of North America and Africa, large ungulates produce substantial volumes of fibrous and nitrogen-rich dung containing many bacteria and protozoa. Insect **coprophages** (dung-feeding organisms) utilize this resource in a number of ways. Certain higher flies – such as the Scathophagidae, Muscidae (notably the worldwide house fly *Musca domestica*, Australian *M. vetustissima* and widespread tropical buffalo fly *Haematobia irritans*), Faniidae and Calliphoridae – oviposit or larviposit into freshly laid dung. Development can be completed before the medium becomes too desiccated. Within the dung medium, predatory fly larvae (notably other species of Muscidae) can seriously limit survival of coprophages. However, in the absence of predators or disturbance of the dung, nuisance-level populations of flies can be generated from larvae developing in dung in pastures.

The insects primarily responsible for disturbing dung, and thereby limiting fly-breeding in the medium, are scarab dung beetles, belonging to the family Scarabaeidae. Not all larvae of scarabs use dung: some ingest general soil organic matter, whilst others are herbivorous on plant roots. However, many are coprophages. In Africa, where there are many large herbivores that produce large volumes of dung, several thousand species of scarabs are coprophagous, showing a wide variety of strategies to obtain the resource. Many are able to detect dung as it is produced by the herbivore, and from the time it falls to the ground invasion is very rapid and many individuals may arrive: for example, up to many thousands for a fresh elephant dropping. Most dung beetles excavate networks of tunnels immediately beneath or beside the pad, and pull down pellets of dung (Fig. 8.3). Other beetles excise a chunk of dung and move it some distance to a dug-out chamber,

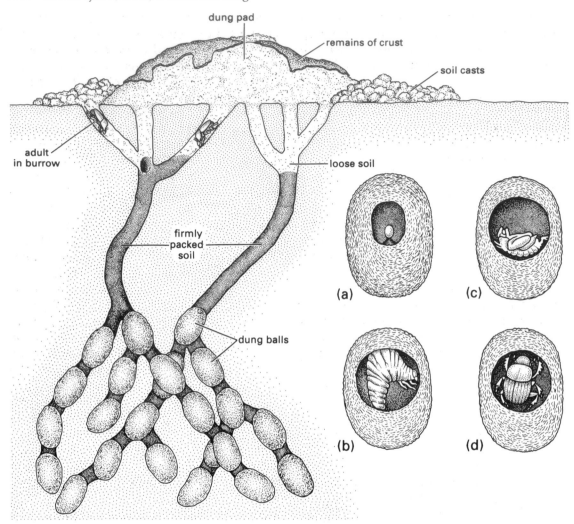

dung pad

remains of crust

soil casts

adult
in burrow

loose soil

firmly
packed
soil

dung balls

(a)   (c)

(b)   (d)

Fig. 8.3 A pair of dung beetles of *Onthophagus gazella* (Coleoptera: Scarabaeidae) filling in the tunnels that they have excavated below a dung pad. The inset shows an individual dung ball within which beetle development takes place: (a) egg; (b) larva that feeds on the dung; (c) pupa; (d) adult just prior to emergence. (After Waterhouse, 1974.)

often within a network of tunnels. This movement from pad to nest chamber may occur either by head-butting an unformed lump or by rolling moulded spherical balls over the ground to the burial site. The female lays eggs into the buried pellets and the larvae develop within the faecal food ball, eating fine and coarse particles. The adult scarabs may also feed on dung but only on the fluids and finest particulate matter. Some scarabs are generalists and utilize virtually any dung encountered, while others specialize according to texture, wetness, pad size, fibre content, geographic area and climate, so that a range of scarab activities ensures that all dung is buried within a few days at most.

In Australia, a continent in which native large herbivores are absent, native dung beetles are unable to handle the volume and texture of dung produced by introduced domestic cattle, horses and sheep. As a result, dung lies around in pastures for prolonged periods, reducing the quality of pasture and allowing the development of prodigious numbers of nuisance flies. A programme to introduce alien dung beetles from Africa and Mediterranean Europe has been successful in accelerating dung burial in many regions.

8.4 INSECT–CARRION INTERACTIONS

In places where ants are important components of the fauna, the corpses of invertebrates are discovered and removed rapidly by ants, which scavenge widely and efficiently. In contrast, vertebrate corpses (carrion) support a wide diversity of organisms, many of which are insects. These form a **succession** – a non-seasonal, directional and continuous sequential pattern of populations of species colonizing and being eliminated as carrion decay progresses. The nature and timing of the succession depends upon the size of the corpse, seasonal and ambient climatic conditions, and the surrounding non-biological (edaphic) environment, such as soil-type. The organisms involved in the succession vary according to whether they are upon or within the carrion, in the substrate immediately below the corpse or in the soil at an intermediate distance below or away from the corpse. Furthermore, each succession will comprise different species in different geographic areas, even in places with similar climates. This is because few species are very widespread in distribution, and each biogeographic area will have its own specialist carrion faunas. However, the broad taxonomic categories of cadaver specialists are similar worldwide.

The first stage in carrion decomposition, **initial decay**, involves only micro-organisms already present in the body, but within a few days the second stage called **putrefaction** begins. About two weeks later, amidst strong odours of decay, the third **black putrefaction** stage begins, followed by a fourth, **butyric fermentation** stage, in which the cheesy odour of butyric acid is present. This terminates in an almost dry carcass; and in the fifth stage slow **dry decay** completes the process, leaving only bones.

The typical sequence of corpse **necrophages**, saprophages and their parasites is often referred to as following 'waves' of colonization. The first wave involves certain blow flies (Diptera: Calliphoridae) and house flies (Muscidae) that arrive within hours or a few days at most. The second wave is of sarcophagids (Diptera) and additional muscids and calliphorids that follow shortly thereafter, as the corpse develops an odour. All these flies either lay eggs or larviposit on the corpse. The principal predators on the insects of the corpse fauna are staphylinid, silphid and histerid beetles, and hymenopteran parasitoids may be entomophagous on all the above hosts. At this stage blow-fly activity ceases as their larvae leave the corpse and pupate in the ground. When the fat of the corpse turns rancid, a third wave of species enter this modified substrate, notably more dipterans, such as certain Phoridae, Drosophilidae, and *Eristalis* rat-tailed maggots (Syrphidae) in the liquid parts. As the corpse becomes butyric, a fourth wave of cheese-skippers (Diptera: Piophilidae) and related flies use the body. A fifth wave occurs as the ammonia-smelling carrion dries out, and adult and larval Dermestidae and Cleridae (Coleoptera) become abundant, feeding on keratin. In the final stages of dry decay, some tineid larvae ('clothes moths') feed on any remnant hair.

Immediately beneath the corpse, larvae and adults of the beetle families Staphylinidae, Histeridae and Dermestidae are abundant during the putrefaction stage. However, the normal, soil-inhabiting groups are absent during the carrion phase, and only slowly return as the corpse enters late decay. The rather predictable sequence of colonization and extinction of carrion insects allows forensic entomologists to estimate the age of a corpse, which can have medico-legal implications in homicide investigations (section 14.6).

8.5 INSECT–FUNGAL INTERACTIONS

8.5.1 Fungivorous insects

Fungi and, to a lesser extent, slime moulds are eaten by many insects, termed **fungivores** or **mycophages**, which belong to a range of orders. Amongst insects that use fungal resources, Collembola, larval and adult Coleoptera and larval Diptera are numerous. Two feeding strategies can be identified: **microphages** gather small particles such as spores and hyphal fragments or use more liquid media, whereas **macrophages** use the fungal material of fruiting bodies, which must be torn apart with strong mandibles. The relationship between fungivores and the specificity of their fungus feeding varies. Insects that develop as larvae in the fruiting bodies of large fungi are often obligate fungivores, and may even be restricted to a narrow range of fungi, whereas insects that enter such fungi late in development or during actual decomposition are more likely to be saprophages or generalists than specialist mycophages. Longer-lasting macrofungi such as the pored mushrooms, Polyporaceae, have a higher proportion of mono- or oligophagous associates than ephemeral and patchily distributed mushrooms such as the gilled mushrooms, Agaricales.

Smaller and more cryptic fungal food resources are also used by insects but the associations tend to be less well studied. Yeasts are naturally abundant on live and fallen fruits and leaves, and **fructivores** (fruit-eaters) such as larvae of certain nitidulid beetles and drosophilid fruit flies are known to seek and eat yeasts. Apparently fungivorous drosophilids that live in decomposing fruiting bodies of fungi also use yeasts, and specialization on particular fungi may reflect variations in preferences for particular yeasts. The fungal component of lichens is probably used by grazing larval lepidopterans and adult plecopterans.

Amongst the Diptera that utilize fungal fruiting bodies, the Mycetophilidae (fungus-gnats) are diverse and speciose, and many appear to have oligophagous relationships with fungi from amongst a wide range used by the family. The use by insects of subterranean fungal bodies in the form of mycorrhizae and hyphae within the soil is poorly known. The phylogenetic relationships of the Sciaridae (Diptera) to the mycetophilid fungus-gnats and evidence from commercial mushroom farms all suggest that sciarid larvae normally eat fungal mycelia. Other dipteran larvae, such as certain phorids and cecidomyiids, feed on commercial mushroom mycelia and associated micro-organisms, and may also use this resource in nature.

8.5.2 Fungus farming by leafcutter ants

The subterranean ant nests of the genus *Atta* and the rather smaller colonies of *Acromyrmex* are amongst the major earthen constructions in neotropical rainforest. Calculations suggest that the largest nests of *Atta* species involve excavation of some 40 tonnes of soil. Both these genera are members of a tribe of myrmecine ants, the Attini, in which the larvae have an obligate dependence on symbiotic fungi for food. Other genera of Attini have monomorphic workers (of a single morphology) and cultivate fungi on dead vegetable matter, insect faeces such as caterpillar 'frass', flowers and fruit. In contrast, *Atta* and *Acromyrmex*, the more derived genera of Attini, have polymorphic workers of several different kinds or castes (section 11.2) that exhibit an elaborate range of behaviours involving cutting living plant tissues, hence the name 'leafcutter ants'. In *Atta*, the largest worker ants excise sections of live vegetation with their mandibles (Fig. 8.4(a)) and transport the pieces to the nest (Fig. 8.4(b)). During

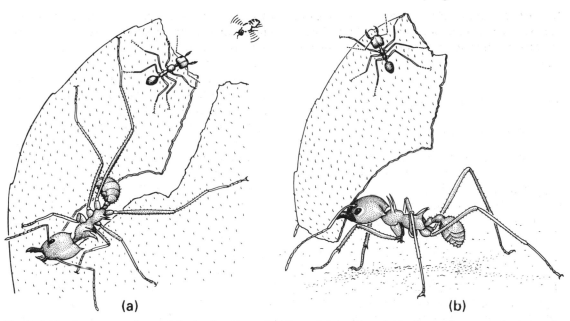

Fig. 8.4 The fungus gardens of the leaf-cutter ant, *Atta cephalotes* (Formicidae), require a constant supply of leaves: (a) a medium-sized worker, called a media, cuts a leaf with its serrated mandibles while a minor worker guards the media from a parasitic phorid fly (*Apocephalus*) that lays its eggs on living ants; (b) a guarding minor hitchhikes on a leaf fragment carried by a media. (After Eibl-Eibesfeldt and Eibl-Eibesfeldt, 1967.)

these processes the working ant has its mandibles full, and may be the target of attack by a particular parasitic phorid fly (illustrated in the top right of Fig. 8.4(a)). The smallest worker is recruited as a defender and is carried on the leaf fragment.

When the material reaches the nest, other individuals lick any waxy cuticle from the leaves and macerate the plant tissue with their mandibles. The mash is then inoculated with a faecal cocktail of enzymes from the hindgut. This initiates digestion of the fresh plant material, which acts as an incubation medium for a fungus, *Attamyces bromatificus*, known only from these 'fungus gardens' of leafcutter ants. Another specialized group of workers tends the gardens by inoculating new substrate with fungal hyphae and removing other species of undesirable fungus in order to maintain a monoculture. Control of alien fungus and bacteria is facilitated by pH

regulation (4.5–5.0) and antibiotics produced by the ants and probably also by the *Attamyces*. In darkness, and at optimal humidity and a temperature close to 25 °C, the cultivated fungal mycelium produces nutritive hyphal bodies called **gongylidia**. These are not sporophores, and appear to have no function other than to provide food for ants in a mutualistic relationship. Gongylidia are easily manipulated by the ants, providing food for adults, and are the exclusive food eaten by larval attine ants. Digestion of fungi requires specialized enzymes that include chitinases produced by ants from their labial glands.

Leafcutter ants dominate the ecosystems in which they occur, with some grassland *Atta* species consuming as much vegetation per hectare as domestic cattle and certain rainforest species estimated to cause up to 80% of all leaf damage and to consume up to 17% of all leaf production. The system is an effective converter of plant cellulose to usable carbohydrate, with at least 45% of the original cellulose of fresh leaves converted by the time the spent substrate is ejected as refuse from the fungus garden. However, fungal gongylidia contribute only a modest fraction of the metabolic energy of the ants, because about 95% of the respiratory requirements of the colony is provided by adults feeding on plant sap.

Leafcutter ants may be termed highly polyphagous, since studies have shown them to utilize between 50% and 70% of all neotropical rainforest plant species. However, since the adults feed on the sap of fewer species, and the larvae are monophagous on fungus, the term polyphagy may be incorrect. The key to the relationship is the ability of the worker ants to harvest from a wide variety of sources, and the ability of the cultivated fungus to grow on a wide range of hosts. Coarse texture and latex production by leaves can discourage attines, and chemical defences may play a role in deterrence. However, leafcutter ants have adopted a strategy that evades plant defensive chemicals that act on the digestive system; they use the the fungus to digest the plant tissue. The ants and fungus co-operate to break down plant defences, with the ants removing protective leaf waxes that deter fungi, and in turn producing carbohydrates from cellulose indigestible to the ants.

8.5.3 Fungus cultivation by termites

The terrestrial microfauna of tropical savannas (grasslands and open woodlands) and some forests of the Afrotropical and Oriental (Indomalayan) zoogeographic regions, can be dominated by a single subfamily of Termitidae, the Macrotermitinae. These termites may form conspicuous above-ground mounds up to 9 m high but more often their nests consist of huge underground structures. Abundance, density and production of macrotermitines may be very high and, with estimates of a live biomass of $10 \, g/m^2$, termites consume more than 25% of all terrestrial litter (wood, grass and leaf) produced annually in some west African savannas.

The litter-derived food resources are ingested, but not digested, by the termites: the food is passed rapidly though the gut and, upon defecation, the undigested faeces are added to comb-like structures within the nest. The combs may be located within many small subterranean chambers or one large central hive or brood chamber. Upon these combs of faeces, a *Termitomyces* fungus develops. The fungi are restricted to Macrotermitinae nests, or occur within the bodies of termites. The combs are constantly replenished and older parts eaten, on a cycle of between five and eight weeks. Fungus action on the termite faecal substrate raises the nitrogen content of the substrate from about 0.3% until in the asexual stages of *Termitomyces* it may reach 8%. These asexual spores (mycotêtes) are eaten by the termites, as well as the nutrient-enriched older comb. Although some species of *Termitomyces* have no sexual stage as is the case for *Attamyces* seen above, others develop above-ground basidiocarps (fruiting bodies, 'mushrooms') at a time that coincides with colony-founding forays of termites from the nest. A new termite colony is inoculated with fungus by means of asexual or sexual spores transferred in the gut of the founder termite(s).

Termitomyces lives as a monoculture on termite-attended combs, but if the termites are removed experimentally or a termite colony dies out, or if the comb is extracted from the nest, many other fungi invade the comb and *Termitomyces* dies. Termite saliva has some antibiotic properties but there is little evidence for *Termitomyces* being able to reduce local competition from other fungi. It seems that *Termitomyces* is favoured in the fungal comb by the remarkably constant microclimate at the comb of a temperature of 30 °C and scarcely varying humidity together with an acid pH of

4.1–4.6. The heat generated by fungal metabolism is regulated appropriately via a complex circulation of air through the passageways of the nest, as illustrated for the above-ground nest of the African *Macrotermes natalensis* in Fig. 11.9.

The origin of the mutualistic relationship between termite and fungus does not seem to derive from joint attack on plant defences, in contrast to the ant–fungus interaction seen in section 8.5.2. Termites are closely associated with fungi, and fungus-infested rotting wood is likely to have been a primitive food preference. The adoption of internal gut symbionts to aid in digestion of lignin and cellulose occurred early in termite evolution, but this involves intake and prolonged retention of large amounts of low nutrient food requiring a substantial gut symbiont fauna. Pressure on the gut is reduced in more derived termites, the Macrotermitinae, by shifting some of the digestion to *Termitomyces* outside the gut. The termite digests complex substances like pectins, chitins and some cellulose, with the aid of gut bacteria. However, the fungus facilitates conversion of indigestible compounds to more nutritious products, allowing the successful utilization of the abundant resource of dead vegetation.

8.6 CAVERNICOLOUS INSECTS

Caves are extensions of the subterranean environment, resembling deep soil habitats in the lack of light and the uniform temperature, but differing in the scarcity of food. Food sources in shallow caves include roots of terrestrial plants, but in deeper caves there is no plant material other than that originating from any stream-derived debris. In many caves nutrient supplies come from fungi and the faeces (guano) of bats and certain cave-dwelling birds, such as swiftlets in the Orient.

Cavernicolous (cave-dwelling) insects include those that seek refuge from adverse external environmental conditions – such as moths and adult flies, including mosquitoes,

that hibernate to avoid winter cold, or aestivate to avoid summer heat and desiccation. **Troglobiont** or **troglobite** insects are restricted to caves, and are often phylogenetically related to soil-dwelling ones. Troglobites are dominated by Collembola (especially the family Entomobryidae); other important groups include the Diplura (especially the family Campodeidae), orthopteroids (including cave crickets, Raphidophoridae) and beetles (chiefly carabids but including fungivorous silphids).

Dermaptera and Blattodea may be abundant in tropical caves, where they are active in guano deposits. In south-east Asian caves one troglobite earwig is ectoparasitic on roosting bats. Associated with cavernicolous vertebrates there are many more conventional ectoparasites such as hippoboscid, nycteribiid and streblid flies, fleas and lice.

8.7 ENVIRONMENTAL MONITORING USING GROUND-DWELLING HEXAPODS

Human activities such as agriculture, forestry and pastoralism have simplified many terrestrial ecosystems. Attempts to quantify the effects of such practices – for the purposes of conservation assessment, classification of land-types and monitoring of impacts – have tended to be phytosociological, emphasizing the use of vegetational mapping data. More recently, data on vertebrate distributions and communities have been incorporated into surveys for conservation purposes.

If the estimates for arthropod diversity given in section 1.3 are correct, then it can be argued that to a close approximation all species are insects. It is rare for data derived from this most diverse group to be available routinely in conservation and monitoring and there are several reasons for this neglect. Firstly, when 'flag-ship' species elicit public reaction to a conservation issue, such as loss of a particular habitat, these organisms are predominantly furry mammals, such as pandas and koalas. Sometimes they are birds,

BOX 8.2
NON-INSECT HEXAPODS (COLLEMBOLA, PROTURA AND DIPLURA)

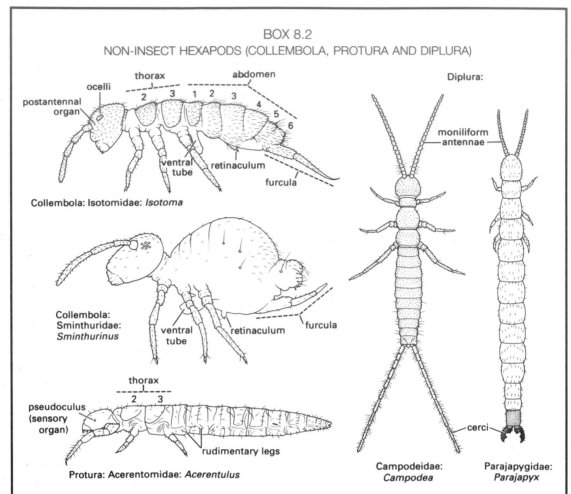

Collembola: Isotomidae: *Isotoma*

Collembola: Sminthuridae: *Sminthurinus*

Protura: Acerentomidae: *Acerentulus*

Campodeidae: *Campodea*

Parajapygidae: *Parajapyx*

The Collembola, Protura and Diplura were once united as the 'Entognatha', based on a similar mouthpart morphology in which mandibles and maxillae are enclosed in folds of the head, except when everted for feeding. Although this condition is probably convergent (section 7.5), it is convenient to treat these classes together here. All have indirect fertilization – males deposit stalked spermatophores which are picked up from the substrate by unattended females. For diagnostic characteristics of the three classes, see sections 7.5.1–3.

Protura (proturans)

Non-insect hexapods, with some 175 species in eight families. Small (< 2 mm long) to very small, delicate, elongate, mostly unpigmented, with fusiform body; head shape conical; thorax poorly differentiated from abdomen; eyes and antennae lacking; mouthparts entognathous, consisting of slender mandibles and maxillae, slightly protruding from pleural fold cavity, maxillary and labial palps present; thorax weakly developed; legs 5-segmented, anterior legs held forward (as shown here for *Acerentulus*, after Nosek, 1973), fulfilling

BOX 8.2—*continued*

antennal sensory function; adult abdomen 12-segmented with gonopore between segments 11 and 12, terminal anus; cerci absent. Immature development **anamorphic** (with segments added posteriorly during development). Proturans are cryptic, found exclusively in soil, moss and leaf litter, but their biology is little known. They are often included in the Ellipura as the sister group to Collembola, but evidence suggests that phylogenetically they are distant to the Collembola, and may be far from any other hexapod.

Collembola (springtails)

Non-insect hexapods, with over 8 000 species in up to 22 recognized families. Small (usually 2–3 mm, up to 12 mm), soft-bodied, with body shape varying from globular to elongate (as illustrated here for *Isotoma* and *Sminthurinus*, after Fjellberg, 1980), pale or often characteristically pigmented grey, blue or black; eyes and/or ocelli often poorly developed or absent; antennae 4- to 6-segmented; entognathous mouthparts comprising elongate maxillae and mandibles enclosed by pleural folds of head, maxillary and labial palps absent; legs 4-segmented; abdomen 6-segmented, with a sucker-like **ventral tube** (the **collophore**), retaining hook and **furcula** (forked jumping organ, usually 3-segmented) on segments 1, 3, 4, respectively, with gonopore on segment 5, anus on segment 6; cerci absent. Amongst hexapods, collembolan eggs uniquely are **microlethical** (lacking large yolk reserves) and **holoblastic** (with complete cleavage). Immature instars similar to adults, developing **epimorphically** (with a constant segment number); maturity attained after five moults, moulting continues for life. Springtails are most abundant in moist litter, where they are major consumers of decaying vegetation, but they also occur in caves, in fungi, commensal with ants and termites, on still water surfaces and in the intertidal zone.

Diplura (diplurans)

Non-insect hexapods, with some 800 species in nine families. Small to medium-sized (2–5 mm, exceptionally to 50 mm), mostly unpigmented and weakly sclerotized; eyes lacking; antennae long, moniliform, multisegmented; mouthparts entognathous, with mandibles and maxillae well developed, tips visible protruding from pleural fold cavity, maxillary and labial palps reduced; legs 5-segmented; thorax little differentiated from abdomen; abdomen 10-segmented, some segments with small styles and protrusible vesicles; gonopore between segments 8 and 9, terminal anus; cerci filiform (as illustrated here for *Campodea*, after Lubbock, 1873) to forceps-like (as in *Parajapyx*, shown here after Womersley, 1939). Immature development epimorphic, with moulting continuing through life; some species gregarious, and females of certain species tend eggs and young. Some diplurans feed on live and decayed vegetation, and japygids are predators.

such as condors, but rarely are they insects. Insects, excepting perhaps some butterflies, seem to lack the necessary charisma in the public perception.

Secondly, invertebrates are difficult to sample in a comparable manner within and between sites. Abundance and diversity fluctuate on a relatively short time scale, in

BOX 8.3
ARCHAEOGNATHA (BRISTLETAILS) AND THYSANURA (SILVERFISH)

The Archaeognatha and Thysanura represent the surviving remnants of a wider radiation of primitively flightless insects. The two apterygote orders are superficially similar but differ in pleural structures and fundamentally in their mouthparts. Thoracic segments subequal and unfused, with poorly developed pleura; abdomen 11-segmented, with styles and often protrusible vesicles on some segments; long, multisegmented, caudal appendix dorsalis mediodorsal on tergum of segment 11, forming an epiproct extension, lying between paired cerci and dorsal to genitalia; in female genitalia, gonapophyses of segments 8 and 9 form an ovipositor; fertilization indirect, by transfer of spermatophore without genital contact; with moulting continuing for life.

Archaeognatha (bristletails)

Primitively wingless insects, with some 350 species in two extant families. Moderate size, elongate and cylindrical; head hypognathous; compound eyes large and dorsally in contact, three ocelli present; antennae multisegmented; mouthparts partially retracted into head, mandibles elongate, **monocondylar** (single-articulated), maxillary palps elongate, 7-segmented; thorax humped; legs with large coxae each bearing style and tarsi 2–3 segmented; abdomen continues contour of thorax, segments 2–9 with ventral muscle-containing styles (representing limbs), segments 1–7 with one or two pairs of protrusible vesicles medial to styles (fully developed only in mature individuals). Paired multisegmented cerci shorter than median

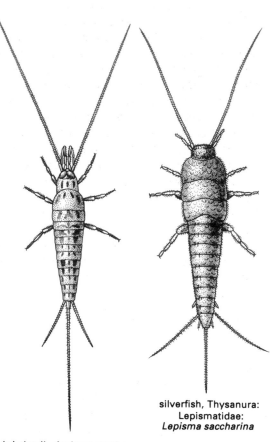

silverfish, Thysanura:
Lepismatidae:
Lepisma saccharina

bristletail, Archaeognatha:
Machilidae: *Petrobius maritima*

caudal appendage (as shown here for *Petrobius maritima*, after Lubbock, 1873). Fertilization indirect, with spermatophores attached to silken lines produced from male gonapophysis. Bristletails are often active nocturnally, feeding on litter, detritus, algae, lichens and mosses, and sheltering beneath bark or in litter during the day. Bristletails can run fast and jump using the arched thorax and flexed abdomen to spring considerable distances.

BOX 8.3—*continued*

Thysanura (Zygentoma; silverfish)

Primitively wingless insects, with some 370 species in four extant families. Moderate size, dorsoventrally flattened, often with silvery scales; head hypognathous to slightly prognathous; compound eyes absent or reduced, comprising isolated ommatidia, one to throo ocelli present, or absent; antennae multisegmented; mouthparts mandibulate, mandibles **dicondylar** (double-articulated), maxillary palps 5-segmented; legs with large coxae and tarsi 2–5 segmented; abdomen continues taper of thorax, at least segments 1–9 but sometimes 2–9 with ventral muscle-containing styles, mature individuals with or without 1 pair of protrusible vesicles medial to styles on segments 2–7, often reduced or absent. Paired elongate multisegmented cerci nearly as long as median caudal appendage (as shown here for *Lepisma saccharina*, after Lubbock, 1873). Fertilization indirect, with flask-shaped spermatophores that females pick up from the substrate. Many silverfish live in litter or under bark, some are subterranean or are cavernicolous, but some species can tolerate low humidity and high temperatures of arid areas. Some species live in mammal burrows, a few are commensals in nests of ants and termites, and several species are familiar synanthropic insects, living in human dwellings. These include *L. saccharina*, *Ctenolepisma longicauda* (silverfishes) and *Lepismodes inquilinus* (the firebrat), which eat materials such as paper, cotton and plant debris, using their own cellulase to digest the cellulose.

response to factors that may be little understood. In contrast, vegetation often shows less temporal variation; and with knowledge of mammal seasonality and of the migration habits of birds, the seasonal variations of vertebrate populations can be taken into account.

Thirdly, arthropods are often more difficult to identify accurately due to the numbers of taxa and some deficiencies in taxonomy mentioned in Chapter 7. Whereas competent mammalogists, ornithologists and field botanists might expect to identify to species level, respectively, all mammals, birds and plants of a geographically restricted area (outside the tropical rainforests), no entomologist could aspire to do so.

None-the-less, aquatic biologists routinely sample and identify all macroinvertebrates in regularly surveyed aquatic ecosystems, for purposes including monitoring of deleterious change in environmental quality (section 9.5). Comparable studies of terrestrial systems, with objectives such as establishment of rationales for conservation and the detection of pollution-induced changes, are undertaken in some countries. The problems outlined above have been addressed in the following ways.

Some charismatic insect species have been highlighted, usually under 'endangered species' legislation designed with vertebrate conservation in mind. Predominantly, these species have been lepidopterans and much has been learnt of the biology of selected species. However, from the perspective of site classification for conservation purposes, community structure of selected soil and litter communities have greater realized and potential value than any single-species study. Sampling problems are alleviated by using a single collection method, often that of pit-fall

BOX 8.4

GRYLLOBLATTODEA, EMBIOPTERA AND ZORAPTERA

These three species-poor orders of hemimetabolous insects are associated with litter habitats and otherwise share only their infrequency of occurrence and the elusiveness of their relationships to the larger hemimetabolous orders.

Grylloblattodea (Notoptera; grylloblattids or rock crawlers)

With 20 species in one family, restricted to western North America and central to eastern Asia, particularly tolerant of cold and high elevations. Moderate size, elongate, cylindrical, soft-bodied, pale, pubescent; head prognathous; compound eyes reduced or absent; ocelli absent; antennae multisegmented; mouthparts mandibulate; quadrate prothorax larger than meso- or metathorax; wings absent; legs cursorial, with large coxae and 5-segmented tarsi; abdomen with 10 visible segments and rudiments of segment 11, with 5- to 9-segmented cerci; female with short ovipositor, male genitalia asymmetrical. Copulation side-by-side (as illustrated here for the commonest Japanese species, *Galloisiana nipponensis*, after Ando, 1982). Eggs may diapause up to a year in moss or damp soil. Nymphs, which resemble adults, develop slowly through eight instars. Active by day and night at low temperatures, saprophagous on dead arthropods and organic material, notably from the surface of ice and snow in spring snow melt, or within caves (including ice caves), but also found in alpine soil and damp places. Grylloblattids may be either primitive orthopterans, or basal on the lineage from which Dictyoptera/Dermaptera is derived.

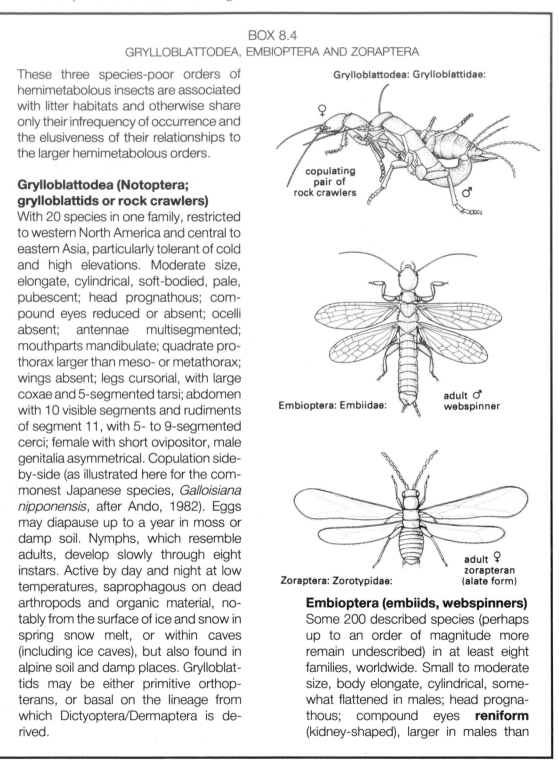

Grylloblattodea: Grylloblattidae:

copulating pair of rock crawlers

Embioptera: Embiidae:

adult ♂ webspinner

Zoraptera: Zorotypidae:

adult ♀ zorapteran (alate form)

Embioptera (embiids, webspinners)

Some 200 described species (perhaps up to an order of magnitude more remain undescribed) in at least eight families, worldwide. Small to moderate size, body elongate, cylindrical, somewhat flattened in males; head prognathous; compound eyes **reniform** (kidney-shaped), larger in males than

BOX 8.4—*continued*

females; ocelli absent; antennae multisegmented; mouthparts mandibulate; quadrate prothorax larger than meso- or metathorax; all females and some males apterous, if present wings (illustrated here for *Embia major*, after Imms, 1913) characteristically soft and flexible, with blood sinus veins stiffened for flight by blood pressure; legs short, with 3-segmented tarsi, basal segment of fore tarsi swollen, containing silk glands, hind femora swollen with strong tibial muscles; abdomen 10-segmented, with rudiments of segment 11, cerci 2-segmented, sensitive; female external genitalia simple, male genitalia complex and asymmetrical. Male holds female with prognathous mandibles and/or asymmetrical cerci during copulation. Immature stages resemble adults except in wings and genitalia. Eggs and early nymphal stages are tended by the female parent. Embioptera live gregariously in silken galleries, spun with tarsal silk glands (present in all instars); they live in leaf litter, beneath stones and in cracks in bark and soil, often around a central retreat. Food comprises litter, moss, bark and dead leaves. Galleries are extended to new food sources, and the safety of the gallery is left only when mature males disperse to new sites, where they mate, do not feed and sometimes are cannibalized by females. Webspinners readily reverse within their galleries, for example, when threatened by a predator. The general morphological resemblance of Embioptera to Plecoptera, and some similarity to the Dermaptera, notably deriving from their prognathy, probably relate to their sister-group relationship to Plecoptera.

Zoraptera
One genus *Zorotypus* comprising some 30 species, worldwide in tropical and warm temperate regions except Australia. Small (<3 mm long), rather termite-like; head hypognathous; compound eyes and ocelli present in winged species, absent in apterous species; antennae moniliform, 9-segmented; mouthparts mandibulate, with 5-segmented maxillary palps and 3-segmented labial palps; subquadrate prothorax larger than similar-shaped meso- and metathorax; wings polymorphic, in some forms apterous in both sexes, in other forms alate with two pairs of wings with reduced venation (as illustrated here for *Zorotypus hubbardi*, after Caudell, 1920), the hind wings smaller, wings shed as in ants and termites; legs with well developed coxae, 2-segmented tarsi, each with two claws; abdomen short and rather swollen, 11-segmented, with 1-segmented cerci; male genitalia asymmetric. Immature stages polymorphic according to wing development. Gregarious, occurring in leaf litter, rotting wood or near termite colonies, eating fungi and perhaps small arthropods. Phylogenetically enigmatic, with relationships to Psocoptera, orthopteroid/blattoids, hemipteroids or plecopteroids.

trapping, but including the extraction of arthropods from litter samples by a variety of means. Pit-fall traps collect mobile terrestrial arthropods by capturing them in containers filled with preserving fluid and sunken level with the substrate. Traps can be aligned along a transect, or dispersed according to a standard quadrat-based sampling regime. Depending on the sample size required, they can be left *in situ* for several days or up to a few

BOX 8.5
DERMAPTERA (EARWIGS)

An order with some 1800 species in ten families, worldwide. Hemimetabolous, small to moderate size (4–25 mm long), elongate, dorsoventrally flattened; head prognathous; compound eyes large, small or absent; ocelli absent; antennae short, annulate, with fewer segments in immature individuals than in adult; mouthparts mandibulate (section 2.3.1; Fig. 2.10); legs relatively short, 3-segmented tarsi with second tarsomeres short; apterous or, if winged, with fore wings small, leathery, smooth, unveined tegmina, hind wings large, membranous, semi-circular (as illustrated here for an adult male of the common European earwig, *Forficula auricularia*) and when at rest, folded fan-like and then longitudinally, protruding slightly from beneath tegmina, with venation dominated by anal fan of branches of A_1 and cross-veins; abdomen with segments telescoped (terga overlapping), with 10 visible segments in male, 8 in female, terminating in prominent cerci modified into forceps, often heavier, larger and more curved in male than in female. Copulation is end-to-end, and male spermatophores may be retained in the female for some months prior to fertilization. Oviparous species lay eggs often in a burrow in debris (Fig. 8.1), guard the eggs and lick them to remove fungus. The female may assist the nymphs to leave the eggs and care for them until the second or third instar, after which she may cannibalize her offspring. The two parasitic tribes Arixeniina and Hemimerina exhibit pseudoplacental viviparity (section 5.9).

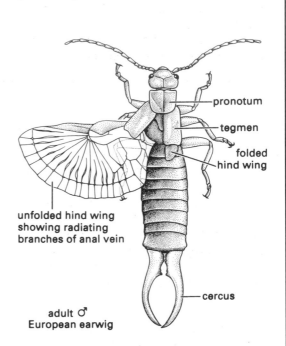

pronotum

tegmen

folded hind wing

unfolded hind wing showing radiating branches of anal vein

cercus

adult ♂
European earwig

Maturity is attained after four or five moults.

Earwigs are mostly cursorial and nocturnal, with most species rarely flying. Feeding is predominantly on dead and decaying vegetable and animal matter, with some predation and some damage to living vegetation, especially in gardens. The Arixeniina are commensals or ectoparasites of bats in Southeast Asia and the Hemimerina are semi-parasites of South African rodents: earwigs in both tribes are blind, apterous and with rod-like forceps. The forceps of free-living earwigs are used for manipulating prey, for defence and offence and in some species for grasping the partner during copulation. The

BOX 8.5—*continued*

common name 'earwig' derives from a supposed predilection for entering ears, but this is unsupported.

A sister-group relationship between Dermaptera and the Dictyoptera (Blattodea, Isoptera and Mantodea) is well supported.

BOX 8.6
BLATTODEA (BLATTARIA; COCKROACHES, ROACHES)

An order of over 3500 species in at least eight families worldwide. Hemimetabolous, small to large (<3– >65 mm), dorsoventrally flattened; head hypognathous; compound eyes moderately large, small or absent in cavernicolous species; ocelli represented by two pale spots; antennae filiform, multisegmented; mouthparts mandibulate; prothorax enlarged, shield-like, often covering head, meso- and metathorax rectangular and subequal; fore wings (Fig. 2.22(c)) sclerotized as tegmina, protecting membranous hind wings, tegmen lacking anal lobe, dominated by branches of veins R and CuA, hind wing with large anal lobe, with many branches in R, CuA and anal sectors, folded fan-like at rest beneath tegmen, wing reduction frequent; legs often spinose (Fig. 2.19), abutting coxae dominating ventral thorax, with tarsi 5-segmented; abdomen with 10 visible segments, subgenital plate (sternum 9), often bearing one or two styles in male, concealing segment 11 represented by paired paraprocts; cerci with one or usually many segments; genitalia of male asymmetrical, female ovipositor valves concealed beneath tergum 10.

adult ♀ cockroach

Mating may involve stridulatory courtship, both sexes may produce sex pheromones, and the female may mount the male prior to end-to-end copulation. Eggs generally are laid in a purse-shaped ootheca comprising two

---

BOX 8.6—*continued*

parallel rows of eggs with a leathery surrounding (section 5.8), which may be carried externally by female (as illustrated here for a female of *Blatella germanica*, after Cornwell, 1968). Certain species demonstrate a range of forms of ovoviviparity in which a variably reduced ootheca is retained within the reproductive tract in a 'uterus' (or brood sac) during embryogenesis, often until nymphal hatching; true viviparity rare. Parthenogenesis occurs in a few species. Nymphs develop slowly, resembling small apterous adults.

Cockroaches are amongst the most familiar insects, due to the cosmopolitan synanthropic habits of some 30 species including *Periplaneta americana* (the American cockroach), *B. germanica* (the German cockroach) and *B. orientalis* (the oriental cockroach). These nocturnal, malodorous, disease-carrying, refuge-seeking, peri-domestic roaches are unrepresentative of the wider diversity of this typically tropical, often diurnal and sometimes arboreal order, with some cavernicolous species. Cockroaches include solitary and gregarious species; and *Cryptocercus* (the woodroach) lives in family groups (Box 11.1) and eats wood, utilizing flagellate internal protozoa to digest cellulose; other xylophagous species use enteric amoebae for this role. Other cockroaches are saprophagous scavengers.

The Blattodea indisputably form the sister group of the Mantodea, with Isoptera being sister group to Blattodea and Mantodea combined.

---

weeks. Depending on the sites surveyed, arthropod collections may be dominated by Collembola, Formicidae and Coleoptera, particularly ground beetles (Carabidae), Tenebrionidae, Scarabaeidae and Staphylinidae, with some terrestrial representatives of many other orders.

Taxonomic difficulties can be alleviated by selecting (from amongst the organisms collected) one or more higher taxonomic groups for species-level identification. The carabids are often selected for study because of the diversity of species sampled, the pre-existing ecological knowledge and availability of taxonomic keys to species-level, although these are largely restricted to temperate northern hemisphere taxa.

Studies to date are ambivalent concerning correlations between species diversity (including taxon richness) established from vegetational survey and those from terrestrial insect trapping. Evidence from the well-documented British biota suggests that vegetational diversity does not predict insect diversity. However, a study in more natural, less human-impacted, environments in southern Norway showed congruence between carabid faunal indices and those obtained by vegetational and bird survey. Further studies are required into the relationships (if any) between terrestrial insect richness and diversity and data obtained by conventional biological survey of selected plants and vertebrates.

FURTHER READING

Collins, N.M. and Thomas, J.A. (eds) (1991) *The Conservation of Insects and their Habitats*, Academic Press, London.

Dallai, R. (1991) Are Protura really insects? in *The Early Evolution of Metazoa and the Significance of Problematic Taxa*, (eds A.M. Simonetta and S. Conway Morris), Cambridge University Press,

Cambridge, New York, Port Chester, Melbourne, Sydney, pp. 263–9.

Dindal, D.L. (ed.) (1990) *Soil Biology Guide*, John Wiley & Sons, Inc., New York, Chichester, Brisbane, Toronto, Singapore.

Eisenbeis, G. and Wichard, W. (1987) *Atlas on the Biology of Soil Arthropods*, 2nd edn, Springer-Verlag, Berlin, Heidelberg, New York, London, Paris, Tokyo.

Lussenhop, J. (1992) Mechanisms of microarthropod–microbial interactions in soil. *Advances in Ecological Research*, **23**, 1–33.

Refseth, D. (1980) Ecological analysis of carabid communities – potential use in biological classification for nature conservation. *Biological Conservation*, **17**, 131–41.

Stork, N.E. (ed.) (1990) *The Role of Ground Beetles in Ecological and Environmental Studies*, Intercept, Andover, Hampshire.

Villani, M.G. and Wright, R.J. (1990) Environmental influences on soil macroarthropod behavior in agricultural systems. *Annual Review of Entomology*, **35**, 249–69.

Wilding, N., Collins, N.M., Hammond, P.M. and Webber, J.F. (eds) (1989) *Insect–Fungus Interactions*, Academic Press, London, San Diego, New York, Berkeley, Boston, Sydney, Toronto, Tokyo.

World Conservation Monitoring Centre (1992) Chapter 11, Soil macrofauna, in *Global Biodiversity: Status of the Earth's Living Resources*, (ed. B. Groombridge), Chapman & Hall, London, Glasgow, New York, Tokyo, Melbourne, Madras, pp. 103–15.

9

AQUATIC INSECTS

A caddisfly larva in its retreat

Almost every inland waterbody, whether a river, stream, seepage or lake, supports a biological community within it. The most familiar components of aquatic communities are often the vertebrates, notably fish and amphibians. However, at least at the macroscopic level, invertebrates provide the highest number of individuals and species, biomass and production. In general, the insects dominate freshwater aquatic systems in which only nematodes can approach insects in terms of species numbers, biomass and productivity. Crustaceans may be abundant, but rarely diverse in species, in saline (especially temporary) inland waters. Some representatives of nearly all orders of insects live in water, and there have been many invasions of freshwater from the land. Insects have been much less successful in marine environments, where there has been very limited diversification, although several unrelated insect groups occur.

This chapter surveys the successful insects in aquatic environments and considers the variety of mechanisms they use to obtain scarce oxygen from the water. Some of their morphological and behavioural modifications to life in water are described, including how they resist water movements. The use of aquatic insects in biological monitoring of water quality is reviewed and the few insects of the marine and inter-tidal zones are discussed. Lastly, taxonomic boxes summarise information on mayflies (Ephemeroptera), dragonflies and damselflies (Odonata), stoneflies (Plecoptera), caddisflies (Trichoptera) and other orders of importance in aquatic ecosystems.

9.1 TAXONOMIC DISTRIBUTION AND TERMINOLOGY

The orders of insects that are almost exclusively aquatic in their immature stages are the Ephemeroptera (mayflies; Box 9.1), Odonata (damselflies and dragonflies; Box 9.2), Plecoptera (stoneflies; Box 9.3) and Trichoptera (caddisflies; Box 9.4). Amongst the major insect orders, a substantial number of Hemiptera, Coleoptera and Diptera have at least some aquatic stages; and in the less speciose, minor orders two families of Megaloptera and some Neuroptera develop in freshwater (Box 9.5). Some Hymenoptera parasitize aquatic prey but these, together with certain collembolans, orthopteroids and other predominantly terrestrial frequenters of damp places, are considered no further in this chapter.

Aquatic entomologists usually (correctly) restrict use of the term **larva** to the immature (i.e. post-embryonic and prepupal) stages of holometabolous insects; **nymph** (or **naiad**) is used for the pre-adult hemimetabolous insects, in which the wings develop externally. However, for the odonates the terms larva, nymph and naiad have been used interchangeably, perhaps because the sluggish, non-feeding, internally reorganizing final-instar odonate has been likened to the pupal stage of a holometabolous insect. Increasingly, some authors, especially North Americans, use 'larva' for the immature stages of all aquatic insects. Here we use the terms in their strict sense, except that immature odonates are referred to as larvae.

Some aquatic adult insects, including notonectid bugs and dytiscid beetles, can use terrestrial oxygen when submerged. Other adult insects are fully aquatic, such as several naucorid bugs, hydrophilid and elmid beetles that can remain submerged for extended periods and can obtain respiratory oxygen from the water. However, by far the greatest proportion of the adults of aquatic insects are aerial, and it is only their nymphal or larval (and often pupal) stages that live permanently below the water surface, where oxygen must be obtained whilst out of direct contact with the atmosphere. The ecological division of life history allows the exploitation of two different habitats, although there are a few insects that remain aquatic throughout their lives. Exceptionally, *Helichus*, a genus of dryopid beetles, has terrestrial larvae and aquatic adults.

9.2 THE EVOLUTION OF AQUATIC LIFESTYLES

Theories concerning the origin of wings in insects (section 7.4) have different implications regarding the evolution of aquatic lifestyles. The paranotal theory suggests that the 'wings' originated in adults of a terrestrial insect for which immature stages may have been aquatic or terrestrial. Some proponents of the preferred exite–endite theory speculate that the progenitor of the pterygotes had aquatic immature stages. Support for the latter hypothesis appears to come from the two extant basal groups of Pterygota (mayflies and odonates) being aquatic, in contrast to the primarily terrestrial extant apterygotes, but the aquatic nature of these groups cannot have been primary, since the spiracular system indicates a preceding terrestrial stage.

Whatever the origins of the aquatic mode of life, all proposed phylogenies demonstrate that it must have been adopted, adopted and lost, and readopted in several lineages, through geological time. The multiple independent adoption of aquatic lifestyles is particularly evident in the Coleoptera and Diptera, with aquatic taxa distributed amongst many families across each of these orders. In contrast, all species of Ephemeroptera and Plecoptera are aquatic. In the Odonata, the only exceptions to an almost universally aquatic lifestyle are the terrestrial larvae of a few species.

Movement from land to water causes physiological problems, the most important of which is the requirement for oxygen. The following section considers the physical properties of oxygen in air and water, and the mechanisms by which aquatic insects obtain an adequate supply.

9.3 AQUATIC INSECTS AND THEIR OXYGEN SUPPLIES

9.3.1 The physical properties of oxygen

Oxygen comprises 200 000 ppm (parts per million) in air, but in aqueous solution only about 15 ppm is attained in saturated cool water. Anaerobic respiration can provide energy at the cellular level but it is inefficient, providing 19 times less energy per unit of substrate respired than aerobic respiration does. Although insects such as bloodworms (chironomid midge larvae) survive extended periods of almost anoxic conditions, most aquatic insects must obtain oxygen from their surroundings in order to function effectively.

The proportions of gases in air and dissolved in water vary according to their solubilities: the amount is inversely proportional to temperature and salinity, and proportional to pressure, decreasing with elevation. In **lentic** (standing) waters, diffusion through water is very slow; it would take years for oxygen to diffuse several metres from the surface in still water. This slow rate, combined with the oxygen demand from microbial breakdown of submerged organic matter, can completely deplete the oxygen on the bottom (**benthic anoxia**). However, the oxygenation of surface waters by diffusion is enhanced by turbulence, which increases the surface area, forces aeration and mixes the water. If this turbulent mixing is prevented, such as in a deep lake with a small surface area, or one with extensive sheltering vegetation or when there is extended ice cover, anoxia can be prolonged or permanent. Living under these circumstances, benthic insects must tolerate wide annual and seasonal fluctuations in oxygen availability.

Oxygen levels in **lotic** (flowing) conditions can reach 15 ppm, especially if the water temperature is low. Oxygen can exceed equilibrium concentrations when photosynthesis generates locally abundant oxygen, such as in macrophyte and algae-rich pools in sunlight. However, as this vegetation respires at night oxygen is consumed, leading to a decline in dissolved oxygen. In these circumstances it is the diurnal range of oxygen tensions that aquatic insects must cope with.

9.3.2 Gaseous exchange in aquatic insects

The gaseous exchange systems of insects depend upon oxygen diffusion, which is rapid through air, slow through water and even slower across the cuticle. Although the insect cuticle is extremely impermeable, diffusion across the body surface may suffice for the smallest aquatic insects, such as some early-instar larvae or all instars of some dipteran larvae. Larger aquatic insects, with respiratory demands equivalent to spiraculate air breathers, require either augmentation of gaseous exchange areas or some other means of obtaining increased oxygen, since the reduced ratio of surface area to volume (Chapter 7) precludes dependence upon cutaneous gas exchange.

Aquatic insects demonstrate a range of mechanisms to cope with the lower oxygen in aqueous solutions compared with the atmosphere. As with their air-breathing relatives, aquatic insects may have open tracheal systems with spiracles, either polypneustic (8–10 spiracles opening on the body surface) or oligopneustic (1–2 pairs of open, often terminal spiracles) or, less commonly, closed tracheal systems that lack direct external connection.

9.3.3 Oxygen uptake with a closed tracheal system

Simple cutaneous gaseous exchange in a closed tracheal system only suffices for the smallest aquatic insects, such as early-instar caddisflies (Trichoptera). For larger insects, cutaneous exchange can account for a substantial part of oxygen uptake, but other mechanisms are needed.

One of the most widespread means of increasing surface area for gaseous exchange

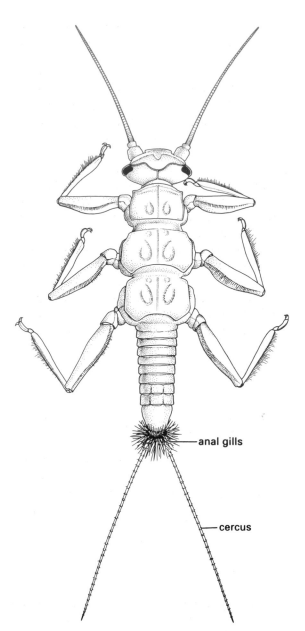

Fig. 9.1 A stonefly nymph (Plecoptera: Gripopterygidae) showing filamentous anal gills.

is by **gills** – tracheated cuticular lamellar extrusions from the body. These are usually abdominal (ventral, lateral or dorsal) or caudal, but may be located on the mentum, maxillae, neck, at the base of the legs, around the anus in some Plecoptera (Fig. 9.1) or even within the rectum, as in dragonfly larvae. Tracheal gills are found in the immature stages of Odonata, Plecoptera, Trichoptera, aquatic Megaloptera and Neuroptera, some aquatic Coleoptera, a few Diptera and pyralid lepidopterans, and probably reach their greatest morphological diversity in the Ephemeroptera.

In interpreting these structures as gills it is important to demonstrate that they do function in oxygen uptake. In experiments with larvae of *Lestes* (Odonata: Lestidae), the huge caudal gill-like lamellae of some individuals were removed by being broken at the site of natural autotomy. Both gilled and ungilled individuals were subjected to low oxygen environments in closed-bottle respirometry and survivorship was assessed. The three caudal lamellae of this odonate met all criteria for gills, namely:

- Large surface area.
- Moist and vascular.
- Able to be ventilated.
- Responsible normally for 20–30% of oxygen uptake.

However, as temperature rose and dissolved oxygen fell, gills accounted for increased oxygen uptake until the maximum uptake reached 70%. At this high level, the pro portion equalled the proportion of gill surface to absolute body surface area. At low temperatures and with dissolved oxygen at the environmental maximum of 9 ppm, the gills of the lestid accounted for very little oxygen uptake; cuticular uptake was presumed to be dominant. When *Siphlonurus* mayfly nymphs were tested similarly, at 12–13°C the gills accounted for 67% of oxygen uptake, which was proportional to their fraction of total surface area of the body.

Another means of extracting dissolved oxygen is through the use of respiratory pigments. These pigments occur almost universally in vertebrates and they are also found in some invertebrates and even in plants and protists. Amongst the aquatic insects some larval chironomids (bloodworms) and a few notonectid bugs possess haemoglobin. Vertebrate haemoglobins have a low affinity for oxygen; that is, oxygen is obtained from high oxygen environments and unloaded in an acid (carbonic acid from dissolved carbon dioxide) environment – the Bohr effect. Where environmental oxygen concentrations are consistently low, as in the almost anoxic and often acidic sediments of lakes, the Bohr effect would be counterproductive. In contrast to vertebrates, the haemoglobins of chironomids therefore have a high affinity for oxygen. Chironomid midge larvae can saturate their haemoglobins through undulating their bodies within their silken tubes or substrate burrows to permit oxygenated water to flow over the cuticle. Oxygen is unloaded when undulations stop, or when recovery from anaerobic respiration is needed. The respiratory pigments allow a much more rapid oxygen release than is available by diffusion alone.

9.3.4 Oxygen uptake with an open spiracular system

For aquatic insects with open spiracular systems, there is a range of possibilities for obtaining oxygen. Many immature stages of Diptera can obtain atmospheric oxygen like the mosquito larva and pupa, which suspend themselves from the water meniscus (Fig. 9.2). There are direct connections between the atmosphere and the spiracles in the terminal respiratory siphon of the larva, or in the thoracic respiratory organ of the pupa. The insect that uses atmospheric oxygen is independent of low dissolved oxygen levels, such as occur in rank or stagnant waters. This is particularly prevalent amongst larvae of flies such as ephydrids, one species of which can live in oil tar ponds, and certain pollution-tolerant hover flies (Syrphidae), the 'rat-tailed maggots'.

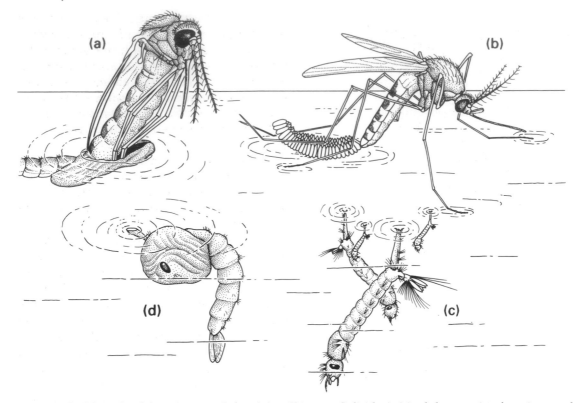

Fig. 9.2 The life cycle of the mosquito *Culex pipiens* (Diptera: Culicidae): (a) adult emerging from its pupal exuviae at the water surface; (b) adult female ovipositing, with her eggs adhering together as a floating raft; (c) larvae obtaining oxygen at the water surface via their siphons; (d) pupa suspended from the water meniscus, with its respiratory horn in contact with the atmosphere. (After Clements, 1992.)

Several other larval Diptera and psephenid beetles have cuticular modifications surrounding the spiracular openings which function as gills that allow an increase in extraction rate of dissolved oxygen without spiracular contact with the atmosphere.

An unusual source of oxygen is the vascular system of roots and stems of aquatic macrophytes. Aquatic insects including the immature stages of some mosquitoes, hover flies and *Donacia* (a genus of chrysomelid beetles) can use this source. In *Mansonia* mosquitoes the spiracle-bearing larval respiratory siphon and pupal thoracic respiratory organ are modified for piercing the plant tissues.

Temporary air stores (**compressible gills**) are common means of storing and extracting

oxygen. Many adult dytiscid, gyrinid, helodid, hydraenid and hydrophilid beetles and both nymphs and adults of many belostomatid, corixid, naucorid and pleid hemipterans use this method of enhancing gaseous exchange. The gill is a bubble of stored air, which is in contact with the spiracles by various means, including sub-elytral retention in adephagan water beetles (Fig. 9.3) and fringes of specialized hydrofuge hairs on the body and legs of some polyphagan water beetles. When the insect dives from the surface, air is trapped in a bubble in which all gases start at atmospheric equilibrium. As the submerged insect respires, oxygen is used up and the carbon dioxide produced is lost by its high solubility in water. Within the bubble, as

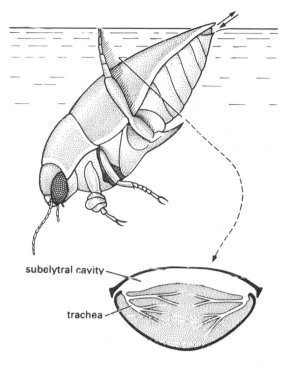

subelytral cavity

trachea

Fig. 9.3 A male water beetle of *Dytiscus* (Coleoptera. Dyticcidae) replenishing its store of air at the water surface. Below is a transverse section of the beetle's abdomen showing the large air store below the elytra and the tracheae opening into this air space. The tarsi of the fore legs are dilated to form adhesive pads that are used to hold the female during copulation. (After Wigglesworth, 1964.)

bubble. However, the available oxygen varies according to the amount of exposed surface area of the bubble and the prevailing water temperature. At low temperatures the metabolic rate is lower, more gases remain dissolved and the gill is long-lasting. Conversely, at higher temperatures metabolism is higher, less gas is dissolved and the gill is less effective.

A further modification of the air-bubble gill, the **plastron**, allows some insects to use permanent air stores, termed an 'incompressible gill'. Water is held away from the body surface by hydrofuge hairs or a cuticular mesh, leaving a permanent gas layer in contact with the spiracles. Most of the gas is relatively insoluble nitrogen but in response to metabolic use of oxygen a gradient is set up and oxygen diffuses from water into the plastron. Most insects with such a gill are relatively sedentary since the gill is not very effective in responding to high oxygen demand. This mode of oxygen extraction is used by adults of some curculionid, dryopid, elmid, hydraenid and hydrophilid beetles, nymphs and adults of naucorid bugs and pyralid moth larvae.

9.3.5 Behavioural ventilation

A consequence of the slow diffusion rate of oxygen through water is the development of an oxygen-depleted layer of water that surrounds the gaseous uptake surface, whether it be the cuticle, gill or spiracle. Aquatic insects exhibit a variety of ventilation behaviours that disrupt this oxygen-depleted layer. Cuticular gaseous diffusers undulate their bodies in tubes (Chironomidae), cases (young caddisfly nymphs) or under shelters (young lepidopteran larvae) to produce fresh currents across the body. This behaviour continues even in later-instar caddisflies and lepidopterans in which gills are developed. Many ungilled aquatic insects select their positions in the water to allow maximum aeration by current flow. Some dipterans, such as blepharicerids

the partial pressure of oxygen drops, more diffuses in from solution in water but not rapidly enough to prevent continued depletion in the bubble. Meanwhile as the proportion of nitrogen in the bubble increases, it diffuses outwards, causing diminution in the size of the bubble. This contraction in size gives rise to the term compressible gill. When the bubble has become too small, it is replenished by returning to the surface.

The longevity of the bubble depends upon the relative rates of consumption of oxygen and of gaseous diffusion between the bubble and the surrounding water. A maximum of eight times more oxygen can be supplied from the compressible gill than was in the original

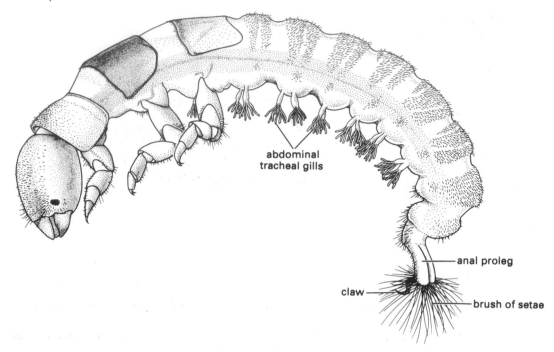

abdominal
tracheal gills

anal proleg

claw

brush of setae

Fig. 9.4 The net spinning larva of the caddisfly *Cheumatopsyche* (Trichoptera: Hydropsychidae).

and deuterophlebids, are found only in torrents; ungilled simuliids, plecopterans and caseless caddisfly larvae (Fig. 9.4) are found commonly in high-flow areas. The very few sedentary aquatic insects with gills, notably black-fly (simuliid) pupae, some adult dryopid beetles and the immature stages of a few lepidopterans, maintain locally high oxygenation by positioning themselves in areas of well-oxygenated flow. For mobile insects, swimming actions, such as leg movements, prevent the formation of a low-oxygen boundary layer.

Although most gilled insects use natural water flow to bring oxygenated water to them, they may also undulate their bodies, beat their gills, or pump water in and out of the rectum, as in anisopteran larvae. In the lestid zygopteran larvae (section 9.3.3), ventilation is assisted by 'pull-downs' (or 'push-ups') that effectively move oxygen-reduced water away from the gills. In contrast, when dissolved oxygen is reduced through an increase in temperature, *Siphlonurus* nymphs elevate the frequency and increase the percentage of time spent beating gills.

9.4 THE AQUATIC ENVIRONMENT

The two different aquatic physical environments, the lotic (flowing) and lentic (standing), place very different constraints on the organisms living therein. In the following sections, these conditions are highlighted and some of the morphological and behavioural modifications of aquatic insects are discussed.

9.4.1 Lotic adaptations

In lotic systems, the velocity of flowing water is a major influence on:

- substrate type, with boulders deposited in fast flow and fine sediments in slow flow areas;
- transport of particles, either as a food source

Fig. 9.5 Dorsal and lateral views of the larva of a species of water penny (Coleoptera: Psephenidae).

for filter feeders, or, during peak flows, as scouring agents; and

- maintenance of high levels of dissolved oxygen.

A stream or river contains heterogeneous microhabitats, with riffles (shallower, stony, fast-flowing sections) interspersed with deeper, natural pools. Areas of erosion of the banks alternate with those where sediments are deposited, and there may be areas of unstable, shifting sandy substrates. The banks may have trees (a vegetated **riparian** zone) or be unstable, with mobile deposits that change with every flood. Typically, where there is riparian vegetation, there will be local accumulations of drifted **allochthonous** (external to the stream) material such as leaf packs and wood. In parts of the world where extensive pristine forested catchments remain, the courses of streams are often periodically blocked by naturally fallen trees. Where the stream is open to light, and nutrient levels allow, **autochthonous** growth of plants and macroalgae (**macrophytes**) will occur and aquatic flowering plants may be abundant, as in chalk streams.

Characteristic insect faunas inhabit these various substrates, many with particular morphological modifications. Those that live in strong current (**rheophilic** species) tend to be dorsoventrally flattened (Fig. 9.5), sometimes with laterally projecting legs. These are not strictly adaptations to strong currents, being found in many aquatic insects, but they do afford avoidance or minimization of exposure, by allowing the insect to remain within a boundary layer of calmer water close to the surface of the substrate. However, the detail of hydraulic flow in natural waters is much more complex than was earlier believed, and the relationship between body shape, streamlining and current velocity is not a simple one.

The cases constructed by many rheophilic caddisflies assist in streamlining or otherwise modifying the effects of flow. The variety of shapes of the cases must act as ballast against displacement (Fig. 9.6). Several aquatic larvae have suckers, or convergently evolved structures that allow the insect to stick to quite smooth exposed surfaces, such as rock-faces on waterfalls and cascades. Silk is widely produced, allowing maintenance of position

(a)

(b)   **(c)**   **(d)**

Fig. 9.6 Portable larval cases of representative families of caddisflies (Trichoptera): (a) Helicopsychidae; (b) Philorheithridae; (c) and (d) Leptoceridae.

in fast flow. Black-fly larvae (Simuliidae) spin a silken pad attached to a rock surface, to which the posterior claws of the larva are attached. Others, including hydropsychid caddisflies and chironomid midges, use silk in constructing retreats. Some spin silken mesh nets to trap food brought into proximity by the stream flow.

Many lotic insects are smaller than their counterparts in standing waters. Their size, together with flexible body design, allows them to live amongst the cracks and crevices of boulders, stones and pebbles in the bed (**benthos**) of the stream, or even in unstable, sandy substrates. Another means of avoiding the current is to live in accumulations of leaves (leaf packs) or to mine in immersed wood –

substrates that are used by many beetles and specialist dipterans, such as crane-fly larvae (Diptera: Tipulidae).

Two behavioural strategies are more evident in running waters than elsewhere. The first is the strategic use of the current to allow **drift** from an unsuitable location, with the possibility of finding a more suitable patch. Predatory aquatic insects frequently drift to locate aggregations of prey. Many other insects, such as stoneflies and mayflies, notably *Baetis* (Ephemeroptera: Baetidae), may show a diurnal periodic pattern of drift. 'Catastrophic' drift is a behavioural response to physical disturbance, such as pollution or severe flow episodes. An alternative response of burrowing deep into the substrate (the **hyporheic** zone), is another particularly lotic behaviour. In the hyporheic zone, the vagaries of flow regime, temperature and perhaps predation can be avoided, although food and oxygen availability may be diminished.

9.4.2 Lentic adaptations

With the exception of wave action at the shores of larger bodies of water, the effects of water movement cause little or no difficulties for aquatic insects that live in lentic environments. However, oxygen availability is more of a problem and lentic taxa show a greater variety of mechanisms for enhanced oxygen uptake.

The lentic water surface is used by many more species (the **neustic** community of **semi-aquatic** insects) than the lotic surface, because the physical properties of surface tension in standing water that can support an insect are disrupted in turbulent flowing water. Water-striders (Hemiptera: Gerromorpha: Gerridae and Veliidae) are amongst the most familiar neustic insects that exploit the surface film (Box 9.5). They use hydrofuge (water-repellent) hair piles on the legs and venter to avoid breaking the film. Water-striders move with a rowing motion and they locate prey items (and in some species, mates)

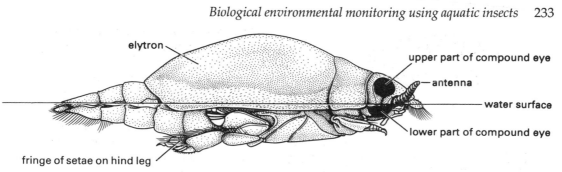

Fig. 9.7 The whirligig beetle, *Gyretes* (Coleoptera: Gyrinidae), swimming on the water surface. The divided compound eye allows the beetle to see both above and below water simultaneously; the hydrofuge hairs on the margin of the elytra repel water. (After White *et al.*, 1984.)

by detecting vibratory ripples on the water surface. Certain staphylinid beetles use chemical means to move around the meniscus, discharging from the anus a detergent-like substance that locally releases surface tension and propels the beetle forward. Some elements of this community can be found in still-water areas of streams and rivers, and related species of Gerromorpha can live in estuarine and even oceanic water surfaces (section 9.7).

Underneath the meniscus of standing water, larvae of many mosquitoes feed (Fig 2.16), and hang suspended by their respiratory siphons (Fig. 9.2) as do certain crane flies and stratiomyiids (Diptera). Whirligig beetles (Gyrinidae) (Fig. 9.7) also are able to straddle the interface between water and air, with an upper unwettable surface and the lower wettable one. Uniquely, each eye is divided such that the upper part can observe the aerial environment and the lower half can see under water.

Between the water surface and the benthos, planktonic organisms live in a zone divisible into an upper **limnetic** zone (that light reaches) and the deeper **profundal** zone. The most abundant planktonic insects belong to *Chaoborus* (Diptera: Chaoboridae); these 'phantom midges' undergo diurnal vertical migration, and their predation on *Daphnia* is discussed in section 12.4. Other insects such as diving beetles (Dytiscidae) and many hemipterans, such as Corixidae, dive and swim actively through this zone in search of prey. The profundal zone generally lacks planktonic insects but may support an abundant benthic community, predominantly of chironomid midge larvae, most of which possess haemoglobin. The profundal benthic zone of some deep lakes, such as Lake Baikal in Siberia, even supports some midges, although at eclosion the pupa may have to rise more than 1 km to the water surface.

In the **littoral** zone, in which light reaches the benthos and macrophytes can grow, insect diversity is at its maximum. Many differentiated microhabitats are available and physico-chemical parameters are less restricting than in the dark, cold and perhaps anoxic conditions of the deeper waters.

9.5 BIOLOGICAL ENVIRONMENTAL MONITORING USING AQUATIC INSECTS

Aquatic insects form communities that vary with their geographic location, according to historical biogeographic and ecological processes. Within a more restricted area, such as a single lake or river drainage, the community structure derived from within this pool of locally available organisms is constrained largely by physico-chemical factors of the environment. Amongst the important factors that govern which species live in a particular water body, variations in oxygen availability obviously lead to different insect communities. For example, in low oxygen conditions, perhaps caused by oxygen-demanding sewage pollution, the community is typically species-poor and differs in composition from a comparable well-oxygenated system, as might be found upstream of a pollution site. Similar changes in community structure can be seen in relation to other physico-chemical factors such as temperature, sediment and substrate type and, of increasing concern, pollutants such as pesticides, acidity and heavy metal deposition.

All of these factors, which are generally subsumed under the term 'water quality', can be measured physico-chemically. However, physico-chemical monitoring requires:

- Knowledge of which of the hundreds of substances to monitor.
- Understanding of the synergistic effects when two or more pollutants interact (which often exacerbates or multiplies the effects of any compound alone).
- Continuous monitoring to detect pollutants that may be intermittent, such as nocturnal release of industrial waste products.

The problem is that it is rarely known in advance which of the many substances released into waterways are significant biologically; even with such knowledge, continuous monitoring of more than a few is difficult and expensive. Supposing that these impediments could be overcome, the important question remains: what are the biological effects of pollutants? Organisms and communities, exposed to aquatic pollutants, integrate multiple present and immediate past environmental effects. Increasingly, insects are used in the description and classification of aquatic ecosystems and in the detection of deleterious effects of human activities. For the latter purpose, aquatic insect communities (or a subset of the animals that comprise an aquatic community) are used as surrogates for humans: their observed responses give early warning of damaging changes.

In this **biological monitoring** of aquatic environments, the advantages of using insects include:

- The ability to select amongst the many insect taxa in any aquatic system, according to the resolution required.
- The availability of many ubiquitous or widely distributed taxa, allowing elimination of non-ecological reasons why a taxon might be missing from an area.
- The functional importance of insects in aquatic ecosystems, ranging from secondary producers to top predators.
- The ease and lack of ethical constraints in sampling aquatic insects, giving numbers of individuals and taxa that can be handled, yet be informative.
- The ability to identify most aquatic insects to a meaningful level.
- The predictable and easily detectable responses of many aquatic insects to disturbances, such as particular types of pollution.

Typical responses observed when aquatic insect communities are disturbed include:

- Increased abundance of certain mayflies, such as Caenidae with protected abdominal gills, and caddisflies including filter-feeders such as Hydropsychidae (like the one illustrated in the vignette for this chapter), as particulate material, including sediment, increases.
- Increase in numbers of haemoglobin-possessing bloodworms (Chironomidae) as dissolved oxygen is reduced.
- Loss of stonefly nymphs (Plecoptera) as water temperatures increase.
- Substantial reduction in diversity with pesticide run-off.
- Increased abundance of a few species but general loss of diversity with elevated nutrient levels (organic enrichment, or **eutrophication**).

More subtle community changes can be observed in response to less overt pollution sources, but it can be difficult to separate environmentally induced changes from natural variations in community structure.

9.6 INSECTS OF TEMPORARY WATERBODIES

In a geological time scale, all waterbodies are temporary. Lakes fill with sediment, become marshes and eventually dry out completely. Erosion reduces the catchments of rivers and their courses change. These historical changes are slow compared with the lifespan of insects and have little impact on the aquatic fauna, apart from a gradual alteration in environmental conditions. However, in certain parts of the world, waterbodies may fill and dry on a much shorter time scale. This is particularly evident where rainfall is very seasonal or intermittent, or where high temperatures cause elevated evaporation rates. Rivers may run during periods of predictable seasonal rainfall, such as the 'winterbournes' on chalk downland in southern England that flow only during, and immediately following, winter rainfall. Others may flow only intermittently after unpredictable heavy rains, such as streams of the arid zone of central Australia and deserts of the western United States. Temporary bodies of standing waters may last for as little as a few days, as in water-filled footprints of animals, rocky depressions, pools beside a falling river, or in impermeable clay-lined pools filled by flood or snow-melt.

Even though temporary, these habitats are very productive and teem with life. Aquatic organisms appear almost immediately such habitats form. Amongst the macroinvertebrates, crustaceans are numerous and many insects thrive in ephemeral waterbodies. Some insects lay eggs into a newly formed aquatic habitat within hours of the filling, and it seems that gravid females of these species are transported to such sites over long distances, associated with the frontal meteorological conditions that bring the rainfall. An alternative to colonization by the adult is the deposition by the female of desiccation-resistant eggs into the dry site of a future pool.

This behaviour is seen in some odonates and many mosquitoes, especially of the genus *Aedes*. Development of the diapausing eggs is induced by environmental factors that include wetting, perhaps requiring several consecutive immersions (section 6.8).

A range of adaptations is shown amongst insects living in ephemeral habitats compared with their relatives in permanent waters. Firstly, development to the adult is often more rapid, perhaps because of increased food quality and lowered inter-specific competition. Secondly, development may be staggered or asynchronous, with some individuals reaching maturity very rapidly, thereby increasing the possibility of at least some adult emergence from a short-lived habitat. Associated with this is a greater variation in size of adult insects from ephemeral habitats – with metamorphosis hastened as a habitat diminishes. Certain larval midges (Diptera: Chironomidae and Ceratopogonidae) can survive drying of an ephemeral habitat by resting in cocoons lined with silk or mucus amongst the debris at the bottom of a pool. In this condition, limited desiccation of the body can be tolerated and development is continued when the next rains fill the pool.

The most remarkable specialization involves a larval chironomid midge, *Polypedilum vanderplankl*, that lives in West Africa on granite outcrops in temporary pools, such as those that form in depressions made by native people when grinding grain. The larvae do not form cocoons when the pools dry, but their bodies lose water until they are almost completely dehydrated. In this condition of **cryptobiosis** (alive but with all metabolism ceased), the larvae can tolerate temperature extremes, with artificially imposed temperatures in dry air, from $-27\,°C$ up to more than $100\,°C$. On wetting, the larvae revive rapidly, feed and continue development until the onset of another cycle of desiccation or pupation and emergence.

Persistent temporary pools develop a fauna of predators, including immature beetles, bugs and odonates, which are the offspring of aerial colonists. These colonization events are important in the genesis of faunas of newly flowing intermittent rivers and streams. In addition, immature stages present in remnant water beneath the stream bed may move into the main channel, or colonists may be derived from permanent waters with which the temporary water connects. It is a frequent observation that novel flowing waters are colonized initially by a single species, often otherwise rare, that rapidly attains high population densities and then declines rapidly with the development of a more complex community, including predators.

Temporary waters are often saline, because evaporation concentrates salts, and this type of pool develops communities of specialist saline-tolerant organisms. However, few if any species of insect living in saline inland waters also occur in the marine zone – nearly all of the former are derived from freshwater relatives.

9.7 INSECTS OF THE MARINE, INTERTIDAL AND LITTORAL ZONES

The estuarine and subtropical and tropical mangrove zones are transitions between fresh and marine waters. Here, the extremes of the truly marine environment, such as wave and tidal actions, and some osmotic effects, are ameliorated. Mangroves and 'saltmarsh' communities (such as *Spartina*, *Sarcocornia*, *Halosarcia* and *Sporobolus*) support a complex phytophagous insect fauna on the emergent vegetation. In intertidal substrates and tidal pools, biting flies (mosquitoes and biting midges) are abundant and may be diverse. At the littoral margin, species of any of four families of hemipterans stride on the surface, some venturing onto the open water. A few other insects, including some *Bledius* staphylinid beetles, cixid fulgoroid bugs and root-feeding *Pemphigus* aphids, occupy the zone of prolonged inundation by saltwater. This fauna is restricted compared with freshwater and terrestrial ecosystems.

Splash-zone pools on rocky shores have salinities that vary from rain-water dilution and solar concentration. They can be occupied by many species of corixid bugs and several larval mosquitoes and crane flies. Flies and beetles are diverse on sandy and muddy marine shores, with some larvae and adults feeding along the strandline, often aggregated on and under stranded seaweeds.

Within the intertidal zone that lies between high and low neaptide marks, the period of tidal inundation varies with the location within the zone. The insect fauna of the upper level is indistinguishable from the strandline fauna. At the lower end of the zone, in conditions that are almost fully marine, crane flies, chironomid midges and species of several families of beetles can be quite abundant. The female of a remarkable New Zealand and Australian marine trichopteran (Chathamiidae: *Philanisus plebeius*) lays its eggs in a starfish coelom. The early-instar caddisflies feed on starfish tissues, but later free-living instars construct cases of algal fragments.

Three lineages of chironomid midges are amongst the few insects that have diversified in the marine zone. *Telmatogeton* (Fig. 6.11) is common in mats of green algae, such as *Ulva*, and occurs worldwide, including on many isolated oceanic islands. In Hawaii the genus has re-invaded freshwater. The ecologically convergent *Clunio* also is found worldwide. In some species adult emergence from marine rockpools is synchronized by the lunar cycle to coincide with the lowest tides. A third lineage, *Pontomyia*, ranges from intertidal to oceanic, with larvae found at depths of up to 30 m on coral reefs.

The only insects on open oceans are pelagic water-striders (*Hylobates*) that have been sighted hundreds of kilometres from shore out in the Pacific Ocean. The distribution of these insects coincides with mid-oceanic accumulations of flotsam, where food of

BOX 9.1
EPHEMEROPTERA (MAYFLIES)

A small order of no more than a few thousand species, with highest diversity in temperate areas. Adult with mouthparts reduced; large compound eyes, especially in male, and three ocelli; antennae filiform, sometimes multisegmented; thorax, particularly mesothorax, enlarged for flight, with large triangular fore wings and smaller hind wings (as illustrated here for an adult male of the ephemerid *Ephemera danica*, after Stanek, 1969, and Elliott and Humpesch, 1983), sometimes much reduced or absent; elongate fore legs in male used to seize female during mating flight; abdomen 10-segmented, typically with three long, multisegmented, caudal filaments consisting of a pair of lateral cerci and usually a median terminal filament. Nymphs with 12–45 aquatic instars; fully developed mandibulate mouthparts; developing wings visible in older nymphs (as shown over for a leptophlebiid nymph); closed tracheal system lacking spiracles, with abdominal lamellar gills on some segments, sometimes elsewhere, including on maxillae and labium; with three usually filiform caudal filaments consisting of paired cerci and a variably reduced (rarely absent) median terminal filament. Penultimate instar or subimago (subadult) fully winged, flying or crawling.

Nymphs graze on periphyton (algae, diatoms, aquatic fungi) or collect fine detritus; few predatory on other aquatic organisms. Development takes from 16 days to longer than one year in cold and high-latitude waters; some species are multivoltine. Subimago and adult are

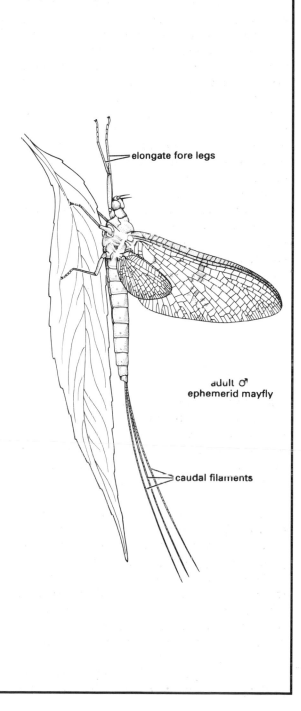

elongate fore legs

adult ♂
ephemerid mayfly

caudal filaments

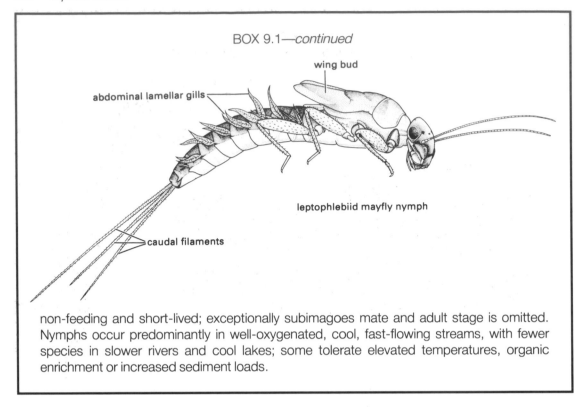

BOX 9.1—*continued*

wing bud

abdominal lamellar gills

leptophlebiid mayfly nymph

caudal filaments

non-feeding and short-lived; exceptionally subimagoes mate and adult stage is omitted. Nymphs occur predominantly in well-oxygenated, cool, fast-flowing streams, with fewer species in slower rivers and cool lakes; some tolerate elevated temperatures, organic enrichment or increased sediment loads.

terrestrial origin supplements a diet of chironomids.

Physiology is unlikely to be a factor restraining diversification in the marine environment because so many different taxa are able to live in inland saline waters and in various marine zones. When living in highly saline waters, submerged insects can alter their osmoregulation to reduce chloride uptake and increase the concentration of their excretion through Malpighian tubules and rectal glands. In the pelagic water-striders, which live on the surface film, contact with saline waters must be quite limited.

Since physiological adaptation appears to be a surmountable problem, explanations for the failure of insects to diversify in the sea must be sought elsewhere. The most likely is that the insects originated well after other invertebrates such as the Crustacea and Mollusca had already dominated the sea. The advantages to terrestrial (including freshwater) insects of internal fertilization and flight are superfluous in the marine environment, where gametes can be shed directly into the sea and dispersal is aided by the tide and oceanic currents. Indeed, of the few successful marine insects, many have modified their wings or lost them altogether.

BOX 9.2
ODONATA (DAMSELFLIES AND DRAGONFLIES)

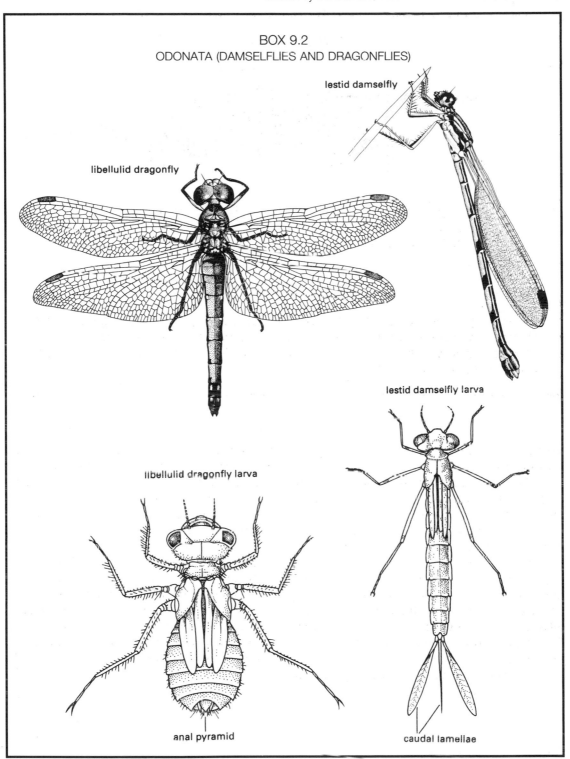

lestid damselfly

libellulid dragonfly

lestid damselfly larva

libellulid dragonfly larva

anal pyramid

caudal lamellae

BOX 9.2—*continued*

A small order of conspicuous insects, comprising a few thousand species, with about one third belonging to suborder Zygoptera (damselflies), two thirds to suborder Anisoptera (dragonflies) and two oriental species in a third suborder, Anisozygoptera. Adult medium to large ($<$ 2– $>$ 15 cm long); mobile head with large, multifaceted compound eyes and three ocelli, short antennae and mandibulate mouthparts; enlarged thorax to accommodate flight muscles of two pairs of elongate membranous wings; slender 10-segmented abdomen. Zygopteran adult with eyes widely separated and fore and hind wings equal in shape with narrow bases (as illustrated in top right figure for an Australian lestid, *Austrolestes*, after Bandsma and Brandt, 1963); anisopteran adult with eyes either contiguous or slightly separated, wings with characteristic closed cells called triangle (T) and hypertriangle (ht) (Fig. 2.22(b)) and hind wings considerably wider at base than fore wings (as illustrated in top left figure for a libellulid dragonfly, *Sympetrum*, after Gibbons, 1986). Larvae with variable number of up to 15 aquatic instars; fully developed mandibulate mouthparts, including extensible, grasping labium or 'mask' (Fig. 12.4); developing wings visible in older nymphs; closed tracheal system lacking spiracles, but specialized gas exchange surfaces present on abdomen as external gills (Zygoptera) or internal folds in rectum (Anisoptera). Zygopteran larva (such as the lestid illustrated on the lower right, after CSIRO, 1970) slender, with head wider than thorax and apex of abdomen with three (rarely two) elongate tracheal gills (caudal lamellae); anisopteran larva (such as the libellulid illustrated on the lower left, after CSIRO, 1970) more stoutly built with head rarely much broader than thorax, and abdominal apex characterized by anal pyramid consisting of three short projections and a pair of cerci in older larvae.

Larvae are predatory on other aquatic organisms. Adults predatory on terrestrial, aerial prey. At metamorphosis (Fig. 6.8), pharate adult moves to water/land surface where atmospheric gaseous exchange commences and then it crawls from water, anchors terrestrially and imago emerges from cuticle of final-instar larva; imago long-lived, active aerial. Larvae occur in all waterbodies, particularly well-oxygenated, standing waters but elevated temperatures, organic enrichment or increased sediment loads are tolerated by many species.

BOX 9.3
PLECOPTERA (STONEFLIES)

adult gripopterygid stonefly

A minor and often cryptic order, with perhaps 2000 species worldwide, predominantly in temperate and cool areas. Hemimetabolous, with adults resembling winged nymphs. Adult mandibulate with filiform antennae, bulging compound eyes and two or three ocelli; thoracic segments subequal; fore and hind wings membranous and similar, except hind wing broader, with folded wings partly wrapping abdomen and extending beyond abdominal apex (as illustrated for an adult of the Australian gripopterygid, *Illiesoperla*), although aptery and brachyptery frequent; legs unspecialized, tarsi 3-segmented; abdomen soft, 10-segmented, with vestiges of segments 11 and 12 as paraprocts, cerci and epiproct, a combination of which serves as male accessory copulatory structures sometimes in conjunction with abdominal sclerites of segments 9 and 10. Nymphs with 12–24, perhaps as many as 33, aquatic instars; with fully developed mandibulate mouthparts; wing pads first visible in half-grown nymphs; closed tracheal system with simple or plumose gills on basal abdominal segments or near anus (Fig. 9.1), sometimes extrusible from anus, or on mouthparts, neck, thorax, or lacking; cerci usually multisegmented, no median terminal filament.

Nymphs may be omnivores, detritivores, herbivores or predators. Adults feed on algae, lichen, higher plants and/or rotten wood; some may not eat. Mature nymphs crawl to water's edge where adult emergence takes place. Nymphs occur predominantly on stony/gravelly substrates in cool water, mostly in well-aerated streams, with fewer species in lakes. Generally very intolerant of pollution.

BOX 9.4
TRICHOPTERA (CADDISFLIES)

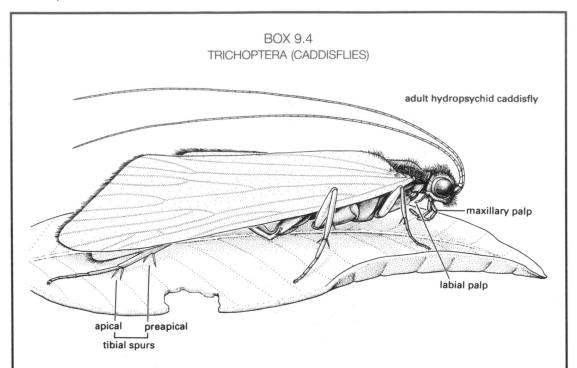

adult hydropsychid caddisfly

maxillary palp

labial palp

apical   preapical

tibial spurs

An order of over 7000 species worldwide. Holometabolous, with adult moth-like (as illustrated here for a hydropsychid), usually covered in hairs; setal warts (setose protuberances) often on dorsum of head and thorax; head with reduced mouthparts, but with 3- to 5-segmented maxillary and 3-segmented labial palps (cf. proboscis of most Lepidoptera), antennae multisegmented, filiform, often as long as or longer than wings, large compound eyes and two or three ocelli; prothorax smaller than meso- or metathorax; wings haired or sometimes scaled, distinguished from Lepidoptera by different wing venation, including anal veins looped in fore wings and no discal cell; abdomen typically 10-segmented, with male terminalia more complex (often with claspers) than in female.

Larva with five to seven aquatic instars, with fully developed mouthparts; three pairs of thoracic legs with at least 5 segments, without the ventral prolegs characteristic of lepidopteran larvae; abdomen terminating in hook-bearing prolegs; closed tracheal system, with tracheal gills often on most or all 9 abdominal segments (Fig. 9.4), sometimes associated with thorax or anus; gas exchange also cuticular, enhanced by ventilatory undulation in tubular case. Pupa aquatic, in silken retreat or case, with large functional mandibles to chew free from pupal case or cocoon; free legs with setose mid-tarsi to swim to water surface; gills coincide with larval gills; eclosion through pharate adult swimming to water surface where pupal cuticle splits and exuviae used as floating platform.

Predominantly univoltine, with development longer than one year at high latitudes and elevations. Larva free-living, net-spinning, or makes saddle-, purse- or tube-case (Fig. 9.6); diverse feeding habits, including predators, filterers and/or shredders of organic matter, and some grazers on macrophytes. Net-spinners restricted to flowing waters, with case-makers frequent also in standing waters. Adults may ingest nectar or water but often do not feed.

BOX 9.5
OTHER AQUATIC ORDERS

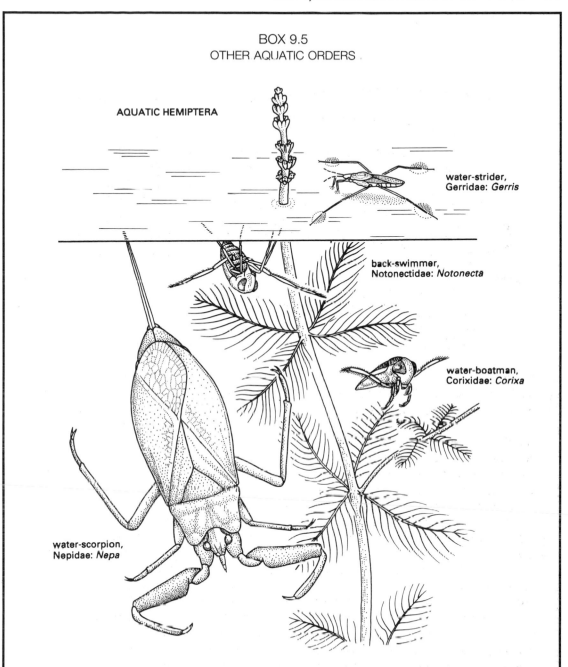

AQUATIC HEMIPTERA

water-strider,
Gerridae: *Gerris*

back-swimmer,
Notonectidae: *Notonecta*

water-boatman,
Corixidae: *Corixa*

water-scorpion,
Nepidae: *Nepa*

Hemiptera (bugs)

Hemimetabolous, with about 4000 aquatic and semi-aquatic (including marine) species in about 20 families worldwide, belonging to three heteropteran infraorders (Gerromorpha, Leptopodomorpha and Nepomorpha). Possessing subordinal characteristics (Box 10.8) of

BOX 9.5—*continued*

mouthparts modified as a rostrum (beak) and fore wings as hemelytra; spiraculate with various gaseous exchange mechanisms; nymphs with one, adults with two or more tarsal segments; antennae 3- to 5-segmented, inconspicuous in aquatic groups but obvious in semi-aquatic ones; often with reduction, loss and/or polymorphism of wings. Nymphs with five, rarely four instars, often univoltine. Gerromorphs (water-striders, represented here by *Gerris*) scavenge or predatory on water surface; diving taxa are either predatory (e.g. back-swimmers (Notonectidae) such as *Notonecta*, water-scorpions (Nepidae) such as *Nepa*, giant water bugs (Belostomatidae) or phytophagous detritivores (some water-boatmen (Corixidae) such as *Corixa*).

Megaloptera (alderflies, dobsonflies, fishflies)

Holometabolous, with about 300 species in two families worldwide. Adult (illustrated in Box 12.3) with unspecialized mandibulate mouthparts, including strong mandibles; unequal wings with large plaited anal field on hind wing that infolds when wings at rest over back; soft abdomen. Larva prognathous, with well-developed mouthparts, including 3-segmented labial palps (similar-looking gyrinid beetle larvae have 1–2-segmented palps); spiraculate with gills consisting of 4- to 5-segmented (Sialidae) or 2-segmented lateral filaments on abdominal segments; abdomen terminating in an unsegmented median caudal filament (Sialidae) or a pair of anal prolegs (as shown here for a species of *Archichauliodes* (Corydalidae)). Pupa beetle-like (Fig. 6.7(a)), except mobile on free legs and with pupal head similar to that of larva, including functional mandibles. Larvae (sometimes called **hellgrammites**) have 10–11 instars and take at least one year, usually two or more, to develop. Pupation is away from water, often in chambers in damp soil under stones, or in damp timber. Larvae predatory in lotic and lentic waters, intolerant of pollution.

Megaloptera: Corydalidae: *Archichauliodes*

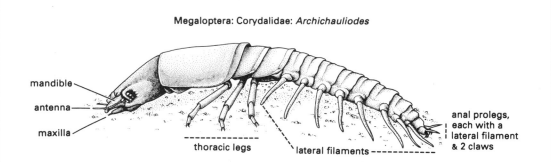

mandible
antenna
maxilla
thoracic legs
lateral filaments
anal prolegs, each with a lateral filament & 2 claws

Neuroptera (lacewings)

Holometabolous, predominantly terrestrial predators (Box 12.3), but spongillaflies (Sisy-ridae) have aquatic larvae. Sisyrid larva with elongate stylet-like mandibles and filamentous antennae, paired ventral abdominal gills, without terminal prolegs; pupa with functional mandibles; adults small, soft-bodied with subequal wings lacking anal lobe on hind wing. Eggs are laid in trees overhanging running water, and hatching larvae drop into water,

BOX 9.5—*continued*

where they seek out and feed upon sponges by sucking out the living cells; three larval instars, with rapid development, may be polyvoltine; pupation in silken cocoon out of water.

Coleoptera (beetles)
A diverse order with over 5000 aquatic species (but comprising less than 2% of the world's described beetle species). About ten families are exclusively aquatic as larva and adult, an additional few are predominantly aquatic as larvae and terrestrial as adults or *vice versa*, and several more have sporadic aquatic representation. Holometabolous; adult diagnostically with mesothoracic wings modified as rigid elytra (Fig. 2.22(d); Box 10.10); gaseous exchange usually by temporary or permanent air stores. Larvae very variable, all with distinct sclerotized head, strongly developed mandibles, 2–3 segmented antenna; three pairs of jointed thoracic legs, lacking abdominal prolegs; open peripneustic (nine pairs of spiracles) tracheal system, but variably reduced spiracle number in most aquatic larvae, some with lateral and/or ventral abdominal gills, sometimes hidden beneath terminal sternite. Pupation terrestrial (except some Psephenidae); pupa lacking functional mandibles.

Diptera (true flies)
Another major order in which aquatic larvae are typical of many Nematocera, with over 10 000 aquatic species in several families, including the speciose Chironomidae (non-biting midges), Ceratopogonidae (biting midges), Culicidae (mosquitoes) and Simuliidae (black flies). Holometabolous; adults terrestrial, aerial (diagnosed in Box 14.4). Larva commonly vermiform, diagnostically with unsegmented prolegs, variably distributed on body; primitively with sclerotized head and horizontally operating mandibles, in more derived groups the head is progressively reduced, ultimately (in the **maggot**) with head and mouthparts atrophied to cephalopharyngeal skeleton; open (amphi- or meta-, rarely propneustic) or closed tracheal systems, gaseous exchange cuticular, through spiracular gills or terminal, elongate respiratory siphon with spiracular connection to atmosphere. Three or four (black flies up to ten) larval instars (Fig 6.1); pupation predominantly subaquatic: pupa non-mandibulate, with appendages fused to body; puparium formed in derived groups (few of which are aquatic) from the tanned retained third-instar larval cuticle. Emergence at the water surface may involve use of the cast exuviae as a platform (Chironomidae and Culicidae), or through the adult rising to the surface in a bubble of air secreted within the pupa (Simuliidae).

Development varies from ten days to longer than one year, with many multivoltine species; adults may be ephemeral to long-lived. At least some dipteran species occur in virtually every aquatic habitat from the marine coast, salt lagoons and sulphurous springs to fresh and stagnant waterbodies, and from temporary containers to rivers and lakes; temperatures tolerated include 0 °C for some species and 55 °C for a few others in thermal pools. This tolerance to pollution of certain taxa is utilized in biological indication of water quality.

FURTHER READING

Eriksen, C.H. (1986) Respiratory roles of caudal lamellae (gills) in a lestid damselfly (Odonata: Zygoptera). *Journal of the North American Benthological Society*, **5,** 16–27.

Eriksen, C.H. and Moeur, J.E. (1990) Respiratory functions of motile tracheal gills in Ephemeroptera nymphs, as exemplified by *Siphlonurus occidentalis* Eaton, in *Mayflies and Stoneflies: Life Histories and Biology*, (ed. I.C. Campbell), Kluwer Academic Publishers, Dordecht, pp. 109–118.

Merritt, R. and Cummins, K.W. (eds) (1984) *An Introduction to the Aquatic Insects of North America*, Kendall/Hunt, Dubuque, Iowa.

Resh, V. and Rosenberg, D. (eds) (1984) *The Ecology of Aquatic Insects*, Praeger, New York, Philadelphia, Eastbourne, Toronto, Hong Kong, Tokyo, Sydney.

Rosenberg, D.M. and Resh, V.H. (eds) (1993) *Freshwater Biomonitoring and Benthic Macroinvertebrates*, Chapman & Hall, New York, London.

Ward, J.V. (1992) *Aquatic Insect Ecology*, John Wiley & Sons, Inc., New York, Chichester, Brisbane, Toronto, Singapore.

10

INSECTS AND PLANTS

Leaf eaten by sawfly larvae

Insects and plants share ancient associations that probably date from the Carboniferous, though fossil evidence suggests that the major taxa of **phytophagous** (plant-feeding) insects became diverse only later, in the Cretaceous (Fig. 7.3). At this time angiosperms (flowering plants) dramatically increased in diversity in a radiation that displaced the previously dominant plant groups of the Jurassic period.

Interpreting the early evolution of the angiosperms is contentious, partially because of the paucity of fossilized flowers prior to the period of radiation, and also because of the apparent rapidity of the origin and diversification within the major angiosperm families. However, according to several hypotheses of their phylogeny, the earliest angiosperms may have been insect-pollinated, perhaps by beetles. Many living representatives of primitive families of beetles feed on fungi, fern spores or pollens of other non-angiosperm taxa such as cycads. Since this feeding type preceded the angiosperm radiation, it can be seen as a preadaptation for angiosperm pollination.

The ability of flying insects to transport pollen from flower to flower on different plants is fundamental to cross-pollination. Other than the beetles, the most significant and diverse present-day pollinator taxa belong to three orders – the Diptera (flies), Hymenoptera (wasps and bees) and Lepidoptera (moths and butterflies). Pollinator taxa within these orders are unrepresented in the fossil record until late in the Cretaceous. Although insects probably pollinated cycads and other more primitive plants, speciation in angiosperms may have been promoted by insect pollinators, through pollinator-mediated isolating mechanisms.

Many modern non-insect hexapods and apterygote insects scavenge in soil and litter, predominantly feeding on decaying plant material (Chapter 8). The earliest true insects probably fed similarly. This manner of feeding certainly brings soil-dwelling insects into contact with plant roots and subterranean storage organs, but specialized use of plant aerial parts by sap sucking, leaf chewing and other forms of phytophagy arose later in the phylogeny of the insects. Feeding on living tissues of higher plants presents problems that are experienced neither by the scavengers living in the soil or litter nor by predators. Firstly, a phytophagous insect must be able to gain and retain a hold on the vegetation in order to feed on leaves, stems or flowers. Secondly, the exposed phytophage may be subject to greater desiccation than an aquatic or litter-dwelling insect. Thirdly, a diet of plant tissues (excluding seeds) is nutritionally inferior in protein, sterol and vitamin content compared with food of animal or microbial origin. Last, but not least, plants are not passive victims of phytophages but have evolved a variety of means to deter herbivores. These include physical defences, such as spines, spicules or sclerophyllous tissue, and chemical defences that may repel, poison, reduce food digestibility or otherwise adversely affect insect behaviour or physiology. Despite these barriers, about half of all living insect species are phytophagous and the exclusively plant-feeding Lepidoptera, Curculionidae (weevils), Chrysomelidae (leaf beetles), Agromyzidae (leaf-mining flies) and Cynipidae (gall wasps) are very speciose. Plants represent an abundant resource and insect taxa that can exploit this resource have flourished.

This chapter begins with a consideration of the evolutionary interactions among insects and their plant hosts. It then deals with the vast array of interactions of insects and living plants by grouping them into three categories, defined by the effects of the insects on the plants. Phytophagy (herbivory) includes leaf chewing, sap sucking, seed predation, gall forming and mining the living tissues of plants (section 10.2). The second category of interactions is important to plant reproduction and involves mobile insects that transport pollen between conspecific plants (pollination) or seeds to suitable germination sites (myrmecochory). These interactions are mutualistic because the insects obtain food or some other resource from the plants that they service (section 10.3). The third category of insect–plant interaction involves insects that live in specialized plant structures and provide their host with either defence against herbivores or nutrition or both (section 10.4). Such mutualisms, like the nutrient-producing fly larvae that live unharmed within the pitchers of carnivorous plants, are unusual but provide fascinating opportunities for evolutionary and ecological studies.

The chapter concludes with seven taxonomic boxes that summarize the morphology and biology of the primarily phytophagous orders Orthoptera, Phasmatodea, Thysanoptera, Hemiptera, Psocoptera, Coleoptera and Lepidoptera.

10.1 COEVOLUTIONARY INTERACTIONS BETWEEN INSECTS AND PLANTS

Reciprocal interactions over evolutionary time between phytophagous insects and their food plants, or pollinating insects and the plants they pollinate, have been described as **co-evolution**. This term was coined and broadly defined by Ehrlich and Raven in 1964 from a study of butterflies and their host plants. Several modes of coevolution are now recognized, differing in the emphasis placed on the specificity and reciprocity of the interactions.

Specific or **pair-wise coevolution** refers to the evolution of a trait of one species (such as an insect's ability to detoxify a poison) in response to a trait of another species (such as the elaboration of the poison by the plant), which in turn originally evolved in response to the trait of the first species (that is, the insect's food preference for that plant). This is a strict mode of coevolution, since reciprocal interactions between specific pairs of species are postulated. The outcomes of such co-evolution may be evolutionary 'arms races' between eater and eaten or convergence of traits in mutualisms so that both members of

an interacting pair appear perfectly adapted to each other. Reciprocal evolution between the interacting species may contribute to at least one of the species becoming subdivided into two or more reproductively isolated populations (as exemplified by figs and fig wasps, Box 10.3), thereby acting to generate species diversity.

Another mode, **diffuse** or **guild coevolution**, describes reciprocal evolutionary change among groups, rather than pairs, of species. Here the criterion of specificity is relaxed so that a particular trait in one or more species (of flowering plants, for example) may evolve in response to a trait or suite of traits in several other species (as, for example, in several different and perhaps unrelated pollinating insects).

These are the main modes of coevolution that relate to insect–plant interactions but clearly they are not mutually exclusive. The study of such interactions is bedevilled with the problem that unequivocal demonstration of the occurrence of any kind of coevolution is difficult. Evolution takes place over geological time and hence the selection pressures responsible for changes in 'coevolving' taxa can be inferred only retrospectively, principally from correlated traits of interacting organisms. Specificity of interactions among living taxa can be demonstrated or refuted far more convincingly than can historical reciprocity in the evolution of the traits of these same taxa. For example, it may be possible to show, by careful observation, that a flower bearing its nectar at the bottom of a very deep tube is pollinated exclusively by a particular moth species with a very long proboscis, or a humming-bird with a particular length and curvature of its beak. Specificity of the association between any individual pollinator species and plant is an observable fact, but flower tube depth and mouthpart morphology are merely correlated and only suggestive of coevolution.

10.2 PHYTOPHAGY (OR HERBIVORY)

The majority of plant species support complex faunas of herbivores, each of which may be defined in relation to the range of plant taxa used. Thus **monophages** are specialists that feed on one plant taxon, **oligophages** feed on few, and **polyphages** are generalists that feed on many plant groups. Gall-inducing cynipid wasps (Hymenoptera) exemplify monophagous insects since nearly all species are host-plant specific; furthermore, all cynipid wasps of the tribe Rhoditini induce their galls only on roses (*Rosa*) (Fig. 10.5(d)) and almost all species of Cynipini form their galls only on oaks (*Quercus*) (Fig. 10.5(c)). The monarch or wanderer butterfly, *Danaus plexippus* (Nymphalidae), is an example of an oligophagous insect; its larvae feed on various milkweeds, predominantly species of *Asclepias*. The Chinese wax scale, *Ceroplastes sinensis* (Hemiptera: Coccidae), is truly polyphagous: its recorded host plants belong to about 200 species in at least 50 families.

Many plants appear to have broad-spectrum defences against a very large suite of enemies, including insect and vertebrate herbivores and pathogens. These defences, as mentioned above, are primarily physical or chemical in nature (and are discussed in section 15.6 in relation to host-plant resistance to insect pests). Spines or pubescence on stems and leaves, silica or sclerenchyma in leaf tissue, and leaf shapes that aid camouflage are amongst the physical attributes of plants that may deter some herbivores. Furthermore, in addition to the chemicals considered essential to plant function, most plants contain compounds whose role is assumed to be defensive, although these chemicals may have (or once had) other metabolic functions or may simply be metabolic waste products. Such chemicals are often called **secondary plant compounds**, **noxious phytochemicals** or **allelochemicals**. A huge array exists, including tannins, terpenoid compounds (essential oils), alkaloids, cyanogenic glycosides and sulphur-containing thiols and glucosinolates. The anti-herbivore action of many of these compounds has been demonstrated or inferred. For example, in *Acacia*, the loss of the otherwise widely distributed cyanogenic glycosides in those species that harbour mutualistic stinging ants implies that the secondary plant chemicals do have an anti-herbivore function in the many species that lack ant defences.

In terms of plant defence, secondary plant compounds may act in one of two ways. At a behavioural level, these chemicals may repel an insect or inhibit feeding or oviposition. At a physiological level, they may poison an insect or reduce the nutritional content of its food. However, the same chemicals that repel some insect species may attract others, for either oviposition or feeding (thus acting as kairomones; section 4.3.3(a)). Such attracted insects are said to be adapted to the chemicals of their host plants, by either tolerating, detoxifying or even sequestering them. An example is the wanderer butterfly, *D. plexippus*, which usually oviposits on milkweed plants, many of which contain toxic cardiac glycosides (cardenolides) that the feeding larva can sequester for use as an anti-predator device (sections 13.4.3 and 13.5.1).

Secondary plant compounds have been classified into two broad groups based on their inferred biochemical actions: (a) qualitative or toxic, and (b) quantitative. The former are effective poisons in small quantities (e.g. alkaloids, cyanogenic glycosides), whereas the latter act in proportion to their concentration, being more effective in greater amounts (e.g. tannins, resins, silica). However, for insects that are specialized to feed on particular plants containing either group of secondary plant compound, these chemicals actually can act as phagostimulants.

The observation that some kinds of plants are more susceptible to insect attack than others has also been explained by the relative **apparency** of the plants – large, long-lived trees in clumps are very much more apparent to an insect than small, scattered annual herbs. Apparent plants tend to have quantitative secondary compounds, with high metabolic costs in their production. Unapparent plants often have qualitative or toxic secondary compounds, produced at little metabolic cost. Humans often turn unapparent plants into apparent ones by growing monocultures of annual plants, with corresponding increases in insect damage.

Another consideration is the predictability of resources sought by insects, such as the suggested predictability of the presence of new leaves on a eucalypt or creosote bush in contrast to the erratic spring flush of new leaves on a deciduous tree. However, the question of the predictability (or apparency) of plants to insects is essentially untestable. Furthermore, insects can optimize the use of unpredictably abundant resources by synchronizing their life cycles to environmental cues identical to those used by the plant.

A third correlate of variation in herbivory rates concerns the nature and quantities of resources (i.e. light, water, nutrients) available to plants. One hypothesis is that insect herbivores feed preferentially on stressed plants (e.g. those affected by drought or waterlogging) because stress can alter plant physiology in ways beneficial to insects. Alternatively, insect herbivores may prefer to feed on vigorously growing plants (or plant parts) in resource-rich habitats.

No single theory (model) of herbivory is consistent with all observed patterns of temporal and spatial variation within plant individuals, populations and communities, but all models of current herbivory theory make two assumptions, both of which are difficult to substantiate. These are:

- Damage by herbivores is a dominant selective force on plant evolution.
- Food quality has a dominant influence on the abundance of insects and the damage they cause.

Even the substantial evidence that hybrid plants may incur much higher damage from herbivores than either adjacent parental population is not unequivocal evidence of either assumption. Although selection against hybrids clearly could affect plant evolution, such herbivore preference for hybrids would be expected to constrain rather than promote plant genetic diversification. The food quality of hybrids is arguably higher than that of the parental plants, due to less efficient chemical defences or higher nutritive value of the genetically 'impure' hybrids. It remains unclear whether the overall population abundance of herbivores is altered by the presence of hybrids (or by food quality *per se*) or is merely redistributed among the plants available. The role of natural enemies in regulating herbivore populations is often overlooked in studies of insect–plant interactions.

Many studies have demonstrated that phytophagous insects can impair plant growth, in both the short and the long term. These observations have led to the suggestion that host-specific herbivores may affect the relative abundances of plant species by reducing the competitive abilities of host plants. The occurrence of induced defences (Box 10.1) supports the idea that it is advantageous for plants to deter herbivores. In contrast with this view is the controversial hypothesis that 'normal' levels of herbivory may be advantageous or selectively neutral to plants. Some degree of pruning, pollarding or mowing may increase (or at least not reduce) overall plant reproductive success by altering growth form or longevity and thus life-time seed set. The important evolutionary parameter is life-time reproductive success, although most assessments of herbivore effects on plants involve only measurements of plant production such as biomass and leaf number.

A major problem with all herbivory theories is that they have been founded largely on studies of leaf-chewing insects, since the damage of these insects is easier to measure and factors involved in defoliation are more

BOX 10.1
INDUCED DEFENCES

Many plants contain chemicals that are believed to have a role in deterring some herbivores, or at least reducing their suitability. These are the secondary plant compounds (noxious phytochemicals or allelochemicals) mentioned in section 10.2. Depending on plant species, such chemicals may be present in the foliage at all times, or in only some parts of the plant, or in some parts during particular stages of ontogeny, such as during the growth period of new leaves. Such **constitutive** defences provide the plant with continuous protection, at least against non-adapted phytophagous insects. If defence is costly in energy and if insect damage is intermittent, plants would benefit from being able to turn on their defences only when insect feeding occurs. There is good experimental evidence that, in some plants, damage to the foliage induces chemical changes in the current or subsequent leaves, with adverse affects on insects. This phenomenon is called **induced defence** if the induced chemical response benefits the plant. However, sometimes the induced chemical changes may lead to greater foliage consumption by lowering food quality for herbivores, so that they eat more to obtain the necessary nutrients.

Both short-term (or rapidly induced) and long-term (or delayed) chemical changes have been observed in plants as responses to herbivory. For example, proteinase-inhibitor proteins are produced rapidly by some plants in response to wounds caused by chewing insects. These proteins can significantly reduce the palatability of the plant to some insects. In other plants, the production of phenolic compounds may be increased, either for short or prolonged periods, within the wounded plant part or sometimes the whole plant.

Such induced chemical changes have been demonstrated for some plants. Even when they occur, their functions may not be easy to demonstrate, especially since herbivore feeding is not always deterred. Sometimes induced chemicals may benefit the plant indirectly, not by reducing herbivory but by attracting natural enemies of the insect herbivores (section 4.3.3(c)). Moreover, the results of studies on induced responses may be difficult to interpret due to large variation in foliage quality among and within individual plants, as well as the complication that minor variations in the nature of the damage can lead to different outcomes. In addition, insect herbivore populations in the field are regulated by an array of factors and the effects of plant chemistry may be ameliorated or exacerbated by other conditions.

amenable to experimentation. The effects of sap-sucking, leaf-mining and gall-forming insects may be as important but are generally less well understood, except for some agricultural and horticultural pests such as aphids.

10.2.1 Leaf chewing

The damage caused by leaf-chewing insects is more readily visible than that of, for example, many sap-sucking insects. Furthermore, the insects responsible for loss of leaf tissue are usually easier to identify than the small larvae of species that mine or gall plant parts. By far the most diverse groups of leaf-chewing insects are the Lepidoptera and Coleoptera. Most moth and butterfly caterpillars and many beetle larvae and adults feed on leaves, although plant roots, shoots, stems, flowers in north temperate forests are lepidopteran larvae, such as those of the gypsy moth, *Lymantria dispar* (Lymantriidae). Other important groups of leaf-chewing insects worldwide are the Orthoptera (most species) and Hymenoptera (chiefly the sawflies, such as those depicted in the vignette of this chapter). The stick-insects (Phasmatodea) and psocids (Psocoptera) generally have only minor impacts as leaf chewers, although outbreaks of the spurlegged stick-insect, *Didymuria violescens* (Box 10.6), can result in defoliation of eucalypts in Australia.

High levels of herbivory result in economic losses to forest trees and other plants, and so reliable and repeatable methods of estimating damage are desirable. Most methods rely on estimating leaf area lost due to leaf-chewing insects. This can be done directly by measuring damage on foliage, either by monitoring marked branches or by destructively collecting separate samples over time ('spot sampling'), or indirectly by measuring the production of insect **frass** (faeces). These sorts of measurements have been undertaken in a number of forest types, from rainforests to xeric (dry) forests, in many countries worldwide. Mostly, the herbivory levels are surprisingly uniform. For temperate forests, most values of proportional missing leaf area range from 3 to 17%, with a mean value of 8.8 +/− 5.0% (n = 38) (values from Landsberg and Ohmart, 1989). Data collected from rainforests and mangrove forests reveal similar levels of leaf-area loss (3–15%, mean 8.8 +/− 3.5%). Defoliation levels may be very high and even lead to death of plants during outbreaks, especially when caused by introduced pest species. For some plant taxa, herbivory levels may be high (20–45%) even under natural, non-outbreak conditions.

Levels of herbivory, measured as leaf-area loss, differ among plant populations or communities for a number of reasons. The leaves of different plant species vary in their suitability as insect food because of variations in nutrient content, water content, type and

Fig. 10.1 Christmas beetles of *Anoplognathus* (Coleoptera: Scarabaeidae) on the chewed foliage of a eucalypt tree (Myrtaceae).

and fruits often are eaten as well. Certain Australian adult scarabs, especially species of *Anoplognathus* (Coleoptera: Scarabaeidae) that are commonly called Christmas beetles (Fig. 10.1), can cause severe defoliation of eucalypt trees. The most important foliage-eating pests concentrations of secondary plant compounds, and degree of sclerophylly (toughness). Such differences may occur because of inherent differences among plant taxa or may relate to the maturity and growing conditions of the sampled individual leaves or plants. Communities in which the majority of the constituent tree species belong to different families (such as in many north temperate forests) may suffer less damage from phytophages than communities that are dominated by one or a few genera (such as Australian eucalypt/acacia forests). In the latter systems, specialist insect species may be able to transfer relatively easily to new, closely related plant hosts, so that favourable conditions may result in considerable insect damage to all or most tree species in a given area. In multi-genera tree communities, oligophagous insects are unlikely to switch to species unrelated to their normal hosts. Furthermore, there may be differences in herbivory levels within any given plant population over time due to seasonal and stochastic factors, including variability in weather conditions (which affects insect and plant growth) or plant defences induced by previous insect damage (Box 10.1). Such temporal variation in plant growth and response to insects can strongly influence herbivory estimates that are made over a restricted period.

10.2.2 Plant mining and boring

There is a range of insect larvae that reside within and feed on the internal tissues of living plants. **Leaf-mining** species live between the two epidermal layers of a leaf and their presence can be detected externally after the area that they have fed upon dies, often leaving a thin layer of dry epidermis. This leaf damage appears as tunnels, blotches or blisters (Fig. 10.2). Tunnels may be straight (linear) or convoluted and often widen throughout their course (Fig. 10.2(a)) due to larval growth during development. Generally larvae that live in the confined space between

the upper and lower leaf epidermis are flattened. Their excretory material – frass – is left in the mine as black or brown pellets (Figs 10.2(a)–(e)) or lines (Fig. 10.2(f)).

The leaf-mining habit has evolved independently in only four orders of insects, all of which are holometabolous: the Diptera, Lepidoptera, Coleoptera and Hymenoptera. The commonest types of leaf miners are larval flies and moths. Some of the most prominent leaf mines result from the larval feeding of agromyzid flies (Fig. 10.2(a)–(d)). Agromyzids are virtually ubiquitous; there are about 2500 species which are all exclusively phytophagous. Most are leaf miners, although some mine stems and a few occur in roots or flower-heads. Some anthomyiids and a few other fly species also mine leaves. Most of the lepidopteran leaf miners (Fig. 10.2(e)–(g)) belong to the families Gracillariidae, Gelechiidae, Incurvariidae, Lyonetiidae, Nepticulidae and Tisheriidae. The habits of leaf-mining moth larvae are diverse, with many variations in types of mines, methods of feeding, frass disposal and larval morphology. Generally the larvae are more specialized than those of other leaf-mining orders and are very dissimilar to their non-mining relatives. A number of moth species have habits that intergrade with gall forming and leaf rolling. Leaf-mining Hymenoptera principally belong to the sawfly superfamily Tenthredinoidea, in which most leaf-mining species form blotch mines. Leaf-mining Coleoptera are represented by species of jewel beetles (Buprestidae), leaf beetles (Chrysomelidae) and weevils (Curculionoidea).

Leaf miners can cause economic damage by attacking the foliage of fruit trees, vegetables, ornamental plants and forest trees. The spinach leaf miner (or mangold fly) *Pegomya hyoscyami* (Diptera: Anthomyiidae) causes commercial damage to the leaves of spinach and beet. The larvae of the birch leaf miner, *Fenusa pusilla* (Hymenoptera: Tenthredinidae), produce blotch mines in birch foliage in northeastern North America, where this

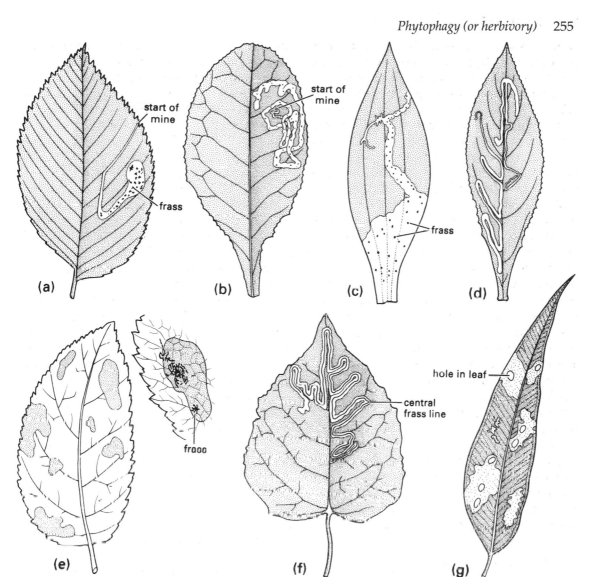

Fig. 10.2 Leaf mines: (a) linear-blotch mine of *Agromyza aristata* (Diptera: Agromyzidae) in leaf of an elm, *Ulmus americana* (Ulmaceae); (b) linear mine of *Chromatomyia primulae* (Agromyzidae) in leaf of a primula, *Primula vulgaris* (Primulaceae); (c) linear-blotch mine of *Chromatomyia gentianella* (Agromyzidae) in leaf of a gentian, *Gentiana acaulis* (Gentianaceae); (d) linear mine of *Phytomyza senecionis* (Agromyzidae) in leaf of a ragwort, *Senecio nemorensis* (Asteraceae); (e) blotch mines of the apple leaf miner, *Lyonetia speculella* (Lepidoptera: Lyonetiidae), in leaf of apple, *Pyrus malus* (Rosaceae); (f) linear mine of *Phyllocnistis populiella* (Lepidoptera: Gracillariidae) in leaf of poplar, *Populus* (Salicaceae); (g) blotch mines of jarrah leaf miner, *Perthida glyphopa* (Lepidoptera: Incurvariidae) in leaf of jarrah, *Eucalyptus marginata* (Myrtaceae). ((a),(e),(f) After Frost, 1959; (b)–(d) after Spencer, 1990.)

sawfly is considered a serious pest. In Australia, certain eucalypts are prone to the attacks of leaf miners, which can cause unsightly damage. The leaf blister sawflies (Hymenoptera: Pergidae: *Phylacteophaga*) tunnel in and blister the foliage of some species of *Eucalyptus* and related genera of Myrtaceae. The larvae of the jarrah leaf miner, *Perthida glyphopa* (Lepidoptera: Incurvariidae), feed in the leaves of jarrah, *Eucalyptus marginata*, causing blotch mines and then holes after the larvae have cut leaf discs for their pupal cases (Fig. 10.2(d)). Jarrah is an important timber tree in Western Australia and the feeding of these leaf miners can cause serious leaf damage in vast areas of eucalypt forest.

Mining sites are not restricted to leaves and some insect taxa display a diversity of habits. For example, different species of *Marmara* (Lepidoptera: Gracillariidae) not only mine leaves but also burrow below the surface of stems, or in the joints of cacti, and a few mine beneath the skin of fruit. One species that typically mines the cambium of twigs extends its tunnels into leaves if conditions are crowded. **Stem mining**, or feeding in the superficial layer of twigs, branches or tree trunks, can be distinguished from **stem boring**, in which the insect feeds deep in the plant tissues.

Stem boring is just one form of **plant boring**, which includes a broad range of habits that can be subdivided according to the part of the plant eaten and whether the insects are feeding on living or on dead and decaying plant tissues. The latter group of saprophytic insects is discussed in section 8.2 and is not dealt with further here. The former group includes larvae that feed in buds, fruits, nuts, seeds, roots, stalks and wood. **Stalk borers**, such as the wheat stem sawflies (Hymenoptera: Cephidae: *Cephus* species) and the European corn borer (Lepidoptera: Pyralidae: *Ostrinia nubilalis*) (Fig. 10.3(a)), attack grasses and more succulent plants, whereas **wood borers** feed in the twigs, stems or trunks of woody plants where they may eat the bark, phloem, sapwood or heartwood. The wood-boring habit is typical of many Coleoptera, especially the larvae of jewel beetles (Buprestidae), longicorn (or longhorn) beetles (Cerambycidae) and weevils (Curculionoidea), and some Lepidoptera (e.g. Hepialidae and

Cossidae, Fig. 1.2) and Hymenoptera. The root-boring habit is well developed in the Lepidoptera, but many moth larvae do not differentiate between the wood of trunks, branches or roots. Many species damage plant storage organs by boring into tubers, corms and bulbs.

The reproductive output of many plants is reduced or destroyed by the feeding activities of larvae that bore into and eat the tissues of fruits, nuts or seeds. **Fruit borers** are notable among the Diptera, especially Tephritidae, such as the apple maggot *Rhagoletis pomonella* and the Mediterranean fruit fly *Ceratitis capitata*); Lepidoptera, for example, some tortricids such as the oriental fruit moth, *Grapholita molesta*, and the codling moth, *Cydia pomonella* (Fig. 10.3(b)); and the Coleoptera, particularly certain weevils, such as the plum curculio, *Conotrachelus nenuphar*. Weevil larvae also are common occupants of seeds and nuts and many species are pests of stored grain (section 10.2.5).

10.2.3 Sap sucking

The feeding activities of insects that chew or mine leaves and shoots cause obvious damage. In contrast, structural damage caused by sap-sucking insects is often inconspicuous since the withdrawal of cell contents from plant tissues usually leaves the cell walls intact. Damage to the plant may be difficult to quantify even though the sap sucker drains plant resources by removing phloem or xylem contents, causing loss of condition such as retarded root growth, fewer leaves or less overall biomass accumulation compared with unaffected plants. These effects may be detectable with confidence only by controlled experiments in which the growth of infested and uninfested plants are compared. Certain sap-sucking insects do cause conspicuous tissue necrosis either by transmitting diseases, especially viral ones, or injecting toxic saliva, whereas others induce obvious tissue

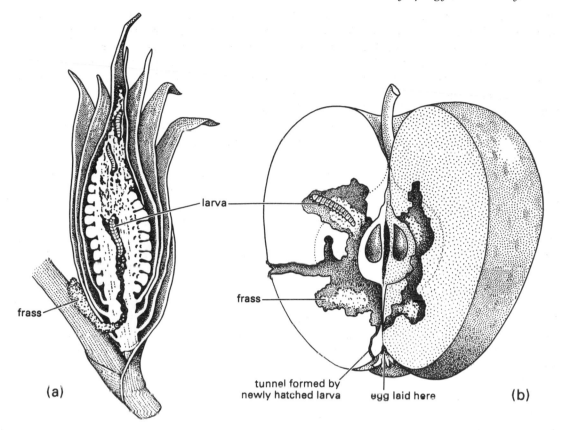

Fig. 10.3 Plant borers: (a) larvae of the European corn borer, *Ostrinia nubilalis* (Lepidoptera: Pyralidae), tunnelling in a corn stalk; (b) a larva of the codling moth, *Cydia pomonella* (Lepidoptera: Tortricidae), inside an apple. (After Frost, 1959.)

distortion or growth abnormalities called galls (section 10.2.4).

Most sap-sucking insects belong to the Hemiptera. All hemipterans have long, thread-like mouthparts consisting of appressed mandibular and maxillary stylets forming a bundle lying in a groove in the labium (Box 10.8). The maxillary stylet contains a salivary canal that directs saliva into the plant and a food canal through which plant juice or sap is sucked up into the insect's gut. Only the stylets enter the tissues of the host plant (Fig. 10.4(a)). They may penetrate superficially into a leaf or deeply into a plant stem or leaf midrib, following either an intracellular or intercellular path, depending on species. The feeding site reached by the stylet tips may be in the parenchyma (e.g. some immature scale insects, many Heteroptera), the phloem (e.g. most aphids, mealybugs, soft scales, psyllids and leafhoppers) or the xylem (e.g. spittlebugs and cicadas). In addition to a hydrolysing type of saliva, many species produce a solidifying saliva that forms a sheath around the stylets as they enter and penetrate the plant tissue. This sheath can be stained in tissue sections and allows the feeding tracks to be traced to the feeding site (Fig. 10.4(b),(c)). The two feeding strategies of hemipterans, stylet-sheath and maccrate-and-flush feeding, are described in section 3.6.2 and the gut specializations of homopterans for dealing with a watery diet are discussed in Box 3.3.

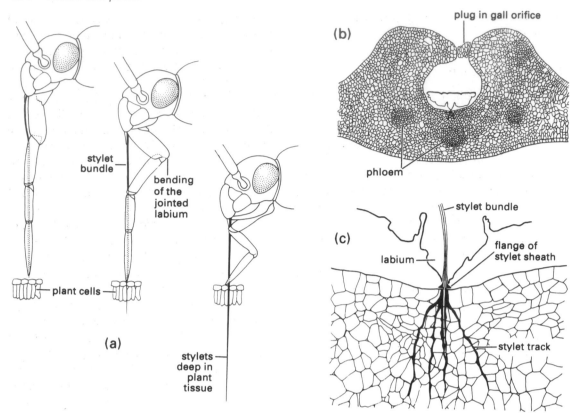

Fig. 10.4 Feeding in phytophagous Hemiptera: (a) penetration of plant tissue by a mirid bug showing bending of the labium as the stylets enter the plant; (b) transverse section through a eucalypt leaf gall containing a feeding nymph of a scale insect, *Apiomorpha* (Eriococcidae); (c) enlargement of the feeding site of (b), showing multiple stylet tracks (formed of solidifying saliva) resulting from probing of the parenchyma. ((a) After Poisson, 1951.)

Many species of plant-feeding Hemiptera are considered serious agricultural and horticultural pests. Loss of sap leads to wilting, distortion or stunting of shoots. Movement of the insect between host plants can lead to the efficient transmission of plant viruses and other diseases, especially by aphids and whiteflies. The sugary excreta (**honeydew**) of phloem-feeding Homoptera, particularly coccoids, is used by black sooty moulds which soil leaves and fruits and can impair photosynthesis.

Most thrips (Thysanoptera) also feed by sucking plant juices. They either rasp the leaf surface and suck up the sap as it exudes from the epidermal cells or penetrate the tissues more deeply using their stylets (Fig. 2.13) to pierce the epidermis and then rupture the cells below. Damaged areas discolour and the leaf, bud, flower or shoot may wither and die. Typically, plant damage is concentrated on rapidly growing tissues, so that flowering and leaf flushing may be seriously disrupted. Some thrips transmit viruses during feeding or inject toxic saliva.

Outside the Hemiptera and Thysanoptera, the sap-sucking habit is rare in extant insects. Many fossil species, however, had a rostrum with piercing-and-sucking mouthparts. For example, palaeodictyopteroids (Fig. 7.2) probably fed by imbibing juices from plant organs.

10.2.4 Gall formation

Insect-induced plant **galls** result from a very specialized type of insect–plant interaction in which the morphology of plant parts is altered, often substantially and characteristically, by the influence of the insect. Generally galls are defined as pathologically developed cells, tissues or organs of plants that have arisen by hypertrophy (increase in cell size) and/or hyperplasia (increase in cell number) due to stimulation from foreign organisms. Some galls are induced by viruses, bacteria, fungi, nematodes and mites, but many more are caused by insects. The study of plant galls is called **cecidology**; gall-causing animals (insects, mites and nematodes) are **cecidozoa**; and galls induced by cecidozoa are referred to as **zoocecidia**. Cecidogenic insects account for about 2% of all described insect species, with perhaps 13 000 species of gall formers.

On a world basis, the principal cecidozoa in terms of number of species represent just three orders of insects – the Hemiptera, Diptera and Hymenoptera. In addition, about 300 species of mostly tropical Thysanoptera (thrips) are associated with galls, although not necessarily as inducers. Some species of Coleoptera (mostly weevils) and microlepidoptera (small moths) induce galls. Most hemipteran galls are elicited by Homoptera, in particular aphids, coccoids and psyllids; their galls are structurally diverse and those of gall-forming eriococcids (Coccoidea: Eriococcidae) often exhibit spectacular sexual dimorphism, with galls of female insects much larger and more complex than those of their conspecific males (Fig. 10.5(a),(b)). Worldwide, there are several hundred gall-forming coccoid species in about seven families, about 350 gall-forming Psyllidae, and perhaps 700 gall-inducing aphid species distributed among the three families Adelgidae, Aphididae and Phylloxeridae.

The Diptera contains the highest number of gall-inducing species, perhaps thousands, but the taxonomy of many dipteran gall formers is inadequate. Most cecidogenous flies belong to one family, the Cecidomyiidae (gall midges), most of which form leaf galls, with some causing stem, bud, flower and sometimes root galls. The other fly family that includes some important cecidogenic species is the Tephritidae, in which gall inducers associated with Compositae may be useful for biological control of some weeds. Three superfamilies of wasps contain large numbers of gall-inducing species: the Cynipoidea contains the gall wasps (Cynipidae, perhaps 2000 species) which are among the best known gall insects in Europe and North America, where hundreds of species form often extremely complex galls, especially on oaks and roses (Fig. 10.5(c),(d)); the Tenthredinoidea has a number of gall-forming sawflies, such as *Pontania* species (Tethredinidae) (Fig. 10.5(g)); and several families of Chalcidoidea include gall-formers, especially species in the Agaonidae (fig wasps; Box 10.3), Eurytomidae and Pteromalidae.

There is enormous diversity in the patterns of development, shape and cellular complexity of insect galls (Fig. 10.5). They range from relatively undifferentiated masses of cells ('indeterminate' galls) to highly organized structures with distinct tissue layers ('determinate' galls). Determinate galls usually have a shape that is specific to each insect species. Some of the most histologically complex and specialized galls are formed by cynipid wasps and eriococcids; these galls have distinct tissue layers or types that may bear little resemblance to the plant part from which they are derived. Among the determinate galls, different shapes correlate with mode of gall formation, which is related to the initial position and feeding method of the insect. Some common types of galls are:

- **Covering galls**, in which the insect becomes enclosed within the gall, either with an

Fig. 10.5 A variety of insect-induced galls: (a) two coccoid galls, each formed by a female of *Apiomorpha munita* (Hemiptera: Eriococcidae) on the stem of *Eucalyptus melliodora*; (b) a cluster of galls, each containing a male of *A. munita* on *E. melliodora*; (c) three oak cynipid galls formed by *Cynips quercusfolii* (Hymenoptera: Cynipidae) on a leaf of *Quercus* sp.; (d) rose bedeguar galls formed by *Diplolepis rosae* (Hymenoptera: Cynipidae) on *Rosa* sp.; (e) a leaf petiole of lombardy poplar, *Populus nigra*, galled by the aphid *Pemphigus spirothecae* (Hemiptera: Aphididae); (f) three psyllid galls, each formed by a nymph of *Glycaspis* sp. (Hemiptera: Psyllidae) on a eucalypt leaf; (g) willow bean galls of the sawfly *Pontania proxima* (Hymenoptera: Tenthredinidae) on a leaf of *Salix* sp. ((c)–(e), (g) After Darlington, 1975.)

opening (ostiole) to the exterior, as in coccoid galls (Fig. 10.5(a),(b)), or without any ostiole, as in cynipid galls (Fig. 10.5(c)).

- **Filz galls**, characterized by their hairy epidermal outgrowths (Fig. 10.5(d)).
- **Roll** and **fold galls**, in which differential growth provoked by insect feeding results in rolled or twisted leaves, shoots or stems that are often swollen, as in many aphid galls (Fig. 10.5(e)).
- **Pouch galls**, which develop as a bulge of the leaf blade, forming an invaginated pouch on one side and a prominent bulge on the other, as in many psyllid galls (Fig. 10.5(f)).
- **Mark galls**, in which the insect egg is deposited inside stems or leaves so that the larva is completely enclosed throughout its development, as in sawfly galls (Fig. 10.5(g)).
- **Pit galls**, in which a slight depression, sometimes surrounded by a swelling, is formed where the insect feeds.
- **Bud** and **rosette galls**, which vary in complexity and cause enlargement of the bud or sometimes multiplication and miniaturization of new leaves, forming a gall resembling a pine cone.

Gall formation may involve two separate processes: (a) initiation, and (b) subsequent growth and maintenance of structure. Usually galls can be stimulated to develop only from actively growing plant tissue; therefore galls are initiated on young leaves, flower buds, stems and roots and rarely on mature plant parts. Some complex galls develop only from undifferentiated meristematic tissue, which becomes moulded into a distinctive gall by the activities of the insect. Development and growth of insect-induced galls (including, if present, the nutritive cells upon which some insects feed) depend upon continued stimulation of the plant cells by the insect. Gall growth ceases if the insect dies or reaches maturity. It appears that gall insects, rather than the plants, control most aspects of gall formation, largely via their feeding activities.

The mode of feeding differs in different taxa as a consequence of fundamental differences in mouthpart structure. The larvae of gall-forming beetles, moths and wasps have biting and chewing mouthparts, whereas nymphal aphids, coccoids, psyllids and thrips have piercing and sucking mouthparts. Larval gall midges have vestigial mouthparts and largely absorb nourishment by suction. Thus these different insects mechanically damage and deliver chemicals (or perhaps genetic material – see below) to the plant cells in a variety of ways.

Little is known about what stimulates gall induction and growth. Wounding and plant hormones (such as cytokinins) appear important in indeterminate galls, but the stimuli are probably more complex for determinate galls. Oral secretions, anal excreta and accessory gland secretions have been implicated in different insect–plant interactions that result in determinate galls. The best-studied compounds are the salivary secretions of Hemiptera. Salivary substances, including amino acids, auxins (and other plant growth regulators), phenolic compounds and phenol oxidases, in various concentrations, may have a role either in gall initiation and growth or in overcoming the defensive necrotic reactions of the plant. Plant hormones, such as auxins and cytokinins, must be involved in cecidogenesis but it is equivocal whether these hormones are produced by the insect, or by the plant as a directed response to the insect, or are incidental to gall induction. In certain complex galls, such as those of eriococcoids and cynipids, it is conceivable that the development of the plant cells is redirected by semi-autonomous genetic entities (viruses, plasmids or transposons) transferred from the insect to the plant. Thus the initiation of such galls may involve the insect acting as a DNA or RNA donor, as in some wasps that parasitize insect hosts (Box 12.1). Unfortunately, in comparison to anatomical and physiological studies of galls, genetic investigations are in their infancy.

The gall-forming habit may have evolved either from plant mining and boring (especially for Lepidoptera, Hymenoptera and certain Diptera) or from sedentary surface feeding (as is likely for Hemiptera, Thysanoptera and cecidomyiid Diptera). It is believed to be beneficial to the insects, rather than a defensive response of the plant to insect attack. All gall insects derive their food from the tissues of the gall, as well as some shelter or protection from natural enemies and adverse conditions of temperature or moisture. The relative importance of these environmental factors to the origin of the galling habit is difficult to ascertain because current advantages of gall living may differ from those gained in the early stages of gall evolution. Clearly most galls are 'sinks' for plant assimilates – the nutritive cells that line the cavity of wasp and fly galls contain higher concentrations of sugars, protein and lipids than ungalled plant cells. Thus one advantage of feeding on gall rather than normal plant tissue is the availability of high quality food. Moreover, for sedentary surface feeders (such as aphids, psyllids and coccoids) galls furnish a more protected microenvironment than the normal plant surface. Some cecidozoa may 'escape' from certain parasitoids and predators that are unable to penetrate galls, particularly those with thick woody walls.

Other natural enemies specialize in feeding on gall-living insects or their galls and often it is difficult to determine which insects were the original inhabitants. Galls are remarkable for the association of an extremely complex community of species, other than the gall causer, belonging to diverse insect groups. These other species may be either parasitoids of the gall former (i.e. parasites that cause the eventual death of their host; Chapter 12) or inquilines ('guests' of the gall former) that obtain their nourishment from tissues of the gall. In some cases, gall inquilines cause the original inhabitant to die through abnormal growth of the gall; this may obliterate the cavity in which the gall former lives or prevent emergence from the gall. If two species are obtained from a single gall or a single type of gall, one of these insects must be a parasitoid or an inquiline. There are even cases of hyperparasitism, in which the parasitoids themselves are subject to parasitization (section 12.3.1).

10.2.5 Seed predation

Plant seeds usually contain higher levels of nutrients than other tissues, providing for the growth of the seedling. This resource is used by specialist seed-eating insects. Notable are many beetles, harvester ants (especially species of *Messor*, *Monomorium* and *Pheidole*) which store seeds in underground granaries, bugs (many Coreidae, Lygaeidae, Pentatomidae, Pyrrhocoridae and Scutelleridae) that suck out the contents of developing or mature seeds and a few moths (such as some Gelechiidae and Oecophoridae).

Harvester ants are ecologically significant seed predators. These are the dominant ants in terms of biomass or colony numbers in deserts and dry grasslands in many parts of the world. Usually the species are highly polymorphic, with the larger individuals possessing powerful mandibles capable of cracking open seeds. Seed fragments are fed to larvae, but probably many harvested seeds escape destruction either by being abandoned in stores or by germinating quickly within the ant nests. Thus seed harvesting by ants, which could be viewed as exclusively detrimental, may carry some benefits to the plant through dispersal and provision of local nutrients to the seedling.

An array of beetles (especially bruchine Chrysomelidae and Curculionidae) develop either entirely within individual seeds or consume several seeds within one fruit. Some seed beetles (Bruchinae or Bruchidae, according to different texts), particularly those attacking leguminous food plants such as peas and beans, are serious pests. Species that eat dried seeds are pre-adapted to be pests of stored products such as pulses and grains. Adult beetles typically oviposit onto the developing ovary or the seeds or fruits and the larvae then mine through the fruit or seed wall or coat. The larvae develop and pupate inside seeds, thus destroying them. Successful development usually occurs only in the final stages of maturity of seeds and there appears to be a 'window of opportunity' for the larvae: a mature seed may have an inpenetrable seed coat but if young seeds are attacked the plant can abort the infected seed or even the whole fruit/pod if little investment has been made in it. Aborted seeds and those shed to the ground (whether mature or not) are less attractive to seed beetles than those retained on the plant, but evidently stored-product pests have no difficulty in developing within cast (i.e. harvested and stored) seeds. The larvae of the granary weevil, *Sitophilus granarius* (Box 10.10), and rice weevil, *S. oryzae*, develop inside dry grains of corn, wheat, rice and other plants.

Plant defence against seed predation includes the provision of protective seed coatings or toxic chemicals (allelochemicals) or both. Another strategy is the synchronous production by a single plant species of an abundance of seeds, often separated by long intervals of time. Seed predators either cannot synchronize their life cycle to the cycle of glut and scarcity, or are overwhelmed and unable to find and consume the total seed production.

10.2.6 Insects as biological control agents for weeds

Weeds are simply plants that are growing where they are not wanted. Some weed species are of little economic or ecological consequence, whereas the presence of others results in significant losses to agriculture or causes detrimental effects in natural ecosystems. Most plants are weeds only in areas outside their native distribution, where suitable climatic and edaphic conditions, usually in the absence of natural enemies, favour their growth and survival. Sometimes exotic plants that have become weeds can be controlled by introducing host-specific phytophagous insects from the area of origin of the weed. This is called classical biological control of weeds and it is analogous to the classical biological control of insect pests explained in detail in section 15.5. Another form of biological control, called augmentation (section 15.5), involves increasing the natural level of insect enemies of a weed and thus requires mass rearing of insects for inundative release. This method of controlling weeds is unlikely to be cost effective for most insect–plant systems. The tissue damage caused by introduced or augmented insect enemies of weeds may limit or reduce vegetative growth (as shown for the weed discussed in Box 10.2), prevent or reduce reproduction or make the weed less competitive than other plants in the environment.

A classical biological control programme involves a sequence of steps that include biological as well as sociopolitical considerations. Each programme is initiated with a review of available data (including taxonomic and distributional information) on the weed, its plant relatives and any known natural enemies, so that the nuisance status of the target weed and a strategy for collecting, rearing and testing the utility of potential insect enemies can be outlined. Regulatory authorities must then approve the proposal to attempt control of the weed. Next, foreign exploration and local surveys must determine the potential control agents attacking the weed in both its native and introduced ranges. The weed's ecology, especially in relation to its natural enemies, must be studied in its native range. The host specificity of potential control agents must be tested, either inside or outside the country of introduction and, in the former case, always in quarantine. The results of these tests will determine whether the regulatory authorities approve the importation of the agents for subsequent release or

BOX 10.2
SALVINIA AND PHYTOPHAGOUS WEEVILS

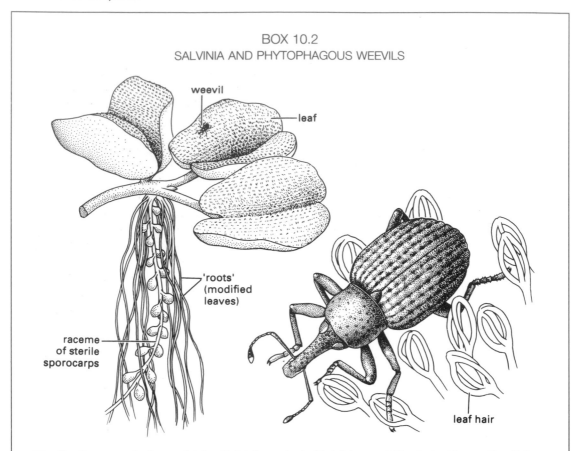

The floating aquatic fern salvinia, *Salvinia molesta* (Salviniaceae) (illustrated here after Sainty and Jacobs, 1981), has spread by human agency since 1939 to many tropical and subtropical lakes, rivers and canals throughout the world. Salvinia colonies consist of ramets (units of a clone) connected by horizontal, branching rhizomes. Growth is favoured by warm, nitrogen-rich water. Conditions suitable for vegetative propagation and the absence of natural enemies in its non-native range have allowed very rapid colonization of large freshwater expanses. Salvinia becomes a serious weed because its thick mats completely block waterways, choking the flow and disrupting the livelihood of people who depend on them for transport, irrigation and food (especially fish, rice, sago palms, etc.). This problem was especially acute in parts of Africa, India, Southeast Asia and Australasia, including the Sepik River in Papua New Guinea. Limited control could be achieved by expensive manual and mechanical removal and herbicides, but by the early 1980s perhaps 2000 km^2 of water surface were covered by this invasive plant. The potential of biological control was recognized in the 1960s, although it was slow to be used (for reasons outlined below) until the 1980s, when outstanding successes were achieved in most areas where it was attempted. Choked lakes and rivers became open water again.

The phytophagous insect responsible for this spectacular control of *S. molesta* is a tiny (2 mm long) weevil (Curculionidae) called *Cyrtobagous salviniae* (shown enlarged in the

BOX 10.2—*continued*

drawing on the right, after Calder and Sands, 1985). Adult weevils feed on salvinia buds, whereas larvae tunnel through buds and rhizomes as well as feeding externally on roots. The weevils are host specific, have a high searching efficiency for salvinia and can live at high population densities without intraspecific interference stimulating emigration. These characteristics allow the weevils to control salvinia effectively.

Initially, biological control of salvinia failed due to unforeseen taxonomic problems with the weed and the weevil. Before 1972, the weed was thought to be *Salvinia auriculata*, which is a South American species fed upon by the weevil *Cyrtobagous singularis*. Even when the correct identity was established as *Salvinia molesta*, it was not until 1978 that its native range was discovered to be southeastern Brazil. Weevils feeding there on *S. molesta* were believed to be conspecific with *C. singularis* feeding on *S. auriculata*. However, after preliminary testing and subsequent success in controlling *S. molesta*, the weevil was recognized as specific to *S. molesta*, new to science and named as *C. salviniae*.

The benefits of control to people living in Africa, Asia and the Pacific are substantial, whether measured in economic terms or as savings in human health and social systems. For example, villages in Papua New Guinea that were abandoned because of salvinia have been re-occupied. Similarly, the environmental benefits of eliminating salvinia infestations are great, since this weed is capable of reducing a complex aquatic ecosystem to a virtual monoculture. The economics of salvinia control have been studied only in Sri Lanka, where a cost:benefit analysis showed returns on investment of 53:1 in terms of cash and 1678:1 in terms of hours of labour. Appropriately, the team responsible for the ecological research that lead to biological control of salvinia was recognized by the award of the UNESCO Science Prize in 1985. Taxonomists made essential contributions by establishing the true identities of the salvinias and weevils.

only for further testing, or refuse approval. After importation, there is a period of rearing in quarantine to eliminate any imported diseases or parasitoids, prior to mass rearing in preparation for field release. Release is dependent on the quarantine procedures being approved by the regulatory authorities. After release, the establishment, spread and effect of the insects on the weed must be monitored. If weed control is attained at the initial release site, the spread of the insects is assisted by manual distribution to other sites.

There have been some outstandingly successful cases of deliberately introduced insects controlling invasive weeds. The control of the water weed salvinia by a *Cyrtobagous* weevil (Box 10.2) and prickly pear cacti, *Opuntia* species, by the larvae of the *Cactoblastis* moth are just two examples. On the whole, however, the chances of successful biological control of weeds by released phytophagous organisms are not high (Fig. 10.6) and vary in different circumstances, often unpredictably.

In general, perennial weeds of uncultivated areas are well suited to classical biological control, since long-lived plants, which are predictable resources, are generally associated with host-specific insect enemies. Cultivation can disrupt these insect populations. In contrast, augmentation of insect enemies of a weed may be best suited to annual weeds of cultivated land, where mass-reared insects

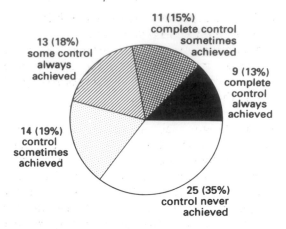

Fig. 10.6 Possible outcomes of releases of alien phytophagous organisms against invasive plants for the biological control of these weeds. The data include 72 weed species for which agents have been introduced and established long enough to permit control assessment. (After Sheppard, 1992, based on data from Julien, 1992.)

could be released to control the plant early in its growing season. Sometimes it is claimed that highly variable, genetically outcrossed weeds are hard to control and that insects 'newly associated' (in an evolutionary sense) with a weed have greater control potential due to their infliction of greater damage. The number of studies for which control assessment is possible are limited and the reasons for variation or failure in control of weeds are diverse. Currently, prediction of the success or failure of control in terms of weed or phytophage ecology and behaviour is unsatisfactory. The interactions of plants, insects and environmental factors are complicated and likely to be case-specific.

In addition to the uncertainty of success of classical biological control programmes, the control of certain weeds can cause potential conflicts of interest. Sometimes the target weed may not be considered a weed by everyone. For example, in Australia, the introduced *Echium plantagineum* (Boraginaceae) is called 'Paterson's curse' by those who consider it an agricultural weed and

'Salvation Jane' by some pastoralists and beekeepers who regard it as a source of fodder for livestock and nectar for bees. A second type of conflict may arise if the natural phytophages of the weed are oligophagous rather than monophagous, as they may feed on a few species other than the target weed. In this case, the introduction of insects that are not strictly host-specific may pose a risk for beneficial and/or native plants in the proposed area of introduction of the control agent. For example, some of the insects that have been introduced into Australia as control agents for *E. plantagineum* also feed on other boraginaceous plants. The risks of damage to such non-target species must be assessed carefully prior to releasing foreign insects for the biological control of a weed. Some introduced phytophagous insects may become pests in their new habitat.

10.3 INSECTS AND PLANT REPRODUCTIVE BIOLOGY

Insects are intimately associated with plants. Agriculturalists, horticulturalists and gardeners are aware of their role in damage and disease dispersal. However, certain insects are vitally important to many plants, assisting in their reproduction, through pollination, or their dispersal, through spreading their seeds.

10.3.1 Pollination

Sexual reproduction in plants involves **pollination** – the transfer of pollen (male germ cells in a protective covering) from the anthers of a flower to the stigma (Fig. 10.7(a)). A pollen tube grows from the stigma down the style to an ovule in the ovary where it fertilizes the egg. Pollen generally is transferred either by an animal pollinator or by the wind. Transfer from anthers to stigma of the same plant (either of the same flower or a different flower) is self-pollination; transfer between flowers on different plants (with different genotypes)

(a)

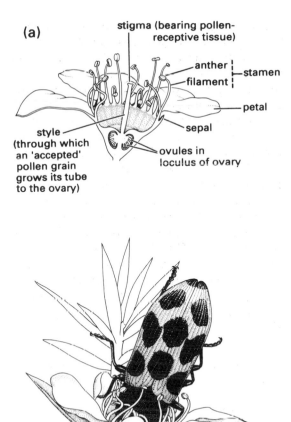

(b)

Fig. 10.7 Anatomy and pollination of a tea-tree flower, *Leptospermum* (Myrtaceae): (a) diagram of a flower showing the parts; (b) a jewel beetle, *Stigmodera* sp. (Coleoptera: Buprestidae), feeding from a flower.

of the same species is cross-pollination. Most flowering plants are pollinated by animals, especially insects, and it is argued that the success of the angiosperms relates to the development of these interactions. The benefits of insect pollination (**entomophily**) over wind pollination (**anemophily**) include:

- Increase in pollination efficiency, including reduction of pollen wastage.

- Successful pollination under conditions unsuitable for wind pollination.
- Maximization of the number of plant species in a given area, since even rare plants can receive conspecific pollen carried into the area by insects.

Within-flower self-pollination also brings some of these advantages, but continued selfing induces deleterious homozygosity and is rarely a dominant fertilization mechanism.

In general it is advantageous to a plant for its pollinators to be specialist visitors that faithfully pollinate the flowers of only one or a few plant species. Pollinator constancy, which may initiate the isolation of small plant populations, is especially prevalent in the Orchidaceae – the most speciose family of vascular plants.

The major **anthophilous** (flower-frequenting) taxa among insects are the beetles (Coleoptera), flies (Diptera), wasps, bees and ants (Hymenoptera), thrips (Thysanoptera), and butterflies and moths (Lepidoptera). These insects visit flowers primarily to obtain nectar or pollen, but even some predatory insects may pollinate the flowers that they visit. Nectar primarily consists of a solution of sugars, especially glucose, fructose and sucrose. Pollen often has a high protein content plus sugar, starch, fat, and traces of vitamins and inorganic salts. In the case of a few bizarre interactions, male hymenopterans are attracted by neither pollen nor nectar but by the resemblance of certain orchid flowers in shape, colour and odour to their conspecific females. In attempting to mate (**pseudocopulate**) with the insect-mimicking flower, the male inadvertently pollinates the orchid with pollen that adhered to his body during previous pseudocopulations. Pseudocopulatory pollination is common among Australian thynnine wasps (Tiphiidae) and occurs in a few other wasp groups, some bees and rarely in ants.

Cantharophily (beetle pollination) may be the oldest form of insect pollination.

Beetle-pollinated flowers are often white or dull in colour, strong-smelling and regularly bowl- or dish-shaped (Fig. 10.7). Beetles visit flowers mostly for pollen, although nutritive tissue or easily accessible nectar may be utilized, and the plant's ovaries are usually well protected from the biting mouthparts of their pollinators. The major beetle families that commonly or exclusively contain anthophilous species are the Buprestidae (jewel beetles; Fig. 10.7(b)), Cantharidae (soldier beetles), Cerambycidae (longicorn or longhorn beetles), Cleridae (checkered beetles), Dermestidae, Lycidae (net-winged beetles), Melyridae (soft-winged flower beetles), Mordellidae (tumbling flower beetles), Nitidulidae (sap beetles) and Scarabaeidae (scarabs).

Myophily (fly pollination) occurs when flies visit flowers to obtain nectar, although hover flies (Syrphidae) feed chiefly on pollen rather than nectar. Fly-pollinated flowers tend to be less showy than other insect-pollinated flowers but may have a strong smell, often malodorous. Flies generally utilize many different sources of food and thus their pollinating activity is irregular and unreliable. However, their sheer abundance and the presence of some flies throughout the year mean that they are important pollinators for many plants. Both dipteran suborders contain anthophilous species. Among the Nematocera, mosquitoes and bibionids are frequent blossom visitors, and predatory midges, principally of *Forcipomyia* species (Ceratopogonidae), are essential pollinators of cocoa flowers. Pollinators are more numerous in the Brachycera, in which at least 30 families are known to contain anthophilous species. Major pollinator taxa are the Bombyliidae (bee flies), Syrphidae and muscoid families.

Many members of the large order Hymenoptera visit flowers for nectar or pollen. The suborder Apocrita, which contains most of the wasps (as well as bees and ants), is more important than the Symphyta (sawflies) in terms of **sphecophily** (wasp pollination). Many pollinators are found in

the superfamilies Ichneumonoidea and Vespoidea. Fig wasps (Chalcidoidea: Agaonidae) are highly specialized pollinators of the hundreds of species of figs (Box 10.3). Ants (Vespoidea: Formicidae) are rather poor pollinators (Box 10.4), although **myrmecophily** (ant pollination) is known for a few plant species.

Bees are regarded as the most important group of insect pollinators. They collect nectar and pollen for their brood as well as for their own consumption. There are over 20 000 species of bees worldwide and all are anthophilous. Plants that depend on **melittophily** (bee pollination) often have bright (yellow or blue), sweet-smelling flowers with nectar guides – lines (often visible only in UV light) on the petals that direct pollinators to the nectar. The main bee pollinator worldwide is the honey bee, *Apis mellifera* (Apidae). The pollination services provided by this bee are extremely important for many crop plants (section 1.2) but in natural ecosystems serious problems can be caused. Honey bees compete with native insect pollinators by depleting nectar and pollen supplies and may disrupt pollination by displacing the specialist pollinators of native plant species.

Most members of the Lepidoptera feed from flowers using a long, thin proboscis. In the speciose Ditrysia (the 'higher' Lepidoptera) the proboscis is retractile (Fig. 2.12), allowing feeding and drinking from sources distant from the head. Such a structural innovation may have contributed to the radiation of this successful group, which contains 98% of all lepidopteran species. Flowers pollinated by butterflies and moths are often regular, tubular and sweet-smelling. **Phalaenophily** (moth pollination) is typically associated with light-coloured, pendant flowers that have nocturnal or crepuscular anthesis (opening of flowers), whereas **psychophily** (butterfly pollination) is typified by red, yellow or blue upright flowers that have diurnal anthesis.

Insect–plant interactions associated with pollination are clearly mutualistic. The plant is

BOX 10.3
FIGS AND FIG WASPS

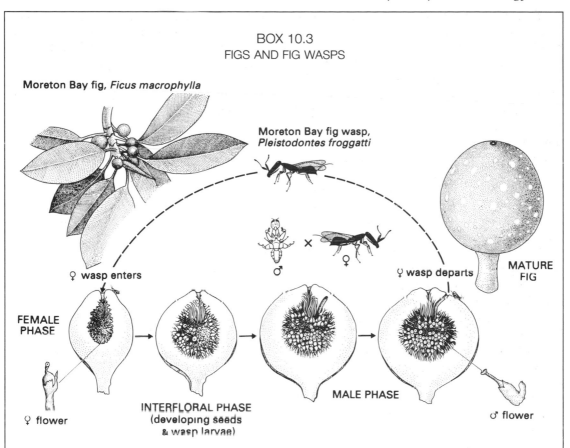

Moreton Bay fig, *Ficus macrophylla*

Moreton Bay fig wasp,
Pleistodontes froggatti

♀ wasp enters

♂   ×   ♀

♀ wasp departs

MATURE
FIG

FEMALE
PHASE

MALE PHASE

♀ flower

INTERFLORAL PHASE
(developing seeds
& wasp larvae)

♂ flower

Figs belong to the large and mostly tropical genus *Ficus* (Moraceae) of about 900 species. Each species of fig (except for the self-fertilizing cultivated edible fig) has a complex obligatory mutualism with usually only one species of pollinator. All of these pollinators are fig wasps belonging to the hymenopteran family Agaonidae, which comprises numerous species in many genera. Each fig tree produces a large crop of 500–1 000 000 fruit (syconia) as often as twice a year, but each fruit requires the action of at least one wasp in order to set seeds. Fig species are either dioecious (with male syconia on separate plants to those bearing the female syconia) or monoecious (with both male and female flowers in the same syconium). The following description of the life cycle of a fig wasp in relation to fig flowering and fruiting applies to monoecious figs, such as *F. macrophylla* illustrated here (after Froggatt, 1907, and Galil and Eisikowitch, 1968).

The female wasp enters the fig syconium via the ostiole (small hole), pollinates the female flowers which line the spheroidal cavity inside, oviposits in some of them (always short-styled ones) and dies. Each wasp larva develops within the ovary of a flower, which becomes a gall flower. Female flowers (usually long-styled ones) that escape wasp oviposition form seeds. About a month after oviposition, wingless male wasps emerge from their seeds and mate with female wasps still within the fig ovaries. Shortly after, the female wasps emerge from their seeds, gather pollen from another lot of flowers within the syconium (which is now in the

---

BOX 10.3—*continued*

male phase) and depart the mature fig to locate another conspecific fig tree in the phase of fig development suitable for oviposition. Different fig trees in a population are in different sexual stages, but all figs on one tree are synchronized. Species-specific volatile attractants produced by the trees allow precise location of another fig tree by the wasps.

For any given fig and wasp pair, reciprocal selection pressures presumably result in matching of fig and fig wasp traits. For example, the sensory receptors of the wasp respond only to the volatile chemicals of its host fig, and the size and morphology of the guarding scales of the fig ostiole allow entry only to a fig wasp of the 'correct' size and shape. It is likely that divergence in a local population of either fig or fig wasp, whether by genetic drift or selection, will induce coevolutionary change in the other. Host specificity provides reproductive isolation among both wasps and figs, so coevolutionary divergence among populations is likely to lead to speciation. The amazing diversity of *Ficus* and Agaonidae may be a consequence of this coevolution.

---

fertilized by appropriate pollen, while the insect obtains food (or sometimes fragrances) supplied by the plant, often specifically to attract the pollinator. It is clear that plants may experience strong selection as a result of insects. In contrast, in most pollination systems the evolution of the pollinators may have been little affected by the plants that they visited. For most insects, any particular plant is just another source of nectar or pollen and even insects that appear to be faithful pollinators over a short observation period may utilize a range of plants in their lifetime. Nevertheless, symmetrical influences do occur in some insect–plant pollination systems, as evidenced by the specializations of each fig wasp species to the fig species that it pollinates (Box 10.3), and by correlations between moth proboscis lengths and flower depths for a range of orchids and some other plants. For example, the Madagascar star orchid, *Angraecum sesquipedale*, which has floral spurs often exceeding 30 cm in length, is pollinated by a giant hawk moth, *Xanthopan morgani praedicta* (Sphingidae), with a proboscis up to about 25 cm long. Only this moth can reach the nectar at the apex of the floral

spurs and, during the process of pushing its head into the flower, it pollinates the orchid.

10.3.2 Myrmecochory – seed dispersal by ants

Many ants are seed predators that harvest and eat seeds (section 10.2.5). Seed dispersal may occur when seeds are accidentally lost in transport or seed stores are abandoned. Some plants have very hard seeds that are inedible to ants and yet many ant species actively collect and disperse them, a phenomenon called **myrmecochory**. These seeds have food bodies, called **elaiosomes**, and special chemical attractants that stimulate ants to collect them. Elaiosomes are seed appendages that vary in size, shape and colour and contain nutritive lipids, proteins and carbohydrates in varying proportions. These structures have diverse derivations from various ovarian structures in different plant groups. The ants, gripping the elaiosome with their mandibles (Fig. 10.8), carry the entire seed back to their nest, where the elaiosomes are removed and typically fed to the ant larvae. The hard seeds are then discarded, intact and viable, either in

---

BOX 10.4

FLOWER-VISITING NON-POLLINATORS

Ants are commonly anthophilous (flower-loving) but rarely pollinate the plants that they visit. Two hypotheses, perhaps acting together, have been postulated to explain the paucity of ant pollination. First, ants are flightless and are often small and are smooth-bodied. They are unlikely to facilitate cross-pollination because the foraging of each worker is confined to one plant, they often avoid contact with the anthers and stigmas, and pollen does not adhere easily to them. Second, the metapleural glands of ants produce secretions that spread over the integument and inhibit fungi and bacteria, but can also affect pollen viability and germination. Some plants have actually evolved mechanisms to deter ants but a few, especially in hot dry habitats, appear to have evolved adaptations to ant pollination.

---

an abandoned gallery of the nest or close to the nest entrance in a refuse pile.

Myrmecochory is a worldwide phenomenon, but is disproportionately prevalent in three plant assemblages: early-flowering herbs in the understory of north temperate mesic forests, perennials in Australian and southern African sclerophyll vegetation, and an eclectic assemblage of tropical plants. Myrmecochorous plants number more than 1500 species in Australia and about 1300 in South Africa, whereas only about 300 species occur in the rest of the world. They are distributed amongst more than 20 plant families and thus represent an ecological rather than a phylogenetic group, although they are predominantly legumes.

This association is of obvious benefit to the ants, for which the elaiosomes represent food, and the mere existence of the elaiosomes is evidence that the plants have become adapted for interactions with ants. The exact advantages to the plants of this relationship are controversial. Myrmecochory may reduce intraspecific or interspecific competition amongst plants by removing seeds to new sites. Seed removal to underground ant nests may provide protection from fire or seed predators, such as some birds, small mammals and other insects. Alternatively, ant

Fig. 10.8 An ant of *Rhytidoponera tasmaniensis* (Hymenoptera: Formicidae) carrying a seed of *Dillwynia juniperina* (Fabaceae) by its elaiosome (seed appendage).

nests may be rich in some or all plant nutrients, making them better microsites for seed germination and seedling establishment. However, no universal explanation for myrmecochory should be expected, since the relative importance of factors responsible for myrmecochory must vary according to plant species and geographic location.

Myrmecochory can be called a mutualism but the association is not characterized by specificity and reciprocity. There is no evidence that any myrmecochorous plant relies on a single ant species to collect its seeds. Similarly, there is no evidence that any ant species has adapted to collect the seeds of one particular myrmecochorous species. Of course, ants that harvest elaiosome-bearing seeds could be called a guild and the myrmecochorous plants of similar form and habitat could also represent a guild; but it is highly unlikely that myrmecochory represents an outcome of diffuse or guild coevolution, since no reciprocity can be inferred. Elaiosomes are just food items to ants, which display no obvious adaptations to myrmecochory. This fascinating form of seed dispersal appears to be the result of plant evolution, due to selection from ants in general, and not of coevolution of plants and specific ants.

10.4 MUTUALISM IN SPECIALIZED PLANT STRUCTURES

A great many insects live within plant structures. Some inhabit bored-out stems, leaf mines or galls, creating their own living spaces by destruction or physiological manipulation. In contrast, some plants have specialized structures or chambers, that house mutualistic insects, but have been formed in the absence of these guests. Two types of these special insect–plant interactions are discussed below.

10.4.1 Ant–plant interactions involving domatia

Domatia or 'little houses' may be hollow stems, tubers, swollen petioles or thorns, which are used by ants either for feeding or as nest sites, or both. True domatia are cavities that form independently of ants, such as in plants grown in glasshouses from which ants are excluded. It may be difficult to recognize true domatia in the field because ants often take advantage of natural hollows and crevices such as tunnels bored by the larvae of beetles or moths. Plants with true domatia are called ant plants or **myrmecophytes**; they are often trees, shrubs or vines of the secondary regrowth or understory of tropical lowland rainforest.

Ants benefit from association with myrmecophytes through provision of shelter for their nests and readily available food resources. Food comes either directly from the plant through food bodies or extrafloral nectaries (Fig. 10.9(a)), or indirectly via honeydew-excreting homopterans living within the domatia (Fig. 10.9(b)). **Food bodies** are small nutritive nodules on the foliage or stems of ant plants. **Extrafloral nectaries** (EFNs) are plant glands that produce sugary secretions (possibly also containing amino acids) attractive to ants and other insects. Plants with EFNs often occur in temperate areas and lack domatia, for example many Australian *Acacia* species, whereas plants with food bodies nearly always have domatia, and some plants have both EFNs and food bodies. Many myrmecophytes, however, lack both of the latter structures and instead the ants 'farm' scale insects or mealybugs (Coccoidea: Coccidae or Pseudococcidae) for their honeydew and possibly cull them to obtain protein. Like EFNs and food bodies, coccoids can draw the ants into a closer relationship with the plant by providing a resource on that plant.

Obviously myrmecophytes receive some benefits from ant occupancy of their domatia. The ants may provide protection from herbivores and plant competitors or supply nutrients to their host plant. Some ants aggressively defend their plant against grazing mammals, remove herbivorous insects and prune or detach other plants, such as epiphytes and vines, that grow on their host. This extremely aggressive caretaking is demonstrated by ants of *Pseudomyrmex* that protect *Acacia* in tropical America. Rather than protection, some myrmecophytes derive mineral nutrients and nitrogen from ant colony waste via absorption through the inner surfaces of the domatia. Such plant 'feeding' by ants, called **myrmecotrophy**, can be documented by following the fate of a radioactive label placed in ant prey. Prey is taken into the domatia and eaten, and the remains are discarded in refuse tunnels; the label ends up

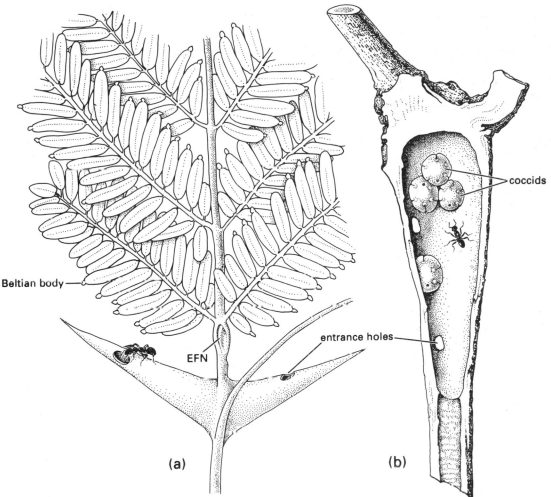

Fig. 10.9 Two myrmecophytes showing the domatia (hollow chambers) that house the ants and the food resources available to the ants: (a) a neotropical bull's-horn acacia, *Acacia sphaerocephala* (Fabaceae), with hollow thorns, Beltian (or food) bodies and an extrafloral nectary (EFN) used by the resident *Pseudomyrmex* ants; (b) a hollow swollen internode of *Kibara* (Monimiaceae) with scale insects of *Myzolecanium kibarae* (Hemiptera: Coccidae), which excrete honeydew that is eaten by the resident ants, *Anonychomyrma scrutator*. ((a) After Wheeler, 1910; (b) after Beccari, 1877.)

in the leaves of the plant. Myrmecotrophy occurs in the epiphytic *Myrmecodia* (Rubiaceae) (Fig. 10.10), species of which occur in the Malaysian and Australian regions, especially in New Guinea.

The majority of ant–plant associations may be opportunistic and unspecialized, although some tropical and subtropical ants (e.g. some *Pseudomyrmex* and *Azteca* species) are totally dependent on their particular host plants (e.g. *Acacia*, *Triplaris* and *Cecropia* species, respectively) for food and shelter. Likewise, if deprived of their attendant ants, these myrmecophytes decline. It is clear that these relationships are obligatorily mutualistic and no doubt others remain to be documented.

Fig. 10.10 A tuber of the epiphytic myrmecophyte, *Myrmecodia beccarii* (Rubiaceae), cut open to show the chambers inhabited by ants. Ants live in smooth-walled chambers and deposit their refuse in warted tunnels, from which nutrients are absorbed by the plant. (After Monteith, 1990.)

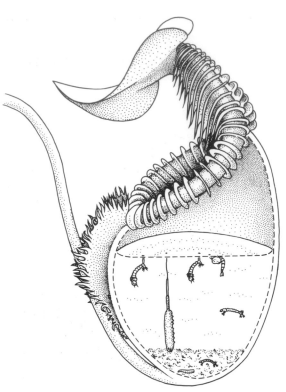

Fig. 10.11 A pitcher of *Nepenthes* (Nepenthaceae) cut open to show fly inquilines in the fluid. Clockwise from the top left: two mosquito larvae, a mosquito pupa, two chironomid midge larvae, a small maggot and a large rat-tailed maggot.

10.4.2 Phytotelmata – plant-held water containers

Many plants support insect communities in structures that retain water. The retention of water in 'containers' such as leaf axils of many bromeliads ('tank-plants'), gingers and teasels, for example, or the rot-holes in trees, appears incidental to the plants. Pitcher plants, however, have a complex architecture which is designed to lure and trap insects so that they are digested in the container liquid (Fig. 10.11).

The pitcher plants are a convergent grouping of the American Sarraceniaceae, Old World Nepenthaceae and Australian endemic Cephalotaceae. They generally live in nutrient-poor soils. Odour, colour and nectar entice insects, predominantly ants, into modified leaves – the 'pitchers'. Exit is prevented by guard hairs and slippery walls: the prey cannot escape and it drowns in the pitcher liquid, which contains digestive enzymes secreted by the plant.

This apparently inhospitable environment provides the home for a few specialist insects that live above the fluid, and many more living as larvae within. The adults of these insects can move in and out of the pitchers with impunity. Mosquito and midge larvae are the most common inhabitants, but other fly larvae of at least 12 families have been reported worldwide, and odonates, spiders and even a stem-mining ant occur in Southeast Asian pitchers. Many of these insect inquilines live in a mutualistic relationship with the plant, digesting trapped prey and micro-organisms and then excreting nutrients in a form readily available to the plant.

BOX 10.5
ORTHOPTERA (GRASSHOPPERS, LOCUSTS, KATYDIDS, CRICKETS)

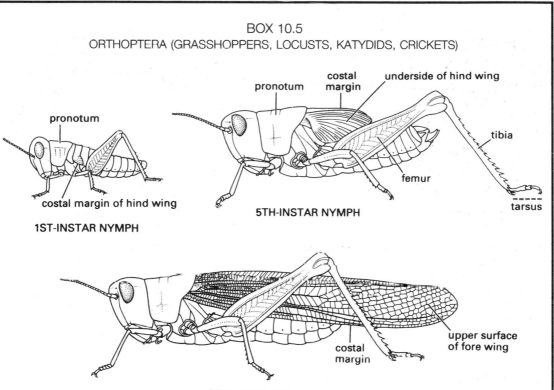

A worldwide order of more than 20 000 species in some 13 families, in two suborders: Caelifera (grasshoppers and locusts) and Ensifera (katydids and crickets). Hemimetabolous; elongate cylindrical, medium-sized to large, with enlarged hind legs for jumping. Hypognathous; mandibulate; compound eyes well developed, ocelli present or absent; antennae multisegmented; prothorax large, pronotum shield-like curving over pleura, mesothorax small, metathorax large; fore wings forming narrow, leathery tegmina, hind wings broad, with numerous longitudinal and cross-veins, folded beneath tegmina by pleating, with aptery and brachyptery frequent; legs often elongate and slender, hind legs large, usually saltatorial, tarsi with 1–4 segments; abdomen with 8–9 annular visible segments, with 2 or 3 terminal segments reduced; female with well-developed appendicular ovipositor (Figs 2.23(b),(c); Box 5.2); cerci 1-segmented.

Courtship may be elaborate and often involves communication by sound production and reception (sections 4.1.3–4). In copulation the male is astride the female; mating is sometimes prolonged for many hours. Ensiferan eggs are laid singly into plants or soil, whereas Caelifera use the ovipositor to bury batches of eggs into soil chambers. Egg diapause is frequent. Nymphs resemble small adults except in lack of development of wings and genitalia, but apterous adults may be difficult to distinguish from nymphs. In all winged species, nymphal wing-pad orientation changes between moults (as illustrated here for a locust); in early instars the wing-pad rudiments are lateral in position with the costal margin

BOX 10.5—*continued*

ventral, until in the penultimate nymphal instar (actually the third last moult) they rotate about their base so that the costal margin is dorsal and the morphological ventral surface is external; the hind wing then overlaps the fore wing (as in the 5th-instar nymph illustrated here). During the moult to the adult, the wings resume their normal position with the costal margin ventral. This wing pad 'rotation', otherwise known only in Odonata, is unique to the Orthoptera amongst the orthopteroid orders.

Caelifera are predominantly day-active, fast-moving, visually acute, terrestrial herbivores, and include some destructive insects such as migratory locusts (section 6.7.5; Fig. 6.13). Ensifera are more often night-active and camouflaged or mimetic predators, omnivores or phytophages.

The Orthoptera form a sister group to the Phasmatodea (Fig. 7.5) and some authors have united the orders, but the different wing bud development, egg morphology and lack of auditory organs in phasmatids warrant their separation.

BOX 10.6
PHASMATODEA (PHASMATIDS, PHASMIDS, STICK-INSECTS OR WALKING STICKS)

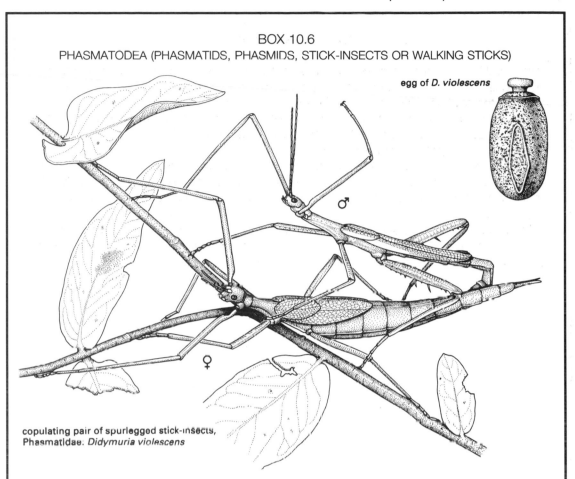

egg of *D. violescens*

♂

♀

copulating pair of spurlegged stick-insects,
Phasmatidae. *Didymuria violescens*

A worldwide, predominantly tropical order of more than 2500 species, conventionally classified in three families (although some workers raise many subfamilies to family rank). Hemimetabolous; elongate cylindrical and stick-like or flattened and often leaf-like, up to > 30 cm in length. Mandibulate; compound eyes anterolaterally placed and relatively small, ocelli only in winged species, often only in male; antennae short to long, with 8–100 segments; prothorax small, mesothorax and metathorax elongate if winged, shorter if apterous; wings, when present, functional in male, often reduced in female, many species apterous in both sexes, fore wings form leathery tegmina, hind wings broad, with network of numerous cross-veins and anterior margin toughened as remigium that protects folded wing; legs elongate, slender, gressorial, with 5-segmented tarsi, can be shed in defence (section 13.3) and may be regenerated in nymphal moult; abdomen 11-segmented, with segment 11 often forming concealed supra-anal plate in males or more obvious segment in females, cerci variably lengthened, 1-segmented; male genitalia concealed, asymmetrical.

In the often prolonged copulation the smaller male is astride the female, as illustrated here for the spurlegged stick-insect, *Didymuria violescens* (Phasmatidae). Eggs often resemble seeds (as shown here in the enlargement of the egg of *D. violescens*, after CSIRO, 1970) and

BOX 10.6—*continued*

are deposited singly, glued on vegetation or dropped to the ground; there may be lengthy egg diapause. Nymphal phasmatids mostly resemble adults except in lack of wing and genitalia development, absence of ocelli and the fewer antennal segments.

Phasmatodea are phytophagous and predominantly resemble (mimic) various vegetational features such as stems, sticks and leaves. In conjunction with crypsis, phasmatids demonstrate an array of anti-predator defences ranging from general slow movement and grotesque and often asymmetrical postures to death-feigning (sections 13.1, 13.2).

The Phasmatodea belong to the orthopteroid assemblage of orders (Fig. 7.5), and are distinguished from the Orthoptera (in which they have sometimes been placed) by their body shape, asymmetrical male genitalia and proventricular structure, and also in the nymphal wing pads, in which there is no 'rotation' during development, as in Orthoptera.

BOX 10.7
THYSANOPTERA (THRIPS)

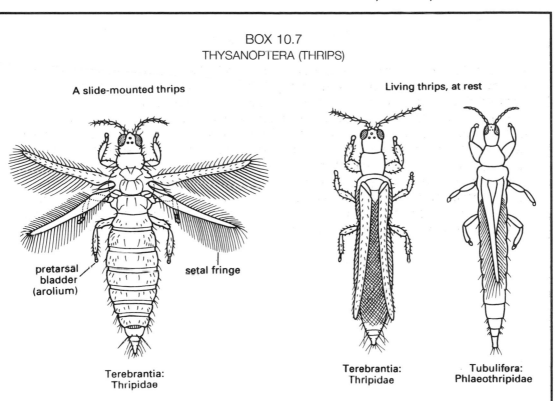

A slide-mounted thrips

Living thrips, at rest

pretarsal bladder (arolium)

setal fringe

Terebrantia: Thripidae

Terebrantia: Thripidae

Tubulifera: Phlaeothripidae

A worldwide order of minute to small insects (from 0.5 mm to a maximum length of 15 mm), comprising about 5000 species in two suborders: Terebrantia with seven families (including the speciose Thripidae) and Tubulifera with one family (speciose Phlaeothripidae). Development intermediate between hemi- and holometabolous; body slender, elongate; head elongate, usually hypognathous; mouthparts (Fig. 2.13(a)) with maxillary laciniae formed as grooved stylets, right mandible atrophied, left mandible formed as stylet, all three stylets form feeding tube; compound eyes small to large, three ocelli in fully winged forms; antennae 4–9-segmented, anteriorly directed; thorax development varies according to presence of wings; fore and hind wings similar, strap-like with long setal fringe (as illustrated on left for a terebrantian thrips, after Lewis, 1973), at rest wings parallel in Terebrantia (middle figure) but overlapping in Tubulifera (right figure), microptery and aptery occur; legs short, gressorial, sometimes with fore legs raptorial, hind legs saltatory, tarsi 1–2-segmented, pretarsus with apical protrusible adhesive ariolum (bladder or vesicle); abdomen 11-segmented (though with only 10 segments visible), male gonitalia concealed, symmetrical; female with cerci absent, ovipositor serrate in suborder Terebrantia, very reduced in suborder Tubulifera.

Eggs are laid into plant tissue (Terebrantia) or into crevices or exposed vegetation (Tubulifera). First- and second-instar nymphs resemble small adults except in wings and genitalia; however, instars 3 to 4 or 5 are resting stages termed 'prepupa' and 'pupa', during which significant tissue reconstruction takes place. Female thrips are diploid, whereas males

BOX 10.7—*continued*

(if present) are haploid, produced from unfertilized eggs. Arrhenotokous parthenogenesis is common; thelytoky is rare (section 5.10).

The primitive feeding mode of thrips was probably fungal feeding. Some species living in litter still eat fungi but thrips are primarily phytophages, including some gall formers, and there are a few predators. Plant-feeding thrips use their single mandibular stylet to pierce a hole through which the maxillary stylets are inserted. The contents of single cells are sucked out one at a time; pollen-feeding thrips similarly remove the contents of individual pollen grains. Thrips may aggregate in flowers, where they may act as pollinators. Subsocial behaviour, including parental care, is exhibited by a few thrips (section 11.1.2(d)).

The Thysanoptera and Hemiptera are sister groups, and the 'lice' (Psocoptera and Phthiraptera) are united as a sister group to Hemiptera + Thysanoptera (Fig. 7.5).

BOX 10.8
HEMIPTERA (BUGS, APHIDS, CICADAS, LEAFHOPPERS, PLANTHOPPERS, PSYLLIDS, SCALE INSECTS, SPITTLEBUGS, WHITEFLIES)

A worldwide order, the most diverse non-endopterygote order with more than 50 000 species in about 100 families. Conventionally divided into two suborders: Heteroptera (bugs) and Homoptera (cicadas, leafhoppers and planthoppers, spittlebugs, aphids, psyllids, scale insects = coccoids, and whiteflies). Some authors recognize three suborders, with 'Homoptera' comprising two suborders: Auchenorrhyncha (cicadas to spittlebugs) defined by their possession of tympanal organs (section 4.1.4) and an aristate antennal flagellum, and Sternorrhyncha (aphids to whiteflies) defined by insertion of labium near or on prosternum and, if winged, by absence of vannus and vannal fold in hind wing. Hemiptera mostly hemimetabolous (Fig. 6.2), compound eyes often large, ocelli present or absent, antennae vary from short with few segments to filiform and multisegmented; mouthparts with mandibles and maxillae modified as needle-like stylets, lying in beak-like, grooved labium, collectively forming a **rostrum** or proboscis; stylet bundle containing two canals, one delivering saliva, the other taking up fluid (as shown here for a pentatomid heteropteran), lacking palps; thorax often with large pro- and mesothorax, small metathorax; both pairs of wings often with reduced venation, sometimes apterous, rarely with only one pair (in male scale insects); legs frequently gressorial, sometimes raptorial, often with complex pretarsal adhesive structures; abdomen variable, cerci lacking.

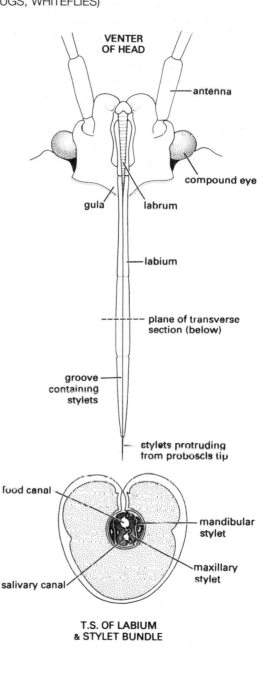

VENTER OF HEAD

antenna

compound eye

gula

labrum

labium

plane of transverse section (below)

groove containing stylets

stylets protruding from proboscis tip

food canal

mandibular stylet

maxillary stylet

salivary canal

T.S. OF LABIUM & STYLET BUNDLE

BOX 10.8—*continued*

Most Heteroptera hold their head horizontally, with rostrum anteriorly distinct from prosternum (although maybe in body contact at coxal bases and on anterior abdomen); wings, at rest, usually folded flat over abdomen (Fig. 5.5) and forewings usually thickened basally and membranous apically to form hemelytra (Fig. 2.22(e)), always with abdominal scent glands; apterous heteropterans identified by rostrum arising from anteroventral head and presence of large gula. Homoptera hold head deflexed with complete length of rostrum appressed to prosternum, directed posteriorly often between coxal bases; wings membranous, resting roof-like over abdomen; apterous homopterans identified by rostrum arising from posteroventral head, lacking gula.

Nymphal Heteroptera (Fig. 6.2) resemble adults except in the lack of development of the wings and genitalia. However, immature Sternorrhyncha show much variation in a range of complex life cycles. Many aphids exhibit parthenogenesis (section 5.10.1), usually alternating with seasonal sexual reproduction. In many sternorrhynchans, notably Aleyrodoidea (whiteflies) and Coccoidea (scale insects), the immature stages may differ greatly from adults, with larviform stages followed by a quiescent, non-feeding 'pupal' stage, in convergently acquired holometaboly.

The primitive feeding mode is piercing and sucking plant tissue (Fig. 10.4). All hemipterans have large salivary glands and an alimentary canal modified for absorption of liquids, with a filter chamber to remove water (Box 3.3). All Homoptera and many Heteroptera rely exclusively on living plant sap. The excretion of large quantities of honeydew by phloem-feeding Sternorrhyncha provide the basis for mutualistic relationships with ants. Many other homopterans exude waxes (Fig. 2.5) that form powdery or plate-like protective covers. Non-phytophagous Heteroptera comprise many predators, some scavengers and a few necrophages, including successful colonizers of aquatic environments (Box 9.5) and some of the few insects to live on the oceans (section 9.7).

Sternorrhyncha may have a separate evolutionary origin to the remaining 'Homoptera' (the Auchenorrhyncha). However, rigorous analysis is lacking and the internal phylogeny of Hemiptera remains unresolved, although evidently Hemiptera and Thysanoptera are sister groups within the Paraneoptera (hemipteroid orders) (Fig. 7.5).

BOX 10.9
PSOCOPTERA (BOOKLICE, BARKLICE OR PSOCIDS)

A worldwide order of common but cryptic small to minute insects, with over 3000 species in 35 families. Hemimetabolous; head large and mobile; compound eyes large; three ocelli present in winged species, absent in apterous ones; antennae usually 13-segmented, filiform; mouthparts with asymmetrical chewing mandibles, rod-shaped maxillary lacinia, reduced labial palps; thorax varies according to presence of wings, pronotum small, meso- and metanotum larger; legs gressorial, slender; wings often reduced or absent (as shown here for the booklouse *Liposcelis entomophilus* (Liposcelidae), after Smithers, 1982), membranous when present, with reduced venation, hind wing coupled to smaller fore wing in flight and at rest, wings held roof-like over abdomen (as shown here for the psocid *Amphigerontia contaminata* (Psocidae), after Badonnel, 1951); abdomen with 10 visible segments, the 11th represented by dorsal epiproct and paired lateral paraprocts; cerci always absent.

Courtship often involves a nuptial dance, followed by spermatozoa transfer via a spermatophore. Eggs are laid in groups or singly onto vegetation or under bark, in sites where nymphs subsequently develop. Parthenogenesis is common, and may be obligatory or facultative. Viviparity is known in at least one genus.

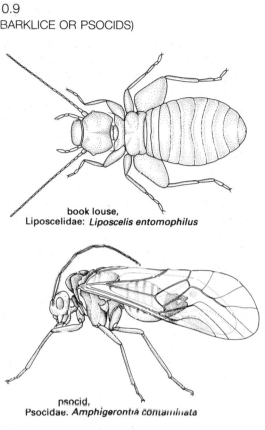

book louse,
Liposcelidae: *Liposcelis entomophilus*

psocid,
Psocidae. *Amphigerontia contaminata*

Adults and nymphs feed on fungi (hyphae and spores), lichens, algae and plant debris. Some species are solitary, others may be communal, forming small groups of adults and nymphs beneath webs.

Psocoptera form the sister group to the Phthiraptera (lice), collectively known as the Psocodea, and belong within the Paraneoptera (hemipteroid orders) (Fig. 7.5).

BOX 10.10
COLEOPTERA (BEETLES)

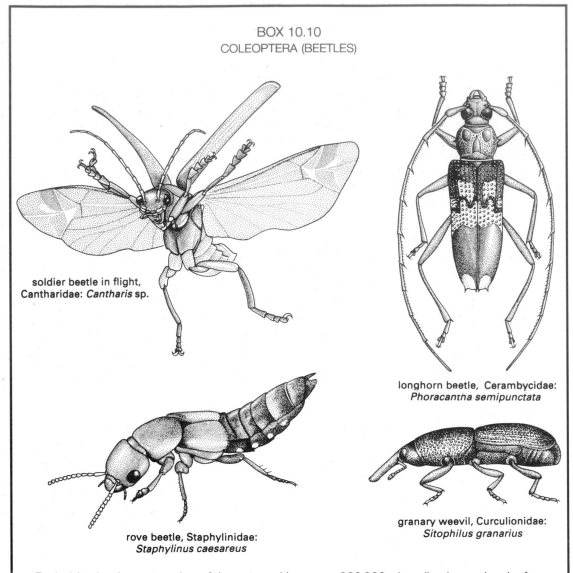

soldier beetle in flight,
Cantharidae: *Cantharis* sp.

longhorn beetle, Cerambycidae:
Phoracantha semipunctata

rove beetle, Staphylinidae:
Staphylinus caesareus

granary weevil, Curculionidae:
Sitophilus granarius

Probably the largest order of insects, with some 300 000 described species in four suborders (Archostemata, Myxophaga, Adephaga and speciose Polyphaga); although family level classification is unstable, some 500 families and subfamilies are recognized. Adults small to very large, usually heavily sclerotized, sometimes even armoured, often compact. Holometabolous; mandibulate; compound eyes well developed (sometimes even meeting medially) to absent, ocelli usually absent; antennae with 11 or frequently fewer segments (exceptionally with 20 segments in male Rhipiceridae); prothorax distinct, large and extending laterally beyond coxae, mesothorax small (at least dorsally), fused to metathorax to form wing-bearing pterothorax; fore wings modified as sclerotized, rigid, elytra (Fig. 2.22(d)) whose movement may assist in lift or may be restricted to opening and

BOX 10.10—*continued*

closing before and after flight, elytra cover hind wings and abdominal spiracles allowing control of water loss, hind wings longer than elytra when extended for flight (as illustrated on the upper left for a soldier beetle, *Cantharis* sp. (Cantharidae), after Brackenbury, 1990), with variably reduced venation much of which is associated with complex pleating to allow wings to be folded longitudinally and transversely beneath elytra even if elytra are reduced in size, as in rove beetles (Staphylinidae) such as *Staphylinus caesareus* (illustrated on the lower left after Stanek, 1969); legs very variably developed, coxae sometimes large and mobile, tarsi primitively 5-segmented although often with reduced number, bearing variously shaped claws and adhesive structures (Fig. 9.3), legs sometimes fossorial (Fig. 8.2(c)) for digging in soil or wood, or modified for swimming (Figs 9.3, 9.7) or jumping; abdomen primitively 9-segmented in female, 10-segmented in male, with at least 1 terminal segment retracted, sterna usually strongly sclerotized, often more so than terga; female with substitutional ovipositor, male external genitalia primitively trilobed (Fig. 2.24(b)); cerci absent.

Larvae exhibit a wide range of morphologies, but most can be recognized by the sclerotized head capsule with opposable mandibles and frequently 5-segmented thoracic legs, and distinguished from similar lepidopteran larvae by the lack of ventral abdominal crochet-bearing prolegs and lack of a median labial silk gland. Similar symphytan wasp larvae have prolegs on abdominal segments 2–7. Beetle larvae vary in body shape and leg structure; some are **apodous** (lacking any thoracic legs; Fig. 6.6(g)), whereas podous larvae may be **campodeiform** (prognathous with long thoracic legs; Fig. 6.6(e)), **eruciform** (grub-like with short legs) or **scarabaeiform** (grub-like but long-legged; Fig. 6.6(f)). Pupation is often in a specially constructed cell or chamber (Fig. 8.1), rarely in a cocoon spun from silk from Malpighian tubules, or exposed as in coccinellids (Fig. 6.7(j)).

Beetles occupy almost every conceivable habitat, including freshwater (Box 9.5), a few marine and intertidal habitats and, above all, in every vegetational microhabitat from external foliage (Fig. 10.1), flowers, buds, stems, bark and roots, to internal sites such as in galls in any living plant tissue or in any kind of dead material in all its various states of decomposition. Saprophagy and fungivory are quite common and dung and carrion are exploited (sections 8.3 and 8.4 respectively). Only a few beetles are parasitic but carnivory is frequent, occurring in nearly all Adephaga and many Polyphaga, including Lampyridae (fireflies) and many Coccinellidae (ladybugs/ladybirds, Fig. 5.6). Herbivorous Chrysomelidae and Curculionidae are widely introduced as biological control agents of weeds and Coccinellidae have been used as biocontrol agents for aphid and coccoid pests of plants. Beetles are significant pests of roots in pastures and crops (especially larval Scarabaeidae), of timber (especially Cerambycidae such as *Phoracantha semipunctata*, illustrated on the upper right after Duffy, 1963) and of stored products (such as the granary weevil, *Sitophilus granarius* (Curculionidae), illustrated on the lower right). The latter tend to be adapted to dry conditions and thrive on stored grains, cereals, pulses and dried animal material such as skins and leather.

Coleoptera are believed to be the sister group of the parasitic Strepsiptera and together these two orders are sister to the neuropteroid orders (Fig. 7.5).

BOX 10.11
LEPIDOPTERA (BUTTERFLIES AND MOTHS)

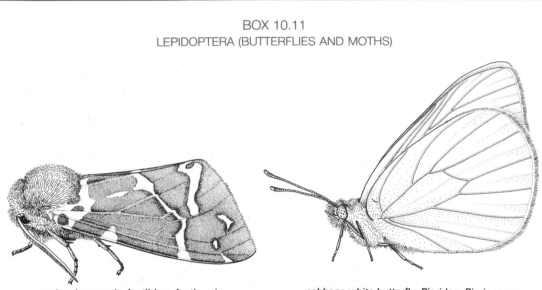

garden tiger moth, Arctiidae: *Arctia caja*   cabbage white butterfly, Pieridae: *Pieris rapae*

One of the major orders in size, with some 140 000 described species in 70 families, and in popularity, with many amateur and professional entomologists studying the order, particularly the butterflies. Three of the four suborders contain few species and lack the characteristic proboscis of the largest suborder Glossata, which contains the speciose series Ditrysia, defined by unique abdominal features especially in the genitalia. Adults very small (some microlepidopterans) to large, with wing span up to 240 mm, always covered in scales. Holometabolous (Fig. 6.3); head hypognathous bearing long, coiled proboscis (Fig. 2.12) formed from greatly elongated maxillary galeae, large labial palps usually present, other mouthparts absent, except mandibles primitively present; compound eyes large, ocelli and/or **chaetosemata** (paired sensory organs lying dorsolateral on head) frequent; antennae multisegmented, often pectinate in moths (Fig. 4.7), knobbed or clubbed in butterflies; prothorax small with paired dorsolateral plates (**patagia**), mesothorax large, bearing scutum and scutellum, and lateral tegula that protects base of fore wing, metathorax small; wings completely covered with double layer of scales (flattened modified macrotrichia), hind and fore wings linked by either frenulum, jugum or simple overlap, venation predominantly of longitudinal veins with few crossveins and some large cells, notably the discal (Fig. 2.22(a)); legs long, gressorial, with five tarsomeres, occasionally fore legs very reduced; abdomen 10-segmented, with segment 1 variably reduced, 9 and 10 modified as external genitalia (Fig. 2.23(a)); internal female genitalia very complex.

Pre-mating behaviour, including courtship, often involves pheromones (Figs 4.8, 4.9). Encounter between the sexes is often aerial but copulation is on the ground or a perch. Eggs are laid on, close to or, more rarely, within a larval host plant. Egg numbers and aggregation are very variable. Diapause is common.

Lepidopteran larvae can be recognized by their sclerotized hypognathous or prognathous head capsule, mandibulate mouthparts, usual six lateral ocelli (Fig. 4.10(a)), short

BOX 10.11—*continued*

3-segmented antennae, 5-segmented thoracic legs with single claws, 10-segmented abdomen with short prolegs on some segments (usually on 3–6 and 10, but may be reduced) (Figs. 6.6(a),(b), 13.5). Silk-gland products are extruded from a characteristic **spinneret** at the median apex of the labial prementum. Pupa usually contained within a silken cocoon, typically adecticous, obtect (a **chrysalis**) (Fig. 6.7(g)–(i)), with only some abdominal segments unfused; exarate in primitive species.

Those adult lepidopterans that feed utilize nutritious liquids such as nectar, honeydew and other seepages from live and decaying plants, and a few species pierce fruit. None suck sap from the vessels of live plants. Many species supplement their diet by feeding on nitrogenous animal wastes. Most larvae feed exposed on higher plants and form the major insect phytophages; a few primitive species feed on non-angiosperm plants, some on fungi; several are predators and others are scavengers, notably amongst the Tineidae (wool moths).

Larvae are often cryptic, particularly when feeding in exposed positions, or warningly coloured (aposematic) to alert predators to their toxicity (Chapter 13). Toxins derived from larval food plants are often retained by adults, which show anti-predator devices including advertisement of non-palatability and defensive mimicry (section 13.5).

Although the butterflies are popularly considered to be distinct from the moths, they form a superfamily Papilionoidea which is sister group to the skippers (Hesperoidea), and together these lie deep within the phylogeny of the Lepidoptera: butterflies are not the sister group to all moths. Butterflies are day-flying whereas most moths are active at night or dusk. In life, butterflies hold their wings together vertically above the body (as shown here on the right for a cabbage white butterfly) in contrast to moths, which hold their wings flat or wrapped around the body (as shown on the left for the garden tiger moth).

Lepidoptera are the sister group of Trichoptera (caddisflies) and together these two orders constitute the Amphiesmenoptera (Fig. 7.5).

FURTHER READING

Armstrong, J.A. (1979) Biotic pollination mechanisms in the Australian flora – a review. *New Zealand Journal of Botany,* **17,** 467–508.

Barbosa, P. and Schultz, J.C. (eds) (1987) *Insect Outbreaks,* Academic Press, San Diego, New York, Berkeley, Boston, London, Sydney, Tokyo, Toronto.

Beattie, A.J. (1985) *The Evolutionary Ecology of Ant–Plant Mutualisms,* Cambridge University Press, Cambridge, London, New York, New Rochelle, Sydney, Melbourne.

Bernays, E.A. (ed.) (1989–1992) *Insect–Plant Interactions, Volumes I–IV,* CRC Press, Inc., Boca Raton, Florida.

Crepet, W.L. and Friis, E.M. (1987) The evolution of insect pollination in angiosperms, in *The Origins of Angiosperms and their Biological Consequences,* (eds E.M. Friis, W.G. Chaloner and P.R. Crane), Cambridge University Press, Cambridge, pp. 181–201.

Farrell, B.D., Mitter, C. and Futuyma, D.J. (1992) Diversification at the insect–plant interface. *BioScience,* **42,** 34–42.

Harley, K.L.S. and Forno, I.W. (1992) *Biological Control of Weeds,* Inkata Press, Melbourne, Sydney.

Haukioja, E. (1990) Induction of defenses in trees. *Annual Review of Entomology,* **36,** 25–42.

Janzen, D.H. (1979) How to be a fig. *Annual Review of Ecology and Systematics,* **10,** 13–51.

Johnson, W.T. and Lyon, H.H. (1992) *Insects that Feed on Trees and Shrubs,* 2nd edn, Comstock Publishing Associates of Cornell University Press, Ithaca and London.

Landsberg, J. and Ohmart, C. (1989) Levels of insect defoliation in forests: patterns and concepts. *Trends in Ecology and Evolution,* **4,** 96–100.

Linhart, Y.B. (1991) Disease, parasitism and herbivory: multidimensional challenges in plant evolution. *Trends in Ecology and Evolution,* **6,** 392–6.

Miller, J.R. and Miller T.A. (eds) (1986) *Insect–Plant Interactions,* Springer-Verlag, New York, Berlin, Heidelberg, London, Paris, Tokyo.

Nilsson, L.A. (1988) The evolution of flowers with deep corolla tubes. *Nature,* **334,** 147–9.

Ohmart, C.P. and Edwards, P.B. (1991) Insect herbivory on *Eucalyptus. Annual Review of Entomology,* **36,** 637–58.

Pellmyr, O. (1992) Evolution of insect pollination and angiosperm diversification. *Trends in Ecology and Evolution,* **7,** 46–9.

Room, P.M. (1990) Ecology of a simple plant–herbivore system: biological control of *Salvinia. Trends in Ecology and Evolution,* **5,** 74–9.

Rosenthal, G.A. and Berenbaum, M.R. (eds) (1992) *Herbivores, their Interactions with Secondary Plant Metabolites,* 2nd edn, Academic Press, San Diego, New York, Boston, London, Sydney, Tokyo, Toronto.

Shorthouse, J.D. and Rohfritsch, O. (eds) (1992) *Biology of Insect-Induced Galls,* Oxford University Press, New York, Oxford.

Thompson, J.N. (1989) Concepts of coevolution. *Trends in Ecology and Evolution,* **4,** 179–83.

Whitham, T.G. (1989) Plant hybrid zones as sinks for pests. *Science,* **244,** 1490–3.

Wiebes, J.T. (1979) Co-evolution of figs and their insect pollinators. *Annual Review of Ecology and Systematics,* **10,** 1–12.

11
INSECT SOCIETIES

Vespid wasp nest

The study of insect social behaviours is a favourite entomological topic and there is a voluminous literature, ranging from the popular to the highly theoretical. The proliferation of some insects, notably the ants and termites, is attributed to the major change from a solitary lifestyle to a social one.

Social insects are ecologically successful and have important effects on human life. Leaf-cutter ants (*Atta* spp.) are the major herbivores in the Neotropics; harvester ants in south-western United States deserts take as many seeds as do mammals. Termites turn over as much or more soil than earthworms in many regions. Numerical dominance of social insects can be astonishing: a Japanese super-colony of *Formica yessensis* was estimated at 306 million workers and over 1 million queens dispersed over $2.7\,km^2$ amongst 45 000 inter-connected nests. In west African savanna, densities of up to 20 million resident ants per hectare have been estimated, and single nomadic colonies of driver ants (*Dorylus* sp.) may attain 20 million workers. Estimates of the value of honey bees in commercial honey production, as well as in pollination of agricultural and horticultural crops, run into hundreds of millions of dollars per annum in the United States alone. Social insects clearly affect our lives.

A broad definition of social behaviour could include all insects that interact in any way with other members of their species, but entomologists limit **sociality** to a more restricted range of **co-operative** behaviours. Amongst the social insects, we can recognize **eusocial** ('true social') insects, which co-operate in reproduction and have division of reproductive effort, and **subsocial** ('below social') insects, which have less strongly developed social habits, falling short of extensive co-operation and reproductive partitioning. **Solitary** insects exhibit no social behaviours.

Eusociality is defined by three traits:

- Division of labour, with a **caste system**

involving sterile individuals assisting those that reproduce.
- Co-operation among colony members in tending the young.
- Overlap of generations capable of contributing to colony functioning.

Eusociality is restricted to all ants and termites and some bees and wasps, such as the vespine paper wasps depicted in the vignette of this chapter (after Blaney, 1976). Subsociality is a more widespread phenomenon, known to have arisen independently in 13 orders of insects, including some cockroaches, embiopterans, thysanopterans, hemipterans, beetles and hymenopterans. As insect lifestyles become better known, forms of subsociality may be found in yet more orders. The term 'presociality' is often used for social behaviours that do not fulfil the strict definition of eusociality. However, the implication that presociality is an evolutionary precursor to eusociality is not always correct and the term is best avoided.

This chapter discusses subsociality before moving on to detailed treatment of eusociality in bees, wasps, ants and termites. It concludes with some ideas concerning the origins and success of eusociality.

11.1 SUBSOCIALITY IN INSECTS

11.1.1 Aggregation

Non-reproductive aggregations of insects, such as the gregarious overwintering of monarch butterflies at specific sites in Mexico and California, are social interactions. Many tropical butterflies form roosting aggregations, particularly in **aposematic** species (distasteful and with warning signals including colour and odour). Aposematic phytophagous insects often form conspicuous feeding aggregations, sometimes using pheromones to lure conspecific individuals to a favourable site (see section 4.3.2(b)). A solitary aposematic insect runs a greater risk

BOX 11.1
CRYPTOCERCUS – ON THE ROAD TO SOCIALITY?

The North American woodroach, *Cryptocercus punctulatus* (illustrated here, after Roth 1982), is found in decaying wood of dead trees in parts of western United States and disjunctly also in the Appalachian Mountains of eastern USA. The only other members of the family Cryptocercidae are two poorly-known species of *Cryptocercus* found in China.

Cryptocercus cockroaches have been given special significance because their diet and digestion resemble those of termites. They have symbiotic cellulose-digesting protozoans living in the hindgut where they digest the extremely nutrient-poor wood that forms the diet. *C. punctulatus* lives in colonies comprising two long-lived mated parents and 15–20 nymphs, predominantly of one age-group. Symbiotic protozoa are acquired by the young by ingestion of cast exuviae (skins) which contain the symbionts in an encysted form (unlike termites, whose symbionts are transferred through faecal ingestion). Parental care involves maintenance of the nest tunnels, defence and feeding the young nymphs on hindgut fluids. As the nymphs grow, they can give a certain amount of nest assistance, but because only one brood is present in the nest at any time nymphs cannot assist in feeding and brood care. The two parents appear to spend up to six years raising a single brood and there is no colony continuity.

This combination of termite-like cellulose-digesting symbionts, and social behaviour involving parental care, suggested that *Cryptocercus* demon-

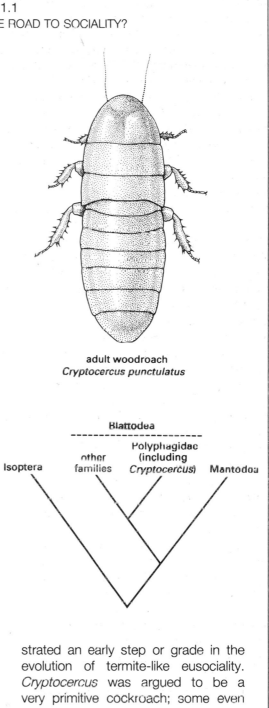

adult woodroach
Cryptocercus punctulatus

strated an early step or grade in the evolution of termite-like eusociality. *Cryptocercus* was argued to be a very primitive cockroach; some even

---

BOX 11.1—*continued*

suggested it was an ancestor of the termites. On the contrary, recent morphological and molecular sequence studies have shown that *Cryptocercus* is evolutionarily distant from the termites, and in fact falls within the cockroach family Polyphagidae (as illustrated in the accompanying cladogram). Therefore *Cryptocercus* cannot show the 'presocial' condition on the evolutionary path to termite eusociality. However, convergent development of social behaviour and acquisition of protozoan gut fauna in *Cryptocercus* and termites may demonstrate the importance of environmental determination of sociality.

---

of being encountered by a naive predator (and being eaten by it) than if it were a member of a conspicuous group. Belonging to a conspicuous social grouping, either of the same or several species, provides benefits by the sharing of protective warning coloration and the education of local predators.

11.1.2 Parental care as a social behaviour

Parental care may be considered to be a social behaviour, although few insects, if any, show a complete lack of parental care: eggs are not deposited randomly. Females select an appropriate oviposition site, affording protection to the eggs and ensuring an appropriate food resource for the hatching offspring. The ovipositing female may protect the eggs in an ootheca, or deposit them directly into suitable substrate with her ovipositor, or modify the environment, as in nest construction. Conventionally, parental care is seen as post-oviposition and post-hatching attention, involving the provision and protection of food resources for the young. A convenient basis for discussing parental care is to distinguish between care with and without nest construction.

(a) Parental care without nesting

For most insects, the highest mortality occurs in the egg and first instar, and many insects tend these stages until the more mature larvae or nymphs can fend better for themselves. The orders of insects in which tending of eggs and young is most frequent are the Blattodea, Orthoptera and Dermaptera (the orthopteroid orders), Embioptera, Psocoptera, Thysanoptera, Hemiptera, Coleoptera and Hymenoptera. There has been a tendency to assume that subsociality in orthopteroids is a precursor of Isoptera eusociality, since the eusocial termites are related to cockroaches. The phylogenetic position and social behaviour, including parental care, of the subsocial cockroach *Cryptocercus* has provoked much speculation on the origin of sociality (Box 11.1).

Egg and early-instar attendance is predominantly a female role, yet paternal guarding is known in some Hemiptera, notably amongst some tropical assassin bugs (Reduviidae) and giant water bugs (Belostomatidae). The female belostomatid oviposits onto the dorsum of the male, which receives eggs in small batches after each copulation. He may even accept eggs from more than one female. He either swims close to the oxygenated surface of the water, or makes deliberate ventilatory movements to create an oxygenated current over the eggs, which die if unattended. There is no tending of belostomatid nymphs, unlike many other hemipterans in which the female (or in some reduviids, the male) often guards at least the early-instar nymphs. In these species, experimental removal of the tending adult leads to increased losses of eggs and nymphs due to parasitization or predation. Other functions of parental care include keeping the eggs free from fungi, maintaining appropriate conditions for egg development, herding the young and sometimes actually feeding them.

In an unusual case, certain treehoppers (Hemiptera: Membracidae) have 'delegated' parental care of their young to ants. Ants obtain honeydew from treehoppers, which are protected from their natural enemies by the presence of the ants. In the presence of protective ants, brooding females may cease to tend a first brood prematurely and raise a second one. Another species of membracid will abandon its eggs in the absence of ants and seek a larger treehopper aggregation, where ants are in attendance, before laying another batch of eggs.

Many wood-mining beetles show advanced subsocial care which verges on the nesting described in the following section and on eusociality. For instance, all Passalidae (Coleoptera) live in communities of larvae and adults, with the adults chewing dead wood to form a substrate for the larvae to feed upon. Some ambrosia beetles (Platypodidae) prepare galleries for their offspring, where the larvae feed on cultivated fungi and are defended by a male that guards the tunnel entrance. Whether or not these feeding galleries are called nests is a matter of semantics.

(b) Parental care with solitary nesting

Nesting is a social behaviour in which the parents use a structure (pre-existing or newly constructed) in which to lay their eggs, and to which they bring food supplies for the young. Nesting, as thus defined, is seen in only five insect orders, of which the nests of eusocial Hymenoptera and the prodigious mounds of the eusocial termites are discussed in sections 11.2.2(b), 11.2.3(b) and 11.2.4(b). Nest-builders amongst the subsocial Orthoptera,

Dermaptera, Coleoptera and subsocial Hymenoptera are discussed below.

Earwigs of both sexes overwinter in a nest. In spring, the male is ejected when the mother starts to tend the eggs. In some species mother earwigs forage and provide food for the young nymphs. Mole crickets and other ground-nesting crickets exhibit somewhat similar behaviour. A greater range of nesting behaviours is seen in the beetles, particularly in the dung beetles (Scarabaeidae) and carrion beetles (Silphidae). Here the food source is short-lived, scattered, but nutrient-rich dung (and carrion) which attracts so many insects that competition is great. Dung beetles that locate a fresh source will remove it from danger of usurpation and desiccation by rapid burial (section 8.3; Box 8.3). Some scarabs roll the dung away from its source; others coat the dung with clay. Both sexes co-operate, but the female is mostly responsible for burrowing and preparation of the larval food source. Eggs are laid on the buried dung and in some species no further interest is taken. In others, parental care is well developed, commonly with maternal attention to fungus reduction, and paternal defence to remove or exclude ants.

Amongst the Hymenoptera, subsocial nesting is restricted to some **aculeate** Apocrita, namely within the superfamilies Chrysidoidea, Vespoidea and Apoidea (Fig. 11.2); these wasps and bees are the most prolific and diverse nest builders amongst the insects. Excepting bees, nearly all these insects are parasitoids: adults attack and immobilize arthropod prey, upon which the young feed. Wasps demonstrate a series of increasingly complex prey handling and nesting strategies, from using the prey's own burrow (e.g. many Pompilidae), to building a simple burrow after prey-capture (a few Sphecidae), to construction of a burrow (which can be termed a nest) prior to prey-capture (most Sphecidae). In bees and masarine wasps, pollen replaces arthropod prey as the food source collected and stored for the larvae. Nest complexity in the aculeates ranges from a single burrow, provisioned with one food item for one developing egg, to linear or radial multicellular nests. The primitive nest site was probably a pre-existing burrow; the construction medium later became soil or sand. Further specializations involved use of plant material (stems, rotten wood, and even solid wood by carpenter bees (Xylocopini)) and free-standing constructions of chewed vegetation (Megachilinae), mud (Eumeninae) and saliva (Colletinae). In some subsocial nesters, many individuals of one species may aggregate, building their nests close together as in mason wasps (Eumeninae).

(c) Parental care with communal nesting

When favourable conditions for nest construction are scarce and scattered throughout the environment, communal nesting may occur. Even under apparently favourable conditions, many subsocial and all eusocial hymenopterans share nests. Communal nesting may stem from daughters choosing to nest in their natal nest, giving enhanced utilization of nesting resources and mutual defence against parasites. However, communal nesting in subsocial species allows 'anti-social' or selfish behaviour: nest and prey usurpation appear to be frequent, and extended time defending the nest against others of the same species may be required. Furthermore, the same cues that lead the wasps and bees to communal nesting sites can easily direct specialized nest parasites to the location. Examples of communal nesting in subsocial species are known or presumed in the Sphecidae, Halictinae, Megachilinae and Andreninae.

After oviposition, female bees and wasps remain in their nests, often until the next generation emerges as adults. They generally guard, but they may also remove faeces and maintain nest hygiene. The supply of food to the nest may be through mass provisioning, as in many communal sphecids and subsocial bees, or replenishment, as seen in the many vespid wasps that return with new prey as their larvae develop.

(d) Subsocial aphids and thrips

Certain aphids belonging to the subfamilies Pemphiginae and Hormaphidinae (Hemiptera: Aphididae) have a sacrificial sterile soldier caste, consisting of some first- or second-instar nymphs that exhibit aggressive behaviour and never develop into adults. A similar phenomenon occurs in other related aphid species, but in this case all nymphs become temporary soldiers which later moult into normal, non-aggressive individuals which reproduce. Soldiers are pseudoscorpion-like, due to body sclerotization and enlarged anterior legs, and will attack intruders using either their frontal horns (anterior cuticular projections; Fig. 11.1) or feeding stylets (mouthparts) as piercing weapons. These modified individuals may defend good feeding sites against competitors or defend their colony against predators. Since the offspring are produced by parthenogenesis, soldiers and normal nymphs from the same mother aphid should be genetically identical, favouring evolution of these non-reproductive and apparently altruistic soldiers (due to increased inclusive fitness via kin selection; section 11.4.1). This unusual aphid polymorphism has led some researchers to claim that the Hemiptera is a third insect order displaying eusociality. Although these few aphid species clearly have a reproductive division of labour, they do not appear to fulfil the other attributes of eusocial insects, since overlap of generations capable of contributing to colony labour is equivocal and the tending of offspring does not occur. Here we consider these aphids to exhibit subsocial behaviour.

A range of subsocial behaviours is seen in a few species of several genera of thrips (Thysanoptera: Phlaeothripidae). At least in the gall thrips, the level of sociality appears to be

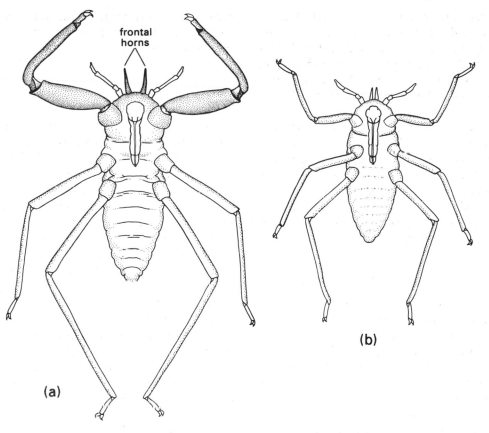

Fig. 11.1 First-instar nymphs of the subsocial aphid *Pseudoregma alexanderi* (Hemiptera: Hormaphidinae): (a) pseudoscorpion-like soldier; (b) normal nymph. (After Miyazaki, 1987.)

similar to that of the aphids discussed above. Thrips sociality is most developed in a bark-dwelling species of *Anactinothrips* from Panama: these thrips live communally, co-operate in brood care and forage with their young in a highly co-ordinated fashion. However, this species has no obvious non-reproductive females and all adults may disappear before the young are fully grown. Evolution of subsocial behaviours in *Anactinothrips* may bring advantages to the young in group foraging, since feeding sites, although stable over time, are patchy and difficult to locate. In several species of Australian gall thrips, adult generations overlap. Females show wing polymorphism and this wing reduction is sometimes associated with enlarged forelegs which could be used for gall defence. Most thrips probably have a haplo-diploid sex determination system. Thus, if single females found a gall in which individuals are polymorphic and multiple generations become established, the high relatedness of the offspring should favour self-sacrificing defence by some individuals (altruism; section 11.4.1). However, eusociality is unlikely in thrips because they do not feed their young and no definitive non-reproductive caste has been demonstrated.

(e) Quasi- and semisociality

Division of reproductive labour is restricted to the subsocial aphids amongst the insect groups discussed above: all females of all the other subsocial insects can reproduce. Within the social Hymenoptera, females show variation in fecundity or reproductive division of labour. The range includes full reproductivity (the subsocial species described above), reduced fecundity (many halictine bees), the laying of only male eggs (workers of *Bombus*), sterility (workers of *Aphaenogaster*) and super-reproductives (queens of *Apis*). This range of female behaviours is reflected in the classification of social behaviours in the Hymenoptera. Thus, in **quasisocial** behaviour, a communal nest consists of members of the same generation, all of which assist in brood rearing, and all females are able to lay eggs even if not necessarily at the same time. In **semisocial** behaviour, the communal nest similarly contains members of the same generation co-operating in brood care but there is division of reproductive labour, with some females (queens) laying eggs whereas their sisters act as workers and rarely lay eggs. This differs from eusociality only in that the workers are sisters to the egg-laying queens, rather than daughters, as is the case in eusociality. As in primitive eusocial hymenopterans there is no morphological (size or shape) difference between queens and workers.

Any or all of the subsocial behaviours discussed above may be evolutionary precursors of eusociality. It is clear that solitary nesting is the primitive behaviour, and that communal nesting and additional subsocial behaviours have arisen independently in many lineages of aculeate hymenopterans.

11.2 EUSOCIALITY IN INSECTS

Eusocial insects have a division of labour in their colonies, with a caste system comprising a restricted reproductive group of one or several **queens**, aided by **workers** (sterile individuals that assist the reproducers) and, in termites and many ants, an additional defensive **soldier** group. There may be further division into subcastes that perform specific tasks. At their most specialized, members of some castes, such as queens and soldiers, may lack the ability to feed themselves. The tasks of workers, therefore, include bringing food to these individuals as well as to the **brood** – the developing offspring.

The primary differentiation is female from male. In eusocial Hymenoptera, which are **haplodiploid** in sex determination, queens control the sex of their offspring. Releasing stored sperm fertilizes haploid eggs which develop into diploid female offspring, whereas unfertilized eggs produce male offspring. At most times of the year, reproductive females (queens or **gynes**) are rare compared with sterile female workers. Males do not form castes and may be infrequent and short-lived, dying soon after mating. In termites (Isoptera), males and females may be equally represented, with both sexes contributing to the worker caste. A single male termite, the **king**, may permanently attend the gyne.

Members of different castes, if derived from a single pair of parents, are close genetically and may be morphologically similar or, as a result of environmental influence, may be morphologically very different (**polymorphic**). Individuals within a caste (or subcaste) often differ behaviourally, in what is termed **polyethism**: either an individual performs different tasks at different times in its life (age polyethism), or individuals within a caste specialize in certain tasks during their lives. The intricacies of social insect caste systems can be considered in terms of the increasing complexity demonstrated in the Hymenoptera, but concluding with the remarkable systems of the termites (Isoptera). The characteristics of these two orders, which contain the majority of the eusocial species, are explained in Boxes 11.3 and 11.4.

11.2.1 The primitively eusocial hymenopterans

Hymenopterans exhibiting primitive eusociality include polistine vespids (paper wasps of

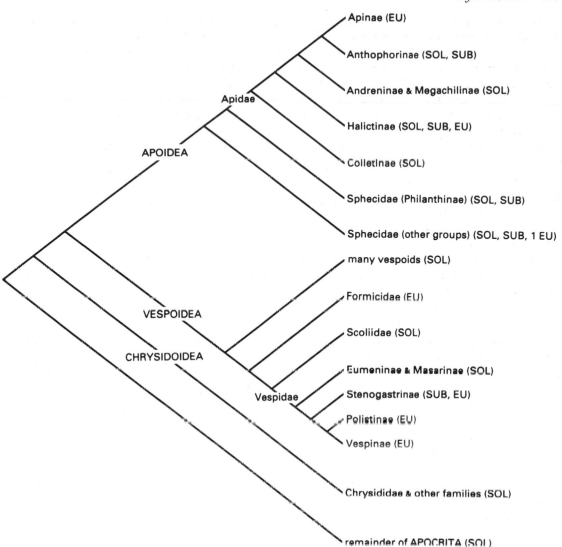

Fig. 11.2 Cladogram showing probable relationships among selected aculeate Hymenoptera to depict the multiple origins of sociality (SOL, solitary; SUB, subsocial; EU, eusocial). The superfamily Apoidea includes the Sphecidae and all bees, here treated as one family, the Apidae, with numerous subfamilies (e.g. Anthophorinae, Colletinae, Halictinae). Traditionally, bees have been classified in several families, a ranking that is unjustified phylogenetically. Probable relationships within non-social wasps and bees are not depicted. (Adapted from several sources including Gauld and Bolton, 1988; Alexander, 1992.)

the genus *Polistes*), stenogastrine wasps (Fig. 11.2) and one sphecid. In these wasps, all individuals are morphologically similar and live in colonies that seldom last more than one year. The colony is founded often by more than one gyne but rapidly becomes **monogynous**, that is, dominated by one queen with other foundresses either departing the nest or remaining but reverting to a worker-like state. The queen establishes a dominance hierarchy

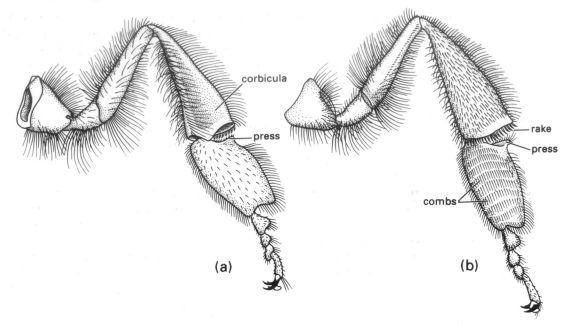

Fig. 11.3 The hindleg of a worker honey bee, *Apis mellifera* (Hymenoptera: Apidae): (a) outer surface showing corbicula, or pollen basket (consisting of a depression fringed by stiff setae), and the press that pushes the pollen into the basket; (b) the inner surface with the combs and rakes that manipulate pollen into the press prior to packing. (After Snodgrass, 1956; Winston, 1987.)

physically by biting, chasing and begging for food. The winning queen gains monopoly rights to egg laying and initiation of cell construction. Dominance may be incomplete, with non-queens laying some eggs: the dominant queen may eat these eggs or allow them to develop as workers to assist the colony. The first brood of females produced by the colony is of small workers, but subsequent workers increase in size as nutrition improves and as worker assistance in rearing increases. Sexual retardation in subordinates is reversible: if the queen dies (or is removed experimentally) either a subordinate foundress takes over or, if none is present, a high-ranking worker can mate (if males are present) and lay fertile eggs. Some other species of primitively eusocial wasps are **polygynous**, retaining several functional queens throughout the duration of the colony, and others are **serially polygynous**, with a succession of functional queens.

Primitively eusocial bees, such as certain species of Halictinae (Fig. 11.2), have a similar breadth of behaviours. In female castes, differences in size between queens and workers range from little or none to there being no overlap in their sizes. Bumble bees (Apidae: *Bombus* species) found colonies through a single gyne, often after a fight to the death between gynes vying for a nest site. The first brood consists only of workers; they are dominated by the queen physically, by aggression and by eating of any worker eggs, and by means of pheromones that modify the behaviour of the workers. In the absence of the queen, or late in the season as the queen's physical and chemical influence wanes, workers can undergo ovarian development. The queen eventually fails to maintain dominance over those workers that have commenced ovarian development, and she is either killed or driven from the nest. When this happens workers are unmated, but they can produce male offspring from their haploid eggs. Gynes are thus derived solely from the fertilized eggs of the queen.

11.2.2 Specialized eusocial hymenopterans – wasps and bees

The highly eusocial hymenopterans comprise the ants (family Formicidae) and some wasps, notably Vespinae, and many bees, including the Apinae (Fig. 11.2). Bees are derived from sphecid wasps and differ from wasps in anatomy, physiology and behaviour in association with their dietary specialization. Most bees provision their larvae with nectar and pollen rather than animal material. Morphological adaptations of bees associated with pollen collection include plumose (branched) hairs, and a widened hind basitarsus adorned with hairs in the form of a brush (**scopa**) or a fringe surrounding a concavity (the **corbicula**, or pollen basket; Fig. 11.3). Pollen collected on the body hairs is groomed by the legs and transferred to either the mouthparts or the scopae or corbiculae. The diagnostic features and the biology of all hymenopterans are dealt with in Box 11.3, which includes an illustration of the morphology of a worker vespine wasp and a worker ant.

(a) Colony and castes in eusocial wasps and bees

The female castes are dimorphic, differing rather strongly in their appearance. Generally, the queen is larger than any worker, as in vespines, such as the European wasps (*Vespula vulgaris* and *V. germanica*), and honey bees (*Apis* species). The typical eusocial wasp queen has a differentially (allometrically) enlarged gaster (abdomen). In worker wasps the bursa copulatrix is small, preventing mating, even though in the absence of a queen their ovaries will develop.

In the vespine wasps, the colony-founding queen or gyne produces only workers in the first brood. Immediately after these are hatched, the queen wasp ceases to forage and devotes herself exclusively to reproduction. As the colony matures, subsequent broods include increasing proportions of males and finally gynes are produced late in the season from larger cells than those from which workers are produced.

The tasks of vespine workers include distribution of protein-rich food to larvae and carbohydrate-rich food to adult wasps, cleaning cells, disposal of dead larvae, ventilation and air-conditioning of the nest by wing-fanning, nest defence by guarding entrances, foraging outside for water, sugary liquids and insect prey, and construction, extension and repair of the cells and inner and outer nest walls with wood pulp, which is masticated to produce paper. Each worker is capable of carrying out any of these tasks but often there is an age polyethism: newly emerged workers tend to remain in the nest engaged in construction and food distribution. A middle-aged foraging period follows, which may be partitioned into wood-pulp collection, predation and fluid-gathering phases. In old age, guarding duties dominate. Since newly recruited workers are produced continuously, the age structure allows flexibility in performing the range of tasks required by an active colony. There are seasonal variations: foraging occupies much of the time of the colony in the founding period, but fewer resources – or a lesser proportion of workers' time – are devoted to these activities in the mature colony. Male eggs are laid in increasing numbers as the season progresses, perhaps by queens, or by workers on whom the influence of the queen has waned.

The biology of the honey bee *Apis mellifera* is extremely well studied because of the economic significance of honey and the relative ease of observing honey-bee behaviour (Box 11.2). Workers differ from queens: they are smaller; they possess wax glands; they have a pollen-collecting apparatus comprising pollen combs and a corbicula on each hind leg; they have a barbed and non-retractible sting; and they have some other features associated with the tasks that they perform. The queen's sting

BOX 11.2
THE DANCE LANGUAGE OF BEES

Honey bees have impressive communication skills. Their ability to communicate forage sites to their nest mates was first recognized when a marked worker provided with an artificial food source was allowed to return to its hive but was then prevented from returning to the food. The rapid appearance of other workers at the food source indicated that information concerning the resource had been transferred within the hive. Subsequent observations using a glass-fronted hive showed that returning foragers often performed a dance on return to the nest. Other workers followed the dancer, made antennal contact and tasted regurgitated food, as depicted here in the upper illustration (after Frisch, 1967). Olfactory communication alone could be discounted by experimental manipulation of food sources and the importance of dancing became recognized. Variations within different dances allow communication and recruitment of workers to close or distant food sources and to food versus prospective nest sites. The purpose and messages associated with three dances – the round, waggle, and dorsoventral abdominal vibrating (DVAV) – have become well understood.

Nearby food is communicated by a simple **round dance**: the incoming worker exchanges nectar and makes tight circles, with frequent reversals, for a few seconds to a few minutes, as shown in the central illustration (after Frisch, 1967). The quality of nectar or pollen from the source is communicated by the vigour of the dance. Although no directionality is conveyed, 89% of 174

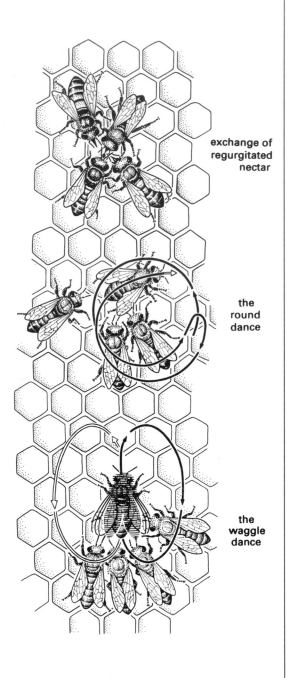

exchange of regurgitated nectar

the round dance

the waggle dance

BOX 11.2—*continued*

workers contacted by the dancer during a round dance were able to find the novel food source within 5 minutes, probably by flying in ever increasing circles until the local source is found.

More distant sources are identified by a **waggle dance** which involves abdomen-shaking during a figure-of-eight circuit, shown in the lower illustration (after Frisch, 1967), as well as food-sharing. Informative characteristics of the dance include the length of the straight part (measured by the number of comb cells traversed), the dance tempo (number of dances per unit time) and the duration of waggling and noise production (buzzing) during the straight line section. Messages conveyed are the energy required to get to the source (rather than absolute distance), quality of the forage, and direction relative to the sun's position (Box 4.2).

The function of the **vibration dance** (DVAV) differs from the round and waggle dances in regulating the daily and seasonal foraging patterns in relation to fluctuating food supply. Workers vibrate their bodies, particularly their abdomens, in a dorsoventral plane, usually whilst in contact with another bee. Vibration dances peak at times of the day and season when the colony needs to be primed for increased foraging and these dances act to recruit workers into the waggle dance area. Vibration dancing with queen contact appears to lessen the inhibitory capacity of the queen, and is used during the period when queen-rearing is taking place. Cessation of this kind of vibration dancing may result in the queen departing with a swarm, or in the mating flight of new queens.

Communication of a suitable site for a new nest differs somewhat from communication of a food source. The returning scout dances without any nectar or pollen and the dance lasts for 15–30 minutes rather than the 1–2 minutes of the forager's dance. At first, several scouts returning from various prospective new sites will all dance, with differences in tempo, angle and duration that indicate the different directions and qualities of the sites, as in a waggle dance. More scouts then fly out to prospect and some sites are rejected. Gradually a consensus is attained, as shown by one dance that indicates the agreed site.

is scarcely barbed and is retractible and reusable, allowing repeated assaults on pretenders to the queen's position. Queens have a shorter proboscis than workers and lack several glands.

Honey-bee workers are more or less monomorphic but exhibit polyethism. Young workers tend to be 'hive bees', engaged in within-hive activities such as nursing larvae and cleaning cells, while older workers are foraging 'field bees'. Seasonal changes are evident: winter bees live for 8–9 months, compared with a life of 4–6 weeks for summer workers. Juvenile hormone (JH) is involved in these behavioural changes, its levels rising from winter to spring, and also in the change from hive bees to field bees. Honey-bee worker activities correlate with seasons, notably in the energy expenditure involved in thermoregulation of the hive.

Caste differentiation in honey bees, as in eusocial hymenopterans generally, is largely **trophogenic**, that is determined by the quantity and quality of larval diet. In species that provision each cell with enough food to allow the egg to develop to the pupa and adult without further replenishment, differences in the food quantity and quality provided to each

Fig. 11.4 Development of the honey bee, *Apis mellifera* (Hymenoptera: Apidae), showing the factors that determine differentiation of the queen-laid eggs into drones, workers and queens (on left) and the approximate developmental times (in days) and stages for drones, workers and queens. (After Winston, 1987.)

cell determine how the larva will develop. In honey bees, although cells are constructed according to the type of caste that is to develop within them, the caste is determined neither by the egg laid by the queen, nor by the cell itself, but by food supplied by workers to the developing larva (Fig. 11.4). The type of cell guides the queen as to whether to lay fertilized or unfertilized eggs, and identifies to the worker which type of rearing (principally food) to be supplied to the occupant. Food given to future queens is known as 'royal jelly' and differs from worker food in having a high sugar content and being composed predominantly of mandibular gland products – pantothenic acid and biopterin. Eggs and larvae up to three days old can differentiate into queens or workers according to upbringing. However, by the third day a potential queen has been fed royal jelly at up to ten times the rate of less rich food supplied to a future worker. If a future queen is transferred to a worker cell for further development at this stage, she will become an intercaste – a worker-like queen. In contrast, if a three-day-old larva reared as a worker is transferred into a queen cell, it

develops into a queen-like worker, still retaining the pollen baskets, barbed stings and mandibles of a worker. After four days of appropriate feeding, the castes are fully differentiated and transfers between cell types result in either retention of the early determined outcome or failure to develop.

Trophogenic effects cannot always be separated from endocrine effects, since nutritional status is linked to corpora allata activity. It is clear that JH levels correlate with polymorphic caste differentiation in eusocial insects but there seems to be much specific and temporal variation in JH titres and no common pattern of control is yet evident.

The queen maintains control over the workers' reproduction principally through pheromones. The mandibular glands of queens produce a compound identified as (E)–9-oxodec–2-enoic acid (abbreviated to 9-ODA) but the intact queen inhibits worker ovarian development more effectively than this active compound. A second pheromone has been found in the gaster of the queen which, together with a second component of the mandibular gland, effectively inhibits ovarian development. Queen recognition by the rest of the colony involves a pheromone disseminated by attendant workers that contact the queen and then move about the colony as messenger bees. Secondly, as the queen moves around on the comb whilst ovipositing into the cells, she leaves a trail of **footprint pheromone**. Production of queens takes place in cells that are distant from the effects of the queen's pheromonal control, as occurs when nests become very large. Should the queen die, the volatile pheromone signal dissipates rapidly and the workers become aware of the absence. Honey bees have very strongly developed chemical communication, with specific pheromones associated with mating, alarm and orientation as well as colony recognition and regulation. Physical threats are rare, and are used only by young gynes towards workers.

Males, termed **drones**, are produced throughout the life of the honey-bee colony, either by the queen or perhaps by workers with developed ovaries. Males contribute little to the colony, living only to mate; their genitalia are ripped out after copulation and they die.

(b) Nest construction in eusocial wasps

The founding of a new colony of eusocial vespid wasps takes place in spring, following the emergence of an overwintering queen. After her departure from the natal colony the previous autumn, the new queen mates but her ovarioles remain undeveloped during the temperature-induced winter quiescence (facultative diapause). As spring temperatures rise, queens leave hibernation and feed on nectar or sap, and the ovarioles grow. The resting site, which may be shared by several overwintering queens, is not a prospective site for the foundation of a new colony. Each queen scouts individually for a suitable cavity and fighting may occur if sites are scarce.

Nest construction begins with the use of the mandibles to scrape wood fibres from sound or, more rarely, rotten wood. The wasp returns to the nest site using visual cues, carrying the wood pulp (masticated with water and saliva) in the mandibles. This pulpy paper is applied to the underside of a selected support at the top of the cavity. From this initial buttress, the pulp is formed into a descending pillar, upon which is suspended ultimately the embryonic colony of 20–40 cells (Fig. 11.5). The first two cells, rounded in cross-section, are attached and then an umbrella-like envelope is formed over them. The envelope is elevated by about the width of the queen's body above the cells, allowing the queen to rest there, curled around the pillar. The developing colony grows by the addition of further cells, now hexagonal in cross section and wider at the open end, and by either extension of the envelope or construction of a new one. The queen forages only for building material at the start of nest construction. As the larvae develop from the first cells, both liquid and insect prey are sought to nourish them, although wood pulp continues to be collected for further cell construction. This first embryonic phase of the life of the colony ceases as the first workers emerge.

As the colony grows, further pillars are added, providing support to more lateral areas where brood-filled cells are aligned in **combs** (series of adjoining cells in parallel rows). The early cells and envelopes become overgrown, and their materials may be reused in later construction. In a subterranean nest, the occupants may have to excavate soil and even small stones to allow colony expansion, resulting in a mature nest (Fig. 11.5) which may contain as many as 12 000 cells. The colony has some independence from external temperature, since thoracic heating through wing beating and larval feeding can raise temperature, and high temperature can be lowered by directional fanning or by evaporation of liquid applied to the pupal cells.

At the end of the season, males and gynes (potential queens) are produced and are fed with larval saliva and prey brought into the

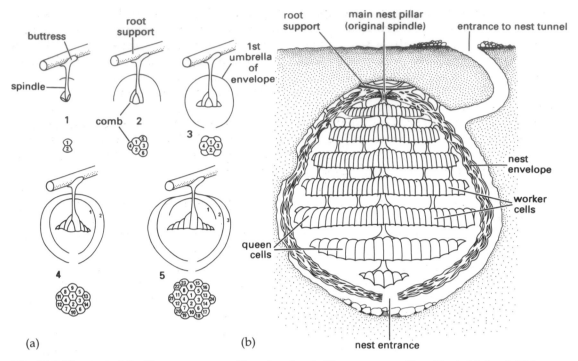

Fig. 11.5 The nest of the European wasp, *Vespula vulgaris* (Hymenoptera: Vespidae): (a) five initial stages of nest construction by the queen (the embryonic phase of the colony's life); (b) a mature nest. (After Spradbery, 1973.)

nest by workers. As the old queen fails and dies, and gynes emerge from the nest, the colony declines rapidly and the nest is destroyed as workers fight and larvae are neglected. Potential queens and males mate away from the nest, and the mated female seeks a suitable overwintering site.

(c) Nesting in honey bees

In honey bees, initiation of new colonies is triggered when the old one gets too crowded. When a bee colony becomes too large and the population density too high, a founder queen, accompanied by a swarm of workers, seeks a new nest site. Since workers cannot survive for long on the honey reserves carried in their stomachs, the need to find a suitable site is imperative. Scouts may have started the search several days before formation of the swarm: when a suitable cavity is found, the scout returns to the cluster and communicates the direction and quality of the site by a dance (Box 11.2). Optimally, a new site should be beyond the foraging territory of the old nest but not so distant that energy is expended in long-distance flight. Bees from temperate areas select enclosed nest sites in cavities of about 40 litres in volume, whereas more tropical bees choose smaller cavities or nest outside.

Following consensus over the nest site, workers start building a nest using wax. **Wax** is unique to social bees and is produced by workers that metabolize honey in fat cells located close to the wax glands. These are modified epidermal cells which lie beneath **wax mirrors** (overlapping plates) ventrally on the fourth to seventh abdominal segments. Flakes of wax are extruded beneath each wax mirror and protrude slightly from each segment of a worker actively producing wax. Wax is quite malleable at the ambient nest temperature of 35 °C, and it is manipulated for cell construction by mixing with saliva. At nest foundation, workers may already have wax protruding from the abdominal wax glands. They start to construct combs of back-to-back hexagonal cells in a parallel series, or comb. Combs are separated from one another by pillars and bridges of wax. A thick cell base of wax is extended into a thin-walled cell of remarkably constant dimensions, despite a series of workers being involved in construction. In contrast to other social insects such as the vespids described above, cells do not hang downwards but are angled at about 13° above the horizontal, thereby preventing loss of honey. The precise orientation of the cells and comb derives from the bees' ability to detect gravity through the proprioceptor hair plates at the base of their necks. Although removal of the hair plates prevents cell construction, worker bees transported in the 1984 *Challenger* space-shuttle flight could construct serviceable cells under conditions of weightlessness.

Unlike most other bees, honey bees do not chew up and reuse wax: once a cell is constructed it is permanently part of the nest, and cells are reused after the brood has emerged or the food contents have been used. Cell sizes vary, with small cells used to rear workers and larger ones for drones (Fig. 11.4). Later in the life of the nest, elongate conical cells in which queens are reared are constructed at the bottom and sides of the nest. The brood develops and pollen is stored in lower and more central cells, whereas honey is stored in upper and peripheral cells. Honey is formed by workers primarily from nectar taken from flowers, but can also be derived from extra-floral nectaries or insect-produced honeydew. Workers carry nectar to the hive in honey stomachs, from which it may be fed directly to the brood and to other adults. More often it is converted to honey by enzymatic digestion of the sugars to simpler forms and reduction of the water content by evaporation, before storage in wax-sealed cells until

required to feed adults or larvae. It has been calculated that in 66 000 bee-hours of labour, 1 kg of beeswax can be formed into 77 000 cells which can support the weight of 22 kg of honey. An average colony requires about 60–80 kg of honey per annum.

When the lower temperatures of a temperate winter arrive, honey bees do not hibernate (unlike wasps). Colonies are active all winter, but foraging is curtailed and no brood is reared. Stored honey is used as an energy source to allow activity and heat generation within the nest. As external temperatures drop, the workers cluster together, heads inwards, forming an inactive layer of bees on the outside with warmer and more active feeding bees on the inside. Despite the prodigious stores of honey and pollen, mortality may be high in extreme cold, or if winter is prolonged.

Bee hives are artificial constructions which resemble feral honey bee nests in some dimensions, notably the distance between the combs. When given wooden frames separated by an invariable natural spacing interval of 9.6 mm, honey bees construct their combs within the frame without formation of the internal waxen bridges needed to separate the combs of a feral nest. This width between combs is approximately the space required for bees to move unimpeded on both combs. The ability to remove frames allows the apiculturalist (bee keeper) to examine and remove the honey, and replace the frames in the hive. The ease of construction allows building of several ranks of boxes.

The hives can be transported to suitable locations without damaging the combs. Although the apiculture industry has developed through commercial production of honey, lack of native pollinators in monocultural agriculture has led to increasing reliance on the mobility of hive bees to ensure pollination of crops as diverse as oilseed, nuts, soybeans, fruits, clover, alfalfa and other fodder crops. In the United States alone, the 1990 value of the pollination duties provided

BOX 11.3
HYMENOPTERA (WASPS, BEES, ANTS, SAWFLIES AND WOOD WASPS)

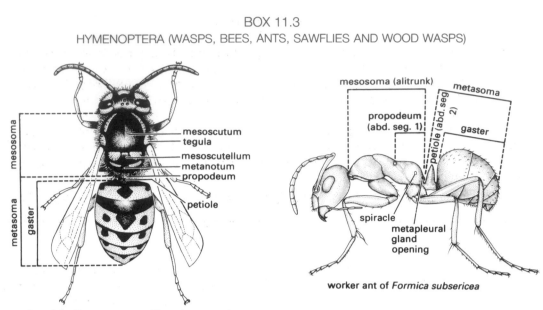

worker of the European wasp, *Vespula germanica*

worker ant of *Formica subsericea*

An order of at least 100 000 described species of holometabolous neopterans, classified in two suborders, the Symphyta (wood wasps and sawflies) and Apocrita (wasps, bees and ants); within the Apocrita, the aculeate taxa (Chrysidoidea, Vespoidea and Apoidea) form a monophyletic group (Fig. 11.2). Adults minute (e.g. Trichogrammatidae, Fig. 15.3) to large (0.15–120 mm long), slender (e.g. many Ichneumonidae) to robust (e.g. the bumble bee depicted in the vignette of Chapter 1); head hypognathous or prognathous; mouthparts range from generalized mandibulate in Symphyta to sucking and chewing with mandibles often used for killing and handling prey, defence and nest building in Apocrita; compound eyes often large; ocelli present, reduced or absent; antennae long, multisegmented, often prominently held forward or recurved dorsally; three conventional segments in symphytan thorax, but in Apocrita abdominal segment 1 (**propodeum**) included in thoracic tagma and then called **mesosoma** (or in ants, **alitrunk**) (as illustrated for workers of the wasp *Vespula germanica* and a *Formica* ant); wings with reduced venation, hind wing with rows of hooks (**hamuli**) along leading edge that couple with hind margin of fore wing in flight; abdominal segment 2 (and sometimes also 3) of Apocrita forms a constriction, or **petiole**, followed by remainder of abdomen, or **gaster**; female genitalia include an ovipositor, comprising three valves and two major basal sclerites, which may be long and highly mobile allowing valves to be directed vertically between legs (Fig. 5.8), ovipositor of aculeate Hymenoptera modified as a sting associated with venom apparatus (Fig. 13.10).

Eggs of endoparasitic species are often deficient in yolk, and sometimes each may give rise to more than one individual (polyembryony; section 5.10.3). Symphytan larvae are eruciform (caterpillar-like) (Fig. 6.6(c) and the vignette of Chapter 10) with three pairs of thoracic legs with apical claws and some abdominal legs; most are phytophagous. Apocritan larvae are apodous (Fig. 6.6(i)), with the head capsule frequently reduced but with by over 1 million bee colonies available for pollinator rental is estimated to be well in excess of US$10 million. Yield losses of over 90% of fruit, seed and nut crops would occur without honey-bee pollination. The role of the many species of eusocial native bees is little recognized, but may be important in areas of natural vegetation.

11.2.3 Specialized hymenopterans – ants

Ants (Formicidae) form a well defined, highly specialized group within the superfamily Vespoidea (Fig. 11.2). The morphology of a worker ant of *Formica* is illustrated in Box 11.3.

(a) Colony and castes in ants

All ants are social and their species are polymorphic. There are two major female castes, the reproductive queen and the workers, usually with complete dimorphism between them. Many ants have monomorphic workers, but others have distinct subcastes called **minor**, **media** or **major** workers, according to their size. Although workers may form clearly different morphs, there is more often a gradient in size. Workers are never winged but wings are present in queens, which shed them after mating, and in most males, which die after mating. Winged individuals are called **alates**. Polymorphism in ants is accompanied by polyethism: the queen's role is restricted to oviposition and the workers perform all other tasks. If workers are monomorphic, there may be temporal or age polyethism, with young workers undertaking internal nurse duties and older ones foraging outside the nest. If workers are polymorphic, the subcaste with the largest individuals – the major workers – usually has a defensive or soldier role.

The workers of certain ants, such as the fire ants (*Solenopsis*), have reduced ovaries and are irreversibly sterile. In others, workers have functional ovaries and may produce some or all of the male offspring by laying haploid (unfertilized) eggs. In some species, when the queen is removed the colony continues to produce gynes from fertilized eggs previously laid by the queen, and males from eggs laid by workers. The inhibition of the queen on her daughter workers is quite exceptional in the African weaver ant (*Oecophylla longinoda*). A mature colony of up to half a million workers, distributed amongst as many as 17 nests, is completely prevented from reproduction by a single queen. Workers, however, do produce male offspring in nests that lie outside the influence (or territory) of the queen. Queens

...iction of reproductive eggs ... may allow the laying of ...ic eggs which are fed to the ...e larvae. By this means the ... prevents any reproductive ...directs much of the protein ...vards her own offspring.

Caste ...ntiation is largely trophogenic (diet determined), involving biased allocation of volume and quality of food given to the larvae. A high protein diet promotes differentiation of gyne/queen and a less rich, more dilute diet leads to differentiation of workers. The queen generally inhibits development of gynes indirectly by modifying the feeding behaviour of workers towards female larvae that have the potential to differentiate as either gynes or workers. In *Myrmica*, large, slowly developing larvae will become gynes, which means that stimulation of rapid development and early metamorphosis of small larvae, or food deprivation and irritating of large larvae by biting to accelerate development, both induce differentiation as workers. When queen influence wanes, either through the increased size of the colony or because the inhibitory pheromone is impeded in its circulation throughout the colony, gynes are produced at some distance from the queen. There is also a role for juvenile hormone (JH) in caste differentiation. JH tends to induce queen development during egg and larval stages, and induces production of major workers from already differentiated workers.

According to a seasonal cycle, ant gynes mature to winged reproductives, or alates, and remain in the nest in a sexually inactive state until external conditions are suitable for departing the nest. At the appropriate time they make their nuptial flight, mate and attempt to found a new colony.

(b) Nesting in ants

The subterranean soil nests of *Myrmica* and the mounds of plant debris of *Formica* are typical temperate ant nests. Colonies are founded when a mated queen sheds her wings and overwinters, sealed into a newly dug nest that she will never leave. In spring, the queen lays some eggs and feeds the hatched larvae by **stomodeal** or **oral trophallaxis**, that is, regurgitation of liquid food from her internal food reserves. Colonies develop slowly whilst worker numbers build up, and a nest may be many years old before alates are produced.

Colony foundation by more than one queen, known as **pleometrosis**, appears to be quite widespread and the digging of the initial nest may be shared, as in the honeypot ant *Myrmecocystus mimicus*. In this species and others, multi-queen nests may persist as polygynous colonies, but monogyny commonly arises through dominance of a single queen, usually following rearing of the first brood of workers. Polygynous nests are often associated with opportunistic use of ephemeral resources, or persistent but patchy resources.

Amongst the best known complex nest constructions of ants are the woven nests of *Oecophylla* species (Fig. 11.6). These African and Asian/Australian weaver ants have extended territories that are continually explored by workers for any leaf that can be bent. A remarkable collaborative construction effort follows, in which leaves are manipulated into a tent shape by linear ranks of workers, often involving 'living chains' of ants that bridge wide gaps between the leaf edges. Another group of workers take larvae from existing nests and carry them delicately between their mandibles to the construction site. There, larvae are induced to produce silk threads from their well-developed silk glands and a nest is woven to link the framework of leaves.

Living plant tissues provide a location for nests of ants such as *Pseudomyrmex ferruginea*, which nests in the expanded thorns of the Central American bull's-horn *Acacia* trees (Fig. 10.9(a)). Cases such as these demonstrate mutualism involving plant defence (section

Fig. 11.6 Weaver ants of *Oecophylla* making a nest by pulling together leaves and binding them with silk produced by larvae held in the mandibles of worker ants. (After CSIRO, 1970; Hölldobler, 1984.)

10.4.1), with plants benefiting from the deterrence of phytophagous animals by ants.

Foraging efficiency of ants can be very high. A typical mature colony of European red ants (*Formica polyctena*) is estimated to harvest about 1 kg of arthropod food per day. The legionary or army and driver ants are popularly known for their voracious predatory activities. These ants, which predominantly belong to the subfamilies Ecitoninae and Dorylinae, alternate cyclically between sedentary (**statary**) and migratory or nomadic phases. In the latter phase, a nightly **bivouac** is formed, which often is no more than an exposed cluster of the entire colony. Each morning, the millions-strong colony moves *in toto*, bearing the larvae. The advancing edge of this massive group raids and forages on a wide range of terrestrial arthropods, and group predation allows even large prey items soldier of *Coptotermes*   soldier of *Nasutitermes*

BOX 11.4
ISOPTERA (TERMITES)

A small order of some 2 300 described species of hemimetabolous neopterans, living socially with polymorphic caste systems of reproductives, workers and soldiers (section 11.2.4(a); Fig. 11.7). All stages small to moderate-sized (even winged reproductives usually <20 mm long); head hypognathous or prognathous; mouthparts typically blattoid, mandibulate, varying between castes; soldiers often with bizarre development of mandibles or with nasus (as illustrated on the left for the mandibulate *Coptotermes* and on the right for the nasute *Nasutitermes*, after Harris, 1971); compound eyes frequently reduced; antennae long, multisegmented with variable number of segments; wings membranous with restricted venation, fore and hind wings similar, except in *Mastotermes*, which has complex venation and an expanded hind wing anal lobe; a pair of 1–5-segmented cerci terminally in all castes; external genitalia absent, except in *Mastotermes* in which female has a reduced blattoid ovipositor and male a membranous copulatory organ; gonads poorly developed in adult soldiers and worker castes.

Internal anatomy dominated by a convoluted alimentary tract including an elaborated hindgut, containing symbiotic cellulose-digesting protozoa in all species except the Termitidae (section 8.5.3 discusses fungal culture by Macrotermitinae). Food exchange between individuals (trophallaxis) is the sole means of replenishment of symbionts to young and newly moulted individuals and is one explanation of the universal eusociality of termites.

Nests may be galleries or more complex structures within wood, such as rotting timber or even a sound tree, or above-ground nests (termitaria) such as prominent earth mounds (Figs 11.8 and 11.9). Termites feed predominantly on cellulose-rich material; many harvest grasses and return the food to subterranean or above-ground nest mounds.

The Isoptera is the sister group of the Blattodea plus Mantodea (Fig. 7.5 and Box 11.1).

to be overcome. After some two weeks of nomadism, a statary period commences during which the queen lays 100 000–300 000 eggs in a statary bivouac. This is more sheltered than a typical overnight bivouac, perhaps within an old ant nest or beneath a log. The eggs hatch after some three weeks, during which period the larvae of the previous oviposition period complete their development to emerge as new workers, which stimulates the next migratory period.

Not all ants are predatory: certain ants are specialist harvesters of grain and seeds (myrmecochory; section 10.3.2) and others feed almost exclusively on insect-produced honey-dew (section 10.4.1). One of the most extra-ordinary examples is seen in the honeypot ants. Workers of these species return to the nest with crops filled with honeydew, which is fed by oral trophallaxis to selected workers called **repletes**. The abdomen of repletes is so distensible that they become virtually immo-bile 'honey pots' (Fig. 2.4) which act as food reserves for all in the nest.

11.2.4 Isoptera (termites)

All termites (Isoptera) are eusocial. Their diagnostic features and biology are summar-ized in Box 11.4.

(a) Colony and castes in termites

In contrast to the adult and female-only castes of holometabolous eusocial Hymenoptera, the castes of the hemimetabolous Isoptera involve immature stages and equal represen-tation of the sexes. Before castes are discussed further, terms for termite immature stages must be clarified. Termitologists refer to the developmental instars of reproductives as nymphs, more properly called brachypterous nymphs, and the instars of sterile lineages as larvae, although strictly the latter are apterous nymphs.

The termites may be divided into two groups: the 'lower' and 'higher' termites. The species-rich higher termites (Termitidae) differ from lower termites in the following manner:

- The Termitidae digest cellulose, using either predigestion by cultivated fungus, incorporation of cellulase in the diet, or the use of anaerobic bacteria for internal diges-tion. This contrasts with the dominance of symbiotic internal flagellates in lower ter-mites.
- Termitidae have a more elaborate and rigid caste system. For example, in most lower termites there is little or no distension of the

queen's abdomen, whereas termitid queens undergo extraordinary **physogastry**, in which the abdomen is distended to 500–1000% of its original size (Fig. 11.7).

All termite colonies contain a pair of **primary reproductives** – the queen and king, which are former alate (winged) adults from an established colony. Upon loss of the primary reproductives, potential replacement reproductives occur. These individuals, called **supplementary reproductives** or **neotenics**, are arrested in their development, either with wings present as buds (brachypterous neo-tenics) or without wings (apterous neotenics or **ergatoids**), and can take on the repro-ductive role if the primary reproductives die. In some species, neotenics are always present in small numbers.

In contrast to these reproductive or poten-tially reproductive castes, the colony is domi-nated numerically by sterile termites that function as workers and soldiers of both sexes. Soldiers have distinctive heavily sclero-tized heads, with large mandibles or with a strongly-produced snout (**nasus**) through which sticky defensive secretions are ejected. Two classes, major and minor soldiers, may occur in some species. Workers are unspecial-ized, weakly pigmented and poorly sclero-tized, giving rise to the popular name of 'white ants'.

Caste differentiation pathways are por-trayed best in the more rigid system of the higher termites (Termitidae), which can then be contrasted with the greater plasticity of the lower termites. In *Nasutitermes exitiosus* (Ter-mitidae, subfamily Nasutitermitinae; Fig. 11.7), two different developmental pathways exist, one leads to reproductives and the other (which is further subdivided) gives rise to sterile castes. This differentiation may occur as early as the first larval stage, although some castes may not be morphologically recogniz-able until later moults. The reproductive pathway (on the left in Fig. 11.7) is relatively constant between termite taxa and is typically

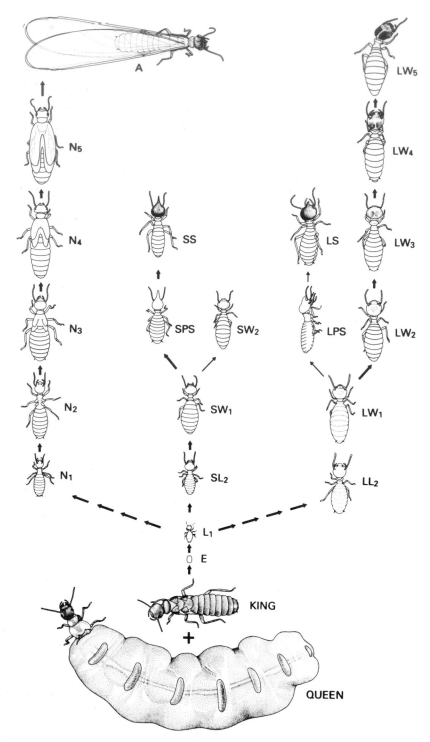

Fig. 11.7 Developmental pathways of the termite *Nasutitermes exitiosus* (Isoptera: Termitidae). Heavy arrows indicate the main lines of development, light arrows the minor lines. A, alate; E, egg; L, larva; LL, large larva; LPS, large presoldier; LS, large soldier; LW, large worker; N, nymph; SL, small larva; SPS, small presoldier; SS, small soldier; SW, small worker. The numbers indicate the stages. (Pathways based on Watson and Abbey, 1985.)

exopterygote, giving rise to alates – the winged reproductives that leave the colony, mate, disperse and found new colonies. In *N. exitiosus* no neotenics are formed; replacement for lost primary reproductives comes from amongst alates retained in the colony. Other *Nasutitermes* show great developmental plasticity.

The sterile (neuter) lineage is complex and variable between different termite species. In *N. exitiosus*, two categories of second-instar larvae can be recognized according to size differences probably relating to sexual dimorphism, although which sex belongs to which size category is unclear. In both lineages a subsequent moult produces a third-instar nymph of the worker caste, either small or large according to the pathway. These third-instar workers have the potential (**competency**) to develop into soldiers (via an intervening **presoldier** instar) or remain as workers through a variable number of further moults. The sterile pathway of *N. exitiosus* involves larger workers continuing to grow at successive moults, whereas the small worker ceases to moult beyond the fourth instar. Those that moult to become presoldiers and then soldiers develop no further.

The lower termites are more flexible, exhibiting more routes to differentiation. Lower termites have no true worker caste, but employ a functionally equivalent 'child-labour' **pseudergate** caste composed of either nymphs whose wing buds have been eliminated (regressed) by moulting or, less frequently, brachypterous nymphs or even undifferentiated larvae. Unlike the 'true' workers of the higher termites, pseudergates are developmentally plastic and retain the capacity to differentiate into other castes by moulting. In lower termites, differentiation of nymphs from larvae, and reproductives from pseudergates, may not be possible until a relatively late instar is reached. If there is sexual dimorphism in the sterile line, the larger workers are often male, but workers may be monomorphic. This may be through

the absence of sexual dimorphism or, more rarely, because only one sex is represented. Moults in species of lower termites may give:

- morphological change within a caste;
- no morphological advance (stationary moult);
- change to a new caste (such as a pseudergate to a reproductive);
- saltation to a new morphology, missing a normal intermediate instar;
- supplementation, adding an instar to the normal route;
- reversion to an earlier morphology (such as a pseudergate from a reproductive); or
- a presoldier from any nymph, late-instar larva or pseudergate.

Instar determination is impossibly difficult in the light of these moulting potentialities. The only inevitability is that a presoldier must moult to a soldier.

Certain unusual termites lack soldiers. Even the universal presence of only one pair of reproductives has exceptions; multiple primary queens cohabit in some colonies of some Termitidae.

Individuals in a termite colony are derived from one pair of parents. Therefore genetic differences existing between castes must be either sex-related or due to differential expression of the genes. Gene expression is under complex multiple and synergistic influences entailing hormones (including neurohormones), external environmental factors and interactions between colony members. Termite colonies are very structured and have high homeostasis – caste proportions are restored rapidly after experimental or natural disturbance by recruitment of individuals of appropriate castes and elimination of individuals excess to colony needs. Homeostasis is controlled by several pheromones that act specifically upon the corpora allata and more generally on the rest of the endocrine system. In the well-studied *Kalotermes*, primary reproductives inhibit differentiation of supplementary reproductives and alate nymphs.

Presoldier formation is inhibited by soldiers but stimulated through pheromones produced by reproductives.

Pheromones that inhibit reproduction are produced inside the body by reproductives and disseminated to pseudergates by **proctodeal trophallaxis**, that is, by feeding on anal excretions. Transfer of pheromones to the rest of the colony is by oral trophallaxis. This was demonstrated experimentally in a *Kalotermes* colony by removing reproductives and dividing the colony into two halves with a membrane. Reproductives were reintroduced, orientated within the membrane such that their abdomens were directed into one half of the colony, their heads into the other. Only in the 'head-end' part of colony did pseudergates differentiate as reproductives: inhibition continued at the 'abdomen-end'. Painting the protruding abdomen with varnish eliminated any cuticular chemical messengers but failed to remove the inhibition on pseudergate development. In contrast, pseudergates became reproductive when the anus was blocked, thereby verifying anal transfer. The inhibitory pheromones produced by both queen and king have complementary or synergistic effects: a female pheromone stimulates the male to release inhibitory pheromone, whereas the male pheromone has a lesser stimulatory effect on the female. Production of primary and supplementary reproductives involves removal of these pheromonal inhibitors produced by functioning reproductives.

Increasing recognition of the role of juvenile hormone (JH) in caste differentiation comes from observations such as the differentiation of pseudergates into soldiers after injection or topical application of JH or implantation of the corpora allata of reproductives. Some of the effects of pheromones on colony composition may be due to JH production by the primary reproductives. Caste determination in Termitidae originates as early as the egg, during maturation in the ovary of the queen. As the queen grows, the corpora allata undergoes hypertrophy and may attain a size 150 times greater than the gland of the alate. The JH content of eggs also varies, and it is possible that high JH level in the egg causes differentiation to follow the sterile lineage. This route is enforced if the larvae are fed proctodeal foods (or trophic eggs) that are high in JH, whereas a low level of JH in the egg allows differentiation along the reproductive pathway. In higher and lower termites, worker and soldier differentiation from the third-instar larva is under further hormonal control, as demonstrated by the induction of individuals of these castes by JH application.

(b) Nesting in termites

In the warmer parts of the temperate northern hemisphere, dry-wood termites are most familiar because of the structural damage that they cause to timber in buildings. Termites are pests of dry and damp wood in the subtropics and tropics, but in these regions termites may be more familiar through their spectacular mound nests. In the timber pests, colony size may be no greater than a few hundred termites, whereas in the mound formers several million individuals may be involved. In all cases, a new nest is founded by a male and female following the nuptial flight of alates. A small cavity is excavated into which the pair seal themselves. Copulation takes place in this royal cell, and egg laying commences. The first offspring are workers which are fed on regurgitated wood or other plant matter, primed with gut symbionts, until they are old enough to feed themselves and enlarge the nest. Early in the life of the colony, production is directed towards workers, with later production of soldiers to defend the colony. As the colony matures, but perhaps not until it is 5–10 years old, production of reproductives commences. This involves differentiation of alate sexual forms at the appropriate season for swarming and foundation of new colonies.

Tropical termites are able to use virtually all

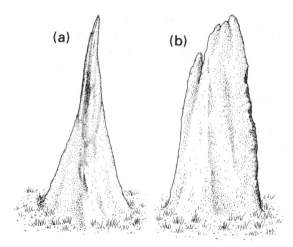

Fig. 11.8 The 'magnetic' mound of the debris-feeding termite *Amitermes meridionalis* (Isoptera: Termitidae) showing: (a) the north/south view, and (b) the east/west view. (After Hadlington, 1987.)

cellulose-rich food sources, from grass tussocks and fungi to living and dead trees, above and below the ground. Workers radiate from the mound, often in subterranean tunnels, less often in above-ground, pheromone-marked trails, in search of materials. In the subfamily Macrotermitinae (Termitidae), fungi are raised in combs of termite faeces within the mound, and the complete culture of fungus and excreta is eaten by the colony (section 8.5.3). These fungus-tending termites form the largest termite colonies known, with estimated millions of inhabitants in some East African species.

The giant mounds of tropical termites mostly belong to species in the Termitidae. As the colony grows through production of workers, the mound is enlarged by layers of soil and termite faeces until mounds as much as a century old attain massive dimensions. Diverse mound architectures characterize different termite species; for example, the 'magnetic mounds' of *Amitermes meridionalis* in northern Australia have a north–south narrow and east–west broad orientation, like a compass (Fig. 11.8). Orientation relates to thermoregulation, since the broad face of the

mound receives maximum exposure to the warming of the early and late sun, with the narrowest face presented to the high and hot midday sun. Aspect is not the only means of temperature regulation: intricate internal design, especially in fungus-farming *Macrotermes* species, allows circulation of air to give microclimatic control of temperature and carbon dioxide (Fig. 11.9).

11.3 INQUILINES OF SOCIAL INSECTS

The abodes of social insects provide many other insects with a hospitable place for their development. The term **inquiline** refers to an organism that shares a home of another. This covers a vast range of organisms that have some kind of obligate relationship with another organism – in this case, a social insect. Complex classification schemes involve categorization of the insect host and the known or presumed ecological relationship between inquiline and host (e.g. myrmecophile, termitoxene). Two alternative divisions appropriate to this discussion involve the degree of integration of the inquiline lifestyle with that of the host: **integrated** inquilines are incorporated into their hosts' social lives by behavioural modification of both parties; **non-integrated** inquilines are adapted ecologically to the nest but do not interact socially with the host. Predatory inquilines may negatively affect the host, whereas other inquilines may merely shelter within the nest, or give benefit, such as by feeding on nest debris.

Integration may be achieved by mimicking the chemical cues used by the host in social communication (such as pheromones), or by tactile signalling that releases social behavioural responses, or both. The term **Wasmannian mimicry** is used to cover some or all of the chemical or tactile mimetic features that allow the mimic to be accepted by a social insect, but the distinction from other forms of mimicry (notably Batesian; Chapter 13) is not always obvious. Wasmannian mimicry may, but need not, include imitation of the body

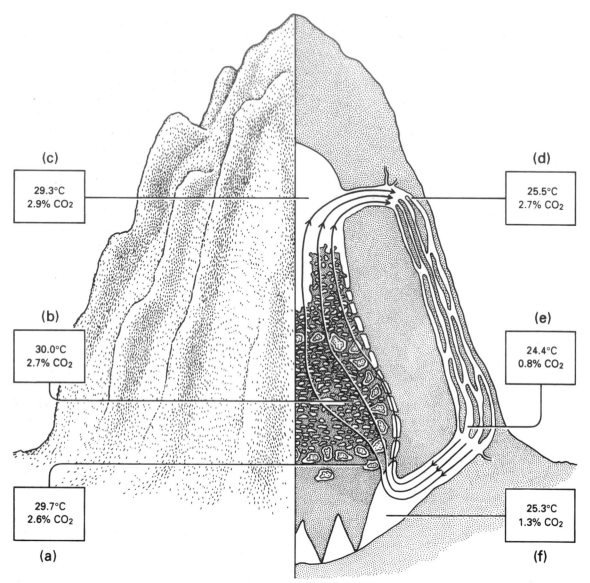

(c)
29.3°C
2.9% CO_2

(d)
25.5°C
2.7% CO_2

(b)
30.0°C
2.7% CO_2

(e)
24.4°C
0.8% CO_2

(a)
29.7°C
2.6% CO_2

(f)
25.3°C
1.3% CO_2

Fig. 11.9 Section through the mound nest of the African fungus-farming termite *Macrotermes natalensis* (Isoptera: Termitidae) showing how air circulating in a series of passageways maintains favourable culture conditions for the fungus at the bottom of the nest and for the termite brood. Measurements of temperature and carbon dioxide are shown in the boxes for the following locations: (a) the fungus combs; (b) the brood chambers; (c) the attic; (d) the upper part of a ridge channel; (e) the lower part of a ridge channel; (f) the cellar. (After Lüscher, 1961.)

form. Conversely, mimicry of a social insect does not necessarily imply inquilinism – the ant mimics shown in Fig. 13.11 are not symbionts but apparently gain some protec-

tion from their natural enemies as a result of their ant-like appearance.

The breaking of the social insect chemical code occurs through the ability of an inquiline to produce appeasement or adoption chemicals – the messengers that social insects use to recognize one another and to distinguish themselves from intruders. Many staphylinid beetles are able to do this; an example is *Atemeles pubicollis*, which lives as a larva in the nest of European *Formica rufa*. The staphylinid larva produces a glandular secretion that induces brood-tending ants to groom the alien. Food is obtained by adoption of the begging posture of an ant larva, in which the larva rears up and contacts the adult ant's mouthparts, provoking a release of regurgitated food. The diet of the staphylinid is supplemented by predation on larvae of ants and of their own species. Pupation and adult eclosion take place in the *Formica rufa* nest. This species of ant ceases activity in winter and during this period the staphylinid seeks alternative shelter. Adult beetles leave the wooded *Formica* habitat and migrate to the more open grassland habitat of *Myrmica* ants. When a *Myrmica* ant is encountered, secretions from the 'appeasement glands' are offered which suppress the aggression of the ant, and then the products of glands on the lateral abdomen attract the ant. Feeding on these secretions appears to facilitate 'adoption', since the ant subsequently carries the beetle back to its nest, where the immature adult overwinters as a tolerated food-thief. In spring, the reproductively mature adult departs for the woods to seek out a *Formica* nest for oviposition.

Amongst the inquilines of termites, many show convergence in shape in terms of physogastry (dilation of the abdomen), seen also in queen termites. In the curious case of *Termitoxenia* and relatives within the family Phoridae (Diptera), the physogastric females from termite nests were the only stage known for so long that published speculation was rife that neither larvae nor males existed. It was suggested that the females hatched directly from huge eggs and were brachypterous throughout their lives, hitching a ride on termites for dispersal. Uniquely amongst the endopterygotes, the flies were believed to be protandrous hermaphrodites, functioning first as males, then as females. The truth is more prosaic: sexual dimorphism in the group is so great that wild-caught flying males had been unrecognized and placed in a different taxonomic group; the females are winged but shed all except the stumps of the anterior veins after mating, before entering the termitarium; although the eggs are large, larval stages are present, but very short lived; and since the post-mated female is **stenogastrous** (with a small abdomen), physogastry develops whilst in the termitarium. Thus *Termitoxenia* is an unconventional fly, well adapted to the rigours of life in a termite nest where eggs are treated by termites as their own, and with attenuation of the vulnerable larval stage, rather than the possessor of a unique suite of life-history features.

Inquilinism is not restricted to non-social insects that abuse the hospitality of social insects. Even amongst the social Hymenoptera, some ants may live as temporary or even permanent social parasites in the nests of other species. A reproductive female inquiline gains access to a host nest and usually kills the resident queen. In some cases, the intruder queen produces workers which eventually take over the nest. In others, the inquiline usurper produces only males and reproductives – the worker caste is eliminated and the nest survives only until the workers of the host species die off.

In a further twist of the complex social lives of ants, some species are slave-makers which capture pupae from the nests of other species and return them to their own nest, where they are reared as slave workers. This phenomenon, known as **dulosis**, occurs in several inquiline species, all of which found their colonies by parasitism.

The phylogenetic relationships between ant hosts and ant inquilines reveals an unexpectedly high proportion of instances in which host and inquiline belong to sister species (i.e. each other's closest relatives) and many more are congeneric close relatives. One possible explanation envisages the situation in which daughter species formed in isolation come into secondary contact after mating barriers have developed. If no differentiation of colony-identifying chemicals has taken place, it is possible for one species to invade the colony of the other, undetected, and parasitization is facilitated.

Non-integrated inquilines are exemplified by hover flies of the genus *Volucella* (Diptera: Syrphidae), the adults of which are Batesian mimics of either *Polistes* wasps or of *Bombus* bees. Female flies appear free to fly in and out of hymenopteran nests, and lay eggs whilst walking over the comb. Hatching larvae drop to the bottom of the nest, where they scavenge on fallen detritus and fallen prey. Another syrphid, *Microdon*, has a myrmecophilous larva so curious that it was described first as a mollusc, then as a coccoid. It lives unscathed amongst nest debris (and perhaps sometimes as a predator on young ant larvae) but the emerged adult is recognized as an intruder. Non-integrated inquilines include many predators and parasitoids whose means of circumventing the defences of social insects are largely unknown.

11.4 EVOLUTION OF EUSOCIALITY

At first impression the complex social systems of hymenopterans and termites bear a close resemblance and it is tempting to suggest a common origin. However, examination of the phylogeny presented in Chapter 7 (Fig. 7.5) shows that these two orders, and the social aphids and thrips, are distantly related and a single evolutionary origin is inconceivable. Thus we will examine the possible routes for the origin of eusociality in Hymenoptera and Isoptera separately.

11.4.1 The origins of eusociality in Hymenoptera

According to estimates derived from the proposed phylogeny of the Hymenoptera,

eusociality has arisen independently in wasps, bees and ants (Fig. 11.2) with multiple origins within wasps and bees. Comparison of life histories between living species with different degrees of social behaviour allows extrapolation to possible historical pathways from solitariness to sociality. Three possible routes have been suggested and, in each case, communal living is seen to provide benefits through sharing the costs of nest construction and defence of offspring.

The first suggestion envisages a monogynous (single queen) subsocial system with eusociality developing through the queen remaining associated with her offspring during increased maternal longevity.

The second scenario involves semisociality and is perhaps applicable only to certain bees: several unrelated females of the same generation associate and establish a colonial nest in which there is some reproductive division of labour, with an association that lasts only for one generation.

The third scenario involves elements of the previous two, with a communal group comprising related females (rather than unrelated) and multiple queens (in a polygynous system), within which there is increasing reproductive division. The association of queens and daughters arises through increased longevity.

These life-history scenarios must be considered in relation to genetic theories concerning eusociality, notably concerning the origins and maintenance by selection of **altruism** (or self-sacrifice in reproduction). Ever since Darwin, there has been debate about altruism: why should some individuals (sterile workers) sacrifice their reproductive potential for the benefit of others?

Four proposals for the origins of the extreme reproductive sacrifice seen in eusociality are group selection, kin selection, maternal manipulation and mutualism. The last three of those proposals are partially or completely compatible with one another. **Group selection**, however, seems incompatible. In this case, selection is argued to operate at the level of the group: an efficient colony with an altruistic division of reproductive labour will survive and produce more offspring compared with one in which rampant individual self-interest leads to anarchy. Whilst this scenario aids in understanding the maintenance of eusociality once it is established, it contributes little if anything to explaining the origins of reproductive sacrifice in non-eusocial or subsocial insects. The concept of group selection operating on pre-eusocial colonies runs counter to the view that selection operates on the genome, and hence the origin of altruistic individual sterility is difficult to accept under group selection. It is amongst the remaining three proposals that the origins of eusociality are more usually sought.

Kin selection stems from recognition that **classical** or **Darwinian fitness** (the direct genetic contribution to the gene pool by an individual through its offspring) is only part of the contribution to an individual's total, or **inclusive** or **extended, fitness**. An additional indirect contribution, termed the **kinship component**, must be included. This is the contribution to the gene pool made by an individual that assists and enhances the reproductive success of its kin. Kin are individuals with similar or identical genotypes derived from the relatedness of their parents. In the Hymenoptera, kin relatedness is enhanced by the haplodiploid sex determination system, in which males are haploid so that sperm (produced by mitosis) contains 100% of their genes. In contrast, the egg (produced by meiosis) is diploid, containing only half the maternal genes. Thus daughter offspring, produced from fertilized eggs, share all their father's genes but only half of their mother's genes. Because of this, full sisters (that is, those with the same father) share on average three-quarters of their genes. Therefore sisters share more genes with each other than they would with their own female offspring (50%). Under these conditions, the inclusive fitness of a sterile

female (worker) is greater than its classical fitness. Since selection operating on an individual should maximize its inclusive fitness, a worker should invest in the survival of her sisters, the queen's offspring, rather than in the production of her own female young.

There are some difficulties with a haplodiploid explanation alone for the origin of eusociality, since altruism does not arise solely from relatedness. Haplodiploidy is universal in hymenopterans and the resultant close relatedness has encouraged repeated eusociality, but eusociality is not universal in the Hymenoptera. Furthermore, eusociality does not occur in other haplodiploid insects, such as thrips, although there may be social behaviour. Other factors involved in promoting eusociality are recognized in Hamilton's rule, which emphasises the ratio of costs and benefits of altruistic behaviour as well as relatedness. The conditions under which selection will favour altruism can be expressed as: $rB - C > 0$, where r is the coefficient of relatedness, B is the benefit gained by the recipient of altruism and C is the cost suffered by the donor of altruism. Thus variation in benefits and costs modify the consequences of a particular degree of relatedness, though these factors are difficult to measure.

Kinship calculations assume that all offspring of a single mother in the colony share an identical father, and this assumption is implicit in the kinship scenario for the origin of eusociality. However, at least in the higher eusocial insects, queens may mate several times with different males. For these eusocial insects, intra-colony relatedness will be less than predicted by the model. This difficulty is unrelated to the origins of eusociality, being more concerned with the maintenance of an already existing eusocial system. Once sterile castes have originated (theoretically under conditions of single paternity), the requirement for high relatedness may be relaxed if sterile workers lack any opportunity to reproduce. Evidently it is the opportunity to help relatives, in combination with high relatedness through haplodiploidy, that predisposes insects to eusociality.

There are two further ideas concerning the origins of eusociality. The first involves **maternal manipulation** of offspring (both behaviourally and genetically), so that, by reducing the reproductive potential of some offspring, the fitness of the parent may be maximized through reproductive success being assured to a few select offspring. The ability of most aculeate females to control the sex of offspring through fertilizing the egg or not, and to vary the size of the offspring through the amount of food supplied, makes maternal manipulation a plausible option for the origin of eusociality.

A further well-supported scenario emphasizes the roles of competition and **mutualism**. This envisages individuals acting to enhance their own classical fitness with contributions to the fitness of neighbours arising only incidentally. Each individual benefits from colonial life through communal defence by shared vigilance against predators and parasites. Thus mutualism (including the benefits of shared defence and nest construction) and kinship encourage the establishment of group living. Differential reproduction within a familial-related colony confers significant fitness advantages on all members through their kinship. In conclusion, the three scenarios of kin selection, female manipulation and mutualism are not mutually exclusive but are compatible in combination: they act in concert to allow evolution of eusociality.

The Vespinae illustrate a trend to eusociality commencing from a solitary existence, with nest-sharing and facultative labour division being a derived condition. Further evolution of eusocial behaviour developed through a dominance hierarchy that arose from female manipulation and reproductive competition among the nest-sharers: the 'winners' are queens and the 'losers' are workers. From this point onwards, individuals act to maximize their fitness and the caste system becomes more rigid. As the queen and colony acquire greater longevity and the number of generations retained increases, short-term monogynous societies (those with a succession of queens) become long-term, monogynous, **matrifilial** (mother–daughter) colonies. Exceptionally, a derived polygynous condition may arise in large colonies or in colonies where queen dominance is relaxed.

11.4.2 The origins of eusociality in Isoptera

In contrast to the haplodiploidy of Hymenoptera, termite sex is determined universally by an XX-XY chromosome system and so there is no genetic predisposition toward kinship-based eusociality. Furthermore, and in contrast with the widespread subsociality of hymenopterans, a lack of any intermediate stages on the route to termite eusociality makes understanding its origin difficult. Subsocial behaviours do occur within the cockroaches and mantids (together forming the sister group to the termites), and it has been proposed that evolution of eusociality in Isoptera can be deduced from the behaviour of *Cryptocercus*-like cockroaches.

However, any suggestion that *Cryptocercus* is placed on a termite lineage is untenable on phylogenetic grounds (Box 11.1) but the environmental conditions of a nutrient-poor food source and adult longevity might have led both convergently to social living. A possible scenario for the origin of the termite social system involves the requirement for internal symbiotic organisms to assist in the digestion of food that is rich in cellulose but poor in nutrients. The need to transfer symbionts in order to replenish supplies that are lost at each moult encourages unusual levels of intra-colony interaction through trophallaxis. Furthermore, transfer of symbionts between members of successive generations requires overlapping generations. Trophallaxis, parental longevity and slow growth induced by the poor diet act together to encourage group cohesion. These factors, together with patchiness of adequate food resources such as rotting logs, can lead to colonial life but do not readily explain altruistic caste origins. When an individual gains substantial benefits from successful foundation of a colony, and with a high degree of intra-colony relatedness (as is found in some termites), eusociality may arise but its origins in termites remain much less clear than for eusocial hymenopterans.

11.5 SUCCESS OF EUSOCIAL INSECTS

Social insects can attain numerical and ecological dominance in some regions. They are most abundant at low latitudes and low elevations, and their activities are most conspicuous either in summer in temperate (or even subarctic and montane) areas, or year-round in subtropical to tropical climates. As a generalization, the most abundant and dominant social insects are the most derived phylogenetically and have the most complex social organization.

Three qualities of social insects are believed to contribute to their competitive advantage, all of which derive from the caste system that allows multiple tasks to be performed. Firstly, the labours of foraging, feeding the queen, caring for offspring and maintenance of the nest are actions that take place simultaneously by different groups rather than sequentially as in solitary insects. Conducting activities in parallel means that one activity does not jeopardize another, and so the nest is not vulnerable to predators or parasites whilst foraging is taking place. Furthermore, mistakes made by individuals have little or no consequence in parallel operations compared with those in serial. Secondly, the ability of the colony to marshall all workers can overcome difficulties that are serious for a solitary insect, such as defence against a much larger or more numerous predator, or construction of a nest under unfavourable conditions.

Thirdly, the specialization of function associated with castes allows some homeostatic regulation, including holding of food reserves in some castes (such as honeypot ants) or in developing larvae, and behavioural control of temperature and other microclimatic conditions within the nest. The ability to vary the proportion of individuals allocated to a particular caste allows appropriate distribution of community resources according to the differing demands of season and colony age. The widespread use of a variety of pheromones allows a high level of control to be exerted, even over millions of individuals. However, within this apparently rigid eusocial system, there is scope for a wide variety of different life histories to have evolved, from the nomadic army ants to the parasitic inquilines.

FURTHER READING

Billen, J. (ed.) (1992) *Biology and Evolution of Social Insects*, Leuven University Press, Leuven.

Breed, M.D., Michener, C.D. and Evans, H.E. (eds) (1982) *The Biology of Social Insects*, Westview Press, Boulder, Colorado.

Carpenter, J (1989) Testing scenarios: wasp social behaviour. *Cladistics*, **5**, 131–144.

Crespi, B.J. (1992) Behavioural ecology of Australian gall thrips (Insecta, Thysanoptera). *Journal of Natural History*, **26**, 769–809.

de Wilde, J. and Beetsma, J. (1982) The physiology of caste development in social insects. *Advances in Insect Physiology*, **16**, 167–246.

Fletcher, D.J.C. and Ross, K.G. (1985) Regulation of reproduction in eusocial Hymenoptera. *Annual Review of Entomology*, **30**, 319–43.

Hardie, J. and Lees, A.D. (1985) Endocrine control of polymorphism and polyphenism, in *Comprehensive Insect Physiology, Biochemistry, and Pharmacology, Volume 8, Endocrinology II*, (eds G.A. Kerkut and L.I. Gilbert), Pergamon Press, Oxford, New York, Toronto, Sydney, Paris, Frankfurt, pp. 441–90.

Hermann, H.R. (ed.) (1979–82) *Social Insects, Volumes I–IV*, Academic Press, New York, London, etc.

Hölldobler, B. and Wilson, E.O. (1990) *The Ants*, Springer-Verlag, Berlin, Heidelberg, London, Paris, Tokyo, Hong Kong.

Itô, Y. (1989) The evolutionary biology of sterile soldiers in aphids. *Trends in Ecology and Evolution*, **4**, 69–73.

Kiester, A.R. and Strates, E. (1984) Social behaviour in a thrips from Panama. *Journal of Natural History*, **18**, 303–14.

Retnakaran, A. and Percy, J. (1985) Fertilization and special modes of reproduction, in *Comprehensive Insect Physiology, Biochemistry, and Pharmacology, Volume I, Embryogenesis and Reproduction,* (eds G.A. Kerkut and L.I. Gilbert), Pergamon Press, Oxford, New York, Toronto, Sydney, Paris, Frankfurt, pp. 231–93.

Robinson, G.E. (1992) Regulation of division of labor in insect societies. *Annual Review of Entomology*, **37**, 637–65.

Tallamy, D.W. and Wood, T.K. (1986) Convergence patterns in subsocial insects. *Annual Review of Entomology*, **31**, 369–90.

Wilson, E.O. (1971) *The Insect Societies*, The Belknap Press of Harvard University Press, Cambridge, Massachusetts.

Wilson, E.O. (1985) The sociogenesis of insect colonies. *Science*, **228**, 1489–95.

12

INSECT PREDATION AND PARASITISM

Scorpionfly feeding on a butterfly pupa

Many insects are phytophagous, feeding directly on primary producers, the algae and higher plants (Chapter 10). These phytophages comprise a substantial food resource for a broad carnivorous group of other organisms, individuals within which may be categorized as follows. A **predator** kills and consumes a number of **prey** animals during its life. Predation involves the interactions in space and time between predator foraging and prey availability, although often it is treated in a one-sided manner as if predation is what the predator does. Animals that live at the expense of another animal (a **host**) that eventually dies as a result are called **parasitoids**, which may live externally (**ectoparasitoids)** or internally (**endoparasitoids**). Those that live at the expense of another animal (also a host) that they do not kill are **parasites**, which likewise can be internal (**endoparasites**) or external (**ectoparasites**). A host attacked by a parasitoid or parasite is **parasitized**. Parasitization is the condition of being parasitized, while **parasitism** describes the relationship between parasitoid or parasite and the host. Predators, parasitoids and parasites, although defined above as if distinct, may not be so clear-cut, since parasitoids may be viewed as specialized predators.

By some estimates, about 25% of insect species are predatory or parasitic in feeding habit in some life-history stage. Representatives from amongst nearly every order of insects are predatory, with adults and immature stages of the Mantodea, the neuropteroid orders (Neuroptera, Megaloptera and Raphidioptera) and the Mecoptera being almost exclusively predatory. The latter orders are considered in Boxes 12.2, 12.3, and 12.5, and the vignette for this chapter depicts a female mecopteran, *Panorpa communis*, feeding on a butterfly pupa. The speciose Hymenoptera have a preponderance of parasitoid taxa almost exclusively using invertebrate hosts, and the uncommon Strepsiptera are unusual in being endoparasites in other insects (Box 12.4). Other parasites that are of medical or veterinary importance, such as lice, adult fleas and many Diptera, are dealt with in Chapter 14.

Insects are amenable to field and laboratory studies of predator–prey interactions since they are unresponsive to human attention, easy to manipulate, may have several generations a year and show a range of predatory and defensive strategies and life histories. Studies of predator–prey and parasitoid–host interactions 'are fundamental to understanding and effecting biocontrol strategies for pest insects. Attempts to model predator–prey interactions mathematically often emphasize parasitoids, since some simplifications can be made. These include the ability to simplify search strategies, because only the adult female parasitoid seeks hosts, and the number of offspring per unit host remain relatively constant from generation to generation.

This chapter shows how predators, parasitoids and parasites **forage**, that is, locate and select their prey or hosts. It looks at morphological modifications of predators for handling prey and how some of the prey defences covered in Chapter 13 are overcome. The means by which parasitoids overcome host defences and develop within their hosts is examined and different strategies of host use by parasitoids are explained. The host use and specificity of ectoparasites is discussed from a phylogenetic perspective. Finally, there is a consideration of the relationships between predator/parasitoid/parasite and prey/host abundances and evolutionary histories.

12.1 PREY/HOST LOCATION

The foraging behaviours of insects, like all other behaviours, comprise a stereotyped sequence of components that leads a predatory or host-seeking insect towards the resource and, on contact, to recognize and use it. Various stimuli along the route elicit an appropriate response, involving either action or inhibition. The foraging strategies of

Fig. 12.1. The basic spectrum of predator foraging and prey defence strategies, varying according to costs and benefits in both time and energy. (After Malcolm, 1990.)

predators, parasitoids and parasites involve trade-offs between profits or benefits (the quality and quantity of resource obtained) and costs, in the form of time expenditure, exposure to suboptimal or adverse environments and the risks of being eaten. Recognition of the time component is important, since all time spent in activities other than reproduction can be viewed, in an evolutionary sense, as time wasted.

In an optimal foraging strategy, the difference between costs and benefits is maximized, through either increasing nutrient gain from prey capture, or reducing effort expended to catch prey, or both. Choices available are:

- Where and how to search.
- How much time to expend in fruitless search in the one area, before moving.
- How much (if any) energy to expend in capture of suboptimal food, once located.

A primary requirement is that the insect be in the appropriate habitat for the resource sought. For many insects this may seem trivial, especially if development takes place in the area which contained the resources used by the parental generation. However, circumstances such as seasonality, climatic vagaries, ephemerality or major resource depletion may necessitate local **dispersal** or perhaps major movement – **migration** – in order to reach an appropriate location.

Even in a suitable habitat, resources are rarely distributed evenly but occur in more or less discrete microhabitat clumps, termed **patches**. Insects show a gradient of responses to these patches. At one extreme, the insect waits in a suitable patch for prey or host organisms to appear. The insect may be camouflaged or apparent, and a trap may be constructed. At the other extreme, the prey or host is actively sought within a patch. The waiting strategy is economically effective, but time-consuming; the active strategy is energy intensive but time-efficient, and trapping lies intermediate between these two (Fig. 12.1). Patch selection is vital to successful foraging.

12.1.1 Sitting and waiting

Sit-and-wait predators find a suitable patch and wait for mobile prey to come within striking range. Since the vision of many insects limits them to recognition of movement rather than precise shape, a sit-and-wait predator may need only to remain motionless in order to be unobserved by its prey. None-the-less, many amongst those that wait have some form of camouflage (**crypsis**). This may be defensive, directed against highly visual predators such as birds, rather than evolved to mislead invertebrate prey. Cryptic predators modelled on a feature that is of no interest to the prey (such as tree bark, lichen, a twig or even a stone) can be distinguished from those that model on a feature of some significance to prey, such as a flower that acts as an insect attractant.

In an example of the latter case, the Malaysian mantid *Hymenopus bicornis* closely resembles the red flowers of the orchid *Melastoma polyanthum* amongst which it rests. Flies are encouraged to land, assisted by the presence of marks resembling flies on the body of the mantid: larger flies that land are eaten by the mantid. In another related example of **aggressive foraging mimicry**, the African flower-mimicking mantid *Idolum* does not rest hidden in a flower but actually resembles one due to petal-shaped, coloured outgrowths of the prothorax and the coxae of the anterior legs. Butterflies and flies that are attracted to this hanging 'flower' are snatched and eaten.

Ambushers include cryptic, sedentary insects such as mantids that prey fail to distinguish from the inert, non-floral plant background. Although these predators rely on the general traffic of invertebrates associated with vegetation, often they locate close to flowers to take advantage of the increased visiting rate of flower-feeders and pollinators.

Odonate larvae, which are major predators in many aquatic systems, are classic ambushers which rest concealed in submerged vegetation or in the substrate awaiting prey to pass. These predators may show dual strategies: if waiting fails to provide food after a fixed period, the hungry insect may change to a more active searching mode. This energy expenditure may bring the predator into an area of higher prey density. In running waters, a disproportionately high number of organisms found drifting passively with the current are predators: this drift constitutes a low-energy means for sit-and-wait predators to relocate, induced by local prey shortage.

Sit-and-wait strategies are not restricted to cryptic and slow-moving predators. Fast-flying, diurnal, visual, rapacious predators such as many robber flies (Diptera: Asilidae) and adult odonates spend much time perched prominently on vegetation. From these conspicuous locations their excellent sight allows them to detect passing flying insects. With rapid and accurately controlled flight, the predator makes only a short foray to capture appropriately-sized prey. This strategy combines energy saving, through not needing to fly incessantly in search of prey, with time efficiency since prey is taken from outside the immediate area of reach of the predator.

Another sit-and-wait technique involving greater energy expenditure is the use of traps to ambush prey. Whilst spiders are the prime exponents of this method, in the warmer parts of the world the pits of certain larval antlions (Neuroptera: Myrmeleontidae; Fig. 12.2(a), (b)) are familiar. The larvae either dig pits directly or form them by spiralling backwards into soft soil or sand. Trapping effectiveness depends upon the steepness of the sides and the diameter and depth of the pit, which vary with species and instar. The larva waits, buried at the base of the conical pit, for passing prey to fall in. Escape is prevented physically by the slipperiness of the slope, and the larva may also flick sand at the prey before dragging it underground to restrict its defensive movements. The location, construction and maintenance of the pit are vitally important to capture efficiency but construction and repair are energetically very expensive. It has been shown experimentally that even starved Japanese antlions (*Myrmeleon bore*) would not relocate their pits to an area where prey was provided artificially. Instead, larvae of this species of antlion reduce their metabolic rate to tolerate famine, even if death by starvation is the result.

In holometabolous ectoparasites, such as fleas and parasitic flies, immature development takes place away from their vertebrate hosts. Following pupation, the adult must locate the appropriate host. Since in many the eyes are reduced or absent, vision cannot be used; and since many ectoparasites are flightless, mobility is restricted. In fleas and some Diptera, in which larval development often takes place in the nest of a host vertebrate, the adult insect waits quiescent in

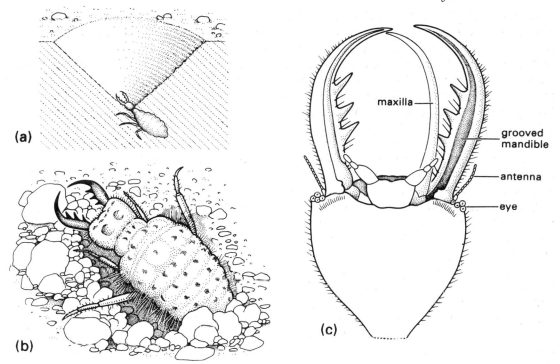

Fig. 12.2. An antlion of *Myrmeleon* (Neuroptera: Myrmeleontidae): (a) larva in its pit in sand; (b) detail of dorsum of larva; (c) detail of ventral view of larval head showing how the maxilla fits against the grooved mandible to form a sucking tube. (After Wigglesworth, 1964.)

the pupal cocoon until the presence of a host is detected. The duration of this quiescent period may be a year or longer, as in the cat flea (*Ctenocephalides canis*) – a familiar phenomenon to humans that enter an empty dwelling that previously housed flea-infested cats. The stimuli to cease dormancy include some or all of the following: vibration, rise in temperature, increased carbon dioxide or another stimulus generated by the host.

In contrast, the hemimetabolan lice spend their lives entirely on a host, with all developmental stages ectoparasitic. Any transfer between hosts is either through phoresy (section 12.1.3) or when host individuals make direct contact, as from mother to young within a nest.

12.1.2 Active foraging

More energetic foraging involves active searching for suitable patches and, once there, for prey or for hosts. Movements associated with foraging and with other locomotory activities, such as seeking a mate, are so similar that the 'motivation' may be recognized only in retrospect, by resultant prey capture or host finding. The locomotory search patterns used to locate prey or hosts are those described for general orientation in section 4.5, and comprise non-directional (random) and directional (non-random) locomotion.

(a) Random or non-directional foraging

The foraging of aphidophagous larval coccinellid beetles and syrphid flies amongst their clumped prey illustrates several features of random food searching. The larvae advance, stop periodically and cast about by swinging their raised anterior bodies from side to side. Subsequent behaviour depends upon whether or not an aphid is encountered. If it is not, motion continues, interspersed with casting and turning at a fundamental frequency. However, if contact is made and feeding has taken place, or if the prey is encountered and lost, searching intensifies with an enhanced frequency of casting and, if the larva is in motion, increased turning or direction-changing. Actual feeding is unnecessary to stimulate this more concentrated search: an unsuccessful encounter is adequate. For early-instar larvae that are very active but have limited ability to handle prey, this stimulus to search intensively near a lost feeding opportunity is important in survival.

Most laboratory-based experimental evidence, and models of foraging based thereon, are derived from single species of walking predators, frequently assumed to encounter a single species of prey randomly distributed within selected patches. Such premises may be justified in modelling grossly simplified ecosystems, such as an agricultural monoculture with a single pest controlled by one predator. Despite the limitations of such laboratory-based models, certain findings appear to have general biological relevance.

An important consideration is that the time allocated to different patches by a foraging predator depends upon the criteria for leaving a patch. Four mechanisms that trigger departure have been recognized:

- A certain number of food items have been encountered (fixed number).
- A certain time has elapsed (fixed time).
- A certain searching time has elapsed (fixed searching time).
- The prey capture rate falls below a certain threshold (fixed rate).

The fixed rate mechanism has been favoured by modellers of optimal foraging, but even this is likely to be a simplification if the forager's responsiveness to prey is non-linear (for example, declines with exposure time) or derives from more than simple prey encounter rate, or prey density. Differences between predator–prey interactions in simplified laboratory conditions and the actuality of the field cause many problems, including failure to recognize variation in prey behaviour that results from exposure to predation (perhaps multiple predators). Furthermore, there are difficulties in interpreting the actions of polyphagous predators, including the causes of predator/parasitoid/parasite behavioural switching between different prey animals or hosts.

(b) Non-random or directional foraging

Several more specific directional means of host-finding can be recognized, including the use of chemicals, sound and light. Experimentally these are rather difficult to establish, and to separate, and it may be that the use of these cues is very widespread, if little understood. Of the variety of cues available, many insects probably use more than one, depending upon distance or proximity to the resource sought. Thus the European sphecid wasp *Philanthus*, which eats only bees, relies initially on vision to locate moving insects of appropriate size. Only bees, or other insects to which bee-odours have been applied experimentally, are captured, indicating a role for odour when near the prey. However, the sting is applied only to actual bees, and not to bee-smelling alternatives, demonstrating a final tactile recognition.

Not only may a step-wise sequence of stimuli be necessary, as seen above, but also appropriate stimuli may have to be present simultaneously in order to elicit appropriate behaviour. *Telenomus heliothidis* (Hymenoptera: Scelionidae), an egg parasitoid of *Heliothis virescens* (Lepidoptera: Noctuidae), will investigate and probe at appropriate-sized round glass beads, emulating *Heliothis* eggs, if they are coated with female moth proteins; but it makes no response to glass beads alone, nor to female moth proteins applied to improperly shaped beads.

Chemicals

The world of insect communication is dominated by chemicals, or pheromones (Chapter 4). Ability to detect kairomones – the chemical odours and messages produced by prey or hosts – allows specialist predators and parasitoids to locate these resources. Certain parasitic tachinid flies and braconid wasps can locate their respective stink bug or coccoid host by tuning to their hosts' long-distance sex attractant pheromones. Several unrelated parasitoid hymenopterans use the aggregation pheromones of their bark and timber beetle hosts. Chemicals emitted by stressed plants, such as terpenes produced by pines when attacked by an insect, act as synomones (communication chemicals that benefit both producer and receiver); for example, certain pteromalid (Hymenoptera) parasitoids locate their hosts, the damage-causing scolytid timber beetles, in this way. Some species of tiny wasps (Trichogrammatidae) that are egg endoparasitoids (Fig. 15.3) are able to locate the eggs of their preferred host moth by the sex attractant pheromones released by the moth. Furthermore, there are several examples of parasitoids that locate their specific insect larval hosts by 'frass' odours – the smells of their faeces. Chemical location is particularly valuable when hosts are concealed from visual inspection, for example, when encased in plant or other tissues.

Chemical detection need not be restricted to tracking volatile compounds produced by the prospective host. Many parasitoids searching for phytophagous insect hosts are attracted initially, and at a distance, to host-plant chemicals in the same manner that the phytophage located the resource. At close range, chemicals produced by the feeding damage and frass of phytophages may allow precise targeting of the host. Once located, the acceptance of a host as suitable is likely to involve similar or other chemicals, judging by

the increased use of rapidly vibrating antennae in sensing the prospective host.

Blood-feeding adult insects locate their hosts using cues which include chemicals emitted by the host. Many female biting flies can detect increased carbon dioxide levels associated with animal respiration and fly upwind towards the source. It is probable that highly host-specific biters are also able to detect subtle odours: for example, human-biting black flies (Diptera: Simuliidae) respond to components of human exocrine sweat glands. Both sexes of tsetse flies (Diptera: Glossinidae) track the odour of exhaled breath, notably carbon dioxide, octanols, acetone and ketones emitted by their preferred cattle hosts.

Sound

The sound signals produced by animals, including those made by insects to attract mates, have been utilized by some parasites to locate their preferred hosts acoustically. The blood-sucking females of *Corethrella* (Diptera: Corethrellidae) locate their favoured host, hylid treefrogs, by following the frogs' calls. Flies of two other dipteran species are known to be attracted by the songs of their hosts: females of the larviparous tachinid *Euphasiopteryx ochracea* locate the male crickets of *Gryllus integer*, and the sarcophagid *Colcondamyia auditrix* finds its male cicada host *Okanagana rimosa* in this manner. This allows quite precise deposition of the parasitic immature stages in, or close to, the hosts in which they are to develop. The acoustic orientation of at least the tachinid may not be species-specific, since experimentally the fly responds phonotactically to certain other orthopteran calls.

Predatory biting midges (Ceratopogonidae) that prey upon swarm-forming flies, such as midges (Chironomidae), appear to use cues similar to those used by their prey to locate the swarm; these may include the sounds produced by wing-beat frequency of the members of the swarm. Vibrations produced by their hosts can be detected by ectoparasites, notably amongst the fleas. There is also evidence that certain parasitoids can detect at close range the substrate vibrations produced by the feeding activity of their hosts. Thus *Biosteres longicaudatus*, a braconid hymenopteran endoparasitoid of a larval tephritid fruit fly (Diptera: *Anastrepha suspensa*) detects vibrations made by the larvae moving and feeding within fruit. These sounds act as a behavioural releaser, stimulating host-finding behaviour as well as acting as a locational cue for their concealed hosts.

Light

The larvae of the Australian and New Zealand cave-dwelling mycetophilid flies, genus *Arachnocampa*, use bioluminescent lures to catch small flies in sticky threads that they suspend from the cave ceiling. Luminescence, as with all communication systems, provides scope for abuse; in this case, the luminescent signal of courtship between beetles is misappropriated. Carnivorous female lampyrids of some *Photurus* species, in an example of aggressive foraging mimicry, imitate the flashing signals of females of up to five other firefly species. The males of these different species flash their responses and are deluded into landing close by the mimetic female, upon which she devours them. The mimicking *Photurus* female will eat the males of her own species, but cannibalism is avoided or reduced since the *Photurus* female is most piratic after mating, at which time she becomes relatively unresponsive to the signals of males of her own species.

12.1.3 Phoresy

Phoresy is a phenomenon in which an individual of one species is transported by an individual of another species. This relationship benefits the carried and does not directly affect the carrier, although in some cases its progeny may be disadvantaged (as we shall see below). Phoresy provides a means of finding a new host or food source. An often observed example involves ischnoceran lice (Phthiraptera) transported by the winged adults of *Ornithomyia* (Diptera: Hippoboscidae). Hippoboscidae are blood-sucking ectoparasitic flies and *Ornithomyia* occurs on many avian hosts. When a host bird dies, lice can reach a new host by attaching themselves by their mandibles to a hippoboscid, which may fly to a new host. However, lice are highly host-specific but hippoboscids are much less so, and chances of any hitch-hiking louse arriving at the appropriate host may not be great.

Amongst the egg-parasitizing hymenopterans (notably the Scelionidae, Trichogrammatidae and Torymidae), some attach themselves to adult females of the host species, thereby gaining immediate access to the eggs at oviposition. *Matibaria manticida* (Scelionidae), an egg parasitoid of the European praying mantid (*Mantis religiosa*), is phoretic, predominantly on female hosts. The adult wasp sheds its wings and may feed on the mantid, and therefore can be an ectoparasite. It moves to the wing bases and amputates the female mantid's wings and then oviposits into the mantid's egg mass whilst it is frothy, before the ootheca hardens. Individuals of *M. manticida* that are phoretic on male mantids may transfer to the female during mating. Certain chalcid hymenopterans (including species of Eucharitidae) have active planidium larvae that actively seek worker ants, on which they attach, thereby gaining transport to the ant nest. Here the remainder of the immature life cycle comprises typical sedentary grubs that develop within ant larvae or pupae.

Females of some egg parasitoids locate and are then phoretic on their female host until it oviposits, at which time the eggs can be ambushed immediately they are laid. Host specificity is thus guaranteed.

An extreme example of phoresy involves the human bot fly *Dermatobia hominis* (Diptera: Cuterebridae) of the neotropical region

(Central and South America), which causes myiasis (section 14.3) of humans and cattle. The female fly, which produces up to 1000 eggs in her lifetime, does not seek out the vertebrate host herself but uses the services of blood-sucking flies, particularly mosquitoes and muscoid flies. She captures a phoretic intermediary and glues around 30 eggs to its body in such a way that flight is not impaired. When the intermediary finds a vertebrate host on which it feeds, an elevation of temperature induces the eggs to hatch rapidly and the larvae transfer to the host, where they penetrate the skin via hair follicles and develop within the resultant wound.

12.2 PREY/HOST ACCEPTANCE AND MANIPULATION

During foraging, there are some similarities in location of prey by a predator and of host by a parasitoid or parasite. When contact is made with the potential prey or host, its acceptability must be established by checking the target's identity, size and age. For example, many parasitoids reject old larvae that are close to pupation. Chemical and tactile stimuli are involved in specific identification, and in subsequent behaviours including biting, ingestion and continuance of feeding. Chemoreceptors on the antennae and ovipositor of parasitoids are vital in chemically detecting host suitability and exact location.

Different manipulations follow acceptance: the predator attempts to eat suitable prey, whereas parasitoids and parasites exhibit a range of behaviours regarding their hosts. A parasitoid either oviposits (or larviposits) directly or subdues and may carry the host elsewhere, for instance to a nest, prior to the offspring developing within or on it. An ectoparasite needs to gain a hold and obtain a meal. The different behavioural and morphological modifications associated with prey and host manipulation are covered in separate sections below, from the perspectives of predator, parasitoid and parasite.

12.2.1 Prey manipulation by predators

When a predator detects and locates suitable prey, it must be captured and restrained in order to be fed upon. Since predation has arisen many times, and in nearly every order, the morphological modifications associated with this lifestyle are highly convergent. Nevertheless, in most predatory insects the principal organs used in capture and manipulation of prey are the legs and mouthparts. Typically, **raptorial** legs of adult insects are elongate and bear spines on the inner surface of at least one of the segments (Fig. 12.3). Prey is captured by closing the spinose segment against another segment, which may itself be spinose, i.e. the femur against the tibia, or the tibia against the tarsus. As well as spines, there may be elongate spurs on the apex of the tibia, and the apical claws may be strongly developed on the raptorial legs. In predators with leg modifications, it is usually the anterior legs that are raptorial, but some hemipterans employ also the mid legs, and scorpionflies (Box 5.1) grasp prey with their hind legs.

Mouthpart modifications associated with predation are of two principal kinds: incorporation of a variable number of elements into a tubular rostrum to allow piercing and sucking of fluids; or development of strengthened and elongate mandibles. Mouthparts modified as a rostrum (Box 10.8) are seen in bugs (Hemiptera) and they function in sucking fluids from plants or from dead arthropods (as in many gerrid bugs) or in predation on living prey, as in many other aquatic insects including species of Nepidae, Belostomatidae and Notonectidae. Amongst the terrestrial bugs, assassin bugs (Reduviidae), which use raptorial fore legs to capture other terrestrial arthropods, are major predators. They inject toxins and proteolytic saliva into captured prey and suck the body fluids through the rostrum. Similar hemipteran mouthparts are used in blood-sucking, as demonstrated by *Rhodnius*, a reduviid that has attained fame for its role in

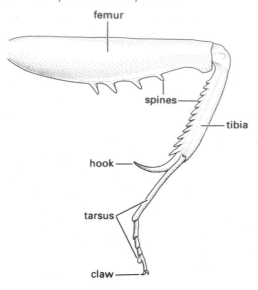

femur

spines

tibia

hook

tarsus

claw

Fig. 12.3 Distal part of the leg of a mantid showing the opposing rows of spines that interlock when the tibia is drawn upwards against the femur. (After Preston-Mafham, 1990.)

experimental insect physiology, and the family Cimicidae, including the bed bug, *Cimex lectularius*.

In the Diptera, mandibles are vital for wound production by the blood-sucking Nematocera (mosquitoes, midges and black flies) but have been lost in the higher flies, some of which have regained the blood-sucking habit. In the stable flies (*Stomoxys*) and tsetse flies (*Glossina*), for example, alternative mouthpart structures have evolved. Some specialized mouthparts of blood-sucking Diptera are described and illustrated in Box 14.4.

Many predatory larvae and some adults have hardened, elongate and apically pointed mandibles capable of piercing durable cuticle. The slender maxilla and sharply pointed and grooved mandible of larval neuropterans (lacewings and antlions) are pressed together to form a composite sucking tube (Fig. 12.2(c)). The composite structure may be straight, as in active pursuers of prey, or curved, as in the sit-and-wait ambushers such

as antlions. Liquid may be sucked or pumped from the prey, using a range of mandibular modifications after enzymatic pre-digestion has liquefied the contents (**extra-oral digestion**).

There are some unusual morphological modifications for predation, such as in the larvae of Chaoboridae (Diptera) which use modified antennae to grasp their planktonic cladoceran prey. Odonate larvae capture passing prey by striking with a highly modified labium (Fig. 12.4), which is projected rapidly outwards by release of hydrostatic pressure, rather than by muscular means.

12.2.2 Host acceptance and manipulation by parasitoids

The two orders with greatest numbers and diversity of larval parasitoids are the Diptera and Hymenoptera. Two basic approaches are displayed once a potential host is located, though there are exceptions. Firstly, as seen in many hymenopterans, it is the adult that seeks out the actual larval development site. In contrast, in many Diptera it is often the first-instar planidium larva that makes the close-up host contact. Parasitic hymenopterans use sensory information from the elongate and constantly mobile antennae to locate precisely even a hidden host. The antennae and specialized ovipositor (Fig. 5.8) bear sensilla that allow host acceptance and accurate oviposition respectively. Modification of the ovipositor as a sting in the aculeate Hymenoptera permits behavioural modifications (section 13.6), including provisioning of the immature stages with a food source captured by the adult and maintained alive in a paralysed state.

Endoparasitoid dipterans, including the Tachinidae, may oviposit (or, in larviparous taxa, deposit a larva) onto the cuticle or directly into the host. In several distantly related families, a convergently-evolved 'substitutional' ovipositor is used (sections 2.5.1

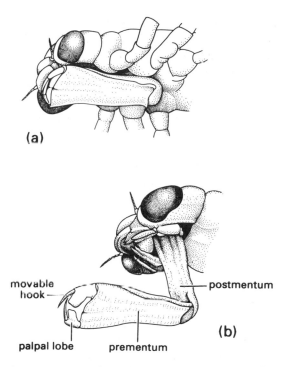

(a)

movable hook —

— postmentum

palpal lobe   prementum   **(b)**

Fig. 12.4 Ventrolateral view of the head of a dragonfly larva (Odonata: Aeshnidae: *Aeshna*) showing the labial 'mask': (a) in folded position; (b) extended during prey capture with opposing hooks of palpal lobe forming claw-like pincers. (After Wigglesworth, 1964.)

and 5.8). Frequently, however, the parasitoid's egg or larva is deposited onto a suitable substrate and the mobile planidium larva is responsible for finding its host. *Euphasiopteryx ochracea*, a tachinid that responds phonotactically to the call of a male cricket, actually deposits larvae around the calling site, and these larvae locate and parasitize not only the vocalist but also other crickets attracted by the call. Hypermetamorphosis, in which the first-instar larva is morphologically and behaviourally different from those of subsequent larval instars, which are sedentary parasitic maggots, is common amongst parasitoids.

Certain parasitic and parasitoid dipterans and some hymenopterans use their aerial flying skills to gain access to a potential host. Some are able to intercept their hosts in flight; others can make rapid lunges at an alert and defended target. Some of the inquilines of social insects (section 11.3) are able to gain access to the nest via an egg laid upon a worker whilst it is active outside the nest. For example, certain phorid flies, lured by ant odours, may be seen darting at ants in an attempt to oviposit on them. A West Indian leafcutter ant (*Atta* sp.) cannot defend itself from such attacks whilst bearing leaf fragments in its mandibles. This problem is frequently addressed (but is unlikely to be completely overcome) by stationing a guard on the leaf during transport; the guard is a small (minima) worker (Fig. 8.4) that uses its jaws to threaten any approaching phorid fly.

The success of attacks of such insects against active and well-defended hosts demonstrates great rapidity in host acceptance, probing and oviposition. This may contrast with the sometimes leisurely manner of many parasitoids of sessile hosts – for example, scale insects, pupae or immature stages that are restrained within confined spaces such as plant tissue and unguarded eggs.

12.2.3 Overcoming host immune responses

Insects that develop within the body of other insects must cope with the active immune responses of the host. An adapted or compatible parasitoid is not eliminated by the cellular immune defences of the host. These defences protect the host by acting against incompatible parasitoids, pathogens and abiotic matter that may invade the host's body cavity. Host immune responses entail mechanisms for recognizing introduced material as non-self, and for inactivating, suppressing or removing the foreign material. The usual host reaction to an incompatible parasitoid is **encapsulation**, that is, surrounding the invading egg or larva by an aggregation of haemocytes (Fig. 12.5). The haemocytes become flattened onto the surface of the parasitoid and phagocytosis

aggregated
haematocytes

Fig. 12.5 Encapsulation of a living larva of *Apanteles* (Hymenoptera: Braconidae) by the haemocytes of a caterpillar of *Ephestia* (Lepidoptera: Pyralidae). (After Salt, 1968.)

commences as the haemocytes build up, eventually forming a capsule that encloses and kills the intruder. This type of reaction rarely occurs when parasitoids infect their normal hosts, presumably because the parasitoid or some factor associated with it alters the host's ability to recognize the parasitoid as foreign or to respond to it. Parasitoids that cope successfully with the host immune system do so in one or more of the following ways:

- Avoidance – for example, ectoparasitoids feed externally on the host (in the manner of predators), egg parasitoids lay into host eggs that are incapable of immune response, and many other parasitoids at least temporarily occupy host organs (such as the brain, a ganglion, a salivary gland or the gut) and thus escape the immune reaction of the host haemolymph.
- Evasion – this includes molecular mimicry (the parasitoid is coated with a substance similar to host proteins and is not recognized as non-self by the host), cloaking (e.g. the parasitoid may insulate itself in a membrane or capsule, either derived from embryonic membranes or host tissues; see also 'subversion' below) and by rapid development in the host.

- Destruction – the host immune system may be blocked by attrition of the host such as by gross feeding that weakens host defence reactions, or by destruction of responding cells (the host haemocytes).
- Suppression – host cellular immune responses may be suppressed by viruses associated with the parasitoids (Box 12.1); often suppression is accompanied by reduction in host haemocyte counts and other changes in host physiology.
- Subversion – in many cases parasitoid development occurs despite host response; for example, physical resistance to encapsulation is known for wasp parasitoids and in dipteran parasitoids the host's haemocytic capsule is subverted for use as a sheath that the fly larva keeps open at one end by vigorous feeding. In many parasitic Hymenoptera, the serosa or trophamnion associated with the parasitoid egg fragments into individual cells that float free in the host haemolymph and grow to form giant cells, or teratocytes, which may assist in overwhelming the host defences.

Obviously the various ways of coping with host immune reactions are not discrete and most adapted parasitoids probably use a combination of methods to enable development within their respective hosts. Parasitoid–host interactions at the level of cellular and humoral immunity are complex and vary greatly among different taxa. Understanding of these systems is still relatively limited but this field of research is producing exciting findings concerning parasitoid genomes and coevolved associations between insects and viruses.

12.3 PREY/HOST SELECTION AND SPECIFICITY

Insects vary in the breadth of food sources they use (Chapters 8 and 10). Some predatory insects are monophagous, utilizing a single species of prey; others are oligophagous,

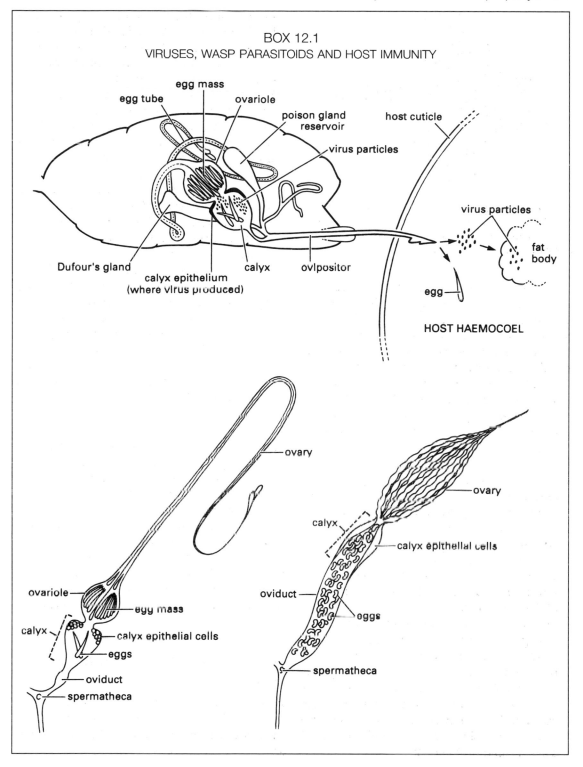

BOX 12.1
VIRUSES, WASP PARASITOIDS AND HOST IMMUNITY

BOX 12.1—*continued*

In certain endoparasitoid wasps in the families Ichneumonidae and Braconidae, the ovipositing female wasp injects the larval host not only with her egg(s), but also with accessory gland secretions and substantial numbers of **virus-like particles** (**VLPs**) or viruses (as depicted in the upper drawing for the braconid *Cardiochiles nigriceps*, after Greany *et al.*, 1984). Many are recognized as viruses belonging to a distinct group, the **polydnaviruses** (**PVs**), which are characterized by the possession of multipartite double-stranded circular DNA. The PVs of braconids (called bracoviruses) differ from the PVs of ichneumonids (ichnoviruses) in morphology, morphogenesis and in relation to their interaction with other wasp-derived factors in the parasitized host. The PVs of different wasp species are generally considered to be distinct viral species. PVs appear to be transmitted between wasp generations through the germline.

It is not clear whether all VLPs are viruses, since the morphology of some VLPs is different to that of typical PVs and some VLPs seem to lack DNA. However, all PVs and VLPs appear to be involved in overcoming the host's immune reaction and are often responsible for other symptoms in infected hosts. For example, the PVs of some wasps can apparently induce most of the changes in growth, development, behaviour and haemocytic activity that are observed in infected host larvae. The PVs of other parasitoids (usually braconids), seem to require the presence of accessory factors, particularly venoms, for the complete prevention of encapsulation of the wasp egg or for fully inducing symptoms in the host. The developing wasp larva probably further reduces the effectiveness of the host immune system by feeding on host haemocytes.

The calyx epithelium of the female reproductive tract is the primary site of replication of PVs (as depicted for the braconid *Cardiochiles nigriceps* in the lower left drawing and for the ichneumonid *Campoletis sonorensis* on the lower right, after Stoltz and Vinson, 1979) and is the only site of VLP assembly (as in the ichneumonid *Venturia canescens*). The lumen of the wasp oviduct becomes filled with PVs or VLPs, which thus surround the wasp eggs. If VLPs or PVs are removed artificially from wasp eggs, encapsulation occurs when the unprotected eggs are then injected into the host. If appropriate PVs or VLPs are injected into the host with the washed eggs, encapsulation is prevented. The physiological mechanism for this protection is not understood, although in the wasp *Venturia*, which coats its eggs in VLPs, it appears that molecular mimicry of a host protein by a VLP protein interferes with the immune recognition process of the lepidopteran host. The VLP protein is similar to a host haemocyte protein involved in recognition of foreign particles.

using few species; and many are polyphagous, feeding on a variety of prey species. As a broad generalization, predators are mostly polyphagous, since a single prey species will rarely provide adequate resources. However, sit-and-wait (ambush) predators, by virtue of their chosen location, may have a restricted diet – for example, antlions may predominantly trap small ants in their pits. Some predators select gregarious prey, such as certain eusocial insects, because the predictable behaviour and abundance of this prey allows monophagy, as long as any chemical defences can be countered by the predator.

The remainder of this section concerns the

more complicated host relations of parasitoids and parasites. In referring to parasitoids and their range of hosts, the terminology of monophagous, oligophagous and poly-phagous is applied, as for phytophages and predators. For parasites, a different, parallel terminology exists: **monoxenous** parasites are restricted to a single host, **oligoxenous** to a few, and **polyxenous** ones avail themselves of many hosts. The following sections discuss first the variety of strategies for host selection by parasitoids, and then the ways in which a parasitized host may be manipulated by the developing parasitoid. The final section considers patterns of host use by parasites, with particular reference to coevolution.

12.3.1 Host use by parasitoids

Parasitoids require only a single individual in which to complete development; they always kill their immature host, and are rarely parasitic in the adult stage. Insect-eating (**entomophagous**) parasitoids show quite a range of strategies for development on their selected hosts. The larva may be ectoparasitic, developing externally, or endoparasitic, developing within the host. Eggs (or larvae) of ectoparasitoids are laid close to or upon the body of the host; the same is sometimes the case with endoparasitoids but the latter more often lay the eggs within the body of the host, using a piercing ovipositor (in hymenopterans) or a substitutional ovipositor (in parasitoid dipterans). Certain parasitoids that feed within host pupal cases or under the covers and protective cases of scale insects and the like are actually ectophages (external feeding), living internal to the protection but external to the insect host body. These different feeding modes give different exposures to the defensive host immune system, which is encountered by endoparasitoids and avoided by ectoparasitoids (section 12.2.3). Ectoparasitoids are often less host-specific than endoparasitoids, since they

have less intimate association with the host compared with the endoparasitoid that must counter the species-specific variations of the host immune system.

Parasitoids may be solitary on their host or gregarious. The number of parasitoids that can develop on a host relates to the size of the host, its post-infected longevity, and the size and biomass of the parasitoid. Development of several parasitoids in one individual host arises commonly through the female ovipositing several eggs on a single host or, less often, by polyembryony, in which a single egg laid by the mother divides and can give rise to numerous offspring (section 5.10.3). Gregarious parasitoids appear able to regulate the clutch size in relation to the quality and size of the host.

Most parasitoids **host discriminate**; that is they can recognize, and generally reject, hosts that are parasitized already, by either themselves, their conspecifics or another species. Parasitized hosts are generally recognizable from a marking pheromone placed inside or externally on the host at the time of oviposition.

Not all parasitoids avoid parasitized hosts. In **superparasitism**, a host receives multiple eggs from either a single individual or several individuals of the same parasitoid species, although the host cannot sustain the total parasitoid burden to maturity. The outcome of multiple oviposition is discussed in section 12.3.2. Theoretical models, some of which have been substantiated experimentally, imply that superparasitism will increase:

- as unparasitized hosts are depleted;
- as parasitoid numbers searching any patch increase;
- in species with high fecundity and small eggs.

Although historically all such instances were deemed to have been 'mistakes', there is some evidence of adaptive benefits deriving from the strategy. Superparasitism is adaptive for

Fig. 12.6 Two examples of the ovipositional behaviour of hymenopteran hyperparasitoids of aphids: (a) endophagous *Alloxysta victrix* (Hymenoptera: Cynipidae) ovipositing into a primary parasitoid inside a live aphid; (b) ectophagous *Asaphes lucens* (Hymenoptera: Pteromalidae) ovipositing onto a primary parasitoid in a mummified aphid. (After Sullivan, 1988.)

individual parasitoids when there is competition for scarce hosts, but avoidance is adaptive when hosts are abundant. Very direct benefits accrue in the case of a solitary parasitoid that uses a host which is able to encapsulate a parasitoid egg (section 12.2.3). Here, the first-laid egg may use all the host haemocytes and a subsequent egg may thereby escape encapsulation.

In **multiparasitism**, a host receives eggs of more than one species of parasitoid. Multiparasitism occurs more often than superparasitism, perhaps because parasitoid species are less able to recognize the marking pheromones placed by species other than their own. Closely related parasitoids may recognize each other's marks, whereas more distantly related species may be unable to do so.

Secondary parasitoids, called **hyperparasitoids**, appear to be able to detect the odours left by a primary parasitoid, allowing accurate location of the site for the development of the hyperparasite at the expense of the primary parasitoid. Some insects are **obligate** hyperparasitoids, developing only within primary parasitoids, whereas others are **facultative** and may develop also as primary parasitoids. Development may be external or internal to the primary parasitoid host, with oviposition into the primary host in the former or into the primary parasitoid in the latter (Fig. 12.6).

External feeding is frequent, and predominantly restricted to the host larval stage: hyperparasitoids of eggs of adult hosts are very rare.

Hyperparasitoids belong to two families of Diptera (certain Bombyliidae and Conopidae), two families of Coleoptera (a few Rhipiphoridae and Cleridae) and notably the Hymenoptera, principally amongst 11 of the 17 families of the superfamily Chalcidoidea and in four subfamilies of Ichneumonidae. However, many parasitoid insects are not attacked by hyperparasitoids, notably the Tachinidae (Diptera) and, amongst the Hymenoptera, the aphidiine Braconidae and the egg parasitoids, Trichogrammatidae, Scelionidae and Mymaridae. Within the Hymenoptera, hyperparasitism has evolved several times, each originating in some manner from primary parasitism, with facultative hyperparasitism demonstrating the ease of the transition. Hymenopteran hyperparasitoids attack a wide range of hymenopteran-parasitized insects, predominantly amongst the Homoptera and Lepidoptera.

Hyperparasitoids often have a broader host range than the frequently oligophagous or monophagous primary parasitoids. However, as with primary parasitoids, endophagous hyperparasitoids seem to be more host-specific than those that feed externally, relating to the greater physiological problems experienced when developing within another living organism. Additionally, foraging and assessment of host suitability of a complexity quite comparable to that of primary parasitoids is known, at least for cynipid hyperparasitoids of aphidophagous parasitoids (Fig. 12.7). Hyperparasitism and the degree of host specificity is fundamental information in biological control programmes (section 15.5.1).

12.3.2 Host manipulation and development of parasitoids

Parasitization may kill or paralyse the host; the developing parasitoid, called an

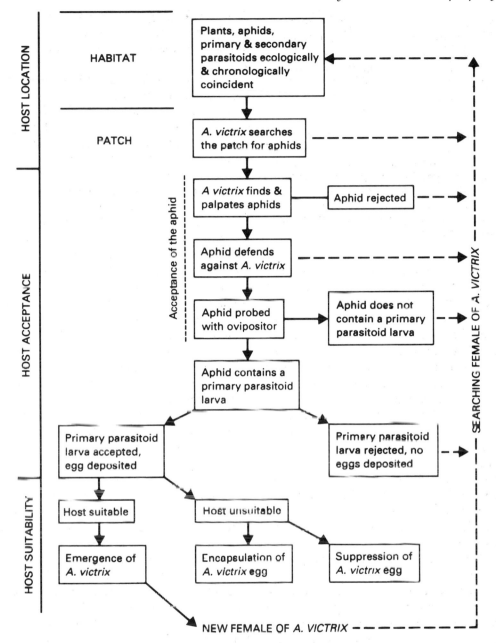

Fig. 12.7 Steps in host selection by the hyperparasitoid *Alloxysta victrix* (Hymenoptera: Cynipidae). (After Gutierrez, 1970.)

idiobiont, develops rapidly, in a situation that differs only slightly from predation. Of greater interest and much more complexity is the **konobiont** parasitoid: it lays its egg(s) in a young host, which continues to grow, thereby providing an increasing food resource. Parasitoid development can be delayed until the host has attained a sufficient size to sustain it.

Host regulation is a feature of konobionts, and certain parasitoids are able to manipulate host physiology, including suppression of its pupation to produce a 'super host'.

Many konobionts respond to hormones of the host. This is demonstrated by: the frequent moulting or emergence of parasitoids in synchrony with the host's moulting or metamorphosis; and/or synchronization of diapause of host and parasitoid. It is uncertain whether, for example, host ecdysteroids act directly on the parasitoid's epidermis to cause moulting, or act indirectly on the parasitoid's own endocrine system to elicit synchronous moulting. Although the specific mechanisms remain unclear, some parasitoids undoubtedly disrupt the host endocrine system, causing either developmental arrest, accelerated or retarded metamorphosis, or inhibition of reproduction in an adult host. This may arise through production of hormones (including mimetic ones) by the parasitoid, or through regulation of the host's endocrine system, or both. In cases of **delayed parasitism**, such as is seen in certain platygastrine and braconid hymenopterans, development of an egg laid in the host egg is delayed for up to a year, until the host is a late-stage larva. Host hormonal changes approaching metamorphosis are implicated in the stimulation of parasitoid development. Specific interactions between the endocrine systems of endoparasitoids and their hosts will play a role in limiting host ranges. Parasitoid-introduced viruses or virus-like particles (Box 12.1) may also modify host physiology and determine host range.

The host is not a passive vessel for parasitoids – as we have seen, the immune system can attack all but the adapted parasitoids. In addition, host quality (size and age) can induce variation in size, fecundity and even the sex ratio of emergent solitary parasitoids. Generally more females are produced from high quality (larger) hosts, whereas males are produced from those of poorer quality, including smaller and superparasitized hosts.

Host aphids reared experimentally on deficient diets (lacking sucrose or iron) produced *Aphelinus* (Hymenoptera: Aphelinidae) parasitoids that developed more slowly, produced more males, and showed lowered fecundity and longevity. The young stages of an endophagous konobiont parasitoid competes with the host tissues for nutrients from the haemolymph. Under laboratory conditions if a parasitoid can be induced to oviposit into an 'incorrect' host (by the use of appropriate kairomones), complete larval development often occurs, showing that haemolymph is adequate nutritionally for development of more than just the adapted parasitoid. Accessory gland secretions (which may include paralysing venoms) are injected by the ovipositing female parasitoid with the eggs, and appear to play a role in regulation of the host's haemolymph nutrient supply to the larva. The specificity of these substances may relate to the creation of a suitable host.

In superparasitism and multiparasitism, larval competition often takes place if the host cannot support all parasitoid larvae to maturity. Depending on the nature of the multiple ovipositions, competition may involve aggression between siblings, other conspecifics or interspecific individuals. Fighting between larvae, especially in mandibulate larval hymenopterans, can result in death and encapsulation of excess individuals. Physiological suppression with venoms, anoxia or food deprivation may occur. Unresolved larval overcrowding in the host can result in a few weak and small individuals emerging, or no parasitoids at all if the host dies prematurely, or resources are depleted before pupation. Gregariousness may have evolved from solitary parasitism in circumstances in which multiple larval development is permitted by greater host size. Evolution of gregariousness may be facilitated when the potential competitors for resources within a single host are relatives – this is particularly so in polyembryony, which produces clonal, genetically identical larvae.

12.3.3 Patterns of host use and specificity in parasites

The wide array of insects that are ectoparasitic upon vertebrate hosts are of such significance to the health of humans and their domestic animals that a complete chapter (Chapter 14) is devoted to them. They will not be considered further here. In contrast to the radiation of ectoparasitic insects using vertebrate hosts and the immense numbers of species of insect parasitoids seen above, there are remarkably few insect parasites of other insects, or indeed of other arthropods.

The largest group of endoparasitic insects using other insects as hosts belong to the Strepsiptera, an order comprising a few hundred exclusively parasitic species (Box 12.4). The characteristically aberrant bodies of their predominantly hemipteran and hymenopteran hosts are termed 'stylopized', so-called for a common strepsipteran genus, *Stylops*. Within the host's body cavity, growth of larvae and pupae of both sexes and of the adult female strepsipteran causes malformations, including displacement of the internal organs. The host's sexual organs degenerate, or fail to develop appropriately.

Although larval Dryinidae (Hymenoptera) develop parasitically part-externally, part-internally in homopterans, virtually all other insect–insect parasitic interactions involve ectoparasitism. The Braulidae is a family of Diptera comprising some aberrant, mite-like flies belonging to a single genus *Braula*, intimately associated with *Apis* (honey bees). Larval braulids scavenge on pollen and wax in the hive, and the adults usurp nectar and saliva from the proboscis of the bee. This association certainly involves phoresy, with adult braulids always found on their hosts' bodies, but whether the relationship is ectoparasitic is open to debate. Likewise, the relationship of several genera of aquatic chironomid larvae with nymphal hosts, such as mayflies, stoneflies and dragonflies, ranges from phoresy to suggested ectoparasitism.

Generally there is little evidence that any of these ecto- and endoparasites using insects show a high degree of specificity at the species level. However, this is not necessarily the case for insect parasites with vertebrate hosts.

The patterns of host specificity and preferences of parasites raise some of the most fascinating questions in parasitology. For example, most orders of mammals bear lice (Phthiraptera), many of which are monoxenic or are found amongst a quite limited range of hosts. Even some marine mammals, namely certain seals, have lice, although whales do not. None of the Chiroptera (bats) harbour lice, despite their apparent suitability and although they are host to many other ectoparasitic insects, including the Streblidae and Nycteribiidae – two families of ectoparasitic Diptera restricted to bats.

Some terrestrial hosts are free of all ectoparasites, others have very specific associations with one or a few guests, whilst in Panama the opossum *Didelphis marsupialis* has been found to harbour 41 species of ectoparasitic insects and mites. Although four or five of these are commonly present, none are restricted to the opossum and the remainder are found on a variety of hosts ranging from distantly related mammals to reptiles, birds and bats.

Some principles concerning the different patterns of distribution of parasites and their hosts can be examined by looking in some detail at cases where close associations of parasites and hosts are expected. The findings can then be related to ectoparasite/host relations in general.

The Phthiraptera are obligate permanent ectoparasites, spending all their lives on their hosts and lacking any free-living stage. Extensive surveys, such as one concerning the lice of neotropical birds that showed an average of 1.1 lice species per bird species across 127 species and 26 families of birds, indicate that lice are highly monoxenous (restricted to one host species). A high level of coevolution between louse and host might be expected and in general related animals have

louse   host

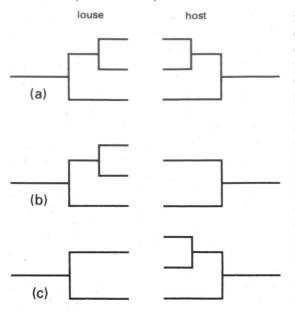

Fig. 12.8 Comparisons of louse and host phylogenetic trees: (a) adherence to Fahrenholz's Rule; (b) independent speciation of the lice; (c) independent speciation of the hosts. (After Lyal, 1986.)

related lice. The widely quoted Fahrenholz's Rule formally states that the phylogenies of hosts and parasites are identical, with every speciation event affecting hosts being matched by a synchronous speciation of the parasites, as shown in Fig. 12.8(a). It follows that:

- Phylogenetic trees of hosts can be derived from the trees of their ectoparasites.
- Ectoparasite phylogenetic trees are derivable from the trees of their hosts. (The potential for circularity of reasoning is evident.)
- The number of parasite species in the group under consideration is identical to the number of host species considered.
- No species of host has more than one species of parasite in the taxon under consideration.
- No species of parasite parasitizes more than one species of host.

Fahrenholz's Rule has been tested for

mammal lice selected from the family Trichodectidae, for which robust phylogenetic trees, derived independently of any host mammal phylogeny, are available. Amongst a sample of these trichodectids, 337 lice species parasitize 244 host species, with 34% of host species parasitized by more than one trichodectid. Several possible explanations exist for these mismatches. Firstly, speciation may have occurred independently amongst certain lice on a single host (Fig. 12.8(b)). This is substantiated, with at least 7% of all speciation events in the sampled Trichodectidae showing this pattern of independent speciation. A second explanation involves secondary transfer of lice species to phylogenetically unrelated host taxa. Amongst extant species, when cases arising from human-induced unnatural host proximity are excluded (accounting for 6% of cases), unmistakable and presumed natural transfers (i.e. between marsupial and eutherian mammal, or bird and mammal) occur in about 2% of speciation events. However, hidden within the phylogenies of host and parasite are speciation events that involve lateral transfer between rather more closely related host taxa, but these transfers fail to match precisely the phylogeny. Examination of the detailed phylogeny of the sampled Trichodectidae shows that a minimum of 20% of all speciation events are associated with distant and lateral secondary transfer, including historical transfers (lying deeper in the phylogenetic trees).

In detailed examinations of relationships between a smaller subset of trichodectids and eight of their pocket gopher (Rodentia: Geomyidae) hosts, substantial concordance was claimed between trees derived from biochemical data for hosts and parasites, and some evidence of cospeciation was found. However, many of the hosts were shown to have two lice species, and unconsidered data show most species of gopher to have a substantial suite of associated lice. Furthermore, a minimum of three instances of lateral transfer (host switching) appeared to have occurred, in all cases between hosts with geographically contiguous ranges. Although many speciation events in these lice 'track' speciation in the host and some estimates even indicate similar ages of host and parasite species, it is evident from the Trichodectidae that strict cospeciation of host and parasite is not the sole explanation of the associations observed.

The reasons why apparently monoxenic lice sometimes do deviate from strict coevolution and cospeciation apply equally to other ectoparasites, many of which show similar variation in complexity of host relationships. Deviations from strict cospeciation arises if host speciation occurs without commensurate parasite speciation (Fig. 12.8(c)). This resulting pattern of relationships is identical to that seen if one of two parasite sister taxa, generated by cospeciation in concert with the host, subsequently became extinct. Frequently a parasite is not present throughout the complete range of its host, resulting perhaps from the parasite being restricted in range by environmental factors independent of those controlling the range of the host. Hemimetabolous ectoparasites, such as lice that spend their entire lives on the host, might be expected to closely follow the ranges of their hosts, but there are exceptions in which the ectoparasite distribution is restricted by external environmental factors. For holometabolous ectoparasites that spend some of their lives away from their hosts, such external factors will be even more influential in governing parasite range. For example, a homeothermic vertebrate may tolerate environmental conditions that cannot be sustained by the free-living stage of a poikilothermic ectoparasite, such as a larval flea. Since speciation may occur in any part of the distribution of a host, host speciation may be expected to occur without necessarily involving the parasite. Furthermore, a parasite may show geographic variation within all or part of the host range that is incongruent with the variation of the host. If either or both

variations leads to eventual species formation, there will be incongruence between parasite and host phylogeny.

Poor knowledge of host and parasite interactions may result in misleading conclusions. A true host may be defined as one which provides the conditions for parasite reproduction to continue indefinitely. When there is more than one true host, there may be a principal (preferred) or exceptional host, depending on the proportional frequencies of ectoparasite occurrence. An intermediate category may be recognized, the sporadic or secondary host, on which parasite development cannot normally take place, but an association arises frequently, perhaps through predator–prey interactions or environmental encounters (such as a shared nest). Small sample sizes and limited biological information can allow an accidental or secondary host to be mistaken for a true host, giving rise to a possible erroneous 'refutation' of cospeciation. Extinctions of certain parasites and true hosts (leaving the parasite extant on a secondary host) will refute Fahrenholz's Rule.

Even assuming perfect recognition of true host specificity and knowledge of the historical existence of all parasites and hosts, it is evident that successful parasite transfers between hosts have taken place throughout the history of host–parasite interactions. Cospeciation is fundamental to host–parasite relations, but the factors encouraging deviations must be considered. Predominantly these concern geographic and social proximity of different hosts, allowing opportunities for parasite colonization of the new host, together with ecological similarity of different hosts, allowing establishment, survival and reproduction of the ectoparasite on the novel host. The results of these factors have been termed **resource tracking**, to contrast with the **phyletic tracking** implied by Fahrenholz's Rule. As with all matters biological, most situations lie somewhere along a continuum between these two extremes and, rather than forcing patterns into one category or the other, interesting questions arise from recognizing and interpreting the different patterns observed.

If all host–parasite relationships are examined, some of the factors that govern host specificity can be identified:

- The stronger the life-history integration with that of the host, the greater the likelihood of monoxeny.
- The greater the vagility (mobility) of the parasite, the more likely it is to be polyxenous.
- The number of accidental and secondary parasite species increases with decreasing ecological specialization and with increase in geographic range of the host, as we saw for the widespread and unspecialized opossum earlier in this section (p.341).

If a single host shares a number of ectoparasites, there may be some ecological or temporal segregation on the host. For example, in haematophagous (blood-sucking) black flies (Simuliidae) that attack cattle, the belly is more attractive to certain species, whereas others feed only on the ears. *Pediculus capitis* and *P. humanus* (Phthiraptera), human head and body lice, are ecologically separated examples of sibling species in which proven reproductive isolation is reflected by only slight morphological differences.

12.4 MODELS OF ABUNDANCES OF PREDATOR/PARASITOIDS AND PREY/HOSTS

There is a substantial body of experimental and theoretical evidence that predators (taken to include parasitoids in this discussion) impose a structure on the ecological communities in which they operate. Experiments show that removal of the most important ('top') predator can give a major shift in community structure, demonstrating that predators control the abundance of subdominant predators and certain prey species.

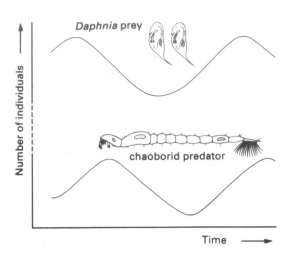

Fig. 12.9 An example of the regular cycling of numbers of predators and their prey: the aquatic planktonic predator *Chaoborus* (Diptera: Chaoboridae) and its cladoceran prey *Daphnia*.

Ecologists have attempted to model the complex relationship between predators and prey, frequently motivated by a desire to understand interactions of biological control agents and target pest species.

Mathematical models may commence from simple interactions between a single monophagous predator and its prey. Experiments and simulations concerning the long term trend in numbers of each show a regular cycling of numbers of predators and prey: when prey are abundant, predator survival is high; as more predators become available, prey abundance is reduced; predator numbers decrease as do those of prey; reduction in predation allows the prey to escape and rebuild numbers. The sinusoidal, time-lagged cycles of predator and prey abundances may occur in some simple natural systems, such as the aquatic planktonic predator *Chaoborus* (Diptera: Chaoboridae) and its cladoceran prey *Daphnia* (Fig. 12.9).

Examination of shorter-term feeding responses using laboratory studies of simple systems shows that predators vary in their responses to prey density, with an early assumption of a linear relationship (increased prey density leading to increased predator feeding) being superseded. A common pattern of functional response of a predator to prey density involves a gradual slowing of the rate of predation relative to increased prey density, until an asymptote is reached. This upper limit beyond which no increased rate of prey capture occurs is due to the time constraints of foraging and handling prey in which there is a finite limit to the time spent in feeding activities, including a recovery period. The rate of prey capture does not depend upon prey abundance alone: individuals of different instars have different feeding rate profiles, and in poikilothermic insects there is an important effect of ambient temperature on activity rates.

The assumptions of predator monophagy usually are biologically unrealistic, and more complex models include multiple prey items. Predator behaviour is based upon optimal foraging strategies involving simulated prey selection varying with changes in proportional availability of different prey items. However, predators may not switch between prey items based upon simple relative numerical abundance. Other factors include differences in prey profitability (nutritional content, ease of handling, etc.), the hunger-level of the predator and perhaps predator learning and development of a search-image for particular prey, irrespective of abundance.

Models of predator prey-foraging and handling in which a more realistic choice between profitable and less profitable prey items are available indicate that:

- Prey specialization ought to occur when the most profitable prey is abundant.
- Predators should switch rapidly from complete dependence on one prey to the other, with partial preference (mixed feeding) being rare.
- The actual abundance of a less abundant prey should be irrelevant to the decision of a predator to specialize on the most abundant prey.

Further problems are associated with the unrealistic assumptions that parasitoid searching behaviour resembles that of a random-searching predator, in being independent of host abundance, the proportion of hosts already parasitized and the distribution of the hosts. As we have seen above, parasitoids are often able to identify and respond behaviourally to already-parasitized hosts. Furthermore, prey (and hosts) are not distributed at random, but occur in patches, and within patches the density is likely to vary. As predators and parasitoids aggregate in areas of high resource density, interactions between predators/parasitoids (**interference**) become significant, perhaps making a profitable area unprofitable.

For a number of reasons, there may be **refuges** from predators and parasitoids within a patch. Amongst California red scale insects (Hemiptera: Diaspididae: *Aonidiella aurantii*) on citrus trees, for example, those on the periphery of the tree may be up to 27 times more vulnerable to two species of parasitoids compared with individual scales in the centre of the tree, which thus may be termed a refuge. The effectiveness of a refuge varies between taxonomic or ecological groups: external leaf-feeding insects support more parasitoid species than leaf-mining insects, which in turn support more than highly concealed insects such as root feeders or those living in structural refuges. These observations have important implications for the success of biological control programmes.

12.5 THE EVOLUTIONARY SUCCESS OF INSECT PREDATION AND PARASITISM

Chapter 10 shows how the development of angiosperms and their colonization by specific plant-eating insects can be seen as an explanation for a substantial diversification of phytophagous insects relative to their non-phytophagous sister taxa. Applying similar reasoning to the Hymenoptera, comparison of the phytophagous suborder Symphyta with its primarily parasitic sister group, the suborder Apocrita, implies that, in this case, adoption of a parasitic lifestyle is associated with a major evolutionary radiation. An explanation may lie in the degree of host restriction: if each species of phytophagous insect were host to a more or less monophagous parasitoid, then we would expect to see a diversification (radiation) of insect parasitoids that corresponded with that of phytophagous insects. Two assumptions need examination in this context: the degree of host specificity and the number of parasitoids harboured by each host.

The question of the degree of monophagy amongst parasites and parasitoids is not answered conclusively. For example, many parasitic hymenopterans are extremely small, and the basic taxonomy and host associations are yet to be fully worked out. There is no doubt that the parasitic hymenopterans are extremely speciose, and show a varying pattern of host specificity from strict monophagy to oligophagy. Amongst parasitoids within the Diptera, the species-rich Tachinidae are relatively general feeders, specializing only in hosts belonging to families or even ordinal groups. Amongst the ectoparasites, lice are predominantly monoxenic, as are many fleas and flies. However, even if several species of ectoparasitic insects were borne by each host species, since the vertebrates are not numerous, ectoparasites contribute relatively little to biological diversification in comparison with the parasitoids of insect and other diverse arthropod hosts.

Concerning the numbers of parasitoid species on each potential host, there is substantial evidence (much of it acquired by the diligence of amateur entomologists) that many hosts support multiple parasitoids. This phenomenon is well known to lepidopterists who endeavour to rear adult butterflies or moths from wild-caught larvae – the frequency and diversity of parasitization is very high. Suites of parasitoid and hyperparasitoid species may attack the same species of host at different seasons, in different locations and in different life-history stages. There are many records of more than ten parasitoid species throughout the range of some widespread lepidopterans, and whilst this is true also for certain well-studied coleopterans, the situation is less clear for other orders of insects.

Finally, some evolutionary interactions between parasites and parasitoids and their hosts may be considered. Firstly, patchiness of potential host abundance throughout the host range seems to provide the opportunity for increased specialization, perhaps leading to species formation within the guild of parasites/parasitoids. This can be seen as a form of niche differentiation, where the total range of a host provides a niche that is ecologically partitioned. Hosts may escape from parasitization within refuges within the range, or by modification of the life cycle, with the introduction of a phase that the parasitoid cannot track. Host diapause may be a mechanism for evading a parasite that is restricted to continuous generations, with an extreme example of escape perhaps seen in the periodic cicada. These species of *Magicicada* grow concealed for many years as nymphs beneath the ground, and the very visible adults appear only every 13 or 17 years. This cycle of a prime number of years may allow avoidance of predators or parasitoids that are able to adapt only to a predictable cyclical life history. Life-cycle shifts in attempts to evade predators may be important in species formation.

Strategies of prey/hosts and predators/parasitoids have been envisaged as evolutionary arms races, with a step-wise sequence of prey/host escape by evolution of successful defences, followed by radiation before the predator/parasitoid 'catches up', in a form of prey/host tracking. An alternative evolutionary model envisages both prey/host and predator/parasitoid evolving defences and circumventing them in virtual synchrony, in an evolutionary stable strategy termed the 'Red Queen' hypothesis (after the description in *Alice in Wonderland* of Alice and the Red Queen running faster and faster to stand still). Tests of each can be devised and models for either can be justified, and it is unlikely that conclusive evidence will be found in the short-term. What is clear is that parasitoids and predators do exert great selective pressure on their hosts or prey, and remarkable defences have arisen, as we shall see in the next chapter.

BOX 12.2
MANTODEA (MANTIDS)

An order of about 2000 species of moderate to large (1–15 cm long) hemimetabolous predators. Males generally smaller than females. Head small, triangular, mobile, with slender antennae, large, widely separated eyes and mandibulate mouthparts; thorax comprising elongate, narrow prothorax and shorter (almost subquadrate) meso- and metathorax; fore wings forming leathery tegmina, anal area reduced; hind wings broad and membranous, with long veins unbranched and many cross-veins; aptery and subaptery frequent; fore legs raptorial (Fig. 12.3 and as illustrated here for a mantid of a *Tithrone* species holding and eating a fly, after Preston-Mafham, 1990), mid and hind legs elongate for walking; abdomen with 10th visible segment bearing variably segmented cerci; ovipositor predominantly internal; external male genitalia asymmetrical.

Eggs are laid in an ootheca produced from accessory gland frothy secretions that harden on contact with the air. Some females guard their ootheca. First-instar nymphs do not feed, but moult immediately. As few as three or as many as 12 instars follow; nymphs resemble adults except for lack of wings and genitalia. Adult mantids are sit-and-wait predators (see the vignette of Chapter 13) which use their fully mobile head and excellent sight to detect prey. Female mantids sometimes consume the male during or after copulation (Box 5.2); males often display elaborate courtship.

Mantodea are undoubtedly the sister group to the Blattodea (cockroaches) (Fig. 7.5), sharing many features such as strong direct flight muscles and weak indirect (longitudinal) flight muscles, asymmetrical male genitalia and multisegmented cerci. Derived features of Mantodea relative to Blattodea involve modifications associated with predation, including leg morphology, elongate prothorax and features associated with visual predation, namely the mobile head with large, separated eyes.

BOX 12.3
NEUROPTERIDA, OR NEUROPTEROID ORDERS: MEGALOPTERA (ALDERFLIES, DOBSONFLIES, FISHFLIES), RAPHIDIOPTERA (SNAKEFLIES) AND NEUROPTERA (LACEWINGS, ANTLIONS, OWLFLIES)

Three small orders with hemimetabolous development, mostly predators, with approximate numbers of described species of 5000 for Neuroptera in about 20 families, 300 in Megaloptera in two widely recognized families, and 200 in Raphidioptera in two families. Adults with multisegmented antennae, large, separated eyes, and mandibulate mouthparts; prothorax may be larger than meso- and metathorax, which are about equal in size; legs may be modified for predation; fore and hind wings quite similar in shape and venation, with folded wings often extending beyond abdomen; abdomen lacking cerci.

Megaloptera are predatory only in the aquatic larval stage (Box 9.5) – adults have strong mandibles, not used in feeding; adults (such as the corydalid, *Archichauliodes guttiferus*, illustrated here) closely resemble neuropterans, except for presence of an anal fold in hind wing.

Raphidioptera (illustrated only in Table 1.3) are terrestrial predators as adults and larvae; adult mantid-like, with elongate prothorax and mobile head used to strike, snake-like, at prey; larva with large prognathous head, sclerotized prothorax slightly longer than membranous meso- and metathorax; pupa mobile.

Adult Neuroptera (illustrated only in Table 1.3) with wings typically characterized by numerous cross-veins and 'twigging' at ends of veins; many are predators but nectar, honeydew and pollen are consumed by some species. Neuropteran larva (Fig. 6.6(d)) are usually specialized, active predators, with prognathous heads and slender, elongate mandibles and maxillae combined to form piercing and sucking mouthparts (Fig. 12.2(c)); all have a blindly ending hindgut. Larval dietary specializations include spider egg masses (for Mantispidae), freshwater sponges (for Sisyridae), or soft-bodied homopterans (for Chrysopidae, Hemerobiidae and Coniopterygidae). Pupation is terrestrial, within shelters spun with silk from Malpighian tubules; pupal mandibles are used to open a toughened cocoon.

The Megaloptera, Raphidioptera and Neuroptera are treated here as separate orders; however, some authorities include the Raphidioptera in the Megaloptera, or all three may be united in the Neuroptera. There is little doubt that the grouping is monophyletic, forming a sister group to the Coleoptera plus Strepsiptera (Fig. 7.5).

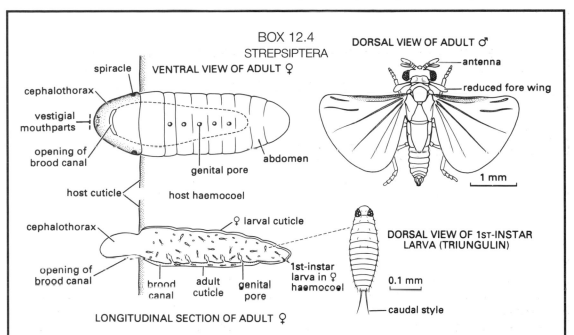

BOX 12.4
STREPSIPTERA

VENTRAL VIEW OF ADULT ♀

DORSAL VIEW OF ADULT ♂

LONGITUDINAL SECTION OF ADULT ♀

DORSAL VIEW OF 1ST-INSTAR LARVA (TRIUNGULIN)

An order of nearly 400 species of highly modified endoparasites with extreme sexual dimorphism. Male (top right figure, after CSIRO, 1970) with large head, bulging eyes with few large facets, without ocelli; antennae flabellate or branched, with 4–7 segments; pro- and mesothorax small; fore wings stubby and without veins, hind wings broad, fan-shaped, with few radiating veins; legs lack trochanters and often also claws; elongate metanotum overlies anterior of tapering abdomen. Female either coccoid-like or larviform, wingless, usually retained in a pharate (cloaked) state, protruding from the host (as illustrated in ventral view and longitudinal section in the left figures, after Askew, 1971). Triungulin (first-instar larva; bottom right figure) with three pairs of thoracic legs, without antennae and mandibles; subsequent instars maggot-like, lacking mouthparts or appendages. Pupa exarate and adectious within puparium formed from last larval instar.

Strepsiptera are ectoparasites of other insects, most commonly of Hemiptera and Hymenoptera. Eggs hatch within female strepsipterans, and active triungulins emerge via a brood passage (shown here on bottom left) and seek out a host, usually in its immature stage. In Stylopidae that parasitize hymenopterans, triungulins leave their host whilst on flowers, and from here seek a suitable adult bee or wasp to gain a ride to the nest, where they enter a host egg or larva.

Entry to the host is via enzymatically softened cuticle, followed by an immediate moult to a maggot-like instar that develops as an endoparasite. The pupa protrudes from the host's body; the male emerges by pushing off a cephalothoracic cap, but the female remains within the cuticle. The virgin female releases pheromones to lure free-flying males, one of which copulates, inseminating through the brood passage on the female cephalothorax.

The Strepsiptera are phylogenetically enigmatic but resemblance of their first-instar triungulins to certain Coleoptera, notably parasitic Rhipiphoridae, indicates a relationship that molecular systematic studies do not refute.

BOX 12.5
MECOPTERA (SCORPIONFLIES, HANGINGFLIES)

An order of about 500 known species in nine families, with common names associated with the two largest families – Bittacidae (hangingflies; Box 5.1) and Panorpidae (scorpionflies, illustrated in Table 1.3 and the vignette of this chapter). Holometabolous; adults with elongate hypognathous rostrum; mandibles and maxillae elongate, slender and serrate; labium elongate; large, separated eyes; filiform, multisegmented antennae; prothorax may be smaller than equally developed meso- and metathorax, each with scutum, scutellum and postscutellum visible; fore and hind wings narrow, similar in size, shape and venation, often reduced or absent; legs may be modified for predation; abdomen 11-segmented, with first tergite fused to metathorax; 1–2-segmented cerci present. Larvae with heavily sclerotized head capsule; mandibulate, with compound eyes; thoracic segments about equal; short thoracic legs with fused tibia and tarsus and single claw; prolegs usually on abdominal segments 1–8, terminal segment (10) either with paired hooks or a suction disk. Pupa (Fig. 6.7(b)) immobile, exarate and mandibulate.

Copulation in certain mecopterans is preceded by elaborate courtship procedures that may involve nuptial feeding (Box 5.1). Oviposition sites vary but known larval development is predominantly in moist litter, or aquatic in Gondwanan Nannochoristidae.

The Mecoptera resemble neuropterans but appear to be the sister group to the Diptera plus Siphonaptera (Fig. 7.5).

FURTHER READING

Askew, R.R. (1971) *Parasitic Insects*, Heinemann, London.

Eggleton, P. and Belshaw, R. (1993) Comparisons of dipteran, hymenopteran and coleopteran parasitoids: provisional phylogenetic explanations. *Biological Journal of the Linnean Society*, **48**, 213–26.

Fleming, J.G.W. (1992) Polydnaviruses: mutualists and pathogens. *Annual Review of Entomology*, **37**, 401–425.

Gauld, I. and Bolton, B. (eds) (1988) *The Hymenoptera*, British Museum (Natural History) in association with Oxford University Press, London.

Hassell, M.P. and Southwood, T.R.E. (1978) Foraging strategies of insects. *Annual Review of Ecology and Systematics*, **9**, 75–98.

Lyal, C.H.C. (1986) Coevolutionary relationships of lice and their hosts: a test of Fahrenholz's Rule, in *Coevolution and Systematics*, (eds A.R. Stone and D.L. Hawksworth), Systematics Association, Oxford, pp. 77–91.

Marshall, A.G. (1981) *The Ecology of Ectoparasitic Insects*, Academic Press, London, New York, Toronto, Sydney, San Francisco.

New, T.R. (1991) *Insects as Predators*, The New South Wales University Press, Kensington.

Ross, K.G. and Matthews, R.W. (eds) (1991) *The Social Biology of Wasps*, Comstock Publishing Associates, Cornell University Press, Ithaca and London.

Salt, G. (1968) The resistance of insect parasitoids to the defence reactions of their hosts. *Biological Reviews of the Cambridge Philosophical Society*, **43**, 200–32.

Salt, G. (1970) *The Cellular Defence Reactions of Insects*, Cambridge University Press, London.

Schmidt, O. and Schuchmann-Feddersen, I. (1989) Role of virus-like particles in parasitoid–host interaction of insects. *Subcellular Biochemistry*, **15**, 91–119.

Stoltz, D. and Whitfield, J.B. (1992) Viruses and virus-like entities in the parasitic Hymenoptera. *Journal of Hymenoptera Research*, **1**, 125–39.

Sullivan, D.J. (1987) Insect hyperparasitism. *Annual Review of Entomology*, **32**, 49–70.

Vinson, S.B. (1984) How parasitoids locate their

hosts: a case of insect espionage, in *Insect Communication*, (ed. T. Williams), Academic Press, London, Orlando, San Diego, New York, Toronto, Montreal, Sydney, Tokyo, pp. 325–348.

Vinson, S.B. (1985) The behavior of parasitoids, in *Comprehensive Insect Physiology, Biochemistry and Pharmacology, Volume 9, Behaviour*, (eds G.A. Kerkut and L.I. Gilbert), Pergamon Press, Oxford, New York, Toronto, Sydney, Paris, Frankfurt, pp. 417–69.

Waage, J. and Greathead, D. (eds) (1986) *Insect Parasitoids*, Academic Press, London.

Whitfield, J.B. (1992) Phylogeny of the non-aculeate Apocrita and the evolution of parasitism in the Hymenoptera. *Journal of Hymenoptera Research*, **1**, 3–14.

13

INSECT DEFENCE

Cryptic mantid among orchid flowers

Fig. 13.1 The basic spectrum of prey defence strategies and predator foraging, varying according to costs and benefits in both time and energy. (After Malcolm, 1990.)

Although some humans eat insects (Chapter 1), many 'western' cultures are reluctant to use them as food. This aversion extends no further than humans. For very many organisms, insects provide a substantial food source: they are nutritious, abundant and diverse, and they are found everywhere. Some animals, termed **insectivores**, rely almost exclusively on a diet of insects; omnivores may eat them opportunistically and many herbivores unavoidably consume insects. Insectivores may be vertebrates or invertebrates, including arthropods – insects certainly eat other insects. Even plants lure, trap and digest insects; for example, pitcher plants (both New World Sarraceniaceae and Old World Nepenthaceae) digest arthropods, predominantly ants, in their fluid-filled pitchers (section 10.4.2), and the flypaper and Venus flytraps (Droseraceae) capture many flies. Insects, however, actively or passively resist being eaten, by means of a variety of protective devices – the insect defences that are the subject of this chapter.

Before continuing, some commonly used terms discussed in Chapter 12 should be reviewed. A predator is an animal that kills and consumes a number of prey animals during its life. Animals that live at the expense of another animal but do not kill it are parasites, which may live internally (endoparasites) or externally (ectoparasites). Parasitoids are those which live at the expense of one animal that dies prematurely as a result. The animal attacked by parasites or parasitoids is a host. All insects are potential prey or hosts to many kinds of predators (either vertebrate or invertebrate), parasitoids or, less often, parasites.

Many defensive strategies exist, including specialized morphology, behaviour, noxious chemicals and responses of the immune system. This chapter deals with aspects of defence that include death feigning, autotomy, crypsis (camouflage), chemical defences, aposematism (warning signals), mimicry and collective defensive strategies. These are directed against a wide range of vertebrates and invertebrates but, since much study has involved insects defending themselves against insectivorous birds, the role of these particular predators will be emphasized. Immunological defence against microorganisms is discussed in Chapter 3 and those used against parasitoids are considered in Chapter 12.

A useful framework for discussion of defence and predation can be based upon the time and energy inputs to the respective behaviours. Thus hiding, escape by running or flight, and defence by staying and fighting involve increasing energy expenditure but diminishing costs in time expended (Fig. 13.1). Many insects will change to another strategy if the previous defence fails: the scheme is not clear-cut and it has elements of a continuum.

BOX 13.1
INDUSTRIAL MELANISM – A BLACK-AND-WHITE CASE?

Britain's long history of enthusiastic amateur lepidopteran studies has allowed the detection of a fascinating case of apparent evolution of a novel pattern in a moth as a response to selective predation. The peppered moth (Geometridae: *Biston betularia*) was known to be a light-coloured moth with a habit of resting cryptically against the pale lichens of tree trunks. In 1848, in the heyday and heartland of the Industrial Revolution in Manchester, England, a **melanic** (dark) morph was reported for the first time. Although the melanic form increased in industrial areas (**industrial melanism**), this change was not followed in detail. Retrospectively, it seemed that these dark *carbonaria* forms of *B. betularia* were differentially successful in eluding bird predation as they rested on lichenless tree trunks blackened by soot. Their pale relatives, which were cryptically protected in clean areas, stood out and were picked off dark trunks by predators. In post industrial Britain, following reduction in smoke emissions and restriction of air pollution, the frequency of melanics has declined. Experimental manipulations showed that moths settled on the appropriately coloured dark or light background, and when moths of each morph were placed on pale and dark tree trunks, as illustrated here, birds were shown to select inappropriately coloured morphs.

Re-examination, stimulated by discovery of pre-industrial melanism in museum-preserved specimens and recent observation of unpredicted proportions of light and dark morphs, has

cast doubts on the simplicity of this often cited example of evolution through natural selection. Firstly, the moths settle primarily under the shoulder (where a tree branch joins the trunk) rather than exposed on the trunk. At this joint, melanic moths are more cryptic than pale ones, even on pale trees in unpolluted areas. Furthermore, if consideration is given to the reflectance of the wings, the pale morph is most cryptic on a grey background of 30% reflectance (100% is pure white and 0% black) rather than white. In appropriately designed choice experiments, moths respond predictably with the pale

BOX 13.1—*continued*

morphs favouring somewhat darker backgrounds than pure white ones. If these factors are included in calculations of selective advantage, the previous curious excess of dark morphs in unpolluted rural areas becomes explicable.

13.1 DEFENCE BY HIDING

In common language a **cryptic** insect, that is, a well-concealed insect that looks very much like its general background environment, is said to 'mimic' its surroundings. However, crypsis is usually excluded from definitions of true mimicry, which is restricted to animal resemblance to another animal that is recognizable by natural enemies (section 13.5).

Insect crypsis can take many forms. The insect may adopt **camouflage**, making it difficult to distinguish from the general background in which it lives, by:

- resembling a uniform coloured background, such as a green geometrid moth on a leaf;
- resembling a patterned background, such as a mottled moth on tree bark (Box 13.1);
- being countershaded – light below and dark above – as in some caterpillars and aquatic insects;
- having a pattern to disrupt the outline, as is seen in many moths that settle on leaf litter; or
- having a bizarre shape to disrupt the silhouette, as demonstrated by some membracid leafhoppers.

In another form of crypsis, termed **masquerade** or **mimesis** to contrast with the camouflage described above, the organism deludes a predator by resembling an inedible object that is a specific feature of its environment – something that is of no inherent interest to a predator. This feature may be an inanimate object, such as the bird dropping resembled by young larvae of some butterflies

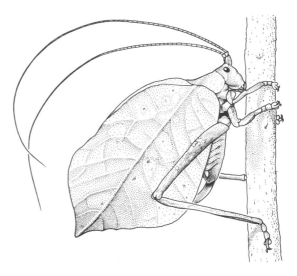

Fig. 13.2 A leaf-mimicking katydid, *Mimetica mortuifolia* (Orthoptera: Tettigoniidae) in which the fore wing resembles a leaf even to the extent of leaf-like venation and spots resembling fungal mottling. (After Belwood, 1990.)

like *Papilio aegeus* (Papilionidae), or an animate object. 'Looper' caterpillars (the larvae of geometrid moths) resemble twigs, membracid homopterans imitate thorns arising from a stem and many stick-insects look very much like sticks and may even move like a twig in the wind. Many insects, notably amongst the lepidopterans and orthopteroids, look like leaves, even to the similarity in venation (Fig. 13.2). These may appear to be dead or alive, mottled with fungus, or even partially eaten as if by a herbivore.

Crypsis is a very common form of insect concealment, particularly in the tropics and

BOX 13.2
AVIAN PREDATORS AS SELECTIVE AGENTS FOR INSECTS

Henry Bates, the first person to suggest a theory for mimicry, argued that it arose through agents selecting among different butterflies, based upon their association of mimetic patterns and unpalatability. Birds, it is contended, are crucial agents in the evolution of insect defences, and examination of the role of avian predators has always been central to these studies.

Two early views appeared to counter the selective importance of birds. The first, that birds do not prey on aposematic monarch butterflies or on either morph of *Biston betularia* (Box 13.1), can be dismissed as due to poor observation. The second, that the total observed insect food items of birds reflected the proportional availability of insect prey, regardless of whether protected or not, was taken to imply that birds could not be selective feeders. Actually, this result is unsurprising if birds exhibit the full range of foraging behaviours shown in Fig. 13.1. The problem is that combining dietary intakes – the results of the many different feeding strategies of insectivorous birds – reveals nothing of the specificity and effectiveness of predator–prey interactions, which relate to prey defence and total prey availability.

Thus winter-roosting monarch butterflies are fed upon by black-backed orioles (Icteridae) that browse selectively on poorly-defended individuals, while black-headed grosbeaks (Fringillidae) appear to be completely insensitive to the toxins. Old World bee-eaters (Meropidae) and neotropical jacamars (Galbulidae) are specialized predators, able to deal with the stings of hymenopterans (the red-throated bee-eater, *Merops bullocki*, is shown here destinging a bee on a branch, after Fry *et al.*, 1992) and toxins of butterflies respectively; a similar suite of birds is able to feed selectively on noxious ants. The ability of these specialist predators to distinguish between varying patterns and edibilities makes them important selective agents in the evolution and maintenance of defensive mimicry.

Birds are observable insectivores for field and laboratory studies: their readily recognizable behavioural responses to unpalatable foods include head-shaking, disgorging of food, tongue extending, bill-wiping, gagging, squawking, and ultimately vomiting. For many birds, a single learning trial with noxious (Class I) chemicals appears to lead to long-term aversion to the particular insect, even with a substantial delay between feeding and illness. However, manipulative studies of bird

---

BOX 13.2—*continued*

diets are complicated by their fear of novelty (neophobia) leading to rejection of startling and frightening displays (section 13.2).

It is unlikely that any insect has completely escaped the attentions of predators and certain birds can overcome even the most severe insect defences. For example, the lubber grasshopper (Acrididae: *Romalea guttata*) is large, gregarious and aposematic and it squirts volatile, pungent chemicals accompanied by a hissing noise when attacked. The lubber is extremely toxic and is avoided by all lizards and birds except one, the loggerhead shrike (Laniidae: *Lanius ludovicanus*), which snatches its prey, including lubbers, and impales them 'decoratively' upon spikes with minimal handling time. These impaled items serve both as food stores and in sexual or territorial displays. *Romalea*, which are emetic to shrikes when fresh, become edible after two days of lardering, presumably by denaturation of the toxins. Impaling behaviour thus proves to be preadaptive, permitting feeding upon an extremely well-defended insect. No matter how good the protection, there is no such thing as total defence in the coevolutionary arms race between prey and predator.

---

amongst nocturnally active insects. It has low energetic costs but relies on the insect being able to select the appropriate background. Experiments with two differently coloured morphs of *Mantis religiosa* (Mantidae), the European praying mantid, have shown that brown and green morphs placed against appropriate and inappropriate coloured backgrounds were fed upon in a highly selectively manner by birds: they removed all 'mismatched' morphs and found no camouflaged ones. Even if the right background is chosen, it may be necessary to orientate correctly: moths with disruptive outlines or with striped patterns resembling the bark of a tree may be concealed only if orientated in a particular direction on the trunk.

In the vignette at the head of this chapter, a nymph of the Indomalayan orchid mantid *Hymenopus coronatus* (Hymenopodidae) is shown blending beautifully with the pink flower spike of an orchid. The crypsis is enhanced by the close resemblance of the femora of the mantid's legs to the flower's petals. Crypsis enables the mantid to avoid detection by its potential prey (flower visitors)

(section 12.1.1) as well as to be concealed from predators.

13.2 SECONDARY LINES OF DEFENCE

Little is known of the learning processes of inexperienced vertebrate predators, such as insectivorous birds. However, studies of the gut contents of birds show that cryptic insects are not immune from predation (Box 13.2). Once found for the first time (perhaps accidentally), birds subsequently seem able to detect cryptic prey via a 'search image' for some element(s) of the pattern. Thus, having discovered that some twigs were caterpillars, American blue jays were observed to continue to peck at sticks in a search for food. Primates can identify stick-insects by one pair of unfolded legs alone, and will attack actual sticks to which phasmatid legs have been affixed experimentally. Clearly, subtle cues allow specialized predators to detect and eat cryptic insects.

Once the deception is discovered, the insect prey may have further defences available in reserve. In the energetically least demanding

Fig. 13.3 The eyed hawk moth, *Smerinthus ocellatus* (Lepidoptera: Sphingidae): top, the brownish fore wings cover the hind wings of a resting moth; bottom, when the moth is disturbed, the black and blue eyespots on the hind wings are revealed. (After Stanek, 1977.)

response, the initial crypsis may be exaggerated, as when a threatened masquerader falls to the ground and lies motionless. This behaviour is not restricted to cryptic insects: even visually obvious prey insects may feign death (**thanatosis**). This behaviour, used by many beetles (particularly weevils), can be successful, as predators lose interest in apparently dead prey or may be unable to locate a motionless insect on the ground. Another secondary line of defence is to take flight and suddenly reveal a flash of conspicuous colour from the hind wings. Immediately on landing the wings are folded, the colour vanishes and the insect is cryptic once more. This behaviour is common amongst certain orthopterans and underwing moths; the colour of the flash may be yellow, red, purple or, rarely, blue.

A third type of behaviour of cryptic insects upon discovery by a predator is the production of a **startle display**. One of the commonest is to open the fore wings and reveal brightly coloured 'eyes' that are usually concealed on the hind wings (Fig. 13.3). Experiments using birds as predators have shown that the more perfect the eye (with increased contrasting rings to resemble true eyes) the better the deterrence. Not all eyes serve to startle: perhaps a rather poor imitation of an eye on a wing may direct pecks from a predatory bird to a non-vital part of the insect's anatomy.

A quite extraordinary type of insect defence is the convergent appearance of part of the body to a feature of a vertebrate, albeit on a much smaller scale. Thus the head of a species of fulgorid bug, commonly called the alligator bug, bears an uncanny resemblance to that of a caiman. The pupa of a particular lycaenid butterfly looks like a monkey head. Some tropical sphingid larvae assume a threat posture which, together with false eyespots that actually lie on the abdomen, give a snake-like impression. These resemblances may deter predators (such as birds that search by 'peering about') by their startle effect, with the incorrect scale of the mimic being overlooked by the predator.

13.3 MECHANICAL DEFENCES

Morphological structures of predatory function, such as the modified mouthparts and spiny legs described in the previous chapter, may also be defensive, especially if a fight ensues. Cuticular horns and spines may deter a predator or have use in fighting, for example, in defence of territory or in combating a rival. For ectoparasitic insects, which are vulnerable to the actions of the host, one line of defence is given by the body shape and sclerotization. Fleas are laterally compressed and biting lice dorsoventrally flattened – shapes that make these insects difficult to dislodge from hairs or feathers. Furthermore, many ectoparasites have resistant bodies, and the heavily sclerotized cuticle of certain beetles must act as a mechanical anti-predator device.

Many insects construct retreats which can deter a predator that fails to recognize the structure as containing anything edible or that is unwilling to eat inorganic material. The cases of caddisfly larvae (Trichoptera), constructed of sand grains, stones or organic fragments (Fig. 9.6) may have originated in response to the physical environment of flowing water but certainly have a defensive role. Similarly, a portable case of vegetable material bound with silk is constructed by the terrestrial larvae of bagworms (Lepidoptera: Psychidae). In both caddisflies and psychids, the case serves to protect during pupation.

Artificial shields are constructed by some insects; for example, the larvae of certain lacewings cover themselves with the sucked-out carcasses of their insect prey or with lichens, and some larvae of chrysomelid beetles use their faeces for the same purpose. These may not act as barriers to a determined predator but will give a first mouthful of inedible material. The waxes and powders secreted by many homopterans (such as scale insects, woolly aphids, whiteflies and fulgorids) may function similarly, and also may entangle the mouthparts of a potential arthropod predator.

Body structures themselves, such as the scales of moths, caddisflies and thrips, can protect since they detach readily to allow the escape of a slightly denuded insect from the jaws of a predator, or the sticky threads of spider webs or the glandular leaves of insectivorous plants such as the sundews. A mechanical defence which seems at first to be maladaptive is **autotomy**, the shedding of limbs, as demonstrated by stick-insects (Phasmatodea) and perhaps crane flies (Diptera: Tipulidae). The upper part of the phasmatid leg has the trochanter and femur fused, with no muscles running across the joint. A special muscle breaks the leg at a weakened zone in response to a predator grasping the leg. Immature stick-insects and mantids can regenerate lost limbs at moulting, and certain autotomized adults can induce an adult moult at which the limb can regenerate.

Secretions of insects can have a mechanical defensive role, acting as a glue or slime that ensnares predators or parasitoids. Certain cockroaches have a permanent slimy coat on the abdomen that confers protection. Lipid secretions from the *cornicles* (also called *siphunculi*) of aphids may gum up predator mouthparts or small parastic wasps. Termite soldiers have a variety of secretions available to them in the form of cephalic glandular products, including terpenes that dry on exposure to air to form a resin. In *Nasutitermes* (Termitidae) the secretion is ejected via the nozzle-like nasus (a pointed snout or rostrum) as a quick-drying fine thread that impairs the movements of a predator such as an ant. This defence counters arthropod predators but is unlikely to deter vertebrates. Mechanical-acting chemicals are only a small selection of the total insect armoury that can be mobilized for chemical warfare.

13.4 CHEMICAL DEFENCES

Chemicals play vital roles in many aspects of insect behaviour. Chapter 4 dealt with the use of pheromones in many forms of communication, including alarm pheromones elicited by the presence of a predator. Similar chemicals, called allomones, play important roles in the defences of many insects, notably amongst many Heteroptera and Coleoptera. The relationship between defensive chemicals and those used in communication may be very close, sometimes with the same chemical fulfilling both roles. Thus a noxious chemical that repels a predator can alert conspecific insects to the predator's presence and may act as a stimulus to action. In the energy/time dimensions shown in Fig. 13.1, chemical defence lies towards the energetically expensive but time efficient end of the spectrum. Chemically-defended insects tend to have high **apparency** to predators, that is, they are usually non-cryptic, active, often relatively large, long-lived and frequently aggregated or social in behaviour. They often signal their distastefulness by **aposematism** – warning signalling usually involving bold colouring but sometimes including odour and perhaps sound production.

13.4.1 Classification by function of defensive chemicals

Amongst the diverse range of defensive chemicals produced by insects, two classes of compounds can be distinguished by their effects on a predator. Class I defensive chemicals are noxious because they irritate, hurt, poison or drug individual predators. Class II chemicals are innocuous, being essentially antifeedant chemicals that merely stimulate the olfactory and gustatory receptors, or aposematic indicator odours. Many insects use mixtures of the two classes of chemicals and, furthermore, Class I chemicals in low concentrations may give Class II effects. Contact by a predator with Class I compounds results in repulsion through, for example, emetic (sickening) properties or induction of pain, and if this unpleasant experience is accompanied by odorous Class II compounds, predators learn to associate the odour with the encounter. This conditioning results in the predator learning to avoid the defended insect at a distance, without the dangers (to both predator and prey) of having to feel or taste it.

Class I chemicals include both **immediate**-acting substances which the predator experiences through handling the prey insect (which may survive the attack), and chemicals with **delayed**, often systemic, effects including vomiting or blistering. In contrast to immediate-effect chemicals sited in particular organs and applied topically (externally), delayed-effect chemicals are distributed more generally within the insect's tissues and haemolymph and are tolerated systemically. Whereas a predator rapidly learns to associate immediate distastefulness with particular prey (especially if it is aposematic), it is

unclear how a predator identifies the cause of nausea some time after the predator has killed and eaten the toxic culprit, and what benefits this action brings to the victim. Experimental evidence from birds shows that these predators are indeed able to associate a particular food item with a delayed effect, perhaps through taste when regurgitating the item. Too little is known of feeding in insects to understand if this applies similarly to predatory insects. Perhaps a delayed poison that fails to protect an individual from being eaten evolved through the education of a predator by a sacrifice, thereby allowing differential survival of relatives (section 13.6).

13.4.2 The chemical nature of defensive compounds

Class I compounds are much more specific and effective against vertebrate than arthropod predators. For example, birds are more sensitive than arthropods to toxins such as cyanides, cardenolides and alkaloids. Cyanogenic glycosides are produced by zygaenid moths (Zygaenidae), *Leptocoris* bugs (Rhopalidae) and *Acraea* and *Heliconius* butterflies (Nymphalidae). Cardenolides are very widespread, occurring notably in monarch butterflies (Nymphalidae), certain cerambycid and chrysomelid beetles, lygaeid bugs, pyrgomorphid grasshoppers and even an aphid. Similarly, a variety of alkaloids are acquired convergently in many coleopterans.

Possession of Class I emetic or toxic chemicals is very often accompanied by aposematism, particularly coloration if directed against visual-hunting diurnal predators. However, visible aposematism is of no use at night, and the sounds emitted by nocturnal moths such as certain Arctiidae when challenged by bats may be aposematic, warning the predator of a distasteful meal.

Class II chemicals tend to be volatile and reactive organic compounds with low molecular weight, such as aromatic ketones, aldehydes, acids and terpenes. Examples include the stink gland products of Heteroptera and the many substances of low molecular weight such as formic acid, emitted by ants. Bitter-tasting but non-toxic compounds like quinones are commonly occurring Class II compounds. Many defensive secretions are complex mixtures that can involve synergistic effects. Thus the carabid beetle *Heluo-morphodes* emits a Class II compound, formic acid, which is mixed with *n*-nonyl acetate that enhances skin penetration of the acid to give a Class I painful effect.

The role of these chemicals in aposematism, warning of the presence of Class I compounds, was considered above. In another role, these Class II chemicals may be used to deter predators such as ants that rely on chemical communication. For example, prey such as certain termites, when threatened by predatory ants, release mimetic ant alarm pheromones, thereby inducing inappropriate ant behaviours of panic and nest defence. In another case, ant-nest inquilines, which might provide prey to their host ants, are unrecognized as potential food because they produce chemicals that appease ants.

Class II compounds alone appear unable to deter many insectivorous birds. For example, blackbirds (Turdidae) will eat notodontid (Lepidoptera) caterpillars that secrete a 30% formic acid solution; many birds actually encourage ants to secrete formic acid into their plumage in an apparent attempt to remove ectoparasites (so-called 'anting').

13.4.3 Sources of defensive chemicals

Many defensive chemicals, notably those of phytophagous insects, are derived from the host plant upon which the larvae (Fig. 13.4) and, less commonly, the adults feed. Frequently a close association is observed between restricted host-plant use (monophagy or oligophagy) and the possession of a chemical defence. An explanation may lie in a coevolutionary 'arms race' in which a plant develops toxins to deter phytophagous

Fig. 13.4 The distasteful and warningly coloured caterpillars of the cinnabar moth, *Tyria jacobaeae* (Lepidoptera: Arctiidae), on ragwort, *Senecio jacobaeae*. (After Blaney, 1976.)

insects. A few phytophages overcome the defences and thereby become specialists able to detoxify or sequester the plant toxins. These specialist herbivores can recognize their preferred host plants, develop on them and use the plant toxins (or metabolize them to closely related compounds) for their own defence.

Although some aposematic insects are closely associated with toxic food plants, certain insects can produce their own toxins. For example, amongst the Coleoptera, meloid beetles synthesize cantharidin, buprestids make buprestin, and some chrysomelids can produce cardiac glycosides. The very toxic staphylinid *Paederus* synthesizes its own blistering agent, paederin. Experimentally it has been shown that certain insects which sequester cyanogenic compounds from plants can still synthesize similar compounds if transferred to toxin-free host plants. If this ability preceded the evolutionary transfer to the toxic host plant, the possession of appropriate biochemical pathways may have preadapted the insect to using them subsequently in defence.

BOX 13.3
INSECT BINARY CHEMICAL WEAPONS

The common name of bombardier beetles (Carabidae: including genus *Brachinus*) derives from the observations by early naturalists that the beetles released volatile defensive chemicals which appeared like a puff of smoke, accompanied by a 'popping' noise resembling gunfire. The spray, released from the anus and able to be directed by the mobile tip of the abdomen, contains *p*-benzoquinone, a deterrent of vertebrate and invertebrate predators. This chemical is not stored; when required, it is produced explosively from components held in paired glands. Each gland is double, comprising a muscular-walled, compressible inner chamber containing a reservoir of hydroquinones and hydrogen peroxide, and a thick-walled outer chamber containing oxidative enzymes. When threatened, the beetle contracts the reservoir, and releases the contents through the newly opened inlet valve into the reaction chamber. Here an exothermic reaction takes place, resulting in the liberation of *p*-benzoquinone at a temperature of 100 °C.

Studies on a Kenyan bombardier beetle, *Stenaptinus insignis*, (illustrated here, after Dean *et al.*, 1990) showed that the discharge is pulsed: the explosive chemical oxidation produces a build-up of pressure in the reaction chamber which closes the one way valve from the reservoir, thereby forcing discharge of the contents through the anus (as shown by the beetle directing its spray at an antagonist in front of it). This relieves the pressure, allowing the valve to open, permitting refilling of the

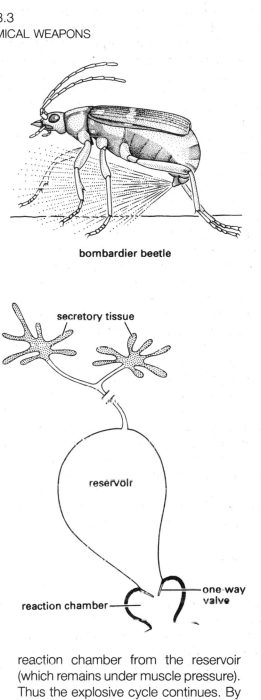

bombardier beetle

secretory tissue

reservoir

reaction chamber——

one way valve

reaction chamber from the reservoir (which remains under muscle pressure). Thus the explosive cycle continues. By this mechanism a high-intensity pulsed

---

BOX 13.3—*continued*

jet is produced by the chemical reaction, rather than requiring extreme muscle pressure. Humans discovered the principles independently and applied them to engineering (as pulse jet propulsion) some millions of years after the bombardier beetles developed the technique.

Amongst the many unusual means of obtaining a defensive chemical, that used by *Photurus* fireflies (Lampyridae) is one of the most bizarre. Many fireflies synthesize deterrent bufadienolides, but *Photurus* females cannot do so. Instead they acquire their supplies by eating male *Photinus* fireflies, which are lured to their deaths by the *Photurus* females that mimic the flashing sexual signal of the *Photinus* female.

13.4.4 Organs of chemical defence

Endogenous defensive chemicals (those synthesized within the insect) are generally produced in specific glands and stored in a reservoir (Box 13.3). Release is through muscular pressure or by evaginating the organ, rather like turning the fingers of a glove inside out. The Coleoptera have developed a wide range of glands, many eversible, that produce and deliver defensive chemicals. Many Lepidoptera use urticating (itching) hairs and spines to inject venomous chemicals into a predator. Venom injection by social insects is dealt with in section 13.6.

In contrast to these endogenous chemicals, exogenous toxins, derived from external sources such as foods, are usually incorporated in the tissues. This makes the complete prey unpalatable but requires the predator to test at close range in order to learn, in contrast to the distant effects of many endogenous compounds. However, the larvae of some swallowtail butterflies (Papilionidae) that feed upon distasteful food plants concentrate the toxins and secrete them into a thoracic pouch

called an **osmeterium**, which is everted if the larvae are touched. The colour of the osmeterium is often aposematic, reinforcing the deterrent effect on a predator (Fig. 13.5). Larval sawflies (Hymenoptera: Pergidae), colloquially called 'spitfires', store eucalypt oils, derived from the leaves that they eat, within a diverticulum of their foregut and ooze this strong smelling, distasteful fluid from their mouths when disturbed (Fig. 13.6).

13.5 DEFENCE BY MIMICRY

The theory of **mimicry** is an interpretation of the close resemblances of unrelated species and was an early application of the theory of evolution. Bates, a naturalist studying in the Amazon in the mid nineteenth century, observed that many similar butterflies, all slow-flying and brightly coloured, seemed to be immune from predators. Although many species were common and related to each other, some were rare and belonged to quite distantly related families. Bates believed that the common species were chemically protected from attack, and this was advertised by their aposematism – high apparency (behavioural conspicuousness) through bright colour and slow flight. The rarer species, he thought, probably were not distasteful but gained protection by their superficial resemblance to the protected ones. On reading the views that Darwin had newly proposed in 1859, Bates realized that his own theory of mimicry involved evolution through natural selection. Poorly protected species gain increased immunity from predation by

osmeterium

head capsule

thoracic legs

ventral prolegs

spiracle

anal proleg

Fig. 13.5 A caterpillar of the orchard butterfly, *Papilio aegeus* (Lepidoptera: Papilionidae), with the osmeterium everted behind its head. Eversion of this glistening, bifid organ occurs when the larva is disturbed; it is accompanied by a pungent smell.

differential survival of subtle variants that more resembled protected species in either appearance, smell, taste, feel or sound. The selective agent is the predator that preferentially eats the inexact mimic. Since that time, mimicry has been interpreted in the light of evolutionary theory, and insects, particularly butterflies, have remained central to mimicry studies.

Understanding mimicry systems requires recognition of three basic components: the **model**, the **mimic** and the **observer**. These components are related to each other through signalling and receiving systems, of which the basic association is the warning signal given by the model (e.g. aposematic colour which warns of a sting or bad taste) perceived by the observer (e.g. a hungry predator). The naive observer must learn of the association between aposematism and subsequent pain or distaste. When learnt, the observer will thereafter avoid the model. The model clearly benefits from this coevolved system, in which the observer can be seen to gain by not wasting time and energy chasing an inedible prey.

Once such a mutually beneficial system has evolved, it is open to manipulation by others. The third component is the mimic: an organism that parasitizes the signalling system through deluding the observer, for example, by false warning colouration. If this provokes

Fig. 13.6 An aggregation of sawfly larvae (Hymenoptera: Pergidae: *Perga*) on a eucalypt leaf. When disturbed, the larvae bend their abdomens in the air and exude droplets of sequestered eucalypt oil from their mouths.

a reaction from the observer similar to true aposematic colouration, the mimic is dismissed as unacceptable food. It is important to realize that the mimic need not be perfect, as long as it can elicit the appropriate avoidance response from the observer. Only a limited subset of the signals given by the model may be required. For example, the black and yellow banding of venomous wasps is an aposematic colour pattern that is displayed by countless species from amongst many orders of insects. The exactness of the match, at least to our eyes, varies considerably. This may be due to subtle differences between several different venomous models, or it may be the inability of the observer to discriminate: if yellow and black banding is all that is required to deter a predator, then there may be little or no selection to refine the mimicry more fully.

13.5.1 Batesian mimicry

In these mimicry triangles, each component has a positive or negative effect on each of the others. In **Batesian mimicry**, an aposematic inedible model has an edible mimic. The model suffers by the mimic's presence because the aposematic signal aimed at the observer is diluted as the chances increase that the observer will taste an edible individual and fail to learn the association between aposematism and distastefulness. The mimic gains from the presence of the protected model and gains from the deception of the observer. Since the mimic's presence is disadvantageous to the model, interaction with the model is negative. The observer benefits by avoiding the noxious model but misses a meal through failure to recognize the mimic as edible.

These Batesian mimicry relationships hold up as long as the mimic remains relatively rare. However, should the model decline (or the mimic become abundant), the protection given to the mimic by the model will wane because the naive observer increasingly encounters and tastes edible mimics. It is evident from studies of butterflies that some palatable mimics attain high density by adopting different models throughout their range. For example, there are up to five mimetic morphs of the mocker swallowtail, *Papilio dardanus*, in Uganda, central Africa, and several more throughout the wide range of the species. Each morph can remain rare relative to the Batesian model but the total population can increase without prejudicing (diluting) the successful mimetic system. In this case, and many other mimetic polymorphisms, the males retain the basic colour pattern of the species and only amongst females in some populations are there mimics of such a variety of models. It may be that there is stabilizing selection on the conservative pattern in males to ensure their recognition by conspecific females for mating purposes.

Larvae of the Old World tropical butterfly *Danaus chrysippus* (Nymphalidae: Danainae) feed predominantly on milkweeds (Asclepiadaceae) from which they can sequester cardenolides that are retained to the aposematic, chemically protected adult stage. A variable but often high proportion of *D. chrysippus* develop on milkweeds lacking bitter and emetic chemicals, and the resulting adult is unprotected. These are intraspecific Batesian **automimics** of their protected relatives. In cases in which there is an unexpectedly high proportion of unprotected individuals, the system may be maintained by parasitoids that preferentially parasitize noxious individuals, perhaps using cardenolides as kairomones in host finding. The situation is further complicated, since unprotected adults, as in many *Danaus* species, actively seek out sources of pyrrolizidine alkaloids from plants to use in production of sex pheromones; these alkaloids also may render the adult less palatable.

13.5.2 Müllerian mimicry

In a contrasting set of relationships, termed **Müllerian mimicry**, the models and mimics are all distasteful and warningly coloured and all benefit from co-existence, since observers learn from tasting any individual. Unlike Batesian mimicry, in which the system will fail as the mimic becomes relatively more abundant, Müllerian mimicry systems gain through enhanced predator learning when the density of component distasteful species increases. 'Mimicry rings' of species may develop, in which organisms from distant families, and even orders, acquire similar aposematic patterns, although the source of

(a)   (b)   (c)

Fig. 13.7 Three nymphalid butterflies that are Müllerian co-mimics in Florida: (a) the monarch (*Danaus plexippus*); (b) the queen (*D. gilippus*); (c) the viceroy (*Limenitis archippus*). (After Brower, 1958.)

protection varies greatly. In the species involved, the warning signal of the co-models differs markedly from that of their close, non-mimetic relatives.

Interpretation of mimicry may be difficult, particularly distinguishing protected from unprotected mimetic components. For example, a century after discovery of one of the seemingly strongest examples of Batesian mimicry, recent studies have cast doubt on the classical interpretation. The system involves two North American danaine butterflies, *Danaus plexippus*, the monarch, and *D. gilippus*, the queen, which are chemically defended models that are mimicked by nymphaline viceroy butterflies (*Limenitis archippus*) (Fig. 13.7). Historically, based on observation of the viceroy's larval food plants and taxonomic affiliation, the butterflies were considered to be palatable, and therefore Batesian mimics. Ideas on palatability were overturned after an experiment in which isolated butterfly abdomens were fed to natural predators (wild-caught red-winged blackbirds). It was impossible that previous exposure to aposematism deterred feeding birds because the aposematically patterned butterfly wings had been removed. Viceroys were found to be as unpalatable as monarchs, and queens were least unpalatable. In Florida populations and

with this particular predator, the system now seems to be Müllerian. The viceroy and monarch are co-models and the queen is a less well chemically protected member that benefits through the asymmetry of its palatability relative to the others. Appropriate experiments to assess palatability, using natural predators and avoiding problems of previous learning by the predator, are uncommon. Other strictly designed tests may show more 'Batesian' systems to be Müllerian.

It has been reasoned that, if all members of a Müllerian mimicry complex are aposematic and distasteful, then an observer (predator) is not deceived by any member and this cannot be mimicry but rather shared aposematism. However, equality of distastefulness across all members of a complex is unlikely. Furthermore, some observers (e.g. specialist predators) may find the least well defended part of the complex to be edible. Thus definitions of mimicry that take into account the range of all possible observers will include Müllerian mimicry.

13.5.3 Mimicry as a continuum

The practicality and even the significance of differentiating two forms of defensive mimicry can be questioned but each gives a different interpretation of the ecology and evolution of the components, and makes quite dissimilar predictions concerning life histories of the participants. For example, the theory of Müllerian mimicry makes predictions that there ought to be certain characteristics:

- Limited numbers of co-modelled aposematic patterns, reducing the number that a predator has to learn.
- Behavioural modifications to 'expose' the pattern to potential predators, such as conspicuous display rather than crypsis, and diurnal rather than nocturnal activity.
- Long post-reproductive life, with prominent exposure to encourage the naive predator to learn of the distastefulness on a post-reproductive individual.

In Batesian mimicry, the model may exhibit characteristics 2 and 3 above. In addition, Batesian mimicry ought to involve mimetic polymorphism, sex-limited (female-only) mimicry and divergence of the model's pattern away from that of the mimic (evolutionary escape). Although all these predictions are met in some mimetic species, there are some exceptions to all of them. Evidently, the rigid demarcation of mimicry into two types is an oversimplification.

Amongst the lycid beetles there are many aposematically odoriferous and warningly coloured species that demonstrate a range of mimetic relationships. The Australian lycid *Metriorrhynchus rhipidius* is protected chemically by odorous methoxy-alkylpyrazine, and by bitter principles and acetylenic antifeedants. Species of *Metriorrhynchus* provide models for mimetic beetles from at least six distantly related families (Buprestidae, Pythidae, Meloidae, Oedemeridae, Cerambycidae and Belidae) and at least one moth. All these mimics are convergent in colour; some have nearly identical alkylpyrazines and distasteful chemicals; others share the alkylpyrazines but have different distasteful chemicals; and some have the odorous chemical but appear to lack

any distasteful chemicals. These aposematically coloured insects form a mimetic series. The oedemerids are clearly Müllerian mimics, modelled precisely on the local *Metriorrhynchus* species and differing only in using cantharidin as an antifeedant. The cerambycid mimics use different repellent odours, whereas the buprestids lack warning odour but are chemically protected by buprestins. Finally pythids and belids are Batesian mimics, apparently lacking any chemical defences. After careful chemical examination, what appears to be a model with many Batesian mimics, or perhaps a Müllerian ring, is revealed to demonstrate a complete range between the extremes of Müllerian and Batesian mimicry.

Although the extremes of these two prominent mimicry systems are well studied, and in some texts appear to be the only systems described, they are but two of the possible permutations involving the interactions of model, mimic and observer. Further complications ensue if model and mimic are of the same species, as in automimicry, or if there is sexual dimorphism and polymorphism. All mimicry systems are complex, interactive and never static, because population sizes change and relative abundance of mimetic species fluctuate so that density-dependent factors play an important role. The defence offered by shared aposematic colouring, and even shared distastefulness, can be circumvented by a specialized predator able to cue on the warning, overcome the defences and eat selected species in the mimicry complex (Box 13.2).

13.6 COLLECTIVE DEFENCES IN GREGARIOUS AND SOCIAL INSECTS

Chemically defended, aposematic insects are often clustered rather than uniformly distributed through suitable habitat. Thus unpalatable butterflies may live in conspicuous aggregations as larvae and as adults; the winter congregation of migratory adult monarch butterflies in California and Mexico is an example. Many chemically defended homopterans aggregate on individual host plants and some vespid wasps congregate conspicuously on the outside of their nests (seen in the vignette of Chapter 11). Orderly clusters occur in the phytophagous larvae of sawflies (Hymenoptera: Pergidae; Fig. 13.6) and some chrysomelid beetles that form defended circles (**cycloalexy**). Some larvae lie within the circle and others form an outer ring with either their heads or abdomens directed outwards, depending upon which end secretes the noxious compounds. These groups often make synchronized displays of head and/or abdomen bobbing, which increase the apparency of the group.

Formation of such clusters is sometimes encouraged by the production of aggregation pheromones by early arriving individuals (section 4.3.2(b)), or may result from the young failing to disperse after hatching from limited egg batches. Benefits to the individual from the clustering of chemically defended insects may relate to the dynamics of predator training. However, these may also involve kin selection in subsocial insects, in which aggregations comprise relatives that benefit at the expense of an individual 'sacrificed' to educate a predator.

The latter scenario for the origin and maintenance of group defence certainly seems to apply to the eusocial Hymenoptera (ants, bees and wasps) as seen in Chapter 11. In these insects, and in the termites (Isoptera), defensive tasks are usually undertaken by morphologically modified individuals called soldiers. In all social insects, excepting the army ants, the focus for defensive action is the nest, and the major role of the soldier caste is to protect the nest and its inhabitants. Nest architecture and location are often a first line of defence, with many nests buried underground, or hidden within trees, with a few easily defendable entrances. Exposed nests, such as those of savanna-zone termites, often have hard, impregnable walls.

Termite soldiers can be male or female; they may have weak sight or be blind, and have enlarged heads (sometimes exceeding the rest of the body length). Soldiers may have well-developed jaws, or be **nasute**, with small jaws but an elongate 'nasus' or rostrum. They may protect the colony by biting, by chemical means or, as in *Cryptotermes*, by **phragmosis** – the blocking of access to the nest with their modified heads. Amongst the most serious adversaries of termites are ants, and complex warfare takes place between the two. Termite soldiers have developed an enormous battery of chemicals, many produced in highly elaborated frontal and salivary glands. For example, in *Pseudacanthotermes spiniger* the salivary glands fill nine-tenths of the abdomen, and *Globitermes sulphureus* soldiers are filled to bursting with sticky yellow fluid used to entangle the predator – and the termite, usually fatally. This suicidal phenomenon is seen also in some *Camponotus* ants which use

Fig. 13.8 Nest guarding by the European ant *Colobopsis truncata*: a minor worker approaching a soldier that is blocking a nest entrance with her plug-shaped head. (After Hölldobler and Wilson, 1990, from Szabó-Patay, 1928.)

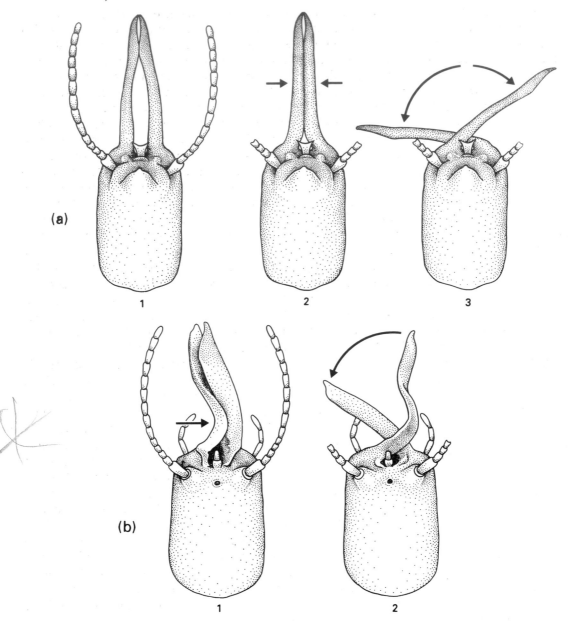

Fig. 13.9 Defence by mandible snapping in termite soldiers: (a) a symmetric snapping soldier of *Termes* in which (1) the long thin mandibles are pressed hard together, and thus (2) bent inwards, before (3) they slide violently across one another; (b) an asymmetric snapping soldier of *Homallotermes* in which force is generated in the flexible left mandible by (1) being pushed against the right one, until (2) the right mandible slips under the left one to strike a violent blow. (After Deligne *et al.*, 1981.)

hydrostatic pressure in the gaster to burst the abdomen and release sticky fluid from the huge salivary glands.

Some of the specialized defensive activities used by termites have developed convergently amongst ants. Thus the soldiers of some formicines, notably the subgenus *Colobopsis*, and several myrmecines show phragmosis, with modifications of the head to allow the blocking of nest entrances (Fig. 13.8). Nest entrances are made by minor workers and are of such a size that the head of a single major worker (soldier) can seal it; in others, such as the myrmecine *Zacryptocerus*, the entrances are larger and a formation of guarding blockers may be required to act as 'gatekeepers'. A further defensive strategy of these myrmecines is for the head to be covered with a crust of secreted filamentous material, such that the head is camouflaged when it blocks a nest entrance on a lichen-covered twig.

Most soldiers use their strongly developed mandibles in colony defence as a means of injuring an attacker. A novel defence in termites involves elongate mandibles that snap against one another, as we might snap our fingers. A violent movement is produced as the pent-up elastic energy is released from the tightly appressed mandibles (Fig. 13.9(a)). In *Capritermes* and *Homallotermes*, the mandibles are asymmetric (Fig. 13.9(b)) and the released pressure results in the violent movement of only the right mandible; the bent left one, which provides the elastic tension, remains immobile. These soldiers can only strike to their left! The advantage of this defence is that a powerful blow can be delivered in a confined tunnel, in which there is inadequate space to open the mandibles wide enough to obtain conventional leverage on an opponent.

Major differences between termite defences and those of social hymenopterans are the restriction of the defensive caste to females in Hymenoptera, and the frequent use of venom injected through an ovipositor modified as a sting (Fig. 13.10). Whereas parasitic hymenopterans use this weapon to immobilize prey, in social aculeate hymenopterans it is a vital weapon in defence against predators. Many subsocial and all social hymenopterans co-operate to sting an intruder *en masse*, thereby escalating the effects of an individual

attack and deterring even large vertebrates. The sting is injected into a predator through a lever (the furcula) acting on a fulcral arm, though fusion of the furcula to the sting base in some ants leads to a less manoeuvrable sting.

Venoms include a wide variety of products, many of which are polypeptides. Biogenic amines such as any or all of histamine, dopamine, adrenaline and noradrenaline (and serotonin in wasps), may be accompanied by acetylcholine, and several important enzymes including phospholipases and hyaluronidases (which are highly allergenic). Wasp venoms have a number of vasopeptides – pharmacologically active kinins that induce vasodilation and relax smooth muscle in vertebrates. Non-formicine ant venoms comprise either similar materials of proteinaceous origin or a pharmacopoeia of alkaloids, or complex mixtures of both types of component. In contrast, formicine venoms are dominated by formic acid.

Venoms are produced in special glands sited on the bases of the inner valves of the ninth segment, comprising free filaments and a reservoir store that may be simple or contain a convoluted gland (Fig. 13.10). The outlet of Dufour's gland enters the sting base ventral to the venom duct. The products of this gland in eusocial bees and wasps are poorly known, but in ants Dufour's gland is the site of synthesis of an astonishing array of hydrocarbons (over 40 in one species of *Camponotus*). These exocrine products include esters, ketones and alcohols and many other compounds used in communication and defence.

The sting is reduced and lost in some social hymenopterans, notably the stingless bees and formicine ants. Alternative defensive strategies have arisen in these groups; thus many stingless bees mimic stinging bees and wasps, and use their mandibles and defensive chemicals if attacked. Formicine ants retain their venom glands, which disperse formic acid through an acidophore, often directed as a spray into a wound created by the mandibles.

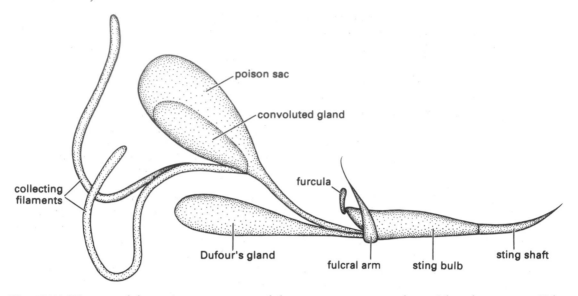

Fig. 13.10 Diagram of the major components of the venom apparatus of a social aculeate wasp. (After Hermann and Blum, 1981.)

Other glands in social hymenopterans produce additional defensive compounds, often with communication roles, and including many volatile compounds that serve as alarm pheromones. These stimulate one or more defensive actions: they may summon more individuals to a threat, marking a predator so that the attack is targeted, or, as a last resort, they may encourage the colony to flee from the danger. Mandibular glands produce alarm pheromones in many insects and also substances that cause pain when they enter wounds caused by the mandibles. The metapleural glands in some species of ants produce compounds that defend against microorganisms in the nest through antibiotic action. Both sets of glands may produce sticky defensive substances and a wide range of pharmacological compounds is currently under study to determine possible human benefit.

Even the best defended insects can be parasitized by mimics (section 13.5), and the best of chemical defences can be breached by a predator (Box 13.2). Although the social insects have some of the most elaborate defences seen in the Insecta, they remain

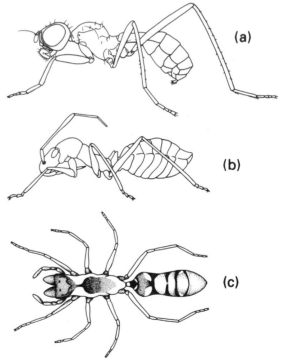

Fig. 13.11 Three ant mimics: (a) a fly (Diptera: Micropezidae: *Badisis*); (b) a bug (Hemiptera: Miridae: Phylinae); (c) a spider (Araneae: Clubionidae: *Sphecotypus*). (After (a) McAlpine, 1990; (b) Atkins, 1980; (c) Oliveira, 1988.)

vulnerable. For example, many insects model themselves on social insects, with representatives of many orders converging morphologically on ants (Fig. 13.11), particularly with regard to the waist constriction and wing loss, and even kinked antennae. The aposematic yellow-and-black patterns of vespid wasps and apid bees provide models for hundreds of mimics throughout the world. Not only are these communication systems of social insects parasitized, but so also are their nests, which provide many parasites and inquilines with a hospitable place for their development (section 11.3).

Defence must be seen as a continuing coevolutionary process, analogous to an 'arms race', in which new defences originate or are modified and then are selectively breached, stimulating improved defences.

FURTHER READING

Eisner, T. (1970) Chemical defense against predation in arthropods, in *Chemical Ecology*, (eds E. Sondheimer and J.B. Simone), Academic Press, New York, pp. 157–217.

Evans, D.L. and Schmidt, J.O. (eds) (1990) *Insect Defenses. Adaptive Mechanisms and Strategies of Prey and Predators*, State University of New York Press, Albany.

Hölldobler, B. and Wilson, E.O. (1990) *The Ants*, Springer Verlag, Berlin, Heidelberg, London, Paris, Tokyo, Hong Kong.

Howlett, R.J. and Majerus, M.E.N. (1987) The understanding of industrial melanism in the peppered moth (*Biston betularia*) (Lepidoptera: Geometridae). *Biological Journal of the Linnean Society*, **30**, 31–44.

Moore, B.P. and Brown, W.V. (1989) Graded levels of chemical defence in mimics of lycid beetles of the genus *Metriorrhynchus* (Coleoptera). *Journal of the Australian Entomological Society*, **28**, 229–33.

Pasteels, J.M., Grégoire, J.-C. and Rowell-Rahier, M. (1983) The chemical ecology of defense in arthropods. *Annual Review of Entomology*, **28**, 263–89.

Ritland, D.B. (1991) Unpalatability of viceroy butterflies (*Limenitis archippus*) and their purported mimicry models, Florida queens (*Danaus gilippus*). *Oecologia*, **88**, 102–8.

Vane-Wright, R.I. (1976) A unified classification of mimetic resemblances. *Biological Journal of the Linnean Society*, **8**, 25–56.

Vane-Wright, R.I. (1981) Mimicry and unknown ecological consequences, in *The Evolving Biosphere*, (ed. P.L. Forey), British Museum (Natural History) and Cambridge University Press, pp. 157–68.

Wickler, W. (1968) *Mimicry in Plants and Animals*, Weidenfeld and Nicolson, London.

Yosef, R. and Whitman, D.W. (1992) Predator exaptations and defensive adaptations in evolutionary balance: no defence is perfect. *Evolutionary Ecology*, **6**, 527–36.

Also papers in (1981) *Biological Journal of the Linnean Society*, **16**, 1–54 (includes a shortened version of classic 1862 paper by H.W. Bates).

14

MEDICAL AND VETERINARY ENTOMOLOGY

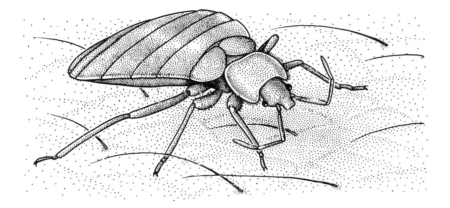

Bed bug on its host

Aside from their impact on agricultural and horticultural crops, insects impinge on us mainly through the diseases they can transmit to humans and their domestic animals. The number of insect species involved is not large but the insects that transmit disease (**vectors**), cause wounds, inject venom or create nuisance have serious social and economic consequences. Thus the study of the veterinary and medical impact of insects is a major scientific discipline.

Medical and veterinary entomology differs from other areas of entomological pursuit and is often much broader in scope. Firstly, the frequent motivation and funding for study is rarely the insect itself, but insect-borne human or animal disease. Secondly, the scientist studying medical and veterinary aspects of entomology must have a wide understanding of not only the insect vector of disease but also the biology of host and parasite. Thirdly, most practitioners do not restrict themselves to insects but have to consider other arthropods, notably ticks, mites and perhaps spiders and scorpions.

For brevity in this chapter, medical entomologists are referred to as those who study all arthropod-borne diseases, including diseases of livestock. The insect, though a vital cog in the chain of disease, need not be the central focus of medical research. Medical entomologists rarely work in isolation but usually function in multi-disciplinary teams that may include medical practitioners and researchers, epidemiologists, virologists and immunologists and ought to include those with skills in insect control.

This chapter deals with entomophobia, followed by allergic reactions, venoms and urtication caused by insects, and then by details of transmission of malaria, an exemplar of insect-borne disease. A review of additional diseases in which insects play an important role is followed by a section on forensic entomology. The chapter is concluded with taxonomic boxes (Boxes 14.2–4) dealing with the Phthiraptera (lice), Siphonaptera (fleas) and medically significant Diptera (flies).

14.1 INSECT NUISANCE AND PHOBIA

Our perceptions of nuisance may be little related to the role of insects in disease transmission. Insect nuisance is often perceived as a product of high densities of a particular species, such as bush flies (*Musca vetustissima*) in rural Australia, or ants and silverfish around the house. Most people have a more justifiable avoidance of filth-frequenting insects such as blow flies and cockroaches, biters such as some ants, and venomous stingers such as bees and wasps. Many serious disease vectors are quite uncommon and have such inconspicuous behaviours (apart from their biting habits) that they may not be perceived as particular nuisances by the lay public.

Harmless insects and arachnids sometimes arouse unwarranted phobic responses (**arachnophobia**, **entomophobia** or **delusory parasitosis**). These cases may cause time-consuming and fruitless inquiry by medical entomologists, when the more appropriate investigations ought to be psychological. None-the-less there certainly are cases in which sufferers of persistent 'insect bites' and persistent skin rashes, for which no physical cause can be established, actually suffer from undiagnosed local or widespread infestation with microscopic mites. In these circumstances, diagnosis of delusory parasitosis, through medical failure to identify the true cause, and referral to psychological counselling is unhelpful, to say the least.

There are some insects that transmit no disease but feed on blood and whose attentions almost universally cause distress: bed bugs. The vignette for this chapter shows *Cimex lectularius* (Hemiptera: Cimicidae), the cosmopolitan common bed bug, whose presence between the sheets indicates poor hygiene conditions.

14.2 VENOMS AND ALLERGENS

14.2.1 Insect venoms

Some people's earliest experiences with insects are memorable for their pain. Although the sting of the females of many social hymenopterans (bees, wasps and ants) can seem quite unprovoked, it is an aggressive defence of the nest. The delivery of venom is through the sting, a modified female ovipositor (Fig. 13.10). The honey-bee sting has backwardly directed barbs, which allows only one use since the bee is fatally damaged when it leaves the sting and accompanying venom sac in the wound as it struggles to retract the sting. In contrast, wasp and ant stings are unbarbed, can be retracted and are capable of repeated use. In some ants, the ovipositor sting is greatly reduced and venom is either sprayed around liberally or can be directed with great accuracy into a wound made by the jaws. The venoms of social insects are discussed in more detail in section 13.6.

14.2.2 Blistering and urtication

Some toxins produced by insects can cause injury to humans even though they are not inoculated through a sting. Blister beetles (Meloidae) contain toxic chemicals – cantharidins – which are released if the beetle is crushed. Cantharidins cause blistering of the skin and, if taken orally, inflammation of the urinary and genital tracts that gave rise to its notoriety (as 'Spanish fly') as a supposed aphrodisiac. Staphylinid beetles of the genus *Paederus* produce potent contact poisons, including paederin, that cause delayed-onset severe blistering and long-lasting ulceration.

Lepidopteran caterpillars, notably of moths, are a frequent cause of skin irritation or **urtication** (a description derived from the reaction to the nettle genus *Urtica*). Some species have hollow spines containing the products of a subcutaneous venom gland, which are released when the spine is broken.

Other species have setae (bristles and hairs) containing toxins which cause intense irritation when the setae contact human skin. Urticating caterpillars include the processionary caterpillars (Notodontidae) and some cup moths (Limacodidae). Processionary caterpillars combine frass, cast larval skins and shed hairs into bags suspended in trees and bushes, in which pupation occurs. If the bag is damaged by contact or by high wind, urticating hairs are widely dispersed.

The pain caused by hymenopteran stings may last a few hours, urtication may last a few days and the most ulcerated beetle-induced blisters may last some weeks. Increased medical significance of these injurious insects comes when repeated exposure leads to allergic disease in some humans.

14.2.3 Insect allergenicity

Insects and other arthropods are often implicated in **allergic disease**, which occurs when exposure to some arthropod allergen (a moderate-sized molecular weight chemical component, usually a protein) triggers excessive immunological reaction in some exposed people or animals. Those who handle insects in their occupations, such as in entomological rearing facilities, tropical fish food production or research laboratories, frequently develop allergic reactions to one or more of a range of insects. Mealworms (beetle larvae of *Tenebrio* species), bloodworms (larvae of *Chironomus* species), locusts and blow flies have all been implicated. Stored products infested with astigmatic mites give rise to allergic diseases such as baker's itch and grocer's itch. The most significant arthropod-mediated allergy arises through the faecal material of house dust mites (*Dermatophagoides pteronyssimus*) which are ubiquitous and abundant in houses throughout the temperate regions of the world. Exposure to naturally occurring allergenic arthropods and their products may be underestimated, although the role of house dust mites in allergy is now well recognized.

The venomous and urticating insects discussed above can cause greater danger when a **sensitized** (previously exposed and allergy-susceptible) individual is bitten again, since anaphylactic shock is possible, with death occurring if the victim is not treated. Individuals showing indications of allergic reaction to hymenopteran stings must take appropriate precautions, including allergen avoidance and carrying adrenalin.

14.3 INSECTS AS CAUSES AND VECTORS OF DISEASE

In tropical and subtropical regions, scientific, if not public, attention is drawn to the rôle of insects in transmitting protozoa, viruses and nematodes. Such pathogens are the causative agents of many important and widespread human diseases, including malaria, dengue, yellow fever, onchocerciasis (river-blindness), leishmaniasis (oriental sore, kala-azar), filariasis (elephantiasis) and trypanosomiasis (sleeping sickness).

The causative agent of disease may be the insect itself, such as the human body or head louse (*Pediculus humanus* and *P. capitis*, respectively) that cause pediculosis, or the mite *Sarcoptes scabiei*, whose skin-burrowing activities cause scabies. In **myiasis** (from *myia*, the Greek word for a fly), the maggots or larvae of blow flies, house flies and their relatives (Diptera: Calliphoridae, Sarcophagidae and Muscidae) can develop in living flesh, either as primary agents or following wounding or damage by other insects, such as ticks and biting flies. If untreated, the animal victim may die. As death approaches and the flesh putrefies through bacterial activity, there may be a third wave of specialist fly larvae and these colonizers are present at death. One particular form of myiasis affecting livestock is known as 'strike'; it is caused in the Americas by the New World screw-worm fly *Cochliomyia hominivorax* (Fig. 6.6(h)) and in the Old World by *Chrysomya bezziana*. The name screw-worm derives from the distinct rings of setae on the maggot resembling the thread of a screw. Virtually all myiases, including screw-worm, can affect humans, particularly under conditions of poor hygiene. Other groups of 'higher' Diptera develop in mammals as endoparasitic larvae in the dermis, intestine or, as in the sheep nostril fly *Oestrus ovis*, in the nasal and head sinuses. In many parts of the world, losses due to fly-induced damage to hides and meat and death caused by myiases may amount to many millions of dollars.

Even more frequent than direct injury due to insects is their action as vectors, transmitting disease-inducing pathogens from one animal or human host to another. This transfer may be by mechanical or biological means. **Mechanical transfer** occurs, for example, when a mosquito transfers myxomatosis from rabbit to rabbit in the blood on its proboscis. Likewise, when a cockroach or house fly acquires bacteria when feeding on faeces it may physically transfer some bacteria from its mouthparts, legs or body to human food, thereby transferring enteric diseases. The causative agent of the disease is passively transported from host to host, and does not increase in the vector. Usually the arthropod is only one of several means of mechanical pathogen transfer, and poor public and personal hygiene often provide additional pathways.

In contrast, **biological transfer** is a much more specific association between insect vector, pathogen and host, and transfer never occurs naturally without all three components. The disease agent replicates (increases) within the vector insect, and there is often quite close specificity between vector and agent. The insect is thus a vital link in biological transfer and efforts to curb disease nearly always involve attempts to reduce vector numbers. In addition, biologically transferred disease may be controlled by seeking to interrupt contact between vector and host, and by direct attack on the pathogen, usually whilst in the host. Disease control comprises a combination of these approaches, each of which requires detailed knowledge of the biology of all three components – vector, pathogen and host.

14.4 GENERALIZED DISEASE CYCLES

In all biologically transferred diseases, a biting (blood-feeding or sucking) adult arthropod, often an insect, particularly a true fly (Diptera), transmits a parasite from human to human, or from animal to human or, more rarely, from human to animal. Some human pathogens (causative agents of human disease such as malaria parasites) can complete their parasitic life-cycles solely within the insect vector and the human host. Human malaria is an example of a disease with a **single cycle** involving *Anopheles* mosquitoes, malaria parasites and humans. Although related malaria parasites occur in animals, notably other primates and birds, these hosts and parasites are not involved in the human malarial cycle. Only a few human insect-borne diseases have single cycles, as in malaria, because these diseases require coevolution of pathogen and vector and *Homo sapiens*. Since *H. sapiens* had an evolutionary recent origin, there has been only a short time for the development of unique insect-borne diseases that require specifically a human rather than any alternative vertebrates for completion of the disease-causing organism's life cycle.

In contrast to single-cycle diseases, many other insect-borne diseases that affect humans include a non-human vertebrate host – for example monkeys in yellow fever, rats in plague, and desert rodents in leishmaniasis. In these cases it is clear that the non-human cycle is primary and the sporadic inclusion of humans in a secondary cycle is not essential to maintain the disease. However, when outbreaks do occur, these diseases can spread in human populations and may involve many cases.

Outbreaks in humans often stem from human actions, such as the spread of people into the natural ranges of the vector and animal hosts which act as disease **reservoirs**. For example, yellow fever in native forested Uganda (central Africa) has a 'sylvan' (woodland) cycle, remaining within canopy-dwelling primates with the exclusively primate-feeding mosquito *Aedes africanus* as the vector. It is only when monkeys and humans coincide at banana plantations close to or within the forest, that *Aedes simpsoni*, a second mosquito vector that feeds on both humans and monkeys, can transfer yellow fever to humans. In another example, *Phlebotomus* sand flies (Psychodidae) depend upon Arabian arid-zone burrowing rodents and, in feeding, transmit *Leishmania* parasites between rodent hosts. Leishmaniasis, a disfiguring ailment, is transmitted to humans when suburban expansion places humans within this rodent reservoir but, unlike yellow fever, there appears to be no change in vector when humans enter the cycle.

In epidemiological terms, the natural cycle is maintained in animal reservoirs: sylvan primates for yellow fever and desert rodents for leishmania. Disease control is complicated by the presence of these reservoirs in addition to a human cycle.

14.5 PATHOGENS

The disease-causing organisms transferred by the insect may be viruses (termed **arboviruses** an abbreviation of *arthropod-borne* viruses), rickettsias, bacteria, protozoa, or filarial nematode worms. Replication of these parasites in both vectors and hosts is required and some complex life cycles have developed, notably amongst the protozoa and filaria. The presence of a parasite in the vector insect (which can be determined by dissection and microscopy or biochemical means) generally appears not to harm the host insect. Transmission can occur when the parasite is at an appropriate developmental stage, and following multiplication or replication (amplification or concentration in the vector). Transfer of parasites from vector to host or *vice versa* takes place when the blood-feeding insect takes a meal from a vertebrate host. The transfer from host to previously uninfected vector is through parasite-infected blood. Transmission to a host by an infected insect is usually by injection along with anticoagulant salivary gland products that keep the wound open during feeding. However, it may also be through deposition of infected faeces close to the wound site.

In the following survey of major arthropod-borne disease, malaria is dealt with in some detail: it is the most devastating and debilitating disease in the world and it illustrates a number of general points concerning medical entomology. Briefer sections review the range of pathogenic diseases involving insects, arranged by phylogenetic sequence of parasite, from virus to filarial worm.

14.5.1 Malaria

(a) The disease

Malaria affects more people, more persistently, throughout more of the world than any other insect-borne disease. Some 120 million new cases arise each year. The World Health Organisation calculated that malaria control during the period 1950–1972 reduced the proportion of the world's population exposed to malaria from 64% to 38% (excluding China). Since then, concern over the unwanted side effects of DDT, resistance of insects to modern pesticides, resistance of malaria parasites to antimalarial drugs, civil unrest in a number of countries and poverty have allowed exposure rates to malaria in many countries to fall back towards the rates of half a century ago. Even in countries such as Australia, in which there is no transmission of malaria, the disease is on the increase among travellers: the number of cases rose from 199 in 1970 to 629 in 1980 and 813 in 1990.

The parasitic protozoans that cause malaria are sporozoans, belonging to the genus *Plasmodium*. Four species are responsible for the human malarias. Others have been described from (but not necessarily causing diseases in) primates, some other mammals, birds and lizards. The vectors of mammalian malaria are always *Anopheles* mosquitoes; other genera are involved in bird plasmodial transmission.

The disease follows a course with a **prepatent period** between infective bite and **patency** – the first appearance of parasites in the erythrocytes (red blood cells). The first clinical symptoms define the end of an **incubation period**, some 9 (*P. falciparum*) to 21 (*P. malariae*) days after infection. Periods of fever followed by severe sweating recur cyclically and follow several hours after synchronous rupture of infected erythrocytes. The spleen is characteristically enlarged.

The four malaria parasites each provoke rather different symptoms:

- *Plasmodium falciparum* causes malignant tertian malaria, which kills many untreated sufferers through, for example, cerebral malaria or renal failure. Fever recurs at 48-hour intervals. The term 'tertian', which is Latin for 'third day', describes the pattern: fever on the first day, normality on the second, and recurrence of fever on the third. *P. falciparum* is limited by a minimum 20 °C isotherm and is thus most common in the warmest areas of the world.
- *Plasmodium vivax* causes benign tertian malaria, which is a less serious disease that rarely kills. It is a more widespread pathogen than *P. falciparum* and has a wider temperature tolerance, extending as far as the 16 °C summer isotherm. Recurrence of fever is every 48 hours, and the disease may persist for up to 8 years, with relapses some months apart.
- *Plasmodium malariae* causes quartan malaria. It is a widespread parasite but more rare than *P. falciparum* or *P. vivax*. If the disease is allowed to persist for an extended period, death occurs through chronic renal failure. Recurrence of fever is at 72 hours ('quartan' means 'fourth day'). It is persistent, with

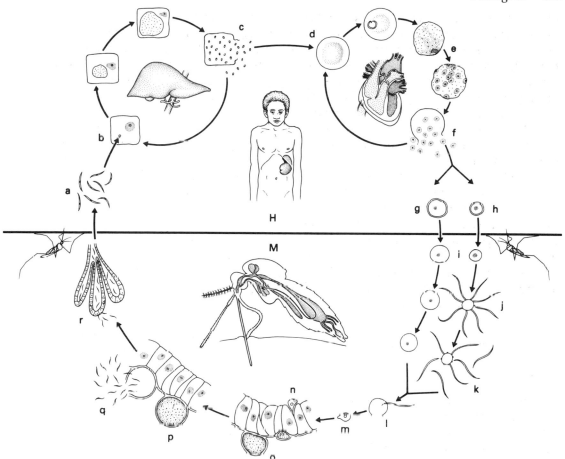

Fig. 14.1. The life cycle of a *Plasmodium*, the malaria pathogen, in human (H) and mosquito (M): (a) injection of sporozoites into human host by feeding mosquito; (b) sporozoite enters liver parenchyma cell; (c) rupture of schizont to release merozoites; (d) merozoites enter erythrocyte; (e) trophozoite becoming schizont; (f) schizont releasing merozoites; (g) female macrogamete formation; (h) male gametocyte formation; (i) gametes lacking erythrocyte coat; (j) flagellate microgamete; (k) exflagellation; (l) zygote formation; (m) ookinete; (n) oocyst forming on mosquito inner midgut wall; (o), (p) oocyst maturation and sporogeny; (q) release of sporozoites; (r) sporozoites entering salivary gland. (Modified after Kettle, 1984; Katz *et al.*, 1989.)

relapses up to half a century after initial attack.

- *Plasmodium ovale* causes a rare tertian malaria with limited pathogenicity and a very long incubation period, with relapse at three-monthly intervals.

(b) The life-cycle of Plasmodium

The malarial cycle (Fig. 14.1) commences with an infected female *Anopheles* mosquito taking a blood meal from a human host. As it feeds, it injects saliva contaminated with the **sporozoite** stage of the *Plasmodium*. The sporozoite circulates in the blood until it reaches the liver, where a **pre-** (or **exo-) erythrocytic schizogonous** cycle takes place in the parenchyma cells of the liver. This leads to the formation of a large **schizont**, containing from 2000 to 40 000 **merozoites**, according to *Plasmodium* species.

The prepatent period of infection that started with an infective bite ends when the merozoites are released and either infect further liver cells or enter the bloodstream and invade the erythrocytes. Invasion occurs when the erythrocyte invaginates to engulf the merozoite, which subsequently feeds as a **trophozoite** within a vacuole. The first and several subsequent **erythrocyte schizogonous** cycles produce a trophozoite that becomes a schizont, which releases from 6 to 16 merozoites, which commence the repetition of the erythrocytic cycle. This synchronous release of merozoites from the erythrocytes liberates parasite products that stimulate the host's cells to release cytokines (a class of immunological mediators), which provoke the fever and illness of a malaria attack. The duration of the erythrocyte schizogonous cycle is the duration of the interval between attacks – 48 hours for tertian malaria and 72 hours for quartan malaria.

After several erythrocyte cycles, some trophozoites do not undergo division but mature to gametocytes, a process that takes 8 days for *P. falciparum* but only 4 days for *P. vivax*. If a female *Anopheles* feeds on an infected human host at this stage in the cycle, she ingests blood containing erythrocytes, some of which contain two types of gametocytes. Within a susceptible mosquito the erythrocyte is disposed of and the two types of gametocytes develop further: half are female gametocytes, which remain large and are termed **macrogametes**; the other half are males, which divide into eight flagellate **microgametes** that rapidly deflagellate, and seek and fuse with a macrogamete to form a **zygote**. All this sexual activity has taken place in a matter of 15 minutes or so while the blood meal passes towards the midgut of the female mosquito. In the midgut, the initially inactive zygote becomes an active **ookinete**, which burrows into the epithelial lining of the midgut to form an **oocyst**.

Asexual reproduction (**sporogony**; Fig. 14.1(m)–(r)) now takes place within the expanding oocyst. In a temperature-dependent process, numerous nuclear divisions give rise to sporozoites. Sporogony does not occur below 16 °C or above 33 °C, which explains the temperature limitations for *Plasmodium* development noted above. The mature oocyst may contain 10 000 sporozoites, which are shed into the haemocoel, whence they migrate into the mosquito's salivary glands. This sporogonic cycle takes a minimum of 8–9 days and produces sporozoites that are active for up to 12 weeks, which is several times the complete life expectancy of the mosquito. At each subsequent feeding, the infective female *Anopheles* injects sporozoites into the next host along with the saliva containing an anticoagulant, and the cycle recommences.

(c) Malaria epidemiology

Malaria exists in many parts of the world but the incidence varies from place to place. As with other diseases, malaria is said to be **endemic** in an area when it occurs at a relatively constant incidence by natural transmission over successive years. Recognized categories of endemicity are based on the incidence and severity of symptoms (spleen enlargement) in both adults and children. An **epidemic** occurs when the incidence in an endemic area rises or a number of cases of the disease occur in a new area. Malaria is said to be in a stable state when there is little seasonal or annual variation in the disease incidence, predominantly transmitted by a strongly **anthropophilic** (human-loving) *Anopheles* vector species. Stable malaria is found in the warmer areas of the world where conditions encourage rapid sporogeny and are usually associated with *P. falciparum* pathogen. In contrast, unstable malaria is associated with sporadic epidemics, often with a short-lived and more **zoophilic** (preferring other animals to humans) vector that may occur in massive numbers. Often ambient temperatures are lower than for areas with stable malaria, sporogeny is slower and the pathogen is more often *P. vivax*.

Disease transmission can be understood only in relation to the potential of each vector to transmit the particular disease. This involves the variously complex relationship between:

- Vector distribution.
- Vector abundance.
- Life expectancy (survivorship) of the vector.
- Predilection of the vector to feed on humans (anthropophily).
- The feeding rate of the vector.
- The vector competence.

With reference to *Anopheles* and malaria, these factors can be detailed:

1. Vector distribution

Anopheles mosquitoes occur almost worldwide, with the exception of cold temperate areas, and there are over 400 known species. The four species of human pathogenic *Plasmodium* are transmitted significantly in nature by only some 30 species of *Anopheles*. Some of these have very local significance; others can be infected experimentally but have no natural role; and perhaps 75% of *Anopheles* are quite **refractory** (intolerant) to malaria. Of the vectorial species, a handful are important in stable malaria, whilst others only become involved in epidemic spread of unstable malaria. Vectorial status can vary across the range of a taxon, an observation that may be due to the hidden presence of **sibling** species that lack morphological differentiation, but differ slightly in biology and have substantially different epidemiological significance, as in the *An. gambiae* complex (Box 14.1).

2. Vector abundance

Anopheles development is temperature-dependent, as in *Aedes aegypti* (Box 6.1), with one or two generations per year in areas where winter temperatures force the hibernation of adult females, but with generation times of perhaps 6 weeks at 16 °C and as short

as 10 days in tropical conditions. Under optimal conditions, with batches of over 100 eggs laid every 2–3 days and a development time of 10 days, 100-fold increases in adult *Anopheles* can take place within 14 days.

Since *Anopheles* larvae develop in water, rainfall significantly governs numbers. The dominant African malaria vector *An. gambiae* (in the restricted sense; Box 14.1) breeds in short-lived pools that require replenishment; increased rainfall obviously increases the number of *Anopheles* breeding sites. On the other hand, rivers where other *Anopheles* species develop in lateral pools or stream-bed pools during a low- or no-flow period will be scoured out by excessive wet-season rainfall. Adult survivorship is clearly related to elevated humidity and, for the female, availability of blood meals and a source of carbohydrate.

3. Vector survival rate

The duration of the adult life of the female infective *Anopheles* mosquito is of great significance in its effectiveness as a disease transmitter. As seen above, there is a minimum period between feeding on an infected host and the sporozoites becoming available for transmission. If a mosquito survives less than 8 or 9 days after an initial infected blood meal, it does not live long enough to transmit malaria. The age of a mosquito can be calculated by finding the physiological age based on the ovarian 'relicts' left by each ovarian cycle (section 6.6.2). With knowledge of this physiological age and the duration of the sporogonic cycle, the proportion of each *Anopheles* vector population of sufficient age to be infective can be calculated. In African *An. gambiae* (in the restricted sense; Box 14.1), three ovarian cycles are completed before infectivity is detected. Maximum transmission of *P. falciparum* to humans occurs in *An. gambiae* which have completed four to six ovarian cycles. Despite these old individuals comprising only 16% of the population, they constitute 73% of infective individuals.

BOX 14.1
ANOPHELES GAMBIAE COMPLEX

☆   *A. bwambae*

⎯⎯⎯⎯   *A. gambiae*

···············   *A. arabiensis*

– – – – –   *A. melas*

— — — —   *A. quadriannulatus*

–·–·–·–·   *A. merus*

In the early days of African malariology, the common and predominantly pool-breeding
Anopheles gambiae was found to be a highly anthropophilic and very efficient vector of
malaria virtually throughout the continent. Subtle variation in morphology and biology
suggested, however, that more than one species might be involved. Initial investigations
allowed morphological segregation of West African *An. melas* and East African *An. merus*;

BOX 14.1—*continued*

both breed in saline waters, unlike the freshwater breeding *An. gambiae*. Reservations remained as to whether the latter belonged to a single species, and studies involving meticulous rearing from single egg masses, cross-fertilization and examination of fertility of thousands of hybrid offspring did reveal discontinuities in the *An. gambiae* gene pool. These were interpreted as supporting four species, a view that was confirmed by banding patterns of the larval salivary gland and ovarian nurse-cell giant chromosomes and also by protein electrophoresis. Even with reliable cytologically determined specimens, traditional morphological features do not allow segregation of the component species of the freshwater members of the *An. gambiae* complex of sibling (or cryptic) species.

   An. gambiae is now restricted to one widespread African taxon; *An. arabiensis* was chosen for a second sibling taxon that in many areas is sympatric with *An. gambiae*; *An. quadriannulatus* is an East and southern African sibling; and *An. bwambae* is a rare and localized taxon from hot mineralized pools in Uganda. The maximum distributional limit of each sibling species is shown here on the map of Africa (data from White, 1985). The siblings differ markedly in their vectorial status: *An. gambiae* and *An. arabiensis* are both **endophilic** (feeding indoors) and highly anthropophilic vectors of malaria and bancroftian filariasis. However, when cattle are present, *An. arabiensis* shows much reduced anthropophily, increased zoophily, and an increased tendency to **exophily** (feeding outdoors) compared with *An. gambiae*. In contrast to these two sibling species, *An. quadriannulatus* is entirely zoophilic and does not transmit disease of medical significance to humans. *An. bwambae* is a very localized vector of malaria that is endophilic if native huts are available.

Clearly adult longevity is important in epidemiological calculations. Raised humidity prolongs adult life and the most important cause of mortality is desiccation.

4. Anthropophily of the vector

To act as a vector, a female *Anopheles* mosquito must feed at least twice – once to gain the pathogenic *Plasmodium* and a second time to transmit the disease. **Host preference** is the term for the propensity of a vector mosquito to feed on a particular host species. In malaria, the host preference for humans (anthropophily) rather than alternative hosts (zoophily) is crucial to human malaria epidemiology. As seen above, stable malaria is associated with vectors that are so strongly anthropophilic that they may never feed on other hosts. In

these circumstances the probability of two consecutive meals being taken from a human is very high, and disease transmission can take place even when mosquito densities are low. In contrast, if the vector has a low rate of anthropophily (a low probability of human feeding) the probability of consecutive blood meals being taken from humans is slight and human malarial transmission by this particular vector is correspondingly low. Transmission takes place only if the vector is present in very high numbers, as occurs in epidemics of unstable malaria.

5. Feeding interval

The frequency of feeding of the female *Anopheles* vector is important in disease transmission. This frequency can be estimated from mark–release–recapture data or from survey of the ovarian-age classes of indoor resting mosquitoes. It is commonly assumed that one blood meal is needed to mature each batch of eggs but some mosquitoes may mature a first egg batch without a meal, whilst some anophelines require two meals. Already infected vectors may experience difficulty in feeding to satiation at one meal, because of blockage of the feeding apparatus by parasites, and may probe many times. This, as well as disturbance during feeding by an irritated host, may lead to feeding on more than one host.

6. Vector competence

Even if an uninfected *Anopheles* feeds on an infectious host, either the mosquito may not acquire a viable infection, or the *Plasmodium* parasite may fail to replicate within the vector. Furthermore, the mosquito may not transmit the infection onwards at a subsequent meal. Thus there is scope for substantial variation, both within and between species, in the **competence** to act as a disease vector.

Allowance must also be made for the human population density, their infective condition and age profiles, since human immunity to malaria increases with age.

The **vectorial capacity** of a given *Anopheles* vector to transmit malaria in a circumscribed human population can be modelled. This involves a relationship between the following:

- The number of female mosquitoes per person.
- The daily biting rate on humans.
- The daily mosquito survival rate.
- The time between mosquito infection and sporozoite production in the salivary glands.
- The vectoral competence.
- Some factor expressing the human recovery rate from infection.

This vectorial capacity must be related to some estimate concerning the biology and prevalence of the parasite when modelling disease transmission, and in monitoring disease control programmes. In malarial studies, the infantile conversion rate (ICR – the rate at which young children develop antibodies to malaria) may be used. In Nigeria (West Africa), the Garki Malaria Project found that over 60% of the variation in the ICR derived from the human-biting rate of the two dominant *Anopheles* species. Only 2.2% of the remaining variation was explained by all other components of vectorial capacity, casting doubt on the value of any measurements other than human-biting rate. This was particularly reinforced by the difficulties and biases involved in obtaining reasonably accurate estimates of many of the vectorial parameters listed above.

14.5.2 Arboviruses

Viruses which multiply in an invertebrate vector and a vertebrate host are termed arboviruses. This definition excludes the mechanically transmitted viruses such as the myxoma virus which causes myxomatosis in rabbits. There is no viral amplification in myxomatosis vectors such as the rabbit flea *Spilopsyllus cuniculi* and, in Australia, *Anopheles* and *Aedes* mosquitoes. Arboviruses are united by their ecologies, notably their ability to replicate in an arthropod. It is an unnatural grouping rather than one based upon virus phylogeny, since arboviruses belong to several virus families. These include some Bunyaviridae, Rheoviridae and Rhabdoviridae and notably many Togaviridae.

The Togaviridae contains two genera. *Alphavirus* includes exclusively mosquito-transmitted viruses, including equine encephalitises. *Flavivirus*, which includes yellow fever, dengue, Japanese encephalitis and further encephalitis viruses, is borne by mosquitoes or ticks.

Yellow fever exemplifies a togavirus life cycle. We have seen something of the African sylvan (forest) cycle in section 14.4, and a similar cycle involving a primate host occurs in Central and South America, though with different mosquito vectors from those in Africa. Although sylvan transmission to humans takes place (as in the Ugandan banana plantations), the disease makes its greatest fatal impact in urban epidemics. The urban and peri-domestic insect vector on both continents is the female of the yellow fever mosquito *Aedes (Stegomyia) aegypti*. This mosquito acquires the virus by feeding on a human yellow fever sufferer in the early stages of disease, from 6 hours pre-clinical to 4 days later. The viral cycle in the mosquito is 12 days, after which the yellow fever virus reaches the mosquito saliva and remains there for the rest of the mosquito's life. With every subsequent blood meal the female mosquito transmits virus-contaminated saliva. Infection results, and yellow fever symptoms develop in the host within a week. Urban disease cycles must originate from infected individuals moving from rural to urban environments. The disease may persist, leading to prolonged outbreaks in the urban environment. Thousands of people have died in past epidemics.

Some South American monkeys may die of yellow fever, whilst African ones are asymptomatic. Perhaps neotropical monkeys have had less time to develop tolerance to the disease. The common vector, *Ae. aegypti*, is a mosquito that may have been transported, together with yellow fever, relatively recently from West Africa to South America, perhaps aboard slave ships. The range of *Ae. aegypti* is greater than that of the disease – the species is present in Australia and Asia, and also in southern United States, where it is expanding. However, only in India are there susceptible but as yet uninfected monkey hosts of the disease.

Other Togaviridae affecting humans and transmitted by mosquitoes cause dengue, dengue haemorrhagic fever, and a number of diseases called encephalites (or encephalitises) since inflammation of the brain occurs in clinical cases. Each togavirus has a preferred mosquito host, frequently an *Aedes (Stegomyia)* species such as *Ae. aegypti* for dengue, and often a *Culex* species for encephalitis. The reservoir hosts for these diseases vary, and include wild birds (at least for encephalitis), with amplification cycles in domestic mammals – for example, pigs for Japanese encephalitis. Horses may also be carriers, giving rise to the name for a subgroup of diseases termed equine encephalites.

A range of togaviruses are transmitted by ixodid ticks, including more viruses that cause encephalitis and haemorrhagic fevers of humans, but more significantly of domestic animals. Bunyaviruses may be tick-borne, notably haemorrhagic diseases of cattle and sheep, particularly when conditions encourage an explosion of tick numbers and disease alters from normal hosts (**enzootic**) to epidemic (**epizootic**) conditions. Mosquito-borne bunyaviruses include African Rift Valley fever, which can produce high mortality amongst African sheep and cattle during mass outbreaks.

Amongst the Reoviridae, bluetongue virus is the best known, most debilitating and most significant economically. The disease, which is virtually worldwide and has many different serotypes, causes tongue ulceration (hence 'bluetongue') and an often terminal fever in sheep. Bluetongue is one of the few diseases in which biting midges of *Culicoides* (Ceratopogonidae) have been clearly established as the sole vectors of an arbovirus of major significance, although many arboviruses have been isolated from these biting flies.

Studies of the epidemiology of arboviruses have been complicated by the discovery that some viruses may persist between generations of vector. La Crosse virus, a bunyavirus which causes encephalitis in the United States, can pass from the adult mosquito through the egg (**transovarial transmission**) to the larva, which overwinters in a near frozen tree-hole. The first emerging female of the spring generation is capable of transmitting La Crosse virus to chipmunk, squirrel or human with her first meal of the year. Transovarial transmission is suspected in other diseases and is substantiated in increasing numbers of cases, including Japanese encephalitis in *Culex tritaenorhynchus* mosquitoes.

14.5.3 Rickettsias and plague

Rickettsias are micro-organisms resembling viruses and bacteria in some features. The genus *Rickettsia* includes virulent pathogens of humans. *R. prowazekii*, which causes endemic typhus, has influenced world affairs as much as any politician, causing deaths of millions of refugees and soldiers in times of social upheaval, such as the years of the Napoleonic invasion of Russia and those following World War I. The vectors of typhus are lice (Box 14.2), notably the body louse *Pediculus humanus*. Infestation of lice indicates insanitary conditions and in western nations, after years of decline, it is resurgent in homeless people. Although both the head louse (*P. capitis*), pubic louse (*Pthirus pubis*) and some fleas can transmit *R. prowazekii* experimentally, they are of little or no epidemiological significance. After the rickettsias of *R. prowazekii* have multiplied in the louse epithelium, they rupture the cells and are voided in the faeces. Since the louse dies, the rickettsias are demonstrated to be rather poorly adapted to the louse host. Human hosts are infected by scratching infected louse faeces (which remain infective for up to two months after deposition) into the itchy site where the louse has fed. There is evidence of low level persistence of rickettsias in those who recover from typhus. These act as endemic reservoirs for resurgence of the disease, and there is some evidence that domestic and a few wild animals may be disease reservoirs. Lice are also vectors of relapsing fever, a spirochaete disease which historically occurred together with epidemic typhus.

Other rickettsial diseases include murine typhus, transmitted by flea vectors, scrub typhus through trombiculid mite vectors and a series of spotted fevers, termed tick-borne typhus, all with tick vectors. Many of these diseases have a wide range of natural hosts – antibodies to the widespread American Rocky Mountain spotted fever (*Rickettsia rickettsi*) are reported from 18 bird and 31 mammal species. Throughout the range of the disease from Virginia to Brazil, several species of ticks with catholic tastes are involved and transmission in this case is through feeding alone, without faecal involvement.

Plague is a rat–flea–rat disease caused by the bacterium *Yersinia pestis*. When humans become involved in plague outbreaks (such as the 'Black Death' of the 14th century in the northern hemisphere) mortality may approach 90% in undernourished people and around 25% in previously well-fed, healthy people. The epidemiological cycle commences amongst rats, with fleas (notably *Xenopsylla cheopis*) naturally transmitting *Y. pestis* between peri-domestic rats, such as the black rat *Rattus rattus*. The plague-bearing fleas may occur on other rodent hosts, including *Rattus norvegicus*, or the bandicoot *Bandicota bengalensis* in India, wild rodents such as ground squirrels (*Citellus* sp.) in the USA and Middle East or gerbils in South Africa and the Middle East. In an outbreak of plague, when the preferred host rats die, the fleas will move on to and in due course kill secondary preference rodent hosts, and eventually move to humans. Plague is a particular problem where rat and flea populations are high, as in overcrowded and insanitary urban conditions.

During its infective lifetime the flea varies in its ability to transmit plague, according to internal physiological changes induced by *Y. pestis*. If the flea takes an infected blood meal, *Y. pestis* increases in the proventriculus and midgut and forms an impassable plug. Further feeding involves a fruitless attempt by the pharyngeal pump to force more blood into the gut, with the result that a contaminated

maxillary palps

tracheae

(a)

(b)

haustellum (labrum, hypopharynx and labium together)

Fig. 14.2 A tsetse fly, *Glossina morsitans* (Diptera: Glossinidae): (a) at the commencement of feeding; (b) fully engorged with blood. Note that the tracheae are visible through the abdominal cuticle in (b). (After Burton and Burton, 1975.)

mixture of blood and bacteria is regurgitated. Even if the blockage is alleviated, the proventriculus now fails to function properly as a one-way valve, and at every subsequent feeding, the flea regurgitates a contaminated mixture of blood and pathogen into each successive host.

Natural variations in the intensity of plague epidemics have been related to the climate of the previous year. Even during plague outbreaks (which lasted for many years) there were periods of fewer cases. These occurred when hot, dry conditions prevented recruitment of fleas, because young fleas are very susceptible to desiccation, and low humidity reduced adult survival in the subsequent year. Bad plague years occurred when a high humidity winter was followed by a high humidity summer, almost independent of prevailing temperature.

14.5.4 Protozoans other than malaria

Protozoans are some of the most important insect-borne pathogens, affecting a substantial proportion of the world's population, particularly in subtropical and tropical areas. In addition to malaria (section 14.5.1), there are two important flagellate protozoans of medical significance.

(a) Trypanosoma

Trypanosoma is a large genus of vertebrate blood parasites that are usually transmitted by blood-feeding 'higher' flies, though trypanosomes that cause Chagas' Disease are transmitted by South American blood-feeding triatomine reduviid bugs, notably *Rhodnius prolixus*. The diseases termed **trypanosomiases** include sleeping sicknesses transmitted to African humans and their cattle by tsetse flies (*Glossina* species) (Fig. 14.2). In this and other diseases, the development cycle of *Trypanosoma* species is complex. Morphological change occurs in the protozoan as it migrates from the tsetse-fly gut, around the posterior free end of the peritrophic membrane, then anteriorly to the salivary gland. Transmission to a human or cattle host is through injection of saliva. Symptoms within the vertebrate depend upon the species of trypanosome: in humans, a vascular and lymphatic infection is followed by an invasion of the central nervous system that gives rise to 'sleeping' symptoms, followed by death.

(b) Leishmania

A second group of trypanosomes are the *Leishmanias*, which are parasites that cause internal visceral or disfiguring external ulcerating diseases of humans and dogs. The vectors are exclusively phlebotomines – small to minute sand flies that can evade mosquito netting and which, in view of their usually very low biting rates, have impressive abilities to transmit disease. Most cycles cause infections in wild animals such as desert and forest rodents, canines and hyraxes. Humans become involved as their homes expand into areas that are naturally home to these animal reservoirs.

14.5.5 Filariases

Two of the five main debilitating diseases transmitted by insects are caused by nematodes, namely filarial worms. The diseases are Bancroftian and Brugian filariases, commonly termed elephantiasis, and onchocerciasis or river-blindness. Other filariases cause minor ailments in humans; and *Dirofilaria immitis* (canine heartworm) is one of the few significant veterinary diseases caused by this type of parasite.

(a) Bancroftian and Brugian filariasis

Two worms, *Wuchereria bancrofti* and *Brugia malayi*, are responsible for more than a hundred million cases of filariasis worldwide. The worms live in the lymphatic system, causing debilitation and oedema, which culminates in the extreme swellings of lower limbs or genitals called elephantiasis. Although the disease is less often seen in the extreme form, the number of sufferers is growing as one major vector, the worldwide peri-domestic mosquito *Culex quinquefasciatus*, increases.

The cycle starts with uptake of small microfilariae with blood taken up by the vector mosquito. The microfilariae move from the mosquito gut through the haemocoel into the flight muscles, where they mature into an infective larva 1.5 mm long. The larvae migrate through the haemocoel into the mosquito head where, when the mosquito next feeds, they rupture the labella and invade the host through the puncture of the mosquito bite. In the human host the larvae mature slowly over many months. The sexes are separate, and pairing of mature worms must take place before further microfilariae are produced. These microfilariae cannot mature without the mosquito phase. Cyclical (nocturnal periodic) movement of microfilariae into the peripheral circulatory system may make them more available to feeding mosquitoes.

(b) Onchocerciasis

Onchocerciasis actually kills no one directly but debilitates millions of people by scarring their eyes, which leads to blindness. The common name of river-blindness refers to the impact of the disease on people living alongside rivers in West Africa and South America, where the insect vectors, *Simulium* black flies (Diptera: Simuliidae), live in flowing waters. The pathogen is a filarial worm, *Onchocerca volvulus*, in which the female is up to 50 mm long and the smaller male 20–30 mm. The adult filariae live in subcutaneous nodules and are relatively harmless. It is the microfilariae that cause the damage to the eye when they invade the tissues and die there. The major black-fly vector has been shown to be one of most extensive complexes of sibling species: *Simulium damnosum* has some 30 cytologically determined sibling species known from West and East Africa. The larvae, which are common filter feeders in flowing waters, are fairly readily controlled, but adults are strongly migratory and re-invasion of previously controlled rivers allows the disease to recur.

14.6 FORENSIC ENTOMOLOGY

Some flies develop in living flesh, with two waves discernible: primary colonizers that cause initial myiases, with secondary myiases developing in pre-existing wounds (section 14.3). A third wave may follow before death. This ecological succession results from changes in the attractiveness of the substrate to different insects. An analogous succession of insects can be seen in a corpse following death, with a somewhat similar course taken whether the corpse is pig, rabbit or human.

BOX 14.2
PHTHIRAPTERA (LICE)

An order of some 3000 species. Highly modified, apterous, dorsoventrally flattened ectoparasites, as in *Werneckiella equi*, the horse louse (Ischnocera: Trichodectidae) illustrated here. Classified in three suborders Amblycera and Ischnocera (the chewing or biting lice, formerly called Mallophaga) and Anoplura (sucking lice). Hemimetabolous; mouthparts mandibulate in Amblycera and Ischnocera and beak-like for piercing and sucking in Anoplura (Fig. 2.14); without maxillary palps in Anoplura; eyes either absent or reduced; antennae either held in grooves (Amblycera) or extended, filiform (and sometimes modified as claspers) in Ischnocera and Anoplura; thoracic segments variably fused, completely fused in Anoplura; legs well developed and stout with strong claw(s) used in grasping host hair or fur. Eggs laid on hair or feathers of the host. Nymphs resemble smaller, less pigmented adults.

Lice are obligate ectoparasites lacking any free-living stage and occurring on all orders of birds and most orders of mammals (with the notable exception of bats). Ischnocera and Amblycera feed on bird feathers and mammal skin, with some few amblycerans feeding on blood. Anoplura feed solely on mammal blood.

The degree of host specificity amongst lice is high and many monophyletic groups of lice occur on monophyletic groups of hosts. However, host speciation and parasite speciation do not match precisely, and historically many transfers have taken place between ecologically proximate but unrelated taxa (section 12.3.3).

VENTRAL VIEW OF HEAD

DORSAL VIEW OF LOUSE

BOX 14.2—*continued*

As with most parasitic insects, some Phthiraptera are involved in disease transmission. *Pediculus humanus*, the human body louse, is one vector of typhus (section 14.5.3). It is notable that the sibling species *P. capitis*, the human head louse (and *Phthirus pubis*, the pubic louse, illustrated on the right in the louse diagnosis in Table 1.3), are insignificant typhus vectors, although often co-occurring with the body louse.

This rather predictable succession in corpses has been used for medico-legal purposes by forensic entomologists as a faunistic method to assess the elapsed time (and even prevailing environmental conditions) since death for human corpses.

The generalized sequence of colonization is as follows. A fresh corpse is rapidly visited by a first wave of *Calliphora* (blow flies) and *Musca* (house flies), which oviposit or drop live larvae on to the cadaver. Their subsequent development to mature larvae (which leave the corpse to pupariate away from the larval site) is temperature-dependent. Given knowledge of the particular species, the larval development times at different temperatures, and the ambient temperature at the corpse, an estimate of the age of a corpse may be made, perhaps accurate to within half a day if quite fresh but with diminishing accuracy with increasing exposure.

As the corpse ages, larvae and adults of *Dermestes* (Coleoptera: Dermestidae) appear, followed by cheese skipper larvae (Diptera: Piophilidae). As the body becomes drier, it is colonized by a sequence of other dipteran larvae, including those of Drosophilidae (fruit flies) and *Eristalis* (Diptera: Syrphidae: the rat-tailed maggot, a hover fly). After some months, when the corpse is completely dry, more species of Dermestidae appear and several species of clothes moth (Lepidoptera: Tineidae) scavenge the desiccated remnants.

This simple outline is confounded by a number of factors including:

- Geography, with different insect species (though perhaps relatives) present in different regions, especially if considered on a continental scale.
- Variation in ambient temperatures, with direct sunlight and high temperatures speeding the succession (even leading to rapid mummification), and shelter and cold conditions retarding the process.
- Variation in exposure of the corpse; burial, even partial, slows the process considerably, and with a very different entomological succession.
- Variation in cause and site of death, with death by drowning and subsequent degree of exposure on the shore giving rise to a different necrophagous fauna from those infesting a terrestrial corpse, with differences between freshwater and marine stranding.

Despite these difficulties, in several cases entomological forensic evidence has proved crucial to criminal investigations.

BOX 14.3
SIPHONAPTERA (FLEAS)

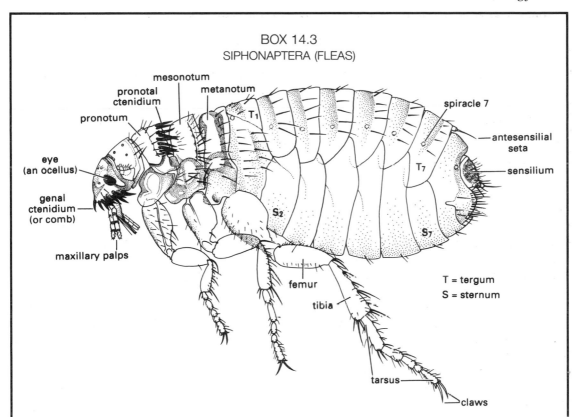

An order of some 2400 species. Highly modified, apterous, and laterally compressed, ectoparasites. Holometabolous; mouthparts (Fig. 2.15) modified for piercing and sucking, without mandibles, with elongate, serrate lacinial blade within sheath formed from maxillary and labial palps, epipharynx forming a long spine; gut with salivary pump to inject saliva into wound and cibarial and pharyngeal pumps to suck up blood; compound eyes absent, ocelli absent to well developed; antennae each lying in deep lateral groove; body with many backwardly-directed setae and spines, some may be grouped into combs (ctenidia) on head (gena) and thorax (especially prothorax); large metathorax houses hind leg muscles; legs long and strong, terminating in strong claws for grasping host hairs.

The large eggs are laid predominantly into the host's nest, where free-living, worm-like larvae (illustrated in Table 1.3) develop on material such as shed skin debris from the host. High temperatures and humidity are required for development by many fleas, including those on the domestic cat (*Ctenocephalides felis*; illustrated here), dog (*C. canis*) and human (*Pulex irritans*). The pupa is exarate and adecticous in a loose cocoon. Both sexes of adult take blood from a host, some species being **monoxenous** (restricted to one host) but many others being **polyxenous**. The plague flea *Xenopsylla cheopis* belongs to the latter group, with polyxeny facilitating transfer of plague from rat to human host (section 14.5.3). Fleas transmit some other diseases of minor significance from other mammals to humans, including murine typhus and tularemia, but, apart from plague, the most common human

BOX 14.3—*continued*

health threat is from allergic reaction to frequent bites from the fleas of pets, *C. felis* and *C. canis*.

Fleas predominantly use mammal hosts, with relatively few birds having fleas, these being derived from many lineages of mammal flea. Some hosts (e.g. *Rattus fuscipes*) have been reported to harbour more than 20 different species of flea, and conversely, some fleas have over 30 recorded hosts, so host specificity is clearly much less than for lice.

BOX 14.4
DIPTERA (FLIES)

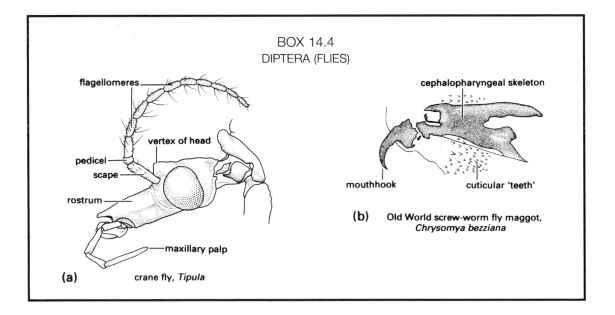

(a) crane fly, *Tipula*

(b) Old World screw-worm fly maggot, *Chrysomya bezziana*

BOX 14.4—*continued*

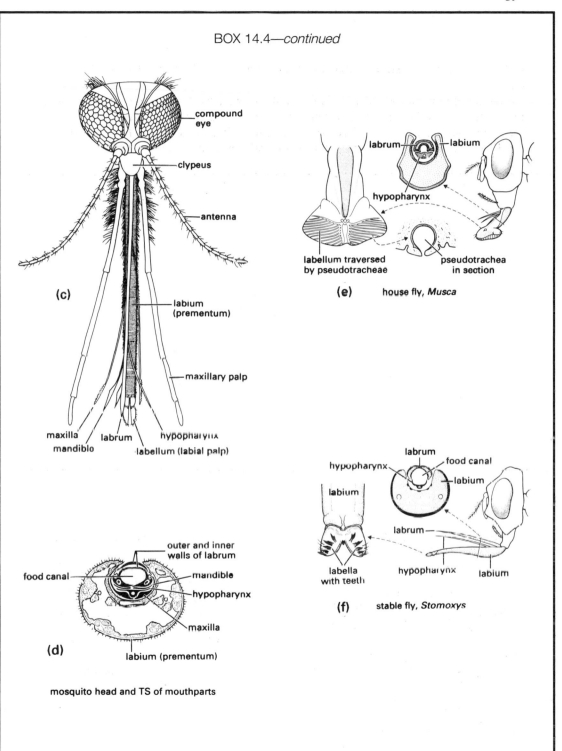

(c)

compound eye

clypeus

antenna

labium (prementum)

maxillary palp

maxilla
mandible
labrum
hypopharynx
labellum (labial palp)

(d)

outer and inner walls of labrum

food canal

mandible

hypopharynx

maxilla

labium (prementum)

mosquito head and TS of mouthparts

(e)   house fly, *Musca*

labrum

labium

hypopharynx

labellum traversed by pseudotracheae

pseudotrachea in section

(f)   stable fly, *Stomoxys*

hypopharynx

labrum

food canal

labium

labium

labrum

labella with teeth

hypopharynx

labium

BOX 14.4—*continued*

An order containing perhaps some 250 000 species, in some 120 families, including several thousand species of medical or veterinary importance. Holometabolous. Adults with variously modified mouthparts, mesothoracic wings and metathoracic halteres (balancers) (Fig. 2.22(f)). Larvae lacking true legs and head structure ranging from a complete sclerotized capsule to acephaly with no external capsule and only an internal skeleton. Pupae adecticous and obtect, or exarate in a puparium.

Two suborders are recognized: Nematocera, comprising crane flies, mosquitoes, midges, gnats and relatives, with slender antennae with upwards of six flagellomeres and 3–5-segmented maxillary palp, illustrated here ((a), after McAlpine, 1981) for a crane fly, *Tipula* (Tipulidae); and Brachycera, containing heavier-built flies including hover flies, bee flies, dung flies and blow flies, with more solid, often shorter, antennae with fewer than seven flagellomeres often with terminal arista (Fig. 2.17(i)), with a 1–2-segmented maxillary palp, illustrated here ((e),(f)) for Muscidae. Within the Brachycera, schizophoran Cyclorrhapha use a ptilinum to aid emergence from the puparium (section 6.2.1(c)).

Fly larvae have a wide variety of habits. Many nematoceran larvae are aquatic (Box 9.5), and brachyceran larvae show a phylogenetic radiation into drier and more specialized larval habits, including phytophagy, predation and parasitization of other arthropods, and myiasis-induction in vertebrates (section 14.3). Myiasis-inducing maggots have a very reduced head but with sclerotized mouthparts known as **mouthhooks** – illustrated here ((b), after Ferrar, 1987) for a third-instar larva of the Old World screw-worm fly *Chrysomya bezziana* (Calliphoridae) – that scrape the host's living flesh.

Adult dipteran mouthparts are illustrated here in frontal view ((c), after Freeman and Bracegirdle, 1971) and transverse section ((d), after Jobling, 1976) for a female mosquito. All dipterans typically have a tubular sucking organ, the **proboscis**, comprising elongate mouthparts (usually including the labrum). A biting-and-sucking type of proboscis appears to be a primitive dipteran feature. Although biting functions have been lost and regained with modifications more than once, blood-feeding is frequent, and leads to the importance of the Diptera as vectors of disease. The blood-feeding female nematocerans (Culicidae, mosquitoes; Ceratopogonidae, biting midges; Psychodidae: Phlebotominae, sand flies; and Simuliidae, black flies) have generally similar mouthparts but differ in proboscis length, allowing penetration of the host to different depths. Mosquitoes can probe deep in search of capillaries, but other blood-feeding nematocerans operate more superficially where a pool of blood is induced in the wound. The labium ends in two sensory **labella**, forming a protective sheath for the functional mouthparts. Enclosed are serrate-edged, cutting mandibles and maxillary lacinia, the curled labrum-epipharynx, and the hypopharynx, all of which are often termed **stylets**. When feeding, the labrum, mandibles and laciniae act as a single unit driven through the skin of the host. The flexible labium remains bowed outside the wound. Saliva, which may contain anti-coagulant, is injected through a salivary duct that runs the length of the sharply pointed and often toothed hypopharynx. Blood is transported up a food canal formed from the curled labrum sealed by either the paired mandibles or the hypopharynx. Capillary blood can flow unaided, but blood must be sucked or pumped from a pool with pumping action from two muscular pumps: the cibarial

BOX 14.4—*continued*

located at the base of the food canal and the pharyngeal in the pharynx between the cibarium and midgut.

Many mouthparts are lost in the 'higher' flies, and are modified for lapping food using pseudotracheae of the labella as 'sponges', illustrated here ((e), after Wigglesworth, 1964) for a house fly, *Musca* (Muscidae). With neither mandibles nor maxillary lacinia to make a wound, blood-feeding cyclorrhaphans often use modified labella, in which the inner surfaces are adorned with sharp teeth, illustrated here ((f), after Wigglesworth, 1964) for a stable fly, *Stomoxys* (Muscidae). Through muscular contraction and relaxation, the labellar lobes dilate and contract repeatedly, creating an often painful rasping of the labellar teeth to give a pool of blood. The hypopharynx applies saliva, and blood is ingested with the aid of a cibarial pump via a feeding tube lying in the labium.

The mouthparts and their use in feeding have implications for disease transmission. Shallow-feeding species such as black flies are more involved in transmission of microfilaria, such as *Onchocerca*, which aggregate just beneath the skin, whilst deeper feeders like mosquitoes transmit pathogens that circulate in the blood. The transmission from fly to host is aided by introduction of saliva into the wound and many parasites aggregate in the salivary glands or ducts. Filariae, in contrast, are too large to enter the wound through this route, and leave the insect host by rupturing the labium or labella during feeding.

FURTHER READING

Dye, C. (1992) The analysis of parasite transmission by bloodsucking insects. *Annual Review of Entomology*, **37**, 1–19.

Kettle, D.S. (1984) *Medical and Veterinary Entomology*, Croom Helm, London and Sydney.

Lane, R.P. and Crosskey, R.W. (eds) (1993) *Medical Insects and Arachnids*, Chapman and Hall, London.

Lehane, M.J. (1991) *Biology of Blood-sucking Insects*, Harper Collins Academic, London.

Smith, K.G.V. (1986) *A Manual of Forensic Entomology*, Trustees of the British Museum (Natural History), London.

15

PEST MANAGEMENT

Biological control of aphids by coccinellid beetles

Pests have no particular ecological significance. They are defined from a purely anthropocentric point of view and insects become pests when they conflict with human welfare, aesthetics or profits. They may be pests either directly through disease transmission (Chapter 14) or indirectly by affecting our domestic animals, cultivated plants or timber reserves. For example, otherwise innocuous insects can provoke severe allergic reaction in sensitized people, and reduction, or even loss, of food-plant yields is a universal result of insect feeding activities and pathogen transmission.

From a conservation perspective, introduced insects become pests when they displace native species, often with ensuing effects on other non-insect species in the community. For example, outside their native range honey bees, *Apis mellifera*, form feral nests and, although they are generalists, may outcompete local insects. Native insects are usually efficient pollinators of a smaller range of native plants than honey bees, and their loss may lead to reduced seed set. Some introduced and behaviourally dominant ants, such as the big-headed ant *Pheidole megacephala* and the Argentine ant *Linepithema humile*, which have displaced or decimated the native insect fauna in Bermuda and Hawaii, seriously threaten endemic species elsewhere. Unfortunately, research on insect pests of relevance to conservation biology is in its infancy in comparison with the vast literature on pests of our crops, garden plants and forest trees.

This chapter deals predominantly with the occurrence and management of insect pests of agriculture, including horticulture or silviculture, and with the control of insects that are of medical and veterinary importance. It commences with a discussion about what constitutes a pest, how damage levels are assessed, and why insects become pests. Next, the effects of insecticides and problems of insecticide resistance are considered, before an overview of integrated pest management

(IPM). The remainder of the chapter discusses the principles and methods of control applied in IPM: chemical control, including insect growth regulators and neuropeptides; biological control using natural enemies (such as the coccinellid beetles shown eating aphids in the vignette of this chapter) and microorganisms; host-plant resistance; mechanical, physical and cultural control; the use of attractants such as pheromones; and finally genetic control of insect pests.

15.1 INSECTS AS PESTS

15.1.1 Assessment of pest status

The pest status of an insect population depends on abundance of individuals as well as the type of nuisance or injury that the insects inflict. **Injury** is the usually deleterious effect of insect activities (mostly feeding) on host physiology, whereas **damage** is the measurable loss of host usefulness, such as yield quality or quantity or aesthetics. Host injury (or insect number used as an injury estimate) does not necessarily inflict detectable damage and even if damage occurs it may not result in appreciable economic loss. Sometimes the damage caused by even a few individual insects is unacceptable, as in fruit infested by codling moth or fruit fly. Other insects must reach high or plague densities before becoming pests, as in locusts feeding on pastures. Most plants tolerate considerable leaf or root injury without significant loss of vigour so that, unless these plant parts are harvestable (e.g. leaf or root vegetables) or are the reason for saleability (e.g. indoor plants), certain levels of insect feeding on these parts should be more tolerable than for fruit which 'sophisticated' consumers normally require to be blemish-free. Often the insect-feeding effects may be merely cosmetic, such as small marks on the fruit surface, and education of the consumers is more desirable than expensive controls. Since market competition demands high standards of appearance for food and other commodities, assessments of pest status often require socio-economic and not just biological judgments.

Pre-emptive measures are sometimes taken to counter the threat of arrival of particular novel insect pests. Generally, however, control becomes economic only when insect numbers cause (or are expected to cause if uncontrolled) financial loss of productivity or marketability greater than the costs of control. Quantitative measures of insect density are used to assess the pest status of different insect species associated with each agricultural crop. In each case, an **economic injury level (EIL)** is determined as the pest density at which the loss caused by the pest equals in value the cost of available control measures or, in other words, the lowest population density that will cause economic damage. The formula for calculating the EIL includes four parameters:

- Costs of control.
- The market value of the crop.
- The yield loss attributable to a unit number of insects.
- The effectiveness of the control.

The formula is: $EIL = C/VDK$, where EIL = pest numbers per production unit (e.g. insects/ha), C = cost of control measure(s) per production unit (e.g. \$/ha), V = market value per unit of product (e.g. \$/kg), D = yield loss per unit number of insects (e.g. kg reduction of crop/n insects), and K = proportionate reduction of insect population caused by control measures.

The calculated EIL will not be the same for different pest species on the same crop or for a particular insect pest on different crops. It may also depend on environmental conditions, such as soil type or rainfall, since these can affect plant vigour and compensatory growth. In most cases, control measures are instigated before the pest density reaches the EIL since there may be a time lag before the measures become effective. The density at which control measures should be applied to prevent an increasing pest population from attaining the EIL is referred to as the **economic threshold** (ET), or 'action threshold'. Although the ET is defined in terms of population density, it actually represents the time for instigation of control measures. It is set explicitly at a different level from the EIL and thus is predictive, with pest numbers being used as an index of the time when economic damage will occur.

Insect pests may be:

- Non-economic, if their populations are never above the EIL (Fig. 15.1(a)).
- Occasional pests, if their population densities exceed the EIL only under special circumstances (Fig. 15.1(b)), such as atypical weather or inappropriate use of insecticides.
- Perennial pests if the general population equilibrium of the pest is close to the economic threshold so that pest population density reaches the EIL frequently (Fig. 15.1(c)).
- Severe or key pests if their numbers, in the absence of controls, always are higher than the EIL (Fig. 15.1(d)). Severe pests must be controlled if the crop is to be grown profitably.

The EIL fails to consider the influence of variable external factors, including the role of natural enemies, resistance to insecticides and the effects of control measures in adjoining fields or plots. Nevertheless, the virtue of the EIL is its simplicity: management depends on the availability of decision rules that can be comprehended and implemented with relative ease. The concept of the EIL was developed primarily as a means for more judicious use of insecticides and its application is largely confined to situations in which control measures are discrete and curative, that is, chemical or microbial insecticides.

The discussion above applies principally to insects that directly damage an agricultural crop. For forest pests, estimation of almost all

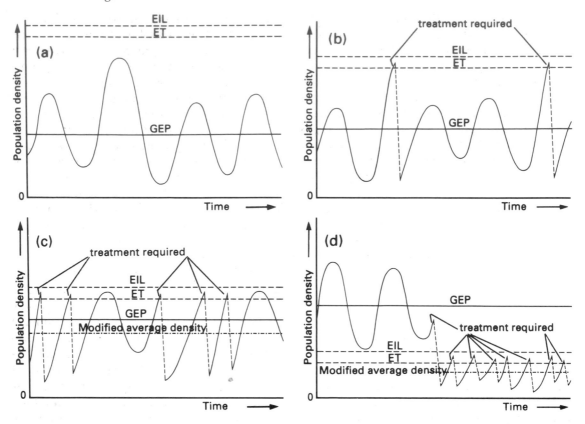

Fig. 15.1 Fluctuations of theoretical insect populations relative to their general equilibrium position (GEP), economic threshold (ET) and economic injury level (EIL). From comparison of the general equilibrium density with the ET and EIL, insect populations can be classified as: (a) non-economic pests if population densities never exceed the ET or EIL; (b) occasional pests if population densities exceed the ET and EIL only under special circumstances; (c) perennial pests if the general population equilibrium is close to the ET so that the ET and EIL are exceeded frequently; or (d) severe or key pests if population densities are always higher than the ET and EIL. In practice, as indicated here, control measures are instigated before the EIL is reached. (After Stern *et al.*, 1959.)

of the components of the EIL are difficult or impossible, and EILs are relevant only to short-term forest products such as Christmas trees. Furthermore, if insects are pests because they can act as vectors (transmitters) of disease of plants or animals, then the economic threshold may be as low as their first appearance. The threat of a virus affecting crops or livestock and spreading via an insect vector requires constant vigilance for the appearance of the vector and the presence of the virus. Precautions may need to be taken

with the first occurrence of either vector or disease symptoms. For economically very serious disease, and often in human health, precautions are taken before any economic threshold is reached, and insect vector and virus population monitoring and modelling are used to estimate when pre-emptive control is required. Calculations such as the vectorial capacity, referred to in Chapter 14, are important in allowing decisions concerning the need and appropriate timing for pre-emptive control measures. However, in human insect-borne disease, such rationales are often replaced by socio-economic ones, in which levels of vector insects that are tolerated in less developed countries or rural areas are perceived as requiring action in developed countries or in urban communities.

A final limitation of the EIL is its unsuitability for multiple pests, since calculations become complicated. However, if injuries from different pests produce the same type of damage, or if effects of different injuries are additive rather than interactive, then the EIL and ET may still apply. The ability to make management decisions for a pest complex (many pests in one crop) is an important part of integrated pest management (section 15.3).

15.1.2 Why insects become pests

Insects may become pests for one or more reasons. First, some previously harmless insects become pests after their accidental or intentional introduction to areas outside their native range, where they escape from the controlling influence of their natural enemies. Such range extensions have allowed many previously innocuous phytophagous insects to flourish as pests, usually following the deliberate spread of their host plants through human cultivation. Second, an insect may be harmless until it becomes a vector of a plant or animal (including human) pathogen. For example, the mosquito vectors of malaria and filariasis occur in Australia but the diseases are currently absent. Third, native insects may become pests if they move from native plants to introduced ones; such host-switching is common for polyphagous and oligophagous insects. For example, the oligophagous Colorado potato beetle switched from other solanaceous host plants to potato, *Solanum tuberosum*, during the nineteenth century (Box 15.1), and the polyphagous larvae of *Helicoverpa* and *Heliothis* (Lepidoptera: Noctuidae) have become serious pests of cultivated cotton and other crops within the native range of the moths.

A fourth, related, problem is that the simplified, virtually monocultural ecosystems in which our food crops, forest trees and livestock are grown create dense aggregations of predictably available resources that encourage the proliferation of specialist and some generalist insects. Certainly the pest status of many native noctuid caterpillars is elevated by the provision of abundant food resources. Moreover, natural enemies of pest insects generally require more diverse habitat or food resources and are discouraged from agro-monocultures. Fifth, in addition to large-scale monocultures, other farming or cultivating methods can lead to previously benign species or minor pests becoming major pests. Cultural practices, such as continuous cultivation without a fallow period, allows build up of insect pest numbers. The inappropriate or prolonged use of insecticides can eliminate natural enemies of phytophagous insects while inadvertently selecting for insecticide resistance in the latter. Released from natural enemies, other previously non-pest species sometimes increase in numbers until they reach economic thresholds. These problems of insecticide use are discussed in more detail below.

15.2 THE EFFECTS OF INSECTICIDES

The chemical insecticides developed during and after the Second World War were initially effective and cheap. Farmers came to rely on the new chemical methods of pest control, which rapidly replaced traditional forms of chemical, cultural and biological control. The 1950s and 1960s were times of an insecticide boom, but use continued to rise and is still the single main pest control method employed today. Although pest populations are suppressed by insecticide use, undesirable effects include:

(1) Selection for insects that are genetically resistant to the chemicals (section 15.2.1).

(2) Destruction of non-target organisms, including pollinators, the natural enemies of the pests and soil arthropods.

(3) **Pest resurgence** – as a consequence of (1) and (2), a dramatic increase in numbers of the targeted pest can occur (e.g. severe outbreaks of cottony-cushion scale as a result of DDT use in California in the 1940s; Box 15.4) and if the natural enemies recover much more slowly than the pest population, the latter can exceed levels found prior to insecticide treatment.

(4) **Secondary pest outbreak** – a combination of suppression of the original target pest and effects (1) and (2) can lead to insects previously not considered pests being released from control and becoming major pests.

(5) Adverse environmental effects, resulting in contamination of soils, water systems and the produce itself with chemicals that accumulate biologically (especially in vertebrates) as the result of biomagnification through food chains.

(6) Dangers to human health either directly from the handling and consumption of insecticides or indirectly via exposure to environmental sources.

Despite increased insecticide use, damage by insect pests has increased disproportionately; for example, insecticide use in the United States increased ten-fold from about 1950 to 1985, whilst the proportion of crops lost to insects about doubled (from 7% to 13%) during the same period. Such figures do not mean that insecticides have not controlled insects because non-resistant insects clearly are killed by chemical poisons. Rather, an array of factors accounts for this imbalance between pest problems and control measures. Human trade has accelerated the spread of pests to areas outside the ranges of their natural enemies. Selection for high crop yield often inadvertently has resulted in crops susceptible to insect pests. Extensive monocultures are commonplace, with reduction in

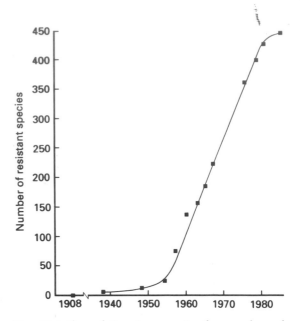

Fig. 15.2 Cumulative increase in the number of arthropod species (mostly insects and mites) known to be resistant to one or more insecticides. (After Georghiou, 1986.)

sanitation and other cultural practices such as crop rotation. Finally, aggressive commercial marketing of chemical insecticides has led to their inappropriate use, especially in developing countries.

15.2.1 Insecticide resistance

Insecticide **resistance** is the result of selection of individuals that are genetically predisposed to survive an insecticide. **Tolerance**, the ability of an individual to survive an insecticide, implies nothing about the basis of survival. Over the past few decades about 500 species of arthropod pests have developed resistance to one or more insecticides (Fig. 15.2). The Colorado potato beetle (Box 15.1) is resistant to almost all the chemicals available for control, and chemically-based pest control of this and many other pests may soon become ineffectual. Many insect species show cross or multiple resistance. **Cross resistance** is the phenomenon of a resistance mechanism

BOX 15.1
THE COLORADO POTATO BEETLE

Leptinotarsa decemlineata (Coleoptera: Chrysomelidae), commonly known as the Colorado potato beetle, is a striking beetle (illustrated here after Stanek, 1969) which has become a major pest of cultivated potatoes in the northern hemisphere. Originally probably native to Mexico, it expanded its host range about a century and a half ago and then spread into Europe from North America in the 1920s. Its present hosts are about 20 species in the family Solanaceae, especially *Solanum* species and in particular *S. tuberosum*, the cultivated potato. Other occasional hosts include *Lycopersicon esculentum*, the cultivated tomato. The adult beetles are attracted by volatile chemicals released by the leaves of *Solanum* species on which they feed and lay eggs. Female beetles live for about two months, in which time each can lay a few thousand eggs. Larvae defoliate potato plants (as illustrated here) resulting in yield losses of up to 100% if damage occurs prior to tuber formation. The Colorado potato beetle is the most important defoliator of potatoes and control measures are necessary if crops are to be grown successfully where it is present.

Insecticides effectively controlled the Colorado potato beetle until it developed resistance to DDT in the 1950s. Since then the beetle has developed resistance to each new insecticide (including synthetic pyrethroids) at progressively faster rates. Currently, many beetle populations are resistant to nearly all registered insecticides. Feeding can be inhibited by applying antifeedants – including neem products

Colorado potato beetle larvae and adult on potato foliage

larva   adult

(Box 15.3) and certain fungicides – to leaf surfaces; however, deleterious effects on the plants and slow suppression of beetle populations has made antifeedants unpopular. Cultural control, via rotation of crops, delays

BOX 15.1—*continued*

infestation of potatoes and can reduce the build-up of early season beetle populations. Diapausing adults mostly overwinter in the soil of fields where potatoes were grown the previous year; they are slow to colonize new fields because much post-diapause dispersal is by walking, but populations of second-generation beetles may or may not be reduced in size compared with those in non-rotated crops. Attempts to produce potato varieties resistant to the Colorado potato beetle have failed to combine useful levels of resistance (either from chemicals or glandular hairs) with a commercially suitable product. Even biological control has been unsuccessful because known natural enemies generally do not reproduce rapidly enough nor individually consume sufficient prey to regulate populations of the Colorado potato beetle effectively, and most natural enemies cannot survive the cold winters of temperate potato-growing areas. However, mass rearing and augmentative releases of certain predators (e.g. two species of pentatomid bugs) and an egg parasitoid (a eulophid wasp) may provide substantial control. Microbial control using sprays of bacterial insecticides can be effective if applications are timed to target the vulnerable early-instar larvae. Two strains of the bacterium *Bacillus thuringiensis* produce toxins which kill the larvae of Colorado potato beetle. Recently, the bacterial genes responsible for producing the toxin of *B. thuringiensis* var. *san diego* (= *B. t.* subsp. *tenebrionis*) have been genetically engineered into potato plants by inserting the genes into another bacterium, *Agrobacterium tumefaciens*, which is capable of inserting its DNA into that of the host plant. Remarkably, these transgenic potato plants are resistant to both adult and larval stages of the Colorado potato beetle. Only time will tell how rapidly the Colorado potato beetle will develop resistance to these 'new' toxins.

---

for one insecticide giving tolerance to another. **Multiple resistance** is the occurrence in a single insect population of more than one defence mechanism against a given compound. Mechanisms of insecticide resistance include:

- Increased behavioural avoidance, since some insecticides such as neem and pyrethroids can repel insects.
- Physiological changes, such as sequestration (deposition of toxic chemicals in specialized tissues), reduced cuticular permeability (penetration) or accelerated excretion.
- Biochemical detoxification (called **metabolic resistance**).

- Increased tolerance due to decreased sensitivity to the presence of the insecticide at its target site (called **target-site resistance**).

The tobacco budworm, *Heliothis virescens* (Lepidoptera: Noctuidae), a major pest of cotton in the United States, exhibits behavioural, penetration, metabolic and target-site resistance. Phytophagous insects, especially polyphagous ones, frequently develop resistance more rapidly than their natural enemies. Perhaps polyphagous herbivores are preadapted to evolve insecticide resistance because they have general detoxifying mechanisms for secondary compounds encountered among their host plants. Resistance has developed under field conditions in some arthropod natural enemies (e.g. some lacewings, parasitic wasps and predatory mites), although few have been tested. In some species, intraspecific variability in insecticide tolerances among populations subjected to differing insecticide doses has been found.

Insecticide resistance in the field is based on relatively few or single genes (monogenic resistance), that is, allelic variants at just one or two loci. Field applications of chemicals designed to kill all individuals leads to rapid evolution of resistance. This is because strong selection favours novel variation, such as a very rare allele for resistance present at a single locus. In contrast, laboratory selection is often weaker, producing polygenic resistance. Single-gene insecticide resistance could also be due to the very specific modes of action of certain insecticides, which allow small changes at the target site to confer resistance.

Management of insecticide resistance requires a programme of controlled use of chemicals with the primary goals of avoiding or slowing the development of resistance in pest populations, causing resistant populations to revert to more susceptible levels, and/or fostering resistance in selected natural enemies. The tactics for resistance management can involve maintaining reservoirs of susceptible pest insects (either in refuges or by immigration from untreated areas) to promote dilution of any resistant genes, varying the dose or frequency of insecticide applications, using less persistent chemicals, and/or applying insecticides as a rotation or sequence of different chemicals or as a mixture. By far the best strategy for retarding the evolution of resistance is to use insecticides only when control by natural enemies fails to curtail economic damage. Resistance monitoring is an integral component of management, as it allows the anticipation of problems and assessment of the effectiveness of operational management tactics.

Recognition of the problems discussed above, cost of insecticides and also a strong consumer backlash against environmentally damaging agronomic practices and chemical contamination of produce, have led to the current development of alternative pest control methods. In some countries and for certain crops, chemical controls increasingly are being integrated with, and sometimes replaced by, other methods.

15.3 INTEGRATED PEST MANAGEMENT (IPM)

Historically, **integrated pest management (IPM)** was first promoted during the 1960s as a result of the failure of chemical insecticides, notably in cotton production in which some regions required at least 12 sprayings per crop. IPM philosophy is to limit economic damage to the crop and simultaneously minimize adverse effects on non-target organisms in the crop and surrounding environment and on consumers of the produce (Box 15.2). Successful IPM requires a thorough knowledge of the biology of the pest insects, their natural enemies and the crop, in order to allow rational use of a variety of cultivation and control techniques under differing circumstances. The factors that regulate populations of insects and other organisms are varied and interrelated in complex ways, so that successful IPM requires an understanding of both population processes (e.g. growth and reproductive capabilities, competition and effects of predation and parasitism) and the effects of environmental factors (e.g. weather, soil conditions, disturbances such as fire, and availability of water, nutrients and shelter), some of which are largely stochastic in nature and may have predictable or unpredictable effects on insect populations.

Despite the economic and environmental advantages of IPM, implementation of IPM programmes has been slow. Reasons include:

(1) Lack of sufficient data on the ecology of many insect pests and their natural enemies.

BOX 15.2
IPM FOR RICE IN INDONESIA

**Pesticide subsidies and usage and rice production
in Indonesia, 1973-1990**

macropterous adult female
of the brown planthopper,
Nilaparvata lugens

An outstandingly successful example of integrated pest management in the tropical world is the Indonesian National IPM Program for rice. Asia's most serious pest of rice (the staple carbohydrate crop) is the brown planthopper, *Nilaparvata lugens* (Hemiptera: Delphacidae; illustrated here after Wilson and Claridge, 1991), which sucks plant sap causing the leaves to go brown and die ('hopperburn'). The rise to pest status of the brown planthopper (BPH) coincided with the widespread cultivation of modern, high-yielding rice varieties and the concomitant increased use of nitrogenous fertilizers and insecticides. It is now clear that indiscriminate insecticide applications released BPH from biological control by natural predators and parasitoids. These include certain spiders that commonly catch both adult and nymphal BPH, veliid bugs (Hemiptera: Veliidae: *Microvelia* and *Mesovelia*) that consume any BPH falling onto the water surface in the rice crop, a mirid egg predator (Hemiptera: Miridae: *Cyrtorhinus*) that locates BPHs by a kairomone they release, and parasitoids of both BPH eggs and nymphs. Predators and parasitoids are killed by chemicals and their populations are slow to recover compared with those of BPH, which rebound when free from enemies – a classic case of pest resurgence. To make matter worse, heavy insecticide use encouraged resistant strains of the BPH.

Until 1986, the major non-insecticidal strategy for control of BPH was the development and use of resistant rice cultivars. Then, with the threat of collapse of the rice industry, Indonesia's President Suharto issued a presidential decree banning 57 of the 63 insecticides

BOX 15.2—*continued*

(mostly toxic organophosphates) used on rice and began diverting millions of dollars from pesticide subsidies to the training of farmers and government field staff in IPM. The Indonesian Government in conjunction with the United Nations Food and Agriculture Organization (FAO) have been responsible for introducing large-scale IPM to Indonesian farmers. IPM training involves active farmer participation in every step of the process, which teaches people how to grow a healthy rice crop, how to recognize and conserve beneficial natural enemies, and to inspect the crop weekly in order to make management decisions based on integrating information from all aspects of crop, pest and natural enemy biology. The benefits to Indonesia of the National IPM Program for rice are enormous. As shown here in the graph (after Anon., 1992), Indonesia saves US$120 million per annum in insecticide subsidies and insecticide use has gone down (with a drop of about 60% in pesticide formulation since 1986) whilst rice production has gone up 13% in the same period. Insecticide-induced pest outbreaks have ceased. Farmers save on production costs and profit from increased yields. Long-term and short-term environmental damage, human occupational poisoning and consumer health risks all are reduced. Similar benefits from large-scale IPM are possible in other Asian countries where rice is grown as a continuous monoculture. The potential of biological control is much higher in tropical than in temperate countries due to high arthropod diversity and year-round activity of natural enemies.

(2) The requirement for knowledge of economic injury levels for each pest of each crop.
(3) The requirement for interdisciplinary research in order to obtain the information mentioned in (1) and (2).
(4) Risks of pest damage to crops associated with IPM strategies.
(5) The apparent simplicity of total insecticidal control combined with the marketing pressures of pesticide companies.
(6) The necessity of training farmers, agricultural extension officers, foresters and others in new principles and methods.

Successful IPM often requires extensive biological research. Such applied research is unlikely to be financed by many industrial companies because IPM may reduce their insecticide market. However, IPM does incorporate the use of chemical insecticides, albeit at a lower level, although its main focus is the establishment of a variety of other methods of controlling insect pests. These usually involve modifying the insect's physical or biological environment; more rarely, they entail changing the genetic properties of the insect (section 15.10). The remainder of this chapter discusses the various principles and methods of insect pest control that could be employed in IPM programmes.

15.4 CHEMICAL CONTROL

Despite the hazards of conventional insecticides, some use is unavoidable but careful chemical choice and application can reduce ecological damage. Carefully timed suppressant doses can be delivered at vulnerable stages of the pest's life cycle or when a pest population is about to explode in numbers. Appropriate and efficient use requires a thorough knowledge of the pest's field biology and an appreciation of the differences among available insecticides.

An array of chemicals has been developed for the purpose of killing insects. The chemicals enter the insect body either by penetrating the cuticle (called contact action or dermal entry), or by inhalation into the tracheal system, or by oral ingestion into the digestive system. Most contact poisons also act as stomach poisons if ingested by the insect, and toxic chemicals that are ingested by the insect after translocation through a host are referred to as **systemic insecticides**. Fumigants used for controlling insects are inhalation poisons. Some chemicals may act simultaneously as inhalation, contact and stomach poisons. Chemical insecticides generally have an acute effect and their mode of action (i.e. method of causing death) is via the nervous system, either by inhibiting acetylcholinesterase (an essential enzyme for transmission of nerve impulses at synapses) or by acting directly on the nerve cells. Most synthetic insecticides (including pyrethroids) are nerve poisons. Other insecticidal chemicals affect the developmental or metabolic processes of insects either by mimicking or interfering with the action of hormones or by affecting the biochemistry of cuticle production.

15.4.1 Insecticides (chemical poisons)

Chemical insecticides may be synthetic or natural products. Natural plant-derived products include:

- **Alkaloids** including nicotine and nornicotine from tobacco.
- **Rotenone** and other **rotenoids** from roots of legumes.
- **Pyrethrins** derived from flowers of *Tanacetum cinerariifolium* (formerly in *Pyrethrum* and then *Chrysanthemum*).
- **Neem** extracts of the tree *Azadirachta indica* (these have a long history of use as insecticides).

Insecticidal alkaloids have been used since the seventeenth century and pyrethrum since at least the early nineteenth. Although nicotine-based insecticides have been phased out for reasons including high mammalian toxicity and limited insecticidal activity, the other categories remain important in pest control. Rotenoids will poison insects and fish, and must be kept out of waterways. Neem derivatives (section 15.4.2; Box 15.3) act as feeding poisons for most nymphs and larvae as well as altering behaviour and disrupting normal development. Pyrethrins (and the structurally related synthetic **pyrethroids**) are especially effective against lepidopteran larvae, kill on contact even at low doses and have low environmental persistence. An advantage of most pyrethrins and pyrethroids, and also neem derivatives, is their much lower mammalian and avian toxicity compared with synthetic insecticides, although pyrethroids are highly toxic to fish. A number of pest insects have already developed resistance to pyrethroids.

The other major classes of insecticides have no natural analogues. These are the synthetic **carbamates** (e.g. aldicarb, carbaryl, carbofuran, methiocarb, methomyl, propoxur), **organophosphates** (e.g. chlorpyrifos, dichlorvos, dimethoate, malathion, parathion, phorate) and **organochlorines** (e.g. aldrin, chlordane, DDT, dieldrin, endosulfan, gamma-BHC (lindane), heptachlor). Certain organochlorines (e.g. aldrin, chlordane, dieldrin and heptachlor) are known as **cyclodienes** because of their chemical structure.

Most synthetic insecticides are **broad spectrum** in action – that is, they have non-specific killing action – and most act on the nervous system of the insect (and, incidentally, of mammals). Organochlorines are chemically stable and thus persistent in the environment; they have a low solubility in water but a moderate solubility in organic solvents, and accumulate in mammalian body fat. Their use is banned in many countries and they are unsuitable for use in IPM. Organophosphates may be highly toxic to mammals but are not stored in fat and, being less environmentally damaging and non-persistent, are suitable for IPM. They usually kill insects by contact or upon ingestion, although some are systemic in action, being absorbed into the vascular system of plants so that they kill most phloem-feeding insects. Non-persistence means that their application must be timed carefully to ensure efficient kill of pests. Carbamates usually act by contact or stomach action, more rarely by systemic action, and have short to medium persistence.

In addition to the chemical and physical properties of insecticides, their toxicity, persistence in the field and method of application are influenced by how they are formulated. **Formulation** refers to what and how other substances are mixed with the active ingredient, and largely constrains the mode of application. Insecticides may be formulated in various ways, including as solutions or emulsions, as unwettable powders that can be dispersed in water, as dusts or granules (i.e. mixed with an inert carrier), or as gaseous fumigants. Formulation may include baits that attract the insects or abrasives that damage the cuticle. The same insecticide can be formulated in different ways according to the application requirements, such as aerial spraying of a crop versus domestic use.

15.4.2 Insect growth regulators

Insect growth regulators (IGRs) are compounds that affect insect growth via interference with metabolism or development. They offer a high level of efficiency against specific stages of many insect pests, with a low level of mammalian toxicity. The two most commonly used groups of IGRs are distinguished by their mode of action. Chemicals that interfere with the normal maturation of insects by disturbing the hormonal control of metamorphosis are fenoxycarb and the **juvenile hormone mimics**, or **juvenoids** (e.g. hydroprene, methoprene). These halt development so that the insect either fails to reach the adult stage or the resulting adults

are sterile and malformed. Since juvenoids deleteriously affect adult rather than immature insects, their use is most appropriate to species in which the adult rather than the larva is the pest, such as fleas, mosquitoes and ants.

The **chitin synthesis inhibitors** (e.g. diflubenzuron, triflumuron) prevent the formation of chitin, a vital component of insect cuticle. Many conventional insecticides cause a weak inhibition of chitin synthesis, but the benzoylphenylureas (also known as acylureas, of which diflubenzuron and triflumuron are examples) strongly inhibit formation of cuticle. Insects exposed to chitin synthesis inhibitors usually die at or immediately after ecdysis. Typically the affected insects shed the old cuticle partially or not all and, if they do succeed in escaping from their exuviae, their body is limp and easily damaged due to the weakness of the new cuticle. IGRs, which are quite persistent indoors, usefully control insect pests in storage silos and domestic premises. Typically, juvenoids are used in urban pest control and inhibitors of chitin synthesis have greatest application in controlling beetle pests of stored grain.

Neem derivatives are another group of growth-regulatory compounds with significance in insect control (Box 15.3). Their ingestion, injection or topical application disrupts moulting and metamorphosis – the effect depends on the insect and the concentration of chemical applied. Treated larvae or nymphs either fail to moult or the moult results in abnormal individuals in the subsequent instar; treated late-instar larvae or nymphs generally produce deformed and non-viable pupae or adults. These physiological effects of neem derivatives are not fully understood but are believed to result from interference with endocrine function; in particular, the main active principle of neem, azadirachtin (AZ), may act as an anti-ecdysteroid by blocking binding sites for ecdysteroid on the protein receptors. AZ may inhibit moulting in insects by preventing the

BOX 15.3
NEEM

The neem tree, *Azadirachta indica* (family Meliaceae), is native to tropical Asia but has been planted widely in the warmer parts of Africa, Central and South America and Australia. It is renowned, especially in India and some areas of Africa, for its anti-insect properties. For example, pressed leaves are put in books to keep insects away and bags of dried leaves are placed in cupboards to deter moths and cockroaches. Extracts of neem seed kernels and leaves act as repellents, antifeedants and growth disruptants. The brown kernels, illustrated here below the complete seeds (after Schmutterer, 1990) are the most important source of the active principles that affect insects. Leaves, also illustrated here (after Corner, 1952), are a secondary source. The main active principle in kernels is **azadirachtin** (**AZ**), a limonoid, but a range of other active compounds is also present. Various aqueous and alcoholic extracts of kernels, neem oil and pure AZ have been tested for their effects on many insects. These neem derivatives can repel, prevent settling and inhibit oviposition, inhibit or reduce food intake and interfere with the regulation of growth, as discussed in section 15.4.2, as well as reduce the fecundity, longevity and vigour of adults. In lepidopteran species, AZ seems to reduce the feeding activity of oligophagous species more than polyphagous ones. The antifeedant (phagodeterrent) action of neem apparently has both a gustatory (regulated by sensilla on the mouthparts) and a non-gustatory component, since injected or topically applied neem

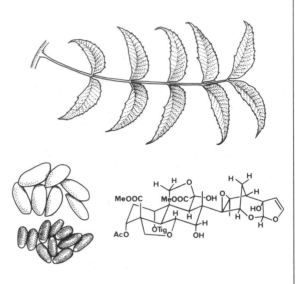

derivatives can reduce feeding even though the mouthparts are not affected directly.

Neem-based products appear effective under field conditions against a broad spectrum of pests, including phytophagous insects of most orders (such as Hemiptera, Coleoptera, Diptera, Lepidoptera and Hymenoptera), stored-product pests, certain pests of livestock, and some mosquito vectors of human disease. Fortunately, honey bees and many predators of insect pests, such as spiders and coccinellid beetles, are less susceptible to neem, making it very suitable for IPM. Furthermore, neem derivatives are non-toxic to warm-blooded vertebrates. Unfortunately, the complex structures of limonoids such as AZ (illustrated here, after Schmutterer, 1990) preclude their economical chemical synthesis, but they are readily available from plant

---

---

usual moult-initiating rise in ecdysteroid titre. Cuticle structures known to be particularly sensitive to ecdysteroids develop abnormally at low doses of AZ.

Two other types of IGRs that currently have little potential in insect control are the anti-juvenile hormone analogues (e.g. precocenes) and ecdysteroid hormones. Anti-juvenile hormones accelerate termination of the immature stages and negate the action of juvenile hormones when the two are administered together, but insufficient activity prevents their practical use. Ecdysteroids are essential to the moulting process of insects and, although their use as IGRs has been considered, they are expensive and have limited insecticidal effect when applied topically or orally.

15.4.3 Neuropeptides and insect control

Insect neuropeptides are small oligopeptides which regulate most aspects of development, metabolism, homeostasis and reproduction. Their diverse functions have been discussed in Chapter 3. Although neuropeptides are unlikely to be used as insecticides *per se*, knowledge of their chemistry and biological actions can be applied in novel approaches to insect control. Neuroendocrine manipulation involves disrupting one or more of the steps of the general hormone process of synthesis, secretion, transport, action and degradation. For example, secretion of a neuropeptide could be interfered with by developing an agent to block or overstimulate at the release site. Alternatively the peptide-mediated response at the target tissue could be blocked or overstimulated by a peptide mimic that penetrates either the gut or the cuticle. Furthermore, the protein nature of neuropeptides makes them amenable to control using recombinant DNA technology and genetic engineering. However, neuropeptides produced by transgenic crop plants or bacteria that express neuropeptide genes may be degraded in the gut of pest insects, and manipulation of insect viruses appears to be more promising for control. Neuropeptide or 'anti-neuropeptide' genes could be incorporated into the genome of insect-specific viruses which would then act as expression vectors of the genes to produce and release the insect hormone(s) within infected insect cells. Baculoviruses have the potential to be used in this way, especially in Lepidoptera. Normally such viruses cause slow or limited mortality in their host insect (section 15.5.2(d)), but their efficacy might be improved by creating an endocrine imbalance that kills infected insects more quickly or increases viral-mediated mortality among infected insects. An advantage of neuroendocrine manipulation is that some neuropeptides may be insect- or arthropod-specific – a property that would reduce deleterious effects on many non-target organisms.

15.5 BIOLOGICAL CONTROL

Regulation of the abundance and distribution of species is influenced strongly by the activities of naturally occurring enemies, namely predators, parasites and parasitoids, pathogens and competitors. In most managed ecosystems these biological interactions are

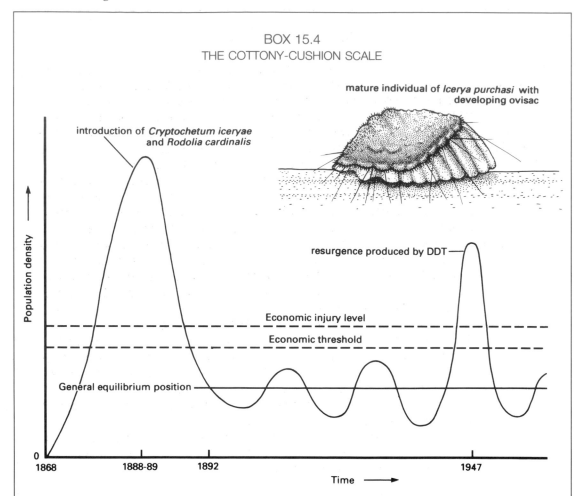

BOX 15.4
THE COTTONY-CUSHION SCALE

introduction of *Cryptochetum iceryae*
and *Rodolia cardinalis*

mature individual of *Icerya purchasi* with
developing ovisac

resurgence produced by DDT

Economic injury level

Economic threshold

General equilibrium position

Population density

0
1868   1888-89   1892   1947

Time

An example of a spectacularly successful classical biological control programme is the control of infestations of the cottony-cushion scale, *Icerya purchasi* (Hemiptera: Margarodidae) in Californian citrus orchards from 1889 onwards, as illustrated in the accompanying graph (after Stern *et al.*, 1959). Control has been interrupted only by DDT use, which killed natural enemies and allowed resurgence of cottony-cushion scale.

The hermaphroditic, self-fertilizing adult of this scale produces a characteristic fluted white ovisac (inset on graph), under which several hundred eggs are laid. This mode of reproduction, in which a single immature individual can establish a new infestation, combined with polyphagy and capacity for multivoltinism in warm climates, makes the cottony-cushion scale a potentially serious pest. In Australia, the insect's country of origin, populations are kept in check by natural enemies, especially ladybird beetles (Coleoptera: Coccinellidae) and parasitic flies (Diptera: Cryptochetidae).

Cottony-cushion scale was first noticed in the United States in about 1868 on a wattle (*Acacia*) growing in a park in northern California. By 1886, it was devastating the new and expanding citrus industry in southern California. Initially the native home of this pest was

BOX 15.4—*continued*

unknown but correspondence between entomologists in America, Australia and New Zealand established Australia as the source. The impetus for introduction of exotic natural enemies came from C.V. Riley, Chief of the Division of Entomology of the United States Department of Agriculture, who arranged for A. Koebele to collect natural enemies in Australia and New Zealand (1888–1889) and ship them to D.W. Coquillett for rearing and release in Californian orchards. Koebele obtained many cottony-cushion scales infected with flies of *Cryptochetum iceryae* and also coccinellids of *Rodolia cardinalis*, the vedalia ladybird. Mortality during several shipments was high and only about 500 vedalia beetles arrived alive; these were bred and distributed to all Californian citrus growers with outstanding results. The vedalia beetles ate their way through infestations of cottony-cushion scale, the citrus industry was saved and biological control became popular.

The parasitic fly was largely forgotten in these early days of enthusiasm for coccinellid predators. Thousands of flies were imported as a result of Koebele's collections but establishment from this source is doubtful. Perhaps the major or only source of the present populations of *C. iceryae* in California was a batch sent in late 1887 by F. Crawford of Adelaide, Australia, to W.G. Klee, the California State Inspector of Fruit Pests, who made releases near San Francisco in early 1888, before Koebele ever visited Australia.

Today both *R. cardinalis* and *C. iceryae* control populations of *I. purchasi* in California. The beetle is dominant in the hot, dry inland citrus areas and the fly most important in the cooler coastal region; interspecific competition can occur if conditions are suitable for both species. The vedalia beetle, and to a lesser extent the fly, were also introduced successfully to many countries worldwide wherever *I. purchasi* became a pest. Both predator and parasitoid have proved to be effective regulators of cottony-cushion scale numbers, presumably due to their specificity and efficient searching ability, aided by the limited dispersal and aggregative behaviour of their target scale insect. Unfortunately, few subsequent biological control programmes involving coccinellids have enjoyed the same success.

severely restricted or disrupted in comparison with natural ecosystems, and certain species escape from their natural regulation to become pests. In **biological control**, deliberate human intervention attempts to restore some balance by introducing or enhancing the natural enemies of target organisms such as insect pests or weed plants.

One advantage of natural enemies is their host specificity but a drawback (shared with other control methods) is that they do not eradicate pests. Thus biological control may not necessarily alleviate all the economic consequences of pests, but control programmes are expected to reduce the abundance of a target pest to below economic threshold levels. In the case of weeds, natural enemies include phytophagous insects; biological control of weeds is discussed in section 10.2.6. Several approaches to biological control are recognized but these categories are not discrete and published definitions vary widely, leading to some confusion. Such overlap is recognized in the following summary of the basic strategies of biological control.

BOX 15.5

TAXONOMY AND BIOCONTROL OF THE CASSAVA MEALYBUG

Cassava (manioc, or tapioca – *Manihot esculenta*) is a staple food crop for 200 million Africans. In 1973 a new mealybug was found attacking cassava in central Africa. Named in 1977 as *Phenacoccus manihoti*, this pest spread rapidly until by the early 1980s it was causing production losses of over 80% throughout tropical Africa. The origin of the mealybug was considered to be the same as the original source of cassava – the Americas. In 1977, what was apparently the same insect was located in Central America and northern South America, along with its parasitic wasps. However, as biological control agents the latter failed to reproduce on the African mealybugs.

Working with existing collections and fresh samples, taxonomists quickly recognized that two closely related mealybug species were involved. The one infesting African cassava proved to be from central South America, rather than further north. When the search for natural enemies was switched to this region, the true *P. manihoti* was eventually found in the Paraguay basin, together with an encyrtid wasp, *Apoanagyrus* (formerly known as *Epidinocarsis*) *lopezi* (J.S. Noyes, personal communication). This wasp gave spectacular biological control when released in Nigeria, and by 1990 had been established successfully in 25 African countries and had spread to more than 2.7 million km^2. The mealybug is now considered to be under almost complete control throughout its range in Africa.

Although an introduction of neotropical origin was clear when the mealybug outbreak first occurred in 1973, the detailed species-level taxonomy was insufficiently refined and the search for the mealybug and its natural enemies was misdirected for three years. The search was redirected thanks to taxonomic research. The savings were enormous: by 1988, the total expenditure on attempts to control the pest was estimated at $14.6 million. In contrast, accurate species identification has led to an annual benefit of an estimated $200 million, and this financial saving may continue indefinitely.

Classical biological control involves the importation and establishment of natural enemies of exotic pests and is intended to achieve control of the target pest with little further assistance. This form of biological control is appropriate when insects that spread or are introduced (usually accidentally) to areas outside their natural range become pests mainly because of absence of natural enemies. Two examples of successful classical biological control are outlined in Boxes 15.4 and 15.5.

Despite the many beneficial aspects of this control strategy, negative environmental impacts can arise through ill-considered introductions of exotic natural enemies. Many introduced agents have failed to control pests; for example, more than 60 predators and parasitoids have been introduced into northeastern North America with little effect on the target gypsy moth, *Lymantria dispar* (Lymantriidae). Some have exacerbated pest problems, whereas others have become pests themselves. Exotic introductions are generally irreversible and non-target species can suffer worse consequences from efficient natural enemies than from chemical insecticides, which are unlikely to cause total extinctions of native insect species.

There are documented cases of introduced biological control agents annihilating native insects. The fly *Bessa remota* (Tachinidae), introduced from Malaysia to Fiji for the control of the coconut moth, *Levuana iridescens* (Zygaenidae), led to the extinction of this endemic species. A number of endemic Hawaiian insects (target and non-target) have become extinct apparently largely as a result of biological control introductions. The endemic snail fauna of Polynesia has been almost completely replaced by accidentally and deliberately introduced alien species. Polyphagous agents have the greatest potential to harm non-target organisms, and native species in tropical and subtropical environments may be especially vulnerable to exotic introductions because, in comparison with temperate areas, biotic interactions can be more important than abiotic factors in regulating their populations. Sadly, the countries and states that may have most to lose from inappropriate introductions are exactly those with the most lax quarantine restrictions and few or no protocols for the release of alien organisms.

Biological control agents that are present already or are non-persistent may be preferred for release. **Augmentation** is the supplementation of existing natural enemies, including **periodic release** of those that do not establish permanently but nevertheless are effective for a while after release. Periodic releases may be made regularly during a season so that the natural enemy population is gradually increased (augmented) to a level at which pest control is very effective. Augmentation or periodic release may be achieved in one of two ways, although in some systems a distinction between the following methods may be inapplicable. **Inoculation** is the periodic release of a natural enemy unable either to survive indefinitely or to track an expanding pest range. Control depends on the progeny of the natural enemies, rather than the original release. **Inundation** resembles insecticide use since control is achieved by the individuals released or applied, rather than by their progeny; control is relatively rapid but short-term. Examples of

inundation include entomopathogens used as microbial insecticides (section 15.5.2) and *Trichogramma* wasps that are mass reared and released into glasshouses. For cases in which short-term control is mediated by the original release and pest suppression is maintained for a period by the activities of the progeny of the original natural enemies, then the control process is neither strictly inoculative nor inundative. Augmentative releases are particularly appropriate for pests that combine good dispersal abilities with high reproductive rates – features that make them unsuitable candidates for classical biological control.

Conservation is another broad strategy of biological control that aims to protect or enhance the activities of natural enemies. In some ecosystems this may involve **preservation** of existing natural enemies through practices that minimize disruption to natural ecological processes. For example, the Indonesian IPM Programme for rice (Box 15.2) aims to encourage management practices, such as reduction or cessation of insecticide use, that cause minimal interference with the predators and parasitoids that are able to control the brown planthopper and other rice pests.

For many crop systems, **environmental manipulation** can greatly enhance the impact of natural enemies in reducing pest populations. Typically this involves altering the habitat available to insect predators and parasitoids to improve conditions for their growth and reproduction by the provision or maintenance of shelter (including overwintering sites), alternative foods or oviposition sites. Similarly, the effectiveness of entomopathogens of insect pests can sometimes be improved by altering environmental conditions at the time of application, such as by spraying a crop with water to elevate the humidity during release of fungal pathogens.

All biological control programmes should be underpinned by sound taxonomic research on both pest and natural enemy species. Failure to invest adequate resources in systematic studies can result in incorrect identifications of the species involved and ultimately has the potential to cost more time and resources than any other step in the biocontrol programme. The value of taxonomy in control programmes is illustrated in Box 15.5 concerning the cassava mealybug in Africa and in Box 10.2 for *Salvinia* control.

The next two subsections cover more specific aspects of biological control by natural enemies. The latter are divided, somewhat arbitrarily, into arthropods (section 15.5.1) and smaller, non-arthropod organisms (section 15.5.2) that are used to control various insect pests. In addition, many vertebrates, especially birds, mammals and fish, are insect predators and their significance as regulators of insect populations should not be underestimated; however, as biological control agents the use of vertebrates is limited because most are dietary generalists and their times and places of activity are not easily manipulated by people. An exception in some circumstances is the mosquito fish *Gambusia*, which has been released in many subtropical and tropical waterways worldwide in an effort to control the immature stages of biting flies, particularly mosquitoes. Although some control has been claimed, competitive interactions have been severely detrimental to small native fishes. Birds are discussed in Chapter 13 as examples of visually hunting predators that influence insect defences.

15.5.1 Arthropod natural enemies

Entomophagous arthropods may be predatory or parasitic. Most predators are either other insects or arachnids, particularly spiders (order Araneae) and mites (subclass Acarina, also called Acari). Predatory mites are important in regulating populations of phytophagous mites, including the pestiferous spider mites (Tetranychidae). Some mites that parasitize immature and adult insects or feed on insect eggs are potentially useful control agents for certain scale insects, grasshoppers

and stored product pests. Spiders, however, have a much greater impact on insect populations than mites and have been recognized as diverse and efficient predators, particularly in tropical ecosystems. The role of spiders may be enhanced in IPM by preservation of existing populations or habitat manipulation for their benefit, but their lack of feeding specificity is restrictive. Predatory beetles (Coleoptera, notably Coccinellidae and Carabidae) and lacewings (Neuroptera: Chrysopidae and Hemerobiidae), have been used successfully in biological control of agricultural pests, but many predatory species are polyphagous and inappropriate for targeting particular pest insects. Entomophagous insect predators may feed on several or all stages (from egg to adult) of their prey and each predator usually consumes several individual prey organisms during its life, with the predatory habit often characterizing both immature and adult instars. The biology of predatory insects is discussed in Chapter 12 from the perspective of the predator.

The other major type of entomophagous insect is parasitic as a larva and free-living as an adult. The larva develops either as an endoparasite within its insect host or externally as an ectoparasite. In both cases the host is consumed and killed by the time the fully-fed larva pupates in or near the remains of the host. Such insects, called parasitoids, are all holometabolous insects and most are wasps (Hymenoptera, especially superfamilies Chalcidoidea, Ichneumonoidea and Platygasteroidea) or flies (Diptera, especially the Tachinidae). The Chalcidoidea contains about 20 families and perhaps 100 000 species, of which most are parasitoids, including egg parasitoids such as the Mymaridae and Trichogrammatidae (Fig. 15.3) and the ecto- and endoparasitic Aphelinidae and Encyrtidae, which are species-rich biocontrol agents of aphids, mealybugs (Box 15.5), scale insects and whiteflies. The Ichneumonoidea includes two families, the Braconidae and Ichneumonidae, which contain numerous parasitoids

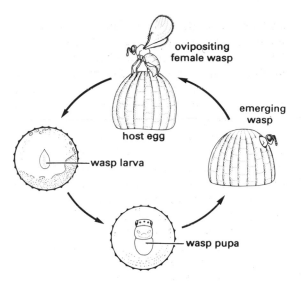

oviposting female wasp

emerging wasp

host egg

wasp larva

wasp pupa

Fig. 15.3 Generalized life cycle of an egg parasitoid. A tiny female wasp of a *Trichogramma* species (Hymenoptera: Trichogrammatidae) oviposits into a lepidopteran egg; the wasp larva develops within the host egg, pupates and emerges as an adult. The full life cycle often takes only one week. (After van den Bosch and Hagen, 1966.)

mostly feeding on insects and often exhibiting quite narrow host specificity. The Platygasteroidea contains the Platygasteridae that are parasitic on insect eggs and larvae and the Scelionidae that parasitize the eggs of insects and spiders. Parasitoids from many of these wasp groups have been utilized for biological control, whereas within the Diptera only the tachinids are commonly used as biocontrol agents.

Parasitoids are often parasitized themselves by secondary parasitoids, called hyperparasitoids (section 12.3.1), which may reduce the effectiveness of the primary parasitoid in controlling the primary host – the pest insect. In classical biological control, great care is usually taken specifically to exclude the natural hyperparasitoids of primary parasitoids and the parasitoids and specialized predators of other introduced exotic natural enemies. However, some highly efficient natural enemies, especially certain predatory

coccinellids, sometimes eliminate their food organisms so effectively that their own populations die out, with subsequent uncontrolled resurgence of the pest. In such cases, limited biological control of the natural enemies of the pest may be warranted. More commonly, exotic parasitoids that are imported free of their natural hyperparasitoids are utilized by indigenous hyperparasitoids in the new habitat, with varying detrimental effects on the biological control programme. Little can be done to solve this problem, except to test the host-switching abilities of some indigenous hyperparasitoids prior to introductions of the natural enemies. Of course, the same problem applies to introduced predators which may become subject to parasitization and predation by indigenous insects in the new area. Such hazards of classical biological control programmes result from the complexities of food webs, which can be unpredictable and difficult to test in advance of introductions.

Some positive management steps can facilitate long-term biological control. For example, there is clear evidence that the numbers and effectiveness of predators and parasitoids can be fostered by providing a stable, structurally and floristically diverse habitat near or within a crop. Habitat stability is naturally higher in perennial systems (e.g. forests, orchards and ornamental gardens) than in annual or seasonal crops (especially monocultures), due to differences in the duration of the crop. In unstable systems, the permanent provision or maintenance of groundcover, hedgerows or strips or patches of cultivated or remnant native vegetation enables natural enemies to survive unfavourable periods such as winter or harvest time and then invade the next crop. Shelter from climatic extremes, particularly during winter in temperate areas, and alternative food resources (when the pest insects are unavailable) are essential to the continuity of predator and parasitoid populations. In particular, the free-living adults of parasitoids generally require quite different food sources from their larvae, such as nectar or pollen from flowering plants. Thus appropriate cultivation practices can contribute significant benefits to biological control. Diversification of agroecosystems can also provide refuges for pests but densities are likely to be low, with damage only significant for crops with low economic injury levels. For these crops, biological control must be integrated with other methods of IPM.

Pest insects must contend not only with predators and parasitoids but also with competitors. Competitive interactions appear to have little regulative influence on most phytophagous insects, but may be important for species that utilize spatially or temporally restricted resources, such as rare or dispersed prey/host organisms, dung or animal carcasses. Interspecific competition can occur within a guild of parasitoids or predators, particularly for generalist feeders and facultative hyperparasitoids, and may inhibit biological control agents.

15.5.2 Microbial control

Micro-organisms are prokaryotes (e.g. bacteria and viruses) or small eukaryotes (e.g. protozoa, fungi and nematodes). Some are pathogenic, usually killing insects, and of these many are host specific to a particular insect genus or family. Infection is from spores, viral particles or organisms that persist in the insect's environment, often in the soil. These pathogens enter insects by several routes. Entry via the mouth (*per os*) is common for viruses, bacteria, nematodes and protozoa. Cuticular or wound entry is typical of fungi and nematodes; the spiracles and anus are other sites of entry. Viruses and protozoa can also infect insects via the female ovipositor or during the egg stage. The micro-organisms then multiply within the living insect but have to kill it to release more infectious spores, particles or (in the case of nematodes) juveniles. Disease is quite common in dense insect populations and under environmental conditions suitable to the micro-organisms. At

low host density, however, disease incidence is often low due to lack of contact between the pathogens and their insect hosts.

Micro-organisms that cause diseases in natural or cultured insect populations can be used as biological control agents in the same way as other natural enemies (section 15.5, above). The usual strategies of control are appropriate, namely:

- Classical biological control, i.e. an introduction of an exotic pathogen (such as the bacterium *Bacillus popilliae* established in the United States for the control of the Japanese beetle *Popillia japonica* (Scarabaeidae)).
- Augmentation via either (a) inoculation (e.g. a single treatment that provides season-long control, as in the fungus *Verticillium lecanii* used against *Myzus persicae* aphids in glasshouses), or (b) inundation, i.e. entomopathogens (such as *Bacillus thuringiensis*) used as microbial insecticides; section 15.5.2(c).
- Conservation of entomopathogens through manipulation of the environment (e.g. raising the humidity to enhance the germination and spore viability of fungi).

Some disease organisms are quite host-specific (e.g. viruses) whereas others, such as fungal and nematode species, often have wide host ranges but possess different strains that vary in their host adaptation. Thus, when formulated as a stable microbial insecticide, different species or strains can be used to kill pest species with little or no harm to non-target insects. In addition to virulence for the target species, other advantages of microbial insecticides include their compatibility with other control methods and the safety of their use (non-toxic and non-polluting). For some **entomopathogens** (insect pathogens) there are further advantages such as the rapid onset of feeding inhibition in the host insect, stability and thus long shelf-life, and often the ability to self-replicate and so persist in target populations. Obviously not all of these advantages apply to every pathogen; many have a slow action on host insects, with efficacy dependent on suitable environmental conditions (e.g. high humidity or protection from sunlight) and appropriate host age or density. The very selectivity of microbial agents can have practical drawbacks as when a single crop has two or more unrelated pest species, each requiring separate microbial control. All entomopathogens are more expensive to produce than chemicals and the cost is even higher if several agents must be used. However, bacteria, fungi and nematodes that can be mass produced in liquid fermenters (*in vitro* culture) are much cheaper to produce than those micro-organisms (viruses and protozoa) requiring living hosts (*in vivo* techniques). Some of the problems with the use of microbial agents are being overcome by research on formulations and mass production methods.

Insects can become resistant to microbial pathogens, as evidenced by the early success in selecting honey bees and silkworms resistant to viral, bacterial and protozoan pathogens. Furthermore, many pest species exhibit significant intraspecific genetic variability in their responses to all major groups of pathogens. The current rarity of significant field resistance to microbial agents probably results from the limited exposure of insects to pathogens rather than any inability of most pest insects to evolve resistance. Of course, unlike chemicals, pathogens do have the capacity to coevolve with their hosts and over time there is likely to be a constant trade-off between host resistance, pathogen virulence and other factors such as persistence.

Each of the five major groups of micro-organisms (viruses, bacteria, protozoa, fungi and nematodes) have different applications in insect pest control. Insecticides based on the bacterium *Bacillus thuringiensis* have been used most widely, but entomopathogenic fungi, nematodes and viruses have specific and often highly successful applications. Although protozoa are responsible for natural disease outbreaks in many insect populations and can be appropriate for classical biological control, they have less potential commercially than other micro-organisms because of their typical low pathogenicity (infections are chronic rather than acute) and the present difficulty of large-scale production for most species.

(a) Nematodes

Nematodes from four families – the Mermithidae, Heterorhabditidae, Steinernematidae and Neotylenchidae – include useful or potentially useful control agents for insects. Mermithid nematodes are large and infect their host singly, eventually killing it as they break through the cuticle. They kill a wide range of insects, but aquatic larvae of black flies and mosquitoes are prime targets for biological control by mermithids.

Heterorhabditids and steinernematids are small, soil-dwelling nematodes associated with symbiotic gut bacteria (of the genus *Xenorhabdus*) which are pathogenic to host insects, killing them by septicaemia. In conjunction with their respective bacteria, nematodes of *Heterorhabditis* and *Steinernema* can kill their hosts within two days of infection. They can be mass produced easily and cheaply and applied with conventional equipment, and they have the advantage of being able to search for their hosts. The infective stage is the third-stage juvenile (or dauer stage), the only stage found outside the host. Host location is an active response to chemical and physical stimuli. Although these nematodes are best at controlling soil pests, some plant-boring beetle and moth pests can be controlled as well.

The Neotylenchidae contains the parasitic *Deladenus siricidicola*, which is one of the biological control agents of the sirex woodwasp, *Sirex noctilio*, a serious pest of forestry plantations of *Pinus radiata* in Australia. The juvenile nematodes infect larvae of *S. noctilio*, leading to sterilization of the resulting adult female wasp. This nematode has two completely different forms – one with a parasitic life cycle completely within the sirex wood-wasp and the other with a number of cycles feeding within the pine tree on the fungus introduced by the ovipositing wasp. The fungal feeding cycle of *D. siricidicola* is used in the mass culture of the nematode so that infective juvenile nematodes are obtained for the purposes of classical biological control.

(b) Fungi

Fungi are the commonest disease organisms in insects: approximately 750 species are known to infect arthropods, although only a few dozen naturally infect agriculturally and medically important insects. When fungal spores contact and adhere to an insect, they germinate and send out hyphae which penetrate the cuticle, invade the haemocoel and cause death either rapidly, due to release of toxins, or more slowly due to massive hyphal proliferation that disrupts insect body functions. The fungus then sporulates, releasing spores that can establish infections in other insects, and thus the fungal disease may spread through the insect population.

Sporulation and subsequent spore germination and infection of entomopathogenic fungi often require moist conditions. Although formulation of fungi in oil improves their infectivity at low humidities, water requirements may restrict the use of some species to particular environments, such as soil, glasshouses or tropical crops. Despite this limitation, the advantage of fungi as control agents is their ability to infect insects by penetrating the cuticle at any developmental stage. This property means that insects of all ages and feeding habits, even sap-sucking, are susceptible to fungal disease. The disadvantages are that fungi can be difficult to mass produce and the storage-life of some fungal products can be limited unless kept at low temperature. Useful species of entomopathogenic fungi belong to genera such as *Beauveria*, *Entomophthora*, *Hirsutella*, *Metarhizium*, *Nomuraea* and *Verticillium*. Many of these fungi overcome their hosts after very little growth in the insect haemocoel, in which case toxins are believed to cause death.

Verticillium lecanii is used commercially to control aphids and scale insects in European glasshouses. *Entomophthora* species are also useful for aphid control in glasshouses. Species of *Beauveria* and *Metarhizium*, known as white and green muscardines respectively (by the colour of the spores), are pathogens of soil pests, such as termites and beetle larvae, and can affect other insects such as spittlebugs of sugarcane and certain moths that live in moist microhabitats. One *Metarhizium* species, *M. flavoviride*, may be useful as a myco-insecticide for locusts.

(c) Bacteria

Bacteria rarely cause disease in insects, although dead insects are frequently invaded by saprophytic bacteria which mask the real cause of death. Relatively few bacteria are used for pest control, but several have proved to be useful entomopathogens against particular pests. *Bacillus popilliae* is an obligate pathogen of scarab beetles (Scarabaeidae) and causes milky disease, named for the white appearance of the body of infected larvae. Ingested spores germinate in the larval gut and lead to septicaemia. Infected larvae and adults are slow to die, which means that *B. popilliae* is unsuitable as a microbial insecticide, but the disease can be transmitted to other beetles by spores that persist in the soil. Thus *B. popilliae* is useful in biological control by introduction or inoculation. Two species of *Serratia* are responsible for amber disease in the scarab *Costelytra zealandica*, a pest of pastures in New Zealand, and have been developed for scarab control. *Bacillus sphaericus* has a toxin which kills mosquito larvae.

The strains of *Bacillus thuringiensis* (usually called Bt) have a broad spectrum of activity against larvae of many species of Lepidoptera,

Coleoptera and aquatic Diptera, but can be used only as inundative insecticides because of lack of persistence in the field. Bt was first isolated from diseased silkworms (*Bombyx mori*) about a century ago, by a Japanese bacteriologist who deduced that a toxin was involved in the pathogenicity of Bt. Shortly afterwards, other Japanese researchers demonstrated that the toxin was a protein present only in sporulated cultures and that it was absent from culture filtrates and thus was not an exotoxin. Many of the isolates of Bt have been commercialized for insect control. Bt is produced in large liquid fermenters and formulated in various ways, including as dusts and granules which can be applied to plants as aqueous sprays. Currently, the largest market for Bt-based products is the North American forestry industry.

Bt forms spores, each containing a protein-aceous inclusion called a **crystal**, which is the source of the toxins that cause most larval deaths. The mode of action of Bt varies among different susceptible insects: in some species, insecticidal action is associated with the toxic effects of the crystal proteins alone (as for some moths and black flies), but in many others (including a number of lepidopterans) the presence of the spore enhances toxicity substantially, and in a few insects death results from septicaemia following spore germination in the insect midgut rather than from the toxins. For insects affected by the toxins, paralysis occurs in mouthparts, gut and often the body, so that feeding is inhibited. Upon ingestion by a larval insect, the crystal is dissolved in the midgut, releasing proteins called delta-endotoxins. These proteins are protoxins which must be activated by midgut proteases before they can interact with gut epithelium and disrupt its integrity, after which the insect ultimately dies. Early-instar larvae generally are more susceptible to Bt than older larvae or adult insects.

Effective control of insect pests by Bt depends on the following factors:

- The insect population must be uniformly young to be susceptible.
- The Bt must be sprayed evenly and the insects must be feeding actively so that they consume a lethal dose.
- The Bt must persist and not be rapidly denatured by UV light.
- The strain and formulation of Bt must be suitable for the insect target.

At present there are about 40 recognized Bt subspecies based on serotype and certain biochemical and host-range data. There is disagreement concerning the basis of the Bt classification scheme since it may be more appropriate to use a system based on the crystal toxin genes, which directly determine the level and range of Bt activity. A recent nomenclature and classification scheme for crystal genes (cry) is based on their phenotype, types of crystal proteins produced and the protein's host range as insecticidal toxins. Toxins are encoded by the *cryI*, *cryII*, *cryIII*, *cryIV* and *cyt* gene classes; *cryI* genes are associated with bipyramidal crystals which are toxic to lepidopteran larvae, *cryII* with cuboidal crystals active against both lepidopteran and dipteran larvae, *cryIII* with flat square crystals toxic to coleopteran larvae, and *cryIV* and *cyt* with variously shaped crystals which kill dipteran larvae. *B. t. israelensis*, for example, has *cryIV* and *cyt* genes, whereas *B. t. tenebrionis* has *cryIII* genes and *B. t. kurstaki* has *cryI* and *cryII* genes. In addition, some cultures of Bt produce exotoxins which are effective against various insects, including larvae of the Colorado potato beetle. The nature and insecticidal effects of the various isolates of Bt are far from simple and further research on the modes of action of the toxins is desirable, especially for understanding the basis of potential and actual resistance to Bt.

Since 1970, Bt products have been used increasingly for control of various Lepidoptera, such as caterpillars on crucifers and in forests. Until 1986, with the exception of one case of resistance in a stored-grain moth, *Plodia interpunctella* (Pyralidae), there were no serious resistance problems. The first insect to show resistance in the field was a major plant pest, the diamondback moth, *Plutella xylostella* (Plutellidae). Watercress growers in Japan and Hawaii complained that Bt's ability to kill this pest had become reduced; by 1989 further reports of resistant moths in Hawaii were confirmed in areas where frequent, high doses of Bt had been used, while in Japan by 1988 an extremely high level of Bt resistance was found in moths in greenhouses where watercress had been grown year-round with a total of 40–50 applications of Bt over three to four years. Moths resistant to Bt were also reported in Thailand, the Philippines and mainland United States, and laboratory studies indicated that other insects could be bred to show differing levels of resistance.

Problems with chemical insecticides have stimulated interest in the use of Bt products as an alternative method of pest control. In addition to conventional applications of Bt, genetic engineering with Bt genes has produced transgenic plants which manufacture their own protective toxins (section 15.6.1). Current optimism has led to the belief that insects are unlikely to develop extremely high levels of Bt resistance in the field due both to instability of resistance and to dilution by immigrants from susceptible populations. Strategies to prevent or slow down the evolution of resistance to Bt are the same as those used to retard resistance to synthetic insecticides. Obviously, the continued success of Bt products and the benefits of technological advances will depend on appropriate use as well as understanding and limiting resistance to the Bt crystal proteins.

(d) Viruses

Many viruses infect and kill insects, but those with potential for insect control are from just three viral groups, all with proteinaceous inclusion bodies which enclose the virions (virus particles). These 'occluded' viral species are considered safe because they have been found only in arthropods and appear unable to replicate in vertebrates or vertebrate cell cultures, although distant relatives of two of these groups have wider host ranges. Many 'non-occluded' viruses that infect insects are considered unsafe for pest control because of their lack of specificity and possible adverse side-effects, such as infection of vertebrates or beneficial insects.

The useful entomopathogenic groups are the nuclear polyhedrosis viruses (NPVs), granulosis viruses (GVs) (both belonging to Baculoviridae – the baculoviruses), the cytoplasmic polyhedrosis viruses (CPVs) (Reoviridae: *Cypovirus*) and the entomopoxviruses (EPVs) (Poxviridae: Entomopoxvirinae). Baculoviruses replicate within the nuclei of the host cells, whereas the CPVs and EPVs replicate in the host cell cytoplasm. Baculoviruses have DNA genomes and are found mostly in endopterygotes, such as moth and beetle larvae, which become infected when they ingest the inclusion bodies with their food. Inclusion bodies dissolve in the high pH of the insect midgut and release the virion(s) (Fig. 15.4). These infect the gut epithelial cells and usually spread to other tissues, particularly the fat body. The inclusion bodies of NPVs are usually very stable and may persist in the environment for years (if protected from UV light, as in the soil), increasing their utility as biological control agents or microbial insecticides. The host specificity of different viruses also influences their potential usefulness as pest control agents; some baculoviruses (such as the *Helicoverpa* NPV) are specific to an insect genus. CPVs have RNA genomes and have been found in more than 200 insect species, mainly of Lepidoptera and Diptera; their inclusion bodies are less stable than those of NPVs. EPVs have large DNA genomes and infect a wide range of hosts in the Orthoptera, Lepidoptera, Coleoptera and Diptera, but individual viral isolates generally have a

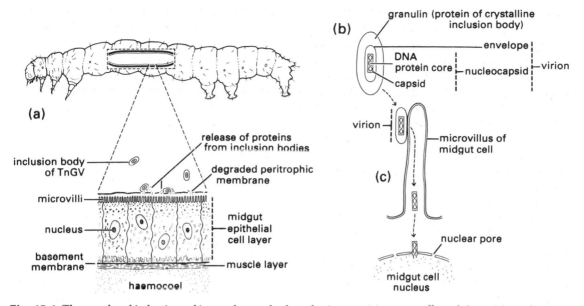

Fig. 15.4 The mode of infection of insect larvae by baculoviruses: (a) a caterpillar of the cabbage looper, *Trichoplusia ni* (Lepidoptera: Noctuidae), ingests the viral inclusion bodies of a granulosis virus (called TnGV) with its food and the inclusion bodies dissolve in the alkaline midgut, releasing proteins that destroy the insect's peritrophic membrane, allowing the virions access to the midgut epithelial cells; (b) a granulosis virus inclusion body with virion in longitudinal section; (c) a virion attaches to a microvillus of a midgut cell, where the nucleocapsid discards its envelope, enters the cell and moves to the nucleus in which the viral DNA replicates. The newly synthesized virions then invade the haemocoel of the caterpillar, where viral inclusion bodies are formed in other tissues (not shown). (After Entwistle and Evans, 1985; Beard, 1989.)

narrow host range. Infection of insect cells follows a similar path to that of baculoviruses.

For certain pests, viral insecticides provide feasible alternatives to chemical controls but several factors may restrict the usefulness of different viruses. Ideally, viral insecticides should be host specific, virulent, kill quickly, persist for a reasonable time in the environment after application and be easy to provide in large amounts. CPVs fulfil these requirements poorly; the others score better, although they are inactivated by UV light within hours or days, they often kill larvae slowly or have a low virulence, and production costs can be large due to the necessity for *in vivo* production. Potency problems may be overcome by genetic engineering to increase either the speed of action or the virulence of

naturally occurring viruses, such as the baculoviruses that infect the heliothine pests (Lepidoptera: Noctuidae: *Helicoverpa* and *Heliothis* species) of cotton. The presence of particular viral proteins appears to enhance the action of baculoviruses; viruses could be altered to produce much more protein or the gene controlling protein production could be added to viruses that lack it. Insect pests that damage valuable crops, such as bollworms of cotton and *Neodiprion* sawflies of coniferous forest trees, are suitable for viral control because substantial economic returns offset the large costs of development (including genetic engineering) and production. The other way in which insect viruses could be manipulated for use against pests is to transform the host plants so that they produce the viral proteins that damage the gut lining of phytophagous insects. This is analogous to the engineering of host-plant resistance by incorporating foreign genes into plant genomes using the crown-gall bacterium as a vector.

15.6 HOST-PLANT RESISTANCE TO INSECTS

Plant resistance to insects consists of inherited genetic qualities that result in a plant being less damaged than another (susceptible) one that is subject to the same conditions but lacks these qualities. Plant resistance is a relative concept since spatial and temporal variations in the environment influence its expression and effectiveness. Generally, the production of plants resistant to particular insect pests is accomplished by selective breeding for resistance traits. The three functional categories of plant resistance to insects are:

- **Antibiosis**, in which the plant is consumed and adversely affects the biology of the phytophagous insect.
- **Antixenosis**, in which the plant is a poor host, deterring any insect feeding.
- **Tolerance**, in which the plant is able to withstand or recover from insect damage.

Antibiotic effects on insects range from mild to lethal, and antibiotic factors include toxins, growth inhibitors, reduced levels of nutrients, sticky exudates from glandular trichomes (hairs), and high concentrations of indigestible plant components such as silica and lignin. Antixenosis factors include plant chemical repellents and deterrents, pubescence (a covering of simple or glandular trichomes), surface waxes, and foliage thickness or toughness – all of which may deter insect colonization. Tolerance involves only plant features and not insect–plant interactions, as it depends only on a plant's ability to outgrow or recover from defoliation or other damage caused by insect feeding. These categories of resistance are not necessarily discrete: any combination may occur in one plant. Furthermore, induced resistance to one type of insect may render a plant susceptible to another or to a disease.

Selecting and breeding for host-plant resistance can be an extremely effective means of controlling pest insects. For example, at the International Rice Research Institute (IRRI), over 100 rice cultivars have been developed with resistance to all of the major insect pests of rice in southern and Southeast Asia. The grafting of American vine rootstocks resistant to grape phylloxera – an aphid (*Phylloxera vittifolae*) that kills susceptible vines – allowed the recovery of the wine industry following a temporary demise caused by phylloxera in many countries worldwide during the late nineteenth century. Some cotton cultivars are tolerant of the feeding damage of certain insects, whereas other cultivars have been developed for their chemicals (such as gossypol) that inhibit insect growth. In general, there are more cultivars of insect-resistant cereal and grain crops than insect-resistant vegetable or fruit crops. The former often have a higher value per hectare and the latter have a low consumer tolerance of any damage; but, perhaps more importantly, resistance factors can be deleterious to food quality.

Conventional methods of obtaining host-plant resistance to pests are not always successful. Despite more than 50 years of intermittent effort, no commercially suitable potato varieties resistant to the Colorado potato beetle (Chrysomelidae: *Leptinotarsa decemlineata*) have been developed. Most attempts to produce potatoes with high levels of toxic glycoalkaloids have stopped, partly because potato plants with high foliage levels of glycoalkaloids often have tubers rich in these toxins, resulting in risks to human health. Breeding potato plants with glandular trichomes may have limited utility due to the ability of the beetle to adapt to different hosts. The most promising resistance mechanism for control of the Colorado potato beetle involves production of transgenic potato plants that express a foreign gene for a bacterial toxin that kills many insect larvae (Box 15.1). Attempts to produce resistance in other vegetables often fail because the resistance factor is incompatible with product quality – poor taste or toxicity might be introduced with the resistance.

15.6.1 Genetic engineering of host resistance

Molecular biologists are using genetic engineering techniques to produce insect-resistant varieties of several crop plants, including tobacco, tomato and potato, that can manufacture foreign antifeedant or insecticidal proteins under field conditions. The genes encoding these proteins are obtained from bacteria or other plants and are inserted into the recipient plant using a modified system from the crown-gall bacterium, *Agrobacterium tumefaciens*. This bacterium can move part of its own DNA into a plant cell during infection because it possesses a tumour-inducing (Ti) plasmid containing a piece of DNA that can integrate into the chromosomes of the infected plant. Ti plasmids can be modified by removal of their tumour-forming capacity, and useful foreign genes (such as insecticidal toxins) can be inserted; then these plasmid vectors are introduced into plant-cell cultures, from which the transformed cells are selected and regenerated as whole plants.

Insect control via resistant transgenic plants has a number of advantages over insecticide-based control methods, including continuous protection (even of plant parts inaccessible to insecticide sprays), elimination of the vagaries and costs of insecticide use, increased environmental and human safety, and lower cost for the development of a new crop variety compared with a new chemical insecticide. Problems with the use of transgenic plants that produce foreign insecticidal proteins include complications concerning registration and patent applications for these new biological entities and, of less pecuniary concern, the potential for the development of resistance in the target insect populations. For example, insect resistance to the toxins of *Bacillus thuringiensis* (Bt) (section 15.5.2(c)) is to be expected after continuous exposure to these proteins in transgenic plant tissue. This problem might be overcome by restricting expression of the toxins to certain plant parts (e.g. the bolls of cotton rather than the whole cotton plant) or to tissues damaged by insects. A specific limitation of the genetic engineering of plants to produce Bt toxins is that maximum Bt activity with some pest insects requires the presence of the spore not just the toxin.

Furthermore, plant resistance based on toxins (allelochemicals) from genes transferred to plants may result in greater pest problems than those that the genetic engineering was attempting to alleviate. First, at low concentrations, many toxins are more active against natural enemies of phytophagous insects than against their pest hosts, adversely affecting biological control. Alkaloids and other allelochemicals ingested by phytophagous insects have been shown to affect development of, or be toxic to, parasitoids that develop within hosts containing them, and also to be toxic or sterilizing to predators. In some insects, allelochemicals sequestered whilst feeding are passed into the eggs with deleterious consequences for egg parasitoids. Second, allelochemicals can increase the tolerance of pests to insecticides by selecting for detoxifying enzymes that lead to cross-reactions to other chemicals. Most other plant resistance mechanisms decrease pest tolerance to insecticides and thus improve the possibilities of using pesticides selectively to facilitate biological control.

15.7 MECHANICAL AND PHYSICAL CONTROL

Mechanical and physical control refers to non-chemical, non-biological methods which destroy pests or make the environment unsuitable for their entry or survival. The best known mechanical method of pest control is the 'fly-swatter' but another example is the shifting and separating procedure used in flour mills to remove insects. Physical control measures are generally limited to confined environments such as glasshouses, food storage structures (e.g. silos) and domestic premises. An obvious method is physical exclusion such as packaging of food products, semi-hermetic sealing of grain silos or provision of mesh screens on glasshouses. In addition, products may be given treatments or stored under controlled conditions of temperature (low or high), atmospheric gas composition (e.g. low oxygen or high carbon dioxide) or low relative humidity, which can kill or reduce reproduction of insect pests. Traps that use long-wave ultraviolet light (e.g. the 'insect-o-cutor') or adhesive surfaces can be effective in domestic or food retail buildings or in glasshouses, but should not be used outdoors because of the likelihood of catching native or introduced beneficial insects.

15.8 CULTURAL CONTROL

Subsistence farmers have utilized cultural methods of pest control for centuries, and many of their techniques are applicable to large-scale as well as small-scale, intensive agriculture. Typically cultural practices involve reducing insect populations in crops by one or a combination of the following techniques: crop rotation; tillage or burning of crop stubble to disrupt pest life cycles; careful timing or placement of plantings to avoid synchrony with pests; destruction of wild plants that harbour pests; cultivation of non-crop plants to conserve natural enemies; and use of pest-free rootstocks and seeds. Intermixed plantings of several crops – called **intercropping** or **polyculture** – may reduce crop apparency (plant apparency hypothesis) or resource concentration for the pests (resource concentration hypothesis), increase protection for susceptible plants growing near resistant plants (associational resistance hypothesis), and/or promote natural enemies

(the natural enemies hypothesis). Recent agroecology research has compared densities of insect pests and their natural enemies in monocultures and polycultures (including di- and tricultures) to test whether the success of intercropping can be explained better by a particular hypothesis; however, the hypotheses are not mutually exclusive and there is some support for each one.

In medical entomology, cultural control methods consist of habitat manipulations, such as draining marshes and removal or covering of water-holding containers to limit larval breeding sites of disease-transmitting mosquitoes, and covering rubbish dumps to prevent access and breeding by disease-disseminating flies. Examples of cultural control of livestock pests include removal of dung that harbours pestiferous flies and simple walk-through traps that remove and kill flies resting on cattle.

15.9 PHEROMONES AND OTHER INSECT ATTRACTANTS

Insects use a variety of chemical odours called semiochemicals to communicate within and between species (Chapter 4). Pheromones are particularly important chemicals used for signalling between members of the same species and these are often mixtures of two, three or more components which, when released by one individual, elicit a specific response in another individual. Other members of the species, for example prospective mates, arrive at the source. Naturally derived or synthetic pheromones, especially sex pheromones, can be used in pest management to misdirect the behaviour and prevent reproduction of pest insects. The pheromone is released from point-source dispensers, often in association with traps that are placed in the crop. The strength of the insect response depends upon dispenser design, placement and density. The rate and duration of pheromone emission from each dispenser depends upon method of release (e.g. from impregnated rubber, microcapsules, capillaries or wicks), strength of formulation, original volume, surface area from which it is volatilized and longevity or stability of the formulation. Sometimes other attractants, such as food baits or oviposition site lures, can be incorporated into a pest management scheme to function in a manner analogous to pheromones, as discussed below.

There are three main uses for insect pheromones (and sometimes other attractants) in horticultural, agricultural and forest management. The first use is in **monitoring**, initially to detect the presence of a particular pest and then to give some measure of its abundance. A trap containing the appropriate pheromone (or other lure) is placed in the susceptible crop and checked at regular intervals for the presence of any male individuals of the pest lured to the trap. The presence of males of the pest (and by inference, females) can be detected even at very low population densities, allowing early recognition of an impending outbreak. Knowledge of the relationship between trap-catch size and actual pest density allows a decision to be made about when the economic threshold for the crop will be reached and thus facilitates the efficient use of control measures, such as insecticide application. Monitoring is an essential part of IPM.

Pheromone mass trapping is another method of using pheromones in pest management and has been used primarily against forest pests. It is one form of **attraction–annihilation** – a more general method in which individuals of the targeted pest species are lured and killed. Lures may be light (e.g. UV) or colour (yellow is a common attractant), or they may be semiochemicals such as pheromones or odours produced by the mating or oviposition site (e.g. dung), host plant, host animal, or empirical attractants (e.g. fruit fly chemical lures). Sometimes the lure, as with methyl eugenol for tephritid fruit flies, is more attractive than any other substance used by the insect. The insects may be attracted into container or sticky traps, onto an electrocutor grid or onto surfaces treated with toxic chemicals or pathogens. The effectiveness of the attraction–annihilation technique appears to be inversely related to the population density of the pest and the size of the infested area. Thus this method is likely to be most effective for control of non-resident insect pests that become abundant through annual or seasonal immigration, or pests that are geographically restricted or always present at low density. Pheromone mass trapping programmes have been undertaken mostly for certain moths, such as gypsy moth (Lymantriidae: *Lymantria dispar*) using their female sex pheromones, and for bark and ambrosia beetles (Scolytidae) using their aggregation pheromones (section 4.3.2(b)). An advantage of this technique for scolytids is that both sexes are caught. Success has been difficult to demonstrate because of the difficulties of designing controlled, large-scale experiments. Nevertheless, mass trapping appears effective in isolated gypsy moth populations and at low scolytid beetle densities. If beetle populations are high, even removal of part of the pest population may be beneficial since in tree-killing beetles there is a positive feedback between population density and damage.

The third method of practical pheromone use involves sex pheromones and is called **mating disruption** (previously sometimes called 'male confusion', which is an inappropriate term). It has been applied very successfully in the field to a number of moth species, such as the pink bollworm (Gelechiidae: *Pectinophora gossypiella*) in cotton and the oriental fruit moth (Tortricidae: *Grapholita molesta*) in stone fruit orchards. Basically, numerous synthetic pheromone dispensers are placed within the crop so that the level of female sex pheromone in the orchard becomes higher than the background. A reduction in the number of males locating female moths means fewer matings and a lowered population in subsequent generations. The exact behavioural or physiological mechanisms responsible for mating disruption are far from resolved but relate to altered behaviour in males and/or females. Disruption of male behaviour may be through **habituation** – temporary modifications within the central nervous system – rather than adaptation of the receptors on the antennae or confusion resulting in the following of false plumes. The high background levels promoted by use of synthetic pheromones also may mask the natural pheromone plumes of the females so that males can no longer differentiate them. Understanding the mechanism of disruption is important for production of the appropriate type of formulation and quantities of synthetic pheromone needed to cause disruption, and thus to control.

All of the above three pheromone methods have been used most successfully for certain moth, beetle and fruit fly pests. Pest control using pheromones appears most effective for species that: (a) are highly dependent on chemical (rather than visual) cues for locating dispersed mates or food sources, (b) have a limited host range, and (c) are resident and relatively sedentary so that locally controlled populations are not constantly supplemented by immigration. Advantages of using pheromone mass trapping or mating disruption include:

- Non-toxicity, leaving fruit and other products free of toxic chemical (insecticide).
- Application may be required only once or a few times per season.
- The confinement of suppression to the target pest (unless predators or parasitoids use the pest's own pheromone for host location).
- Enhancement of biological control (except for the circumstance mentioned in the previous point).

Limitations of pheromone use include:

- High selectivity and therefore no effect on other primary or secondary pests.

- Cost effectiveness only if the target pest is the main pest for which insecticide schedules are designed.
- Requirement that the treated area be isolated or large to avoid mated females flying in from untreated crops.
- Requirement for detailed knowledge of pest biology in the field (especially of flight and mating activity) since timing of application is absolutely critical to successful control if continuous costly use is to be avoided.

The latter two limitations apply also to pest management using chemical or microbial insecticides, since appropriate timing of insecticide applications is particularly important to target vulnerable stages of the pest, to reduce unnecessary and costly spraying, and to minimize detrimental environmental effects.

15.10 GENETIC MANIPULATION OF INSECT PESTS

Cochliomyia hominivorax (Calliphoridae), the New World screw-worm fly, is a devastating pest of livestock in tropical America, laying eggs into wounds, where the larvae cause myiasis (section 14.3) by feeding in the growing suppurating wounds of the living animals, including some humans. The fly was perhaps endemic but also spread seasonally into southern and south-western United States, where substantial economic losses of stock hides and carcasses provoke a continuing control campaign. Since the female of *C. hominivorax* mates only once, control can be achieved by swamping the population with infertile males, so that the first male to arrive and mate with each female is likely to be sterile and the resultant eggs inviable. The **sterile male technique** in the Americas depends upon mass-rearing facilities, sited in Mexico, where billions of screw-worm flies are reared in artificial media of blood and casein. The larvae (Fig. 6.6(h)) drop to the floor of the rearing chambers, where they form a puparium. At a crucial time, after gametogenesis, sterility of the developing adult is induced by gamma-irradiation of the five-day-old puparia. This treatment sterilizes the males. Although the females cannot be separated in the pupal stage and are also released, the irradiation prevents their ovipositing. The released sterile males mix with the wild population so that the fertile proportion diminishes with each mating. Eradication is a theoretical possibility.

The technique is credited with the eradication of screw-worm fly, first from Florida, then from Texas and the western U.S.A., and is in use to limit the problem in Mexico, whence re-invasions of the United States originate. In 1990, when *C. hominivorax* was introduced accidentally to Libya (North Africa), the Mexican facility was able to produce enough sterile flies to prevent the establishment of this potentially devastating pest. The impressive cost/benefit ratio of screw-worm control and eradication using sterile insect technique has induced the expenditure of substantial sums in attempts to control similar economic pests. The lack of success of all other ventures can be attributed to difficulties with one or more of the following:

- Inability to mass-culture the pest.
- Lack of competitiveness of sterile males.
- Release of an inadequate number of males to swamp the females.
- Failure of irradiated insects to mix with the wild population.
- Poor dispersal of the sterile males from the release site, and rapid reinvasion of wild types.

Attempts have been made to introduce deleterious genes into pest species that can be mass cultured and released, with the intention that the detrimental genes spread through the wild population. The reasons for the failure of these attempts are likely to include those cited above for many sterile insect releases, particularly their lack of competitiveness, together with genetic drift and recombination which reduces the genetic effects.

FURTHER READING

Altieri, M.A. (1991) Classical biological control and social equity. *Bulletin of Entomological Research*, **81**, 365–9.

Anderson, T.E. and Leppla, N.C. (eds) (1992) *Advances in Insect Rearing for Research and Pest Management*, Westview Press, Boulder, San Francisco, Oxford.

Baker, R.R. and Dunn, P.E. (eds) (1990) *New Directions in Biological Control, Alternatives for Suppressing Agricultural Pests and Diseases*, Alan R. Liss, Inc., New York.

Beegle, C.C. and Yamamoto, T. (1992) History of *Bacillus thuringiensis* Berliner research and development. *Canadian Entomologist*, **124**, 587–616.

Burges, H.D. (ed.) (1981) *Microbial Control of Pests and Plant Diseases 1970–1980*, Academic Press, London, New York, Toronto, Sydney, San Francisco.

Caltagirone, L.E. (1981) Landmark examples in classical biological control. *Annual Review of Entomology*, **26**, 213–32.

Caltagirone, L.E. and Doutt, R.L. (1989) The history of the vedalia beetle importation to California and its impact on the development of biological control. *Annual Review of Entomology*, **34**, 1–16.

Coats, J.R. (ed.) (1982) *Insecticide Mode of Action*, Academic Press, New York, London, Paris, San Diego, San Francisco, Sao Paulo, Sydney, Tokyo, Toronto.

DeBach, P. and Rosen, D. (1991) *Biological Control by Natural Enemies*, 2nd edn, Cambridge University Press, New York.

Denholm, I. and Rowland, M.W. (1992) Tactics for managing pesticide resistance in arthropods: theory and practice. *Annual Review of Entomology*, **37**, 91–112.

Dent, D. (1991) *Insect Pest Management*, CAB International, Wallingford.

Gerson, U. and Smiley, R.L. (1990) *Acarine Biocontrol Agents*, Chapman & Hall, London, New York, Tokyo, Melbourne, Madras.

Gill, S.S., Cowles, E.A. and Pietrantonio, P.V. (1992) The mode of action of *Bacillus thuringiensis* endotoxins. *Annual Review of Entomology*, **37**, 615–36.

Hadlington, P. and Gerozisis, J. (1985) *Urban Pest*

Control in Australia, New South Wales University Press, Kensington.

Hare, J.D. (1990) Ecology and management of the Colorado potato beetle. *Annual Review of Entomology*, **35**, 81–100.

Herren, H.R. and Neuenschwander, P. (1991) Biological control of cassava pests in Africa. *Annual Review of Entomology*, **36**, 257–83.

Hill, D.S. (1983) *Agricultural Insect Pests of the Tropics and their Control*, 2nd edn, Cambridge University Press, Cambridge.

Hill, D.S. (1987) *Agricultural Insect Pests of Temperate Regions and their Control*, Cambridge University Press, Cambridge.

Howarth, F.G. (1991) Environmental impacts of classical biological control. *Annual Review of Entomology*, **36**, 485–509.

Keeley, L.L. and Hayes, T.K. (1987) Speculations on biotechnology applications for insect neuroendocrine research. *Insect Biochemistry*, **17**, 639–51.

Kerkut, G.A. and Gilbert, L.I. (eds) (1985) *Comprehensive Insect Physiology, Biochemistry and Pharmacology, Volume 12, Insect Control*, Pergamon Press, Oxford, New York, Toronto, Sydney, Paris, Frankfurt.

Kitching, R.L. and Jones, R.E. (eds) (1981) *The Ecology of Pests. Some Australian Case Histories*, CSIRO, Melbourne.

Kirschbaum, J.B. (1985) Potential implication of genetic engineering and other biotechnologies to insect control. *Annual Review of Entomology*, **30**, 51–70.

Kogan, M. (ed.) (1986) *Ecological Theory and Integrated Pest Management*, John Wiley and Sons, New York.

Lockwood, J.A. (1993) Environmental issues involved in biological control of rangeland grasshoppers (Orthoptera: Acridadae) with exotic agents. *Environmental Entomology*, **22**, 503–18.

Maeda, S. (1989) Expression of foreign genes in insects using baculovirus vectors. *Annual Review of Entomology*, **34**, 351–72.

Mallet, J. (1989) The evolution of insecticide resistance: have the insects won? *Trends in Ecology and Evolution*, **4**, 336–40.

Matteson, P.C. (1992) 'Farmer first' for establishing IPM. *Bulletin of Entomological Research*, **82**, 293–6.

Meeusen, R.L. and Warren, G. (1989) Insect control with genetically engineered crops. *Annual Review of Entomology*, **34**, 373–81.

Metcalf, R.L. and Metcalf, R.A. (1992). *Destructive and Useful Insects*, 5th edn, McGraw-Hill.

Parrella, M.P., Heinz, K.M. and Nunney, L. (1992) Biological control through augmentative releases of natural enemies: a strategy whose time has come. *American Entomologist*, **38**, 172–9.

Pedigo, L.P., Hutchins, S.H. and Higley, L.G. (1986) Economic injury levels in theory and practice. *Annual Review of Entomology*, **31**, 341–68.

Ridgway, R.L., Silverstein, R.M. and Inscoe, M.N. (eds) (1990) *Behavior-Modifying Chemicals for Insect Management, Applications of Pheromones and Other Attractants*, Marcel Dekker, Inc., New York and Basel.

Riechert, S.E. and Lockley, T. (1984) Spiders as biological control agents. *Annual Review of Entomology*, **29**, 299–320.

Rousch, R.T. and Tabashnik, B.E. (eds) (1990) *Pesticide Resistance in Arthropods*, Chapman & Hall, New York and London.

Schmutterer, H. (1990) Properties and potential of natural pesticides from the neem tree, *Azadirachta indica*. *Annual Review of Entomology*, **35**, 271–97.

Smith, C.M. (1989) *Plant Resistance to Insects*, John Wiley and Sons, New York, Chichester, Brisbane, Toronto, Singapore.

Strittmatter, G. and Wegener, D. (1993) Genetic engineering of disease and pest resistance in plants: present state of the art. *Zeitshrift für Naturforschung*, **48c**, 673–88.

van den Bosch, R., Messenger, P.S. and Gutierrez, A.P. (1982) *An Introduction to Biological Control*, Plenum Press, New York.

Waterhouse, D.F. and Norris, K.R. (1987) *Biological Control: Pacific Prospects*, Inkata Press, Melbourne. (Plus two supplements, 1989, 1993.)

GLOSSARY OF ENTOMOLOGICAL AND SELECTED TECHNICAL TERMS

Each scientific and technical field has a particular vocabulary, and entomology is no exception. This is not an attempt by entomologists to restrict access to their science. It results from the need for precision in terminology, whilst avoiding, for example, misplaced anthropocentric terms derived from human anatomy. Many terms are derived from Latin or Greek: when competence in these languages was a prerequisite for scholarship (including in the natural sciences), these terms were comprehensible to the educated, whatever their mother tongue. The utility of these terms remains, although fluency in the languages from which they are derived does not. In this glossary we have tried to define the terms we use in a straightforward manner to complement the definitions used on first mention in the main body of the book. A comprehensive glossary of entomological terms is provided by Torre-Bueno (1989).

Bold within the text of an entry indicates cross-reference to another headword. The following abbreviations are used:

adj. adjective
dim. diminutive
pl. plural
sing. singular

Abdomen The third (posterior) major division (**tagma**) of an insect body.

Acanthae Fine, unicellular, cuticular extensions (Fig. 2.6(c)).

Accessory gland A gland subsidiary to a major one; more specifically, a gland opening into the genital chamber (Figs 3.1, 3.20(a),(b)).

Accessory pulsatile organs Valved pumps aiding the circulation of haemolymph, often lying close to the dorsal vessel.

Acrotergite The anterior part of a secondary segment, sometimes large (then called **postnotum**), often reduced (Fig. 2.7).

Acrotrophic ovariole, *see* **Telotrophic ovariole**.

Activation In embryology, the commencement of embryonic development within the egg.

Aculeate Belonging to the aculeate Hymenoptera – wasps (including all bees) in which the ovipositor is modified as a sting.

Adecticous pupa A pupa with immovable mandibles (Fig. 6.7).

Adenotrophic viviparity Viviparity (producing living offspring) in which there is no free-living larval stage; eggs develop within the female uterus, nourished by special **milk glands** until the larvae mature, at which stage they are laid and immediately pupate; occurring only in some Diptera (Hippoboscidae and *Glossina*).

Aedeagus The male copulatory organ, variably constructed (sometimes refers to the penis alone) (Fig. 5.2).

Age-grading Determination of the physiological age of an insect.

Air-sac Thin-walled, dilated sections of the **tracheae** (Figs 2.24(b) and 3.11(b)).

Akinesis A state of immobility, due to lack of any stimuli.

Alary muscles Paired muscles that support the heart.

Alate Possessing wings.

Alinotum The wing-bearing plate on the dorsum of the meso- or metathorax (Fig. 2.18).

Alitrunk The fused thorax and first abdominal segment (**propodeum**) of adult ants (Box 11.3); *see also* **Mesosoma**.

Alkaloids Chemicals found in plants, many with important pharmacological actions.

Allelochemical A chemical functioning in interspecific communication; *see also* **Allomone**; **Kairomone**; **Pheromone**; **Synomone**.

Allomone A communication chemical that benefits the producer by the effect it invokes in the receiver.

Allopatric Non-overlapping geographic distributions of organisms or taxa; *see also* **Sympatric**.

Altruism Behaviour costly to the individual but beneficial to others.

Ametabolous Lacking metamorphosis, i.e. with no change in body form during development to adult, with immature stages lacking only genitalia.

Amnion In embryology, the layer covering the germ band (Fig. 6.5).

Amphitoky (*also* **amphitokous parthenogenesis**; **deuterotoky**) Parthenogenesis in which the female produces offspring of both sexes.

Amplexiform A form of wing-coupling in which there is extensive overlap between the fore and hind wing, but without any specific coupling mechanism.

Anal In the direction or position of the anus, near the anus or on the last abdominal segment.

Anal area The posterior part of the wing, supported by the anal vein(s) (Fig. 2.20).

Anal *or* **vannal fold** A distinctive fold in the anal area of wing (Fig. 2.20).

Anamorphic development Development in which the immature stages have fewer abdominal segments than the adult; *see also* **Epimorphic development**.

Anemophily Wind pollination.

Anoxic Lacking oxygen.

Antecostal suture (*also* **intersegmental groove**) A groove marking the position of the intersegmental fold between the primary segments (Figs 2.7, 2.18).

Antenna (*pl.* antennae) Paired, segmented, sensory appendages, lying usually anterodorsally, on the head (Figs 2.9, 2.10, 2.17); derived from the second head segment.

Antennomere A subdivision of the antenna.

Anterior At or towards the front (Fig. 2.8).

Anthophilous Flower-loving.

Anthropogenic Caused by humans.

Anthropophilic Associated with humans.

Antibiosis A property of an organism that adversely affects the well-being of another organism that consumes it.

Antixenosis In plant resistance, unsuitability of a plant to a feeding insect.

Anus The posterior opening of the digestive tract (Figs 2.23 and 3.13).

Apical At or towards the apex (Fig. 2.8).

Apneustic A respiratory system without functional spiracles; *see also* **Oligopneustic**; **Polypneustic**.

Apocritan Belonging to the suborder of Hymenoptera (Apocrita) in which the first abdominal segment is fused to the thorax; *see also* **Propodeum**.

Apod larva A larva lacking true legs (Fig. 6.6).

Apode (*adj.* apodous) An organism without legs.

Apodeme An ingrowth of the exoskeleton, tendon-like, to which muscles are attached (Figs 2.23 and 3.2(c)).

Apolysial space The space between the old and the new cuticle that forms during **apolysis**, prior to **ecdysis**.

Apolysis The separation of the old from the new cuticle during moulting.

Apophysis (*pl.* apophyses) An elongate apodeme, an internal projection of the exoskeleton.

Aposematic Warning of unpalatability, particularly using colour.

Aposematism A communication system based on warning signals.

Apparency Obviousness (e.g. of a plant to an insect herbivore).

Appendicular ovipositor The true ovipositor formed from appendages of segments 8 and 9; *see also* **Substitutional ovipositor**.

Appendix dorsalis The medial caudal appendage arising from the epiproct, lying above the anus; present in apterygotes, most mayflies and some fossil insects.

Apterous Wingless.

Apterous neotenic, *see* **Ergatoid**.

Arachnophobia Fear of arachnids (spiders and relatives).

Aroliar pad Pretarsal ventral adhesive pad (Fig. 2.19).

Arolium (*pl.* arolia) Pretarsal pad-like or sac-like structure(s) lying between the claws (Fig. 2.19).

Arrhenotokous parthenogenesis The production of haploid male offspring from unfertilized eggs.

Arthrodial membrane Soft, stretchable cuticle, e.g. between segments (Fig. 2.4).

Articular sclerites Separate, small, movable plates that lie between the body and a wing.

Asynchronous muscle A muscle that contracts many times per nerve impulse, as in many flight muscles and those controlling the cicada tymbal.

Atrium A chamber, especially inside a tubular conducting system, such as the tracheal system (Fig. 3.10(a)).

Automimic A condition of **Batesian mimicry** in which palatable members of a species are defended by their resemblance to members of the same species that are chemically unpalatable.

Autotomy The shedding of appendage(s), notably for defence.

Axillary area An area at the wing base bearing the wing articulation (Fig. 2.20).

Axillary plates Two (anterior and posterior) articulating plates which are fused with the veins in an odonate wing; the anterior supports the costal vein, the posterior, the remaining veins; in Ephemeroptera there is only a posterior plate.

Axillary sclerites Three or four sclerites, which together with the **humeral plate** and

tegula comprise the articular sclerites of the neopteran wing base (Fig. 2.21).

Axon A nerve cell fibre that transmits a nerve impulse away from the cell body (Fig. 3.5); *see also* **Dendrite**.

Azadirachtin (AZ) The chemically active principle of the neem tree.

Basal At or towards either the base or the main body, or closer to point of attachment (Fig. 2.8).

Basalare (*pl.* basalaria) A small sclerite, one of the epipleurites that lies anterior to the pleural wing process; an attachment for the direct flight muscles (Fig. 2.18).

Basisternum The main sclerite of the eusternum, lying between the anterior presternum and posterior sternellum (Fig. 2.18).

Batesian mimicry A mimetic system in which a palatable species obtains protection from predation by resembling an unpalatable species; *see also* **Müllerian mimicry**.

Benthos The bottom sediments of aquatic habitats and/or the organisms that live there.

Biological control (*also* **biocontrol**) The human use of selected living organisms to control populations of pest species (plant or animal).

Biological monitoring Using plants or animals to detect changes in the environment.

Biological transfer The movement of a disease organism from one host to another by one or more vectors in which there is a biological cycle of disease.

Bioluminescence The production by an organism of cold light, commonly involving the action of an enzyme (luciferase) on a substrate (luciferin).

Bivoltine Having two generations in one year; *see also* **Univoltine**; **Multivoltine**.

Bivouac An army ant camp during the mobile phase.

Borer The maker of burrows in dead or living tissue.

Brachypterous Having shortened wings.

Brain In insects, the supraoesophageal

ganglion of the nervous system (Fig. 3.6), comprising protocerebrum, deutocerebrum and tritocerebrum.

Broad spectrum insecticide An insecticide with a broad range of targets, usually acting generally on the insect nervous system.

Brood A clutch of individuals that hatch at the same time from eggs produced by one set of parents.

Bud, *see* **Imaginal disc**.

Bursa copulatrix The female **genital chamber** if functioning as a copulatory pouch (Box 5.3); in Lepidoptera, the primary receptacle for sperm (Fig. 5.4).

Bursicon A neuropeptide hormone that controls hardening and darkening of the cuticle after **ecdysis**.

Caecum (*pl.* caeca) A blind-ending tube or sac (Figs 3.1 and 3.13).

Calliphorin A protein produced in the fat body and stored in the haemolymph of larval Calliphoridae (Diptera).

Calyx (*pl.* calyces) A cup-like expansion, especially of the oviduct into which the ovaries open (Fig. 3.20(a)).

Camouflage Crypsis, in which an organism is indistinguishable from its background.

Campaniform sensillum A mechanoreceptor that detects stress on the cuticle, comprising a dome of thin cuticle overlying one neurone per sensillum, located especially on joints (Fig. 4.2(b)).

Cantharophily Plant pollination by beetles.

Cap cell The outermost cell of a sense organ such as a chordotonal organ (Fig. 4.3).

Carbamate A synthetic insecticide.

Cardiopeptide A neuropeptide hormone that stimulates the dorsal vessel ('heart') causing haemolymph movement.

Cardo The proximal part of the maxillary base (Fig. 2.10).

Castes Morphologically distinctive groups of individuals within a single species of social insect, usually differing in behaviour.

Caudal At or towards the anal (tail) end.

Caudal filament One of two or three terminal filaments (Fig. 7.4; Box 9.1).

Caudal lamella One of two or three terminal gills (Box 9.2).

Cavernicolous (*also* **troglodytic; troglobiont**) Living in caves.

Cecidology The study of plant galls.

Cecidozoa Gall-inducing animals.

Cell Of a wing, an area of the wing membrane partially or completely surrounded by veins; *see also* **Closed cell**; **Open cell**.

Cement layer The outermost layer of the cuticle (Fig. 2.1), often absent.

Central nervous system (CNS) In insects, the central series of ganglia extending for the length of the body (Fig. 3.6); *see also* **Brain**.

Cephalic Pertaining to the head.

Cercus (*pl.* cerci) One of a pair of appendages originating from abdominal segment 11 but usually visible as if on segment 10 (Figs 2.23(b) and 3.1).

Cervical sclerite Small sclerite on the membrane between head and thorax (actually the first thoracic segment) (Fig. 2.9).

Chitin The major component of arthropod cuticle, a polysaccharide composed of subunits of acetyl-glucosamine and glucosamine (Fig. 2.2).

Chitin synthesis inhibitor An insecticide that prevents chitin formation.

Chloride cells Osmoregulatory cells found in the epithelium of abdominal gills of aquatic insects.

Chordotonal organs Sense organs (mechanoreceptors) that perceive vibrations, comprising one to several elongate cells called **scolopidia** (Fig. 4.3); examples include the **tympanal organ**, **subgenual organ** and **Johnston's organ**.

Chorion The outermost shell of an insect egg, which may be multilayered, including exo- and endochorion and wax layer (Fig. 5.7).

Chrysalis, *see* **Pupa**.

Cibarium The dorsal food pouch, lying between the hypopharynx and the inner wall of the clypeus, often with muscular pump (Figs 2.14 and 3.14).

Circadian rhythms Repeated periodic behaviour with an interval of about 24 hours.

Clade A group of organisms proposed to be monophyletic, that is, all descendants of one common ancestor.

Cladistics A classification system in which **clades** are the only permissible groupings.

Cladogram A diagram illustrating the branching sequence of purported relationships of organisms, based on the distribution of shared derived features (synapomorphies).

Claval furrow A flexion line on the wing that separates the **clavus** from the **remigium** (Fig. 2.20).

Clavus An area of the wing delimited by the claval furrow and the posterior margin (Fig. 2.22(e)).

Claw (*strictly*, **pretarsal claw**; **unguis**) A hooked structure on the distal end of the pretarsus, usually paired (Fig. 2.19); more generally, any hooked structure.

Closed cell An area of the wing membrane bounded entirely by veins; *see also* **Open cell**.

Closed tracheal system A gas exchange system comprising tracheae and tracheoles but lacking spiracles and therefore closed to direct contact with the atmosphere (Figs 3.11(d) (f)); *see also* **Open tracheal system**.

Clypeus The part of the insect head to which the labrum is attached anteriorly (Figs 2.9, 2.10); it lies below the frons, with which it may be fused in a **frontoclypeus** or separated by a suture.

Coevolution Evolutionary interactions between two organisms, such as plants and pollinators, hosts and parasites; the degree of specificity and reciprocity varies; *see also* **Guild coevolution; Specific coevolution; Phyletic tracking**.

Colleterial glands Accessory glands of the female internal genitalia that produce secretions used to cement eggs to the substrate.

Collophore The ventral tube of Collembola.

Colon The hindgut between the ileum and rectum (Figs 3.1, 3.13).

Comb In a social hymenopteran nest, a layer of regularly arranged cells (Fig. 11.5).

Common (median) oviduct In female insects, the tubes leading from the fused **lateral oviducts** to the **vagina** (Fig. 3.20(a)).

Competency The potential of a termite of one caste to become another, e.g. a worker to become a soldier.

Compound eye An aggregation of **ommatidia**, each acting as a single facet of the eye (Fig. 2.9).

Conjunctiva, conjunctival membrane, *see* **Intersegmental membrane**.

Connective Anything that connects; more specifically, the paired longitudinal nerve cords that connect the ganglia of the central nervous system.

Constitutive defence Part of the normal chemical composition; *see also* **Induced defence**.

Contact poisons Insecticides that poison by direct contact.

Coprophage (*adj.* coprophagous) A feeder on dung or excrement (Fig. 8.3).

Corbicula The pollen basket of bees (Fig. 11.3).

Corium A section of the heteropteran hemelytron (fore wing), differentiated from the clavus and membrane, usually leathery (Fig. 2.22(e)).

Cornea The cuticle covering the eye or ocellus (Fig. 4.10).

Corneagenous cell One of the translucent cells beneath the **cornea** which secretes and supports the corneal lens (Fig. 4.10).

Cornicles (*also* **siphunculi**) A pair of tubes on the abdomen of aphids that discharge defensive secretions, especially alarm pheromones.

Corpora allata (*sing.* corpus allatum) Paired endocrine glands associated with the stomodeal ganglia behind the brain (Fig. 3.8), the source of juvenile hormone.

Corpora cardiaca (*sing.* corpus cardiacum) Paired glands lying close to the aorta and behind the brain (Fig. 3.8), acting as stores and producers of neurohormones.

Cosmopolitan Distributed worldwide (or nearly so).

Costa The most anterior longitudinal wing

vein, running along the costal margin of the wing and ending near the apex (Figs 2.20, 2.21).

Costal fracture A break or weakness in the costal margin in Heteroptera that divides the corium, separating the cuneus from the embolium (Fig. 2.22(e)).

Coxa (*pl.* coxae) The proximal (basal) leg segment (Fig. 2.19).

Crepuscular Active at low light intensities, dusk or dawn; *see also* **Diurnal**; **Nocturnal**.

Crista acustica (*also* **crista acoustica**) The main chordotonal organ of the tibial typanal organ of katydids (Orthoptera: Tettigoniidae) (Fig. 4.4).

Crochets Curved hooks, spines or spinules on **prolegs**.

Crop The food storage area of the digestive system, posterior to the oesophagus (Figs 3.1 and 3.13).

Cross resistance Resistance of an insect to one insecticide providing resistance to a different insecticide.

Cross-veins Transverse wing veins that link the longitudinal veins.

Crypsis Camouflage by resemblance to environmental features.

Cryptic Hidden, camouflaged, concealed.

Cryptobiosis The state of a living organism during which there are no signs of life and metabolism virtually ceases.

Cryptonephric system A condition of the excretory system in which the **Malpighian tubules** form an intricate contact with the rectum, allowing production of dry excreta (Box 3.4).

Crystalline cone A hard crystalline body lying beneath the cornea in an ommatidium (Fig. 4.10(c)).

Crystalline lens A lens lying beneath the cuticle of the **stemma** of some insects (Fig. 4.10(a)).

Ctenidium Comb (Box 14.3).

Cubitus (Cu) The sixth longitudinal vein, lying posterior to the media, often divided into an anterior branched CuA and a posterior unbranched CuP (Fig. 2.21).

Cuneus The distal section of the corium in the heteropteran wing, separated from the embolium by a costal fracture (Fig. 2.22(e)).

Cursorial Running or adapted for running.

Cuticle The external skeletal structure, secreted by the epidermis, composed of **chitin** and protein comprising several differentiated layers (Fig. 2.1).

Cycloalexy Forming aggregations in defensive circles (Fig. 13.6).

Cyclodienes A class of organochlorine insecticides.

Day-degrees A measure of physiological time, the product of time and temperature above a **threshold**.

Deciduous Falling off, detaching (e.g. at maturity).

Decticous pupa An **exarate pupa** in which the mandibles are articulated (Fig. 6.7).

Delayed effect Of a defensive chemical, an effect that appears after a time lapse from first encounter; *see also* **Immediate effect**.

Delayed parasitism Parasitism in which hatching of the parasite (or parasitoid) egg is delayed until the host is mature.

Delusory parasitosis A psychotic illness in which parasitic infection is imagined.

Dendrite A fine branch of a nerve cell (Fig. 3.5).

Dermal gland A unicellular epidermal gland that may secrete cements, wax, etc. and probably pheromones (Figs 2.1 and 2.5).

Determinate growth With a distinctive final, adult instar; *see also* **Indeterminate growth**.

Detritivore (*adj.* detritivorous) An eater of organic detritus of plant or animal origin.

Deuterotoky, *see* **Amphitoky**.

Deutocerebrum The middle part of the insect brain; the ganglion of the second segment, comprising antennal and olfactory lobes.

Developmental (or **growth**) **threshold** The temperature below which no development takes place.

Diapause Delayed development controlled by particular environmental conditions.

Diapause hormone A hormone produced by neurosecretory cells in the suboesophageal ganglion that affects the timing of future development of eggs.

Dicondylar An articulation (as of a mandible) with two points of articulation (condyles).

Diffuse coevolution, *see* **Guild coevolution**

Diploid With two sets of chromosomes.

Direct flight muscles Muscles that power flight that are attached directly to the wing (Fig. 3.4); *see also* **Indirect flight muscles**.

Dispersal Movement of an individual or population away from its birth site.

Distal At or near the furthest end from the attachment of an appendage – opposite to **proximal** (Fig. 2.8).

Diurnal Active during the day; *see also* **Crepuscular**; **Nocturnal**.

Domatia Plant chambers produced specifically to house certain arthropods, especially ants.

Dorsal On the upper surface (Fig. 2.8).

Dorsal closure The embryological process in which the dorsal wall of an embryo is formed by growth of the germ band to surround the yolk.

Dorsal diaphragm The main fibromuscular septum that divides the haemocoel into the **pericardial** and **perivisceral sinuses** (Fig. 3.9).

Dorsal vessel The 'aorta' and 'heart', the main pump for haemolymph; a longitudinal tube lying in the dorsal pericardial sinus (Fig. 3.9).

Dorsum The upper surface.

Drift Passive movement due to water or air currents.

Drone The male bee, especially of honey bees and bumble bees, derived from an unfertilized egg.

Dufour's gland In aculeate hymenopterans, a sac opening into the poison duct near the sting (Fig. 13.10), site of production of pheromones and/or poison components.

Dulosis A slave-like relationship between parasitic ant species and the captured brood from another species.

Dyar's rule An observational 'rule' governing the size increment found between subsequent instars of the same species (Fig. 6.11).

Ecdysis (*adj.* ecdysial) The final stage of moulting, the process of casting off of the cuticle (Fig. 6.10).

Ecdysone The steroid hormone secreted by the prothoracic gland that is converted to 20-hydroxyecdysone, which stimulates the epidermal moulting.

Ecdysteroid The general term for steroids that induce moulting (Fig. 6.9).

Ecdysterone An old term for 20-hydroxyecdysone, the major moult-inducing steroid.

Eclosion The release of the adult insect from the cuticle of the previous instar; sometimes used of hatching from the egg.

Eclosion hormone A neuropeptide with several functions associated with adult eclosion including increasing cuticle extensibility.

Economic injury level (EIL) The level at which economic pest damage equals the costs of their control.

Economic threshold (ET) The pest density at which control must be applied to prevent the **economic injury level** being reached.

Ectoparasite A parasite that lives externally on, and at the expense of, another, which it does not kill.

Ectoparasitoid A parasite that lives externally on, and at the expense of, another, which it kills.

Ectoperitrophic space The space between the **peritrophic membrane** and the midgut wall (Fig. 3.16).

Ectothermia (*adj.* ectothermic) The inability to regulate the body temperature relative to the surrounding environment.

Ejaculatory duct The duct that leads from the fused vas deferens to the gonopore (Fig. 3.20(b)), through which semen or sperm is transported.

Elaiosome A food body forming an appendage on a plant seed (Fig. 10.8).

Elytron (*pl.* elytra) The modified, hardened fore wing of a beetle that protects the hind wing (Fig. 2.22(d)).

Empodium (*pl.* empodia) A central spine or pad on the pretarsus of Diptera.

Encapsulation A reaction of the host to an endoparasitoid in which the invader is surrounded by haemocytes that eventually form a capsule.

Endemic Restricted to a particular geographic area.

Endite An inwardly directed (mesal) appendage or lobe of a limb segment (Fig. 7.4).

Endocuticle The flexible, unsclerotized inner layer of the procuticle (Fig. 2.1); *see also* **Exocuticle**.

Endogenous rhythms Clock or calendar-like activity patterns, commonly **circadian**, unaffected by external conditions.

Endoparasite A parasite that lives internally at the expense of another, which it does not kill.

Endoparasitoid A parasite that lives internally at the expense of another, which it kills.

Endoperitrophic space In the gut, the space enclosed within the peritrophic membrane (Fig. 3.16).

Endophallus The inner, eversible tube of the penis (Fig. 5.2).

Endophilic Indoor loving, as of an insect that feeds inside a dwelling.

Endopterygote Development in which the wings form within pockets of the integument, with eversion taking place only at the larval–pupal moult (as in the monophyletic grouping Endopterygota).

Endothermy (*adj.* endothermic) The ability to regulate the body temperature higher than the surrounding environment.

Energids In an embryo, the daughter nuclei cleavage products and their surrounding cytoplasm.

Entomopathogen A pathogen (disease-causing organism) that attacks insects particularly.

Entomophage (*adj.* entomophagous) An eater of insects.

Entomophily Pollination by insects.

Entomophobia Fear of insects.

Environmental manipulation Alteration of the environment, particularly to enhance natural populations of insect predators and parasitoids.

Enzootic A disease present in a natural host within its natural range.

Epicoxa A basal leg segment (Fig. 7.4), forming the articular sclerites in all extant insects, believed to have borne the exites and endites that fused to form the evolutionary precursors of wings.

Epicranial suture A Y-shaped line of weakness on the vertex of the head where the split at moulting occurs (Fig. 2.10(a)).

Epicuticle The inextensible and unsupportive outermost layer of cuticle, lying outside the **procuticle** (Fig. 2.1).

Epidemic The spread of a disease from its **endemic** area and/or from its normal host.

Epidermis The unicellular layer of ectodermally derived integument that secretes the cuticle (Fig. 2.1).

Epimeron (*pl.* epimera) The posterior division of the pleuron of a thorax, separated from the **episternum** by the **pleural suture** (Fig. 2.18).

Epimorphic development Development in which the segment number is fixed in the embryo before hatching.

Epipharynx The ventral surface of the labrum, a membranous roof to the mouth (Fig. 2.15).

Epipleurite The more dorsal of the sclerites formed when the pleuron is divided longitudinally; *or* one of two small sclerites of a wing-bearing segment – the anterior **basalare** and posterior **subalare** (Fig. 2.18).

Epiproct Dorsal relic of segment 11 (Fig. 2.23).

Episternum (*pl.* episterna) The anterior division of the pleuron, separated from the **epimeron** by the **pleural suture** (Fig. 2.18).

Epistomal suture, *see* **Frontoclypeal suture**.

Epizootic Of a disease, when epidemic (there is an unusually high number of cases and/or deaths).

Ergatoid (*also* **apterous neotenic**) In termites, a supplementary reproductive derived from a worker, held in a state of arrested development and lacking wings, able to replace reproductives if they die; *see also* **Neotenic**.

Erythrocyte A red blood cell.

Erythrocyte schizogonous cycle The stage in the asexual development of a malaria parasite (*Plasmodium* spp.) in which **trophozoites** within the vertebrate host red blood cells divide to form **merozoites** (Fig. 14.1).

Euplantula (*pl.* euplantulae) (*also* **pulvillus**) Pad-like structure on the ventral surface of some tarsomeres of the leg.

Eusocial Exhibiting co-operation in reproduction and division of labour, with overlap of generations.

Eusternum (*pl.* eusterna) The dominant ventral plate of the thorax that frequently extends into the pleural region (Fig. 2.18).

Eutrophication Nutrient enrichment, especially of water bodies.

Evolutionary systematics A classification system in which **clades** and **grades** are recognized.

Exarate pupa A pupa in which the appendages are free from the body (Fig. 6.7), as opposed to being cemented; *see also* **Obtect pupa**.

Excretion The elimination of metabolic wastes from the body, or their internal storage in an insoluble form.

Exite An outer appendage or lobe of a limb segment (Fig. 7.4).

Exocuticle The rigid, sclerotized outer layer of the procuticle (Fig. 2.1); *see also* **Endocuticle**.

Exo-erythrocytic, *see* **Pre-erythrocytic schizogonous cycle**.

Exogenous rhythms Activity patterns governed by variations in the external environment (e.g. light, temperature, etc.).

Exophily (*adj.* exophilic) Out-of-doors loving, used of biting insects that do not enter buildings; *see also* **Endophilic**.

Exopterygote development Development in which the wings form progressively in sheaths that lie externally on the dorsal or dorsolateral surface of the body (as in the paraphyletic grouping Exopterygota).

Exoskeleton The external, hardened cuticular skeleton to which muscles are attached internally.

Extra-oral digestion Digestion that takes place outside the organism, by secretion of salivary enzymes on or into the food, with soluble digestive products being sucked up.

Exuviae (*sing.* and *pl.*) The cuticle cast off after a moult.

Facultative Not compulsory, optional behaviour, such as facultative parasitism, in which a free-living organism may adopt a parasitic mode of life.

Fat body A loose or compact aggregation of cells, mostly **trophocytes**, suspended in the haemocoel, responsible for storage and excretion.

Femur (*pl.* femora) The third segment of an insect leg, following the coxa and trochanter; often the stoutest leg segment (Fig. 2.19).

Fermentation Breakdown of complex molecules by microbes, as carbohydrates by yeast.

File A toothed or ridged structure used in sound production by stridulation through contact with a **scraper**.

Filter chamber Part of the alimentary canal of many hemipterans, in which the anterior and posterior parts of the midgut are in intimate contact; a system in which most fluid bypasses the absorptive midgut (Box 3.3).

Fitness (Darwinian; inclusive, extended) The contribution of an individual to the gene pool through its offspring (Darwinian) or by enhanced success of its kin (inclusive, extended).

Flabellum In bees, the lobe at the tip of the glossae ('tongue') (Fig. 2.11).

Flagellomere One of the subdivisions of a 'multisegmented' (actually multiannulate) antennal flagellum.

Flagellum The third part of an antenna, distal to the scape and pedicel; more generally, any whip or whip-like structure.

Flexion-line A line along which a wing flexes (bends) when in flight (Fig. 2.20).

Fluctuating asymmetry The level of deviation from absolute symmetry in a bilaterally symmetrical organism, argued to be due to variable stress during development.

Fold-lines A line along which a wing is folded when at rest (Fig. 2.20).

Follicle The oocyte and follicular epithelium; more generally, any sac or tube.

Follicular relic Morphological evidence left in the ovary showing that an egg has been laid (or resorbed), which may include dilation of the lumen and/or pigmentation.

Food canal Canal anterior to the cibarium through which fluid food is ingested (Figs 2.11, 2.12).

Forage To seek and gather food.

Fore Anterior, towards the head.

Fore wings The anterior pair of wings, usually on the mesothorax.

Foregut (*also* **stomodeum**) The part of the gut that lies between the mouth and the midgut (Fig. 3.13), derived from the ectoderm.

Forensic entomologist A scientist who studies the role of insects in legal matters.

Formulation The components and proportions of additional substances that accompany an insecticide when prepared for application.

Fossorial Digging, or adapted for digging (Fig. 8.2).

Frass Solid excreta of an insect, particularly a larva.

Frenate coupling A form of wing-coupling in which one or more hind-wing structures (**frenulum**) attach to a retaining structure (**retinaculum**) on the fore wing.

Frenulum Spines or bristles on the costa of the hind wing that lock with the fore-wing retinaculum in flight.

Frons The single medio-anterior sclerite of the insect head, usually lying between the epicranium and the clypeus (Figs 2.9 and 2.10).

Frontoclypeus The combined **frons** and **clypeus**.

Frontoclypeal (*or* **epistomal**) **suture** A groove that runs across the insect's face, often separating frons from clypeus (Fig. 2.9).

Fructivore (*adj.* fructivorous) A fruit-eater.

Fungivore (*adj.* fungivorous) A fungus-eater.

Furcula The abdominal springing organ of Collembola (Box 8.2); with the fulcral arm, the lever of the hymenopteran sting (Fig. 13.10).

Galea The lateral lobe of the maxillary stipes (Figs 2.10, 2.11, 2.12).

Gall An aberrant plant growth produced in response to the activities of another organism, often an insect (Fig. 10.5).

Ganglion (*pl.* ganglia) A nerve centre; in insects, forming fused pairs of white, ovoid bodies lying in a row ventrally in the body cavity, linked by a double nerve cord (Figs 3.1 and 3.6).

Gas exchange The system of oxygen uptake and carbon dioxide elimination.

Gas gills Specific respiratory (gas exchange) surfaces on aquatic insects, often as abdominal lamellae, but may be present almost anywhere on the body.

Gaster The swollen part of the abdomen of aculeate Hymenoptera, lying posterior to the **petiole** (Box 11.3).

Gena (*pl.* genae) Literally, a cheek; on each side of the head, the part lying beneath the compound eye (Fig. 2.9).

Genital chamber A cavity of the female body wall that contains the gonopore (Fig. 3.20(a)), also known as the **bursa copulatrix** if functioning as a copulatory pouch.

Genitalia All ectodermally derived structures of both sexes associated with reproduction (copulation, fertilization and oviposition).

Genus (*pl.* genera, *adj.* generic) The name of the taxonomic category ranked between species and family; an assemblage of one or more species united by one or more derived features and therefore believed to be of a single evolutionary origin (i.e. **monophyletic**).

Germ anlage In embryology, the germ disc that denotes the first indication of a developing embryo (Fig. 6.5).

Germ band In embryology, the postgastrulation band of thickened cells on the ventral gastroderm, destined to form the ventral part of the developing embryo (Fig. 6.5).

Germarium The structure within an **ovariole**

in which the **oogonia** give rise to **oocytes** (Fig. 3.20(a)).

Giant axon A nerve fibre that conducts impulses rapidly from sense organ to muscle.

Gill A gas exchange organ, found in various forms in aquatic insects.

Gizzard, *see* **Proventriculus.**

Glossa (*pl.* glossae) The 'tongue', one of a pair of lobes on the inner apex of the prementum (Fig. 2.11).

Gonapophysis (*pl.* gonapophyses) A valve (part of the shaft) of the **ovipositor** (Fig. 2.23); also in the genitalia of many male insects (Fig. 2.24).

Gonochorism Sexual reproduction in which males and females are separate individuals.

Gonocoxite The base of an appendage, formed of **coxa** plus **trochanter**, of a genital segment (8 or 9) (Figs 2.23, 2.24); *see also* **Valvifer.**

Gonopore The opening of the genital duct (Fig. 3.20), in the unmodified female the opening of the common oviduct, in the male the opening of the ejaculatory duct.

Gonostyle The style (rudimentary appendage) of the ninth segment (Fig. 2.23), often functioning as a male clasper (Fig. 2.24).

Grade A paraphyletic group, one which does not include all descendants of a common ancestor, united by shared primitive features.

Gregarious Forming aggregations.

Gressorial Walking, or adapted for walking.

Growth threshold, *see* **Development threshold.**

Guild (*or* **diffuse**) **coevolution** Concerted evolutionary change that takes place between groups of organisms, as opposed to between two species; *see also* **Specific coevolution.**

Gula A ventromedian sclerotized plate on the head of prognathous insects (Fig. 2.10).

Gyne A reproductive female hymenopteran, a queen.

Habituation Reduction in the response to a stimulus with repeated exposure, through modification of the central nervous system.

Haemocoel (*also* **hemocoel**) The main body cavity of many invertebrates, including insects, formed from an expanded 'blood' system.

Haemocoelous viviparity Viviparity (producing living offspring) in which the immature stages develop within haemocoel of the parent female, as for example in Strepsiptera.

Haemocyte (*also* **hemocyte**) Insect blood cells.

Haemolymph (*also* **hemolymph**) The fluid filling the **haemocoel**.

Hair A cuticular extension, also called a **macrotrichium** or **seta**.

Hair plate A group of sensory hairs that act as a **proprioceptor** for movement of articulating parts of the body (Fig. 4.2(a)).

Haltere The modified hind wing in Diptera, acting as a balancer (Fig. 2.22(f))

Hamuli (*adj.* hamulate) Hooks along the anterior (costal) margin of the hind wing of Hymenoptera which couple the wings in flight by catching on a fold of the fore wing.

Haplodiploidy (*adj.* haplodiploid) A sex determination system in which **diploid** eggs give rise to females, **haploid** ones to males.

Haploid With one set of chromosomes.

Haustellate Sucking, as of mouthparts

Haustellum Sucking mouthparts (Fig. 14.2).

Head The anterior of the three major divisions (**tagmata**) of an insect body.

Hellgrammite The larva of Megaloptera.

Hemelytron (*pl.* hemelytra) The fore wing of Heteroptera, with thickened basal section and membranous apical section (Fig. 2.22(e)).

Hemimetaboly With development in which the body form gradually changes at each moult, with wing buds growing larger at each moult; incomplete metamorphosis; *see also* **Holometaboly.**

Hemocoel, *see* **Haemocoel.**

Hemocyte, *see* **Haemocyte.**

Hemolymph, *see* **Haemolymph.**

Hermaphroditism Having individuals that possess both testes and ovaries.

Heteromorphosis (*also* **hypermetamorphosis**)

Undergoing a major change in morphology between larval instars, as from **triungulin** to grub.

Hind At or towards the posterior.

Hind wings The wings on the metathoracic segment.

Hindgut (*also* **proctodeum**) The posterior section of the gut, extending from the end of the midgut to the anus (Fig. 3.13).

Holoblastic In embryology, a complete cleavage of the egg.

Holometaboly (*adj.* holometabolous) With development in which the body form abruptly changes at the pupal moult; complete metamorphosis, as in the group Holometabola; *see also* **Hemimetaboly**.

Homology (*adj.* homologous) Morphological identity or similarity of a structure in two (or more) organisms due to common evolutionary origin.

Homoplasy Characters acquired convergently or in parallel, rather than by direct inheritance from an ancestor.

Honeydew A watery fluid containing sugars eliminated from the anus of some Homoptera.

Hormone A chemical messenger that regulates some activity at a distance from the endocrine organ that produced it.

Host An organism that harbours another, especially a parasite or parasitoid, either internally or externally.

Host discriminate To choose between different hosts.

Host preference To prefer one host over another.

Host regulation The ability of a parasitoid to manipulate the host's physiology.

Humeral plate One of the articular sclerites of the neopteran wing base (Fig. 2.21); *see also* **Tegula**; **Axillary sclerite**.

Humus Organic soil.

Hydrostatic skeleton Turgid structural support provided by fluid pressure maintained by muscle contractions on a fixed volume of liquid within especially larval insects.

Hypermetamorphosis, *see* **Heteromorphosis**.

Hyperparasite (*adj.* hyperparasitic) A parasite that lives upon another parasite.

Hyperparasitoid A secondary parasitoid that develops upon another parasite or parasitoid.

Hypognathous With the head directed vertically and mouthparts directed ventrally.

Hypopharynx A median lobe of the preoral cavity of the mouthparts (Fig. 2.10).

Hyporheic Living in the substrate beneath the bed of a waterbody.

Idiobiont A parasitoid that prevents its host from developing any further, by paralysis or death; *see also* **Konobiont**.

Ileum The second section of the hindgut, preceding the colon (Figs 3.1, 3.13).

Imaginal disc *or* **bud** Latent adult structures in an immature insect, visible as groups of undifferentiated cells (Fig. 6.4).

Imago (*pl.* imagines, imagos) The adult insect.

Immediate effect Of a defensive chemical, an effect that appears immediately on first encounter; *see also* **Delayed effect**.

Indeterminate growth Without a distinctive final, adult instar – with no definitive terminal moult; *see also* **Determinate growth**.

Indirect flight muscles Muscles that power flight by deforming the thorax rather than directly moving the wings (Fig. 3.4); *see also* **Direct flight muscles**.

Indirect system Of flight, with power for flight coming from regular deformation of the thorax by indirect flight muscles, rather than from muscular connection to the wings.

Induced defence A chemical change, deleterious to herbivores, induced in foliage as a result of feeding damage.

Industrial melanism The phenomenon of dark **morphs** occurring with a higher than usual frequency in areas in which industrial pollution darkens tree trunks and other surfaces upon which insects may rest.

Inhalation poison An insecticide with a fumigant action.

Innate behaviour Behaviour requiring no choice or learning.

Inner epicuticle The innermost of the three layers of epicuticle, with the procuticle beneath it (Fig. 2.1); *see also* **Outer epicuticle**; **Superficial layer**.

Inoculation To infect with a disease by introducing it into the blood; in biological control, to periodically release a natural enemy which breeds so that the progeny provide control.

Inquiline An organism that lives in the home of another, sharing food; in entomology, used particularly of residents in the nests of social insects, or in plant galls induced by another organism; *see also* **Integrated inquiline**; **Non-integrated inquiline**.

Insectivore (*adj.* insectivorous), (*also* **entomophage**) An insect eater.

Instar The growth stage between two successive moults.

Integrated inquiline An inquiline that is incorporated into the social life of the host by behavioural modification of both inquiline and host.

Integrated pest management (IPM) Integration of chemical means of insect control with other methods, notably **biological control** and habitat manipulation.

Integument The epidermis plus cuticle; the outer covering of the living tissues of an insect.

Intercropping Mixed planting of agricultural crops.

Interference In colours, iridescent colours produced by variable reflection of light by narrowly separated surfaces (as scales of Lepidoptera); in intra- and interspecific interactions between predators and parasitoids, a reduction in the profitability of otherwise high resource density.

Intermediate organ A chordotonal organ in the subgenual organ of the foreleg of some orthopterans (Fig. 4.4), associated with the **tympanum** and believed to respond to sound frequencies of 2–14 kHz.

Intermoult period, *see* **Stadium**.

Intersegmental groove, *see* **Antecostal suture**.

Intersegmental membrane (*also* **conjunctiva**) Membrane between segments, particularly of the abdomen (Fig. 2.7).

Intersternite An intersegmental sternal plate posterior to the eusternum, known as the **spinasternum** except on the metasternum (Fig. 2.7).

Inundative release In biological control, swamping a pest with large numbers of control agents, with control deriving from the released organisms rather than any of their progeny.

Johnston's organ A chordotonal (sensory) organ within the antennal pedicel.

Jugate coupling A mechanism for coupling the fore and hind wings in flight by a large jugum of the fore wing overlapping the hind wing.

Jugal area (*also* **jugum**, *pl.* juga) The posterobasal area of the wing, delimited by the jugal fold and the wing margin (Fig. 2.20).

Jugal fold A fold-line of the wing, dividing the jugal area from the anal area (Fig. 2.20).

Juvenile hormone (JH) A 16, 17 or 18 C-chain hormone released by the corpora allata into the haemolymph, involved in many aspects of insect physiology, including modification of the expression of a moult.

Juvenile hormone mimics (*also* **juvenoids**) Synthetic chemicals that mimic the effect of JH on development.

Kairomone A communication chemical that benefits the receiver and is disadvantageous to the producer; *see also* **Allomone**; **Synomone**.

Kinesis (*pl.* kineses) Movement of an organism in response to a stimulus, usually restricted to response to stimulus intensity only.

King The male primary reproductive in termites (Fig. 11.7).

Kinship component An indirect contribution to an individual's inclusive **fitness** derived from increased reproductive success of the individual's kin (relatives) through the altruistic assistance of the individual.

Klinokinesis Movement resembling a 'random walk', in which changes in direction are made when unfavourable stimuli are encountered, with frequency of turning dependent upon the stimulus intensity.

Klinotaxis A movement in a definite direction relative to a stimulus, either directly towards or away from the source.

Konobiont A parasitoid that allows its host to continue to develop; *see also* **Idiobiont**.

Labella (*sing.* labellum) In certain flies, paired lobes at the apex of the proboscis, derived from labial palps (Box 14.4).

Labial palp A 1–5-segmented appendage of the **labium** (Figs 2.9, 2.10).

Labium (*adj.* labial) The 'lower lip', forming the floor of the mouth, often with a pair of palps and two pairs of median lobes (Figs 2.9 and 2.10); derived from the sixth head segment.

Labrum (*adj.* labral) The 'upper lip', forming the roof of the preoral cavity and mouth (Figs 2.9, 2.10); derived from the first head segment.

Lacinia The mesal lobe of the maxillary stipes (Figs 2.9 and 2.10).

Larva (*pl.* larvae) An immature insect after emerging from the egg, often restricted to insects in which there is complete metamorphosis (holometaboly) but sometimes used for any immature insect that differs strongly from its adult; *see also* **Nymph**.

Lateral At, or close to, the side (Fig. 2.8).

Lateral oviducts In female insects, the paired tubes leading from the ovaries to the common oviduct (Fig. 3.20(a)).

Laterosternite The result of a fusion of the eusternum and a pleural sclerite.

Leaf miner A feeder on the mesophyll layer between the upper and lower epidermis of a leaf (Fig. 10.2).

Lentic Of standing water.

Ligation An experimental technique which isolates one part of the body of a living insect from another, usually by tightening a ligature.

Light compass response, *see* **Menotaxis**.

Ligula The glossae plus paraglossae of the prementum of the labium, whether fused or separate.

Litter A layer of dead vegetative matter overlying the soil.

Longitudinal In the direction of the long axis of the body.

Longitudinal muscles Muscles running along the long axis of the body (Fig. 2.7(a)).

Lotic Of flowing water.

Macrogamete The female gametocyte of a malaria parasite, *Plasmodium* spp. (Fig. 14.1); *see also* **Microgamete**.

Macrophage An eater of large particles; *see also* **Microphage**.

Macrotrichium (*pl.* macrotrichia) A trichoid sensillum, also called a **seta** or **hair**.

Maggot A legless larval insect, usually with a reduced head, frequently a fly.

Major worker An individual of the largest-sized worker caste of termites and ants, specialized for defence; *see also* **Media worker**; **Minor worker**.

Malpighian tubules Thin, blind-ending tubules, originating near the junction of mid- and hindgut (Figs 3.1, 3.13), predominantly involved in regulation of salt, water and nitrogenous waste excretion.

Mandible (*adj.* mandibular) The jaws, either jaw-like in shape in biting and chewing (mandibulate) insects (Figs 2.9, 2.10), or modified as narrow stylets in piercing and sucking insects (Fig. 2.13); the first pair of jaws; derived from the fourth head segment.

Mandibulate Possessing mandibles.

Masquerade A form of **crypsis** in which an organism resembles a feature of its environment that is of no interest to a predator; also called **mimesis**.

Mating disruption A form of insect control in which synthetic sex pheromones (usually of the female) are maintained artificially at a higher level than the background, interfering with mate location.

Matrifilial Of eusocial Hymenoptera whose colonies consist of mothers and their daughters.

Maxilla (*pl.* maxillae) The second pair of jaws, jaw-like in chewing insects (Figs 2.9, 2.10), variously modified in others (Fig. 2.13); derived from the fifth head segment.

Maxillary palp A 1–7-segmented sensory appendage borne on the **stipes** of the maxilla (Figs 2.9, 2.10).

Mechanical transfer The movement of a disease organism from one host to another by passive transfer, with no biological cycle in the **vector**.

Meconium The first excreta of a newly emerged adult following the pupal stage.

Media In wing venation, the fifth longitudinal vein, lying between the radius and the cubitus, with a maximum of eight branches (Fig. 2.21).

Media worker An individual of the medium-sized worker caste of termites and ants, *see also* **Major worker**; **Minor worker**.

Medial Towards the middle.

Median At or towards the middle (Fig. 2.8).

Median flexion-line A fold-line that runs longitudinally through the approximate middle of the wing (Fig. 2.20).

Melanic Darkened.

Melanism Darkening due to increased pigmentation.

Melittophily Pollination by bees.

Menotaxis (including **light compass response**) Movement that maintains a constant orientation relative to a light source.

Mentum The ventral fused plate derived from the labium (Fig. 2.16).

Merozoite The third stage in the asexual cycle of a malaria parasite (*Plasmodium* spp.)(Fig. 14.1), derived from division of the **schizont**.

Mesal Nearer to the midline of the body, also **medial**.

Mesenteron, *see* **Midgut**.

Mesosoma The middle of the three major divisions (**tagmata**) of the insect body, equivalent to the thorax but in apocritan Hymenoptera including the **propodeum**; called the **alitrunk** in adult ants (Box 11.3).

Mesothorax The second (and middle) segment of the thorax (Fig. 2.18).

Metabolic resistance The ability to avoid harm by biochemical detoxification of an insecticide.

Metamorphosis The relatively abrupt change in body form between the end of immature development and the onset of the imaginal (adult) phase.

Metasoma In apocritan Hymenoptera, the **petiole** plus **gaster** (Box 11.3).

Metathorax The third (and last) segment of the thorax (Fig. 2.18).

Microgamete The male gametocyte of a malaria parasite, *Plasmodium* spp., flagellate initially, after which flagella are lost (Fig. 14.1); *see also* **Macrogamete**.

Microlethical An egg lacking large yolk reserves.

Microphage A feeder on small particles, such as spores.

Micropyle A minute opening in the chorion of an insect egg (Fig. 5.7), through which sperm enter.

Microtrichium (*pl.* microtrichia) A subcellular cuticular extension, usually several to very many per cell (Fig. 2.6(d)).

Midgut The middle section of the gut, extending from the end of the proventriculus to the start of the ileum (Figs 3.1, 3.13).

Migration Mass movement, usually in search of more appropriate conditions.

Milk glands Specialized accessory glands in certain **adenotrophically viviparous** flies (Hippoboscidae and *Glossina*) that produce secretions fed upon by larvae.

Mimesis Resemblance to an inedible object in the environment; *see also* **Masquerade**.

Mimic (*adj.* mimetic) One of the three components of a mimicry system, the emitter of false signal(s) received by an **observer**; an individual, population or species that resembles a **model**, usually another species or part thereof; *see also* **Automimic**; **Batesian mimicry**; **Müllerian mimicry**.

Mimicry The resemblance of a **mimic** to a **model**, by which the mimic derives protection from predation provided to the model (e.g. by unpalatability).

Minor worker An individual of the smaller-sized worker caste of termites and ants; *see also* **Major worker**; **Media worker**.

Model One of the three components of a mimicry system, the emitter of signal(s) received by the **observer**; the organism resembled by a **mimic**, protected from predation, for example, by distastefulness.

Molar area The grinding surface of the mandible (Fig. 2.10).

Monocondylar An articulation (as of a mandible) with one point of articulation (condyle).

Monogyny A colony of eusocial insects dominated by one queen.

Monophage (*adj.* monophagous) An eater of only one kind of food, used particularly of specialized **phytophages**.

Monophyletic Evolutionarily derived from a single ancestor, recognized by the joint possession of shared derived feature(s).

Monoxene (*adj.* monoxenous) A parasite restricted to one host.

Morph A genetic form or variant.

Motor neurone A nerve cell with an axon that transmits stimuli from an interneurone to muscles (Fig. 3.5).

Moult increment The increase in size between successive instars (Fig. 6.11).

Moulting The formation of new cuticle followed by **ecdysis** (Figs 6.9, 6.10).

Mouthhooks The head skeleton of the maggot larva of higher flies (Box 14.4).

Multiparasitism Parasitization of a host by two or more parasites or parasitoids.

Multiple resistance The concurrent existence in a single insect population of two or more defence mechanisms against insecticide.

Multiporous Having several small openings.

Multivoltine Having several generations in one year; *see also* **Bivoltine**; **Univoltine**.

Müllerian mimicry A mimetic system in which two or more unpalatable species obtain protection from predation by resembling each other; *see also* **Batesian mimicry**.

Mushroom body A cluster of seminal vesicles and accessory gland tubules into a single mushroom-shaped structure, in certain orthopteroid/blattoid insects.

Mycetocyte A cell containing symbiotic microorganisms, scattered throughout the body, particularly within the fat body, or aggregated in organs called **mycetomes**.

Mycetome An organ containing aggregations of **mycetocytes**, usually located in the fat body or gonads.

Mycophage (*adj.* mycophagous) An eater of fungi; *see also* **Fungivore**.

Myiasis Disease or injury caused by feeding of larval flies on live flesh, of humans or other animals.

Myofibrils Contractile fibres that run the length of a muscle fibre, comprising actin sandwiched between myosin fibres.

Myophily Plant pollination by flies.

Myrmecochory The collection and dispersal of seeds by ants.

Myrmecophily Plant pollination by ants.

Myrmecophyte Plants containing **domatia** that house ants; 'ant-plants'.

Myrmecotrophy The feeding of plants by ants, notably through the waste products of an ant colony.

Naiad An alternative name for the immature stages of aquatic hemimetabolous insects; *see also* **Nymph**; **Larva**.

Nasus A nose, the snout of certain termite soldiers (**nasutes**).

Nasute A soldier termite possessing a snout.

Natatorial Swimming.

Necrophage (*adj.* necrophagous) An eater of dead and/or decaying animals.

Neem *Azadirachta indica* (Meliaceae), the neem tree.

Neotenic In termites, supplementary reproductives, arrested in their development, which have the potential to take on the reproductive role should the primary reproductives be lost.

Neoteny (*adj.* neotenous) The retention of juvenile features into the adult stage.

Nephrocytes (*also* **pericardial cells**) Cells that

sieve the haemolymph for products that they metabolize.

Neuroendocrine cells, *see* **Neurosecretory cells**.

Neurohormone The largest class of insect hormones, also known as neuropeptides, being small proteins secreted within different parts of the nervous system (Fig. 5.10).

Neurone The nerve cell, comprising a cell body, **dendrite** and **axon** (Figs 3.5, 4.3).

Neuropeptide, *see* **Neurohormone**.

Neurosecretory cells Modified neurones found throughout the nervous system (Fig. 3.8), producing insect hormones except ecdysteroids and juvenile hormones.

Neuston (*adj*. neustic) The water surface.

Nocturnal Active at night; *see also* **Crepuscular**; **Diurnal**.

Nomenclature The science of naming (of plants and animals).

Non-integrated inquiline An **inquiline** that is adapted ecologically to the nest of the host, but does not interact socially with the host; *see also* **Integrated inquiline**.

Notum (*pl*. nota) A thoracic **tergum**.

Nulliparous A female that has laid no eggs.

Nymph An immature insect after emerging from the egg, usually restricted to insects in which there is incomplete metamorphosis (**hemimetaboly**); *see also* **Larva**.

Obligatory Compulsory, or exclusive; e.g. obligatory diapause is a resting stage that occurs in every individual of each generation of a univoltine insect.

Observer One of the three components of mimicry systems, the receiver of the signal(s) emitted by the **model** and **mimic**.

Obtect pupa A pupa with body appendages fused (cemented) to the body, not free (Fig. 6.7); *see also* **Exarate pupa**.

Occipital foramen The opening of the back of the head.

Occiput The dorsal part of the posterior cranium (Fig. 2.9).

Ocellus (*pl*. ocelli) The 'simple' eye of adult and nymphal insects, typically three in a triangle on the vertex, with one median and two lateral ocelli (Figs 2.9, 2.11); the **stemma** of some holometabolous larvae.

Oenocyte A cell of the haemocoel, epidermis or especially the fat body, probably with many functions, most of which are unclear, but including synthesis of cuticle paraffins and/or haemoglobin.

Oesophagus The foregut that lies posterior to the pharynx, anterior to the crop (Figs 2.14, 3.1, 3.13).

Oligophage (*adj*. oligophagous) An eater of few kinds of food, e.g. several plant species within one genus or one family; used particularly of phytophages.

Oligopneustic A respiratory system with one to two functional spiracles on each side of the body; *see also* **Apneustic**; **Polypneustic**.

Oligopod larva A larva with legs on the thorax and not on the abdomen (Fig. 6.6).

Oligoxene (*adj*. oligoxenous) A parasite restricted to few hosts.

Ommatidium (*pl*. ommatidia) A single element of the compound eye (Fig. 4.10(c)).

Ontogeny The development from egg to adult.

Oocyst A stage in the sexual cycle of a malaria parasite (*Plasmodium* spp.) within the mosquito midgut lining, formed from the **oo kinete** and which undergoes **sporogony** (asexual reproduction) (Fig. 14.1).

Oocyte An immature egg cell formed from the **oogonium** within the ovariole.

Oogonium The first stage in the development in the **germarium** of an egg from a female germ cell.

Ookinete An active sexual stage in the cycle of a malaria parasite (*Plasmodium* spp.) formed from the zygote and which penetrates the body cavity within the mosquito (Fig. 14.1).

Oostatic hormone A peptide hormone, produced either by the ovary or associated neurosecretory tissue, with inhibitory activities on one or more ovarian functions.

Ootheca A protective surrounding for eggs (Box 8.6).

Open cell An area of the wing membrane partially bounded by veins but including part of the wing margin.

Open tracheal system A gas exchange system comprising tracheae and tracheoles and with spiracular contact with the atmosphere (Fig. 3.11(a)-(c)); *see also* **Closed tracheal system**.

Opisthognathous With the head deflexed such that the mouthparts are directed posteriorly, as in many Hemiptera; *see also* **Hypognathous**; **Prognathous**.

Oral (*or* **stomodeal**) **trophallaxis** In social and subsocial insects, the transfer of nutritive fluid from the mouth to another individual; may be mutual or unidirectional.

Organochlorine An organic chemical containing chlorine; a group of compounds including several insecticides.

Organophosphate An organic chemical containing phosphorus; a group of compounds including several insecticides.

Orthokinesis A response to a stimulus in which the rate of response (such as speed of movement) is positively proportional to the intensity of the stimulus.

Osmeterium (*pl.* osmeteria) An eversible tubular pouch on the prothorax of some larval swallowtail butterflies (Lepidoptera: Papilionidae) (Fig. 13.5), used to disseminate volatile toxic, defensive compounds.

Osmoregulation The regulation of water balance, maintaining the homeostasis of osmotic and ionic content of the body fluids.

Ostium (*pl.* ostia) A slit-like opening in the dorsal vessel ('heart') present usually on each thoracic and the first nine abdominal segments, each ostium having a one-way valve that permits flow of haemolymph from the pericardial sinus into the dorsal vessel (Fig. 3.9).

Outer epicuticle The middle of the three layers of epicuticle, with the inner epicuticle beneath it (Fig. 2.1); *see also* **Inner epicuticle**; **Superficial layer**.

Ovarian cycle The length of time between successive ovipositions.

Ovary One of the paired gonads of female insects, each comprising several **ovarioles**.

Ovariole One of several ovarian tubes that form the ovary (Fig. 3.1(a)), each consisting of a **germarium**, a **vitellarium** and a stalk or **pedicel** (Fig. 3.20(a)).

Oviparous Reproduction in which eggs are laid.

Ovipositor The organ used for laying eggs; *see also* **Appendicular ovipositor**; **Substitutional ovipositor**.

Ovoviviparity Retention of the developing fertilized egg within the mother, considered to be a form of **viviparity** but in which there is no nutrition of the hatched young; *see also* **Oviparity**.

Paedogenesis (*adj.* paedogenetic) Reproduction in an immature stage.

Pair-wise coevolution, *see* **Specific coevolution**.

Palp (*also* **palpus**, *pl.* palpi) Finger-like, usually segmented appendage of the maxilla (**maxillary palp**) and labium (**labial palp**) (Figs 2.9, 2.10).

Panoistic ovariole An ovariole that lacks nurse cells; *see also* **Telotrophic ovariole**.

Paraglossa (*pl.* paraglossae) One of a pair of lobes distolateral on the prementum of the labium, lying outside the glossae, but mesal to the labial palp (Fig. 2.10).

Paramere One of a pair of lobes lying lateral to the penis, forming part of the **aedeagus** (Fig. 2.24).

Paranota (*sing.* paranotum, *adj.* paranotal) Postulated lobes of the thoracic terga from which, it has been argued, the wings derive.

Paraphyletic A group (grade) evolutionarily derived from a single ancestor, but which does not contain all descendants, recognized by the joint possession of shared primitive character(s); rejected in **cladistics** but often accepted in **evolutionary systematics**; *see also* **Monophyletic**; **Polyphyletic**.

Paraproct Ventral relics of segment 11 (Fig. 2.23).

Parasite An organism that lives at the expense of another (host) which it does not usually kill; *see also* **Ectoparasite**; **Endoparasite**; **Parasitoid**.

Parasitism The relationship between parasitoid or parasite and host.

Parasitization The condition of being parasitized, by either a parasitoid or a parasite.

Parasitized The state of a host that supports a parasitoid or parasite.

Parasitoid A parasite that kills its host; *see also* **Ectoparasitoid; Endoparasitoid.**

Parous A female that has laid at least one egg.

Parthenogenesis Development from an unfertilized egg; *see also* **Amphitoky; Arrhenotokous parthenogenesis; Paedogenesis; Thelytoky.**

Patch A discrete area of microhabitat.

Patency In the course of the malaria disease, the first appearance of parasites in the red blood cells.

Pedicel The stem or stalk of an organ; the stalk of an ovariole (Fig. 3.20(a)); the second antennal segment (Fig. 2.10); the 'waist' of an ant.

Penis (*pl.* penes), *also* **phallus** The median intromittent organ (Figs 2.24, 3.20(b)), variously derived in different insect orders.

Pericardial cells, *see* **Nephrocytes**

Pericardial sinus The body compartment that contains the **dorsal vessel** ('heart') (Fig. 3.9).

Perineural sinus The ventral body compartment that contains the nerve cord, separated from the perivisceral sinus by the **ventral diaphragm** (Fig. 3.9).

Periodic release The regular release of biological control agents that are effective in control but unable to establish permanently.

Peripheral nervous system The network of nerve fibres and cells associated with the muscles.

Peritreme A sclerotized plate surrounding an orifice, notably around a spiracle.

Peritrophic membrane A thin sheath lining the midgut epithelium of many insects (Fig. 3.16).

Perivisceral sinus The central body compartment, delimited by the **ventral diaphragm** and **dorsal diaphragm.**

Pest resurgence The rapid increase in numbers of a pest following cessation of control measures or resulting from development of **resistance** and/or elimination of natural enemies.

Petiole A stalk; in apocritan Hymenoptera, the narrow second (and sometimes third) abdominal segments that precede the gaster, forming the 'waist' (Box 11.3).

Phalaenophily Plant pollination by moths.

Phallobase In male genitalia, the support for the **aedeagus** (Figs 2.24(b), 5.2).

Phallomere A lobe lateral to the penis.

Phallus, *see* **Penis.**

Pharate Within the cuticle of the previous stage, 'cloaked'.

Pharynx The anterior part of the foregut, anterior to the oesophagus (Figs 2.14, 3.13).

Phenetic A classification system in which overall resemblance between organisms is the criterion for grouping; *see also* **Cladistics; Evolutionary systematics.**

Pheromone A chemical used in communication between individuals of the same species, releasing a specific behaviour or development in the receiver; pheromones play roles in aggregation, alarm, courtship, queen recognition, sex, sex attraction, spacing (epideictic or dispersion) and trail-marking.

Pheromone mass trapping The use of pheromones to lure pest insects, which are then killed

Photoperiod The duration of the light (and therefore also dark) part of the 24-hour daily cycle.

Phragma (*pl.* phragmata) Plate-like **apodemes,** notably those of the antecostal suture of the thoracic segments that support the longitudinal flight muscles (Figs 2.7(d), 2.18).

Phragmosis The closing of a nest opening with part of the body.

Phyletic tracking Strict coevolution in which the phylogenies of each taxon (e.g. host and parasite, plant and pollinator) match precisely.

Phylogeny (*adj.* phylogenetic) Evolutionary history (of a taxon).

Physiological time A measure of development time based upon the amount of heat required rather than calendar time elapsed.

Physogastry Having a swollen abdomen, as in mature queen termites (Fig. 11.7), ants and bees.

Phytophage (*adj.* phytophagous) An eater of plants.

Phytophagy The eating of plants.

Plant resistance A range of mechanisms by which plants resist insect attack; *see also* **Antibiosis; Antixenosis; Tolerance.**

Plasma The aqueous component of haemolymph.

Pleometrosis The foundation of a colony of social insects by more than one queen.

Pleural coxal process The anterior end of the pleural ridge providing reinforcement for the coxal articulation (Fig. 2.18).

Pleural ridge The internal ridge dividing the pterothorax into the anterior **episternum** and posterior **epimeron**.

Pleural suture The externally visible indication of the pleural ridge, running from the leg base to the tergum (Fig. 2.18).

Pleural wing process The posterior end of the pleural ridge providing reinforcement for the wing articulation (Fig. 2.18).

Pleuron (*pl.* pleura; *adj.* pleural; *dim.* pleurite) The lateral region of the body, bearing the limb bases.

Poison glands Accessory glands that produce poison, as in the stings of Hymenoptera (Fig. 13.10).

Pollination The transfer of pollen from male to female flower parts.

Polyculture The cultivation of several crops intermingled.

Polydnaviruses (PVs) A group of viruses found in the ovaries of some parasitic wasps, involved in overcoming host immune responses when injected with the wasp eggs.

Polyethism Within a social insect **caste**, the division of labour either by specialization throughout the life of an individual or by different ages performing different tasks.

Polyembryony The production of more than one embryo (often many) from a single egg, notably in parasitic insects.

Polygyny Social insects that have several queens, either at the same time or sequentially (serial polygyny).

Polymorphic Of a species, existing in two or more genetic variants (morphs).

Polyphage (*adj.* polyphagous) An eater of many kinds of food, e.g. many plant species from a range of families; used particularly of phytophages.

Polyphenism Environmentally induced differences between successive generations, lacking a genetic basis.

Polyphyletic A group evolutionarily derived from more than one ancestor, recognized by the possession of one or more features evolved convergently; rejected in **cladistics** and **evolutionary systematics**.

Polypneustic A respiratory system with at least eight functional spiracles on each side of the body; *see also* **Apneustic; Oligopneustic**.

Polypod larva A type of larva with jointed legs on the thorax and prolegs on the abdomen (Fig. 6.6).

Polytrophic ovariole An ovariole in which several nurse cells remain closely attached to each oocyte as it moves down the ovariole.

Polyxene (*adj.* polyxenous) A parasite with a wide range of hosts.

Pore canals Fine tubules that run through the cuticle and carry epidermally derived compounds to the **wax canals** and thus to the epicuticular surface (Fig. 2.1).

Pore kettle The chamber within a chemoreceptor sensillum that has many pores (slits) in the wall (Fig. 4.6).

Post-coxal bridge The pleural area behind the coxa, often fused with the sternum (Fig. 2.18).

Posterior At or towards the rear (Fig. 2.8).

Posterior cranium The posterior, often horseshoe-shaped, area of the head.

Postgena The lateral part of the occipital arch

posterior to the post-occipital suture (Fig. 2.9).

Postmentum The proximal part of the labium (Figs 2.10, 12.4).

Postnotum The posterior part of a pterothoracic notum, bearing the phragmata that support longitudinal muscles (Figs 2.7, 2.18).

Postoccipital suture A groove on the head that indicates the original head segmentation, separating the postocciput from the remainder of the head (Fig. 2.9).

Postocciput The posterior rim of the head behind the postoccipital suture (Fig. 2.9).

Post-tarsus, *see* **Pretarsus**.

Precosta The most anterior wing vein (Fig. 2.21).

Precoxal bridge The pleural area anterior to the coxa, often fused with the sternum (Fig. 2.18).

Predation Preying on other organisms.

Predator An organism that eats more than one other organism during its life; *see also* **Parasitoid**.

Pre-erythrocytic (*also* **exo-erythrocytic**) **schizogonous cycle** In the asexual cycle of a malaria parasite (*Plasmodium* spp.), the schizogonous cycle within the parenchyma cells of the liver, producing a large schizont (Fig. 14.1).

Pregenital segments The first seven abdominal segments.

Prementum The free distal end of the labium, usually bearing labial palps, glossae and paraglossae (Figs 2.10, 12.4).

Prepatent period In the course of a disease, the time between infection and first symptoms.

Prescutum The anterior third of the **alinotum** (either meso- or metanotum), in front of the **scutum** (Fig. 2.18).

Presoldier In termites, an intervening stage between worker and soldier.

Press The process on the proximal apex of the tarsus of a bee that pushes pollen into the **corbicula** (Fig. 11.3).

Presternum A smaller sclerite of the ** custernum**, lying anterior to the basisternum.

Pretarsus (*pl.* pretarsi), *also* **post-tarsus** The distal segment of the insect leg (Fig. 2.19).

Prey A food item for a predator.

Primary cycle In a disease, the cycle that involves the typical host; *see also* **Secondary cycle**.

Primary reproductives In termites, the king and queen founders of a colony (Fig. 11.7).

Proboscis A general term for elongate mouthparts (Fig. 2.12 and Box 14.4); *see also* **Rostrum**.

Procephalon In embryology, the anterior head formed by fusion of the primitive anterior three segments (Fig. 6.5).

Proctodeal trophallaxis In social and subsocial insects, the transfer of nutritive fluid from the anus to another individual; may be mutual or unidirectional.

Proctodeum (*adj.* proctodeal), *see* **Hindgut**.

Procuticle The thicker layer of cuticle, which in sclerotized cuticle comprises an outer exocuticle and inner endocuticle; lying beneath the thinner epicuticle (Fig. 2.1).

Prognathous With the head horizontal and the mouthparts directed anteriorly; *see also* **Hypognathous**; **Opisthognathous**.

Proleg An unsegmented leg of a larva.

Pronotum The upper (dorsal) plate of the prothorax.

Propodeum In apocritan Hymenoptera, the first abdominal segment if fused with the thorax to form a **mesosoma** (or **alitrunk** in ants) (Box 11.3).

Proprioceptors Sense organs that respond to the position of body organs.

Prothoracic gland The thoracic or cephalic glands (Fig. 3.8) that secrete ecdysteroids (Fig. 5.10).

Prothoracicotropic hormone A neuropeptide hormone secreted by the brain that controls aspects of moulting and metamorphosis via action on the **corpora cardiaca**.

Prothorax The first segment of the thorax (Fig. 2.18).

Protocerebrum The anterior part of the insect brain, the ganglia of the first segment, comprising the ocular and associative centres.

Protraction Withdrawal, the converse of extension.

Protrusible (*or* **exsertile**) **vesicle** A small sac or bladder, capable of being extended or protruded.

Proventriculus (*also* **gizzard**) The grinding organ of the foregut (Figs 3.1, 3.13).

Proximal The part of an appendage closer to or at the body (opposite to distal).

Pseudergate In 'lower' termites the equivalent to the worker caste, comprising immature nymphs or undifferentiated larvae.

Pseudocopulation The attempted copulation of an insect with a flower.

Pseudoplacental viviparity Viviparity (producing living offspring) in which a **microlethical** egg develops via nourishment from a presumed placenta.

Pseudotrachea A ridged groove on the ventral surface of the labellum of some higher Diptera (Box 14.4), used to take up liquid food.

Psychophily Plant pollination by butterflies.

Pterostigma A pigmented (and denser) spot near the anterior margin of the fore and sometimes hind wings (Figs 2.20, 2.21, 2.22(b)).

Pterothorax The enlarged second and third segments of the thorax bearing the wings in pterygotes.

Ptilinum A sac everted from a fissure between the antennae of schizophoran flies (Diptera) that aids puparium fracture at emergence.

Pulvillus (*pl.* pulvilli) (*also* **euplantula**) Ventral pad on some **tarsomeres** of leg (Fig. 2.19).

Pupa (*adj.* pupal) The inactive stage between larva and adult in holometabolous insects; also termed a **chrysalis** in butterflies.

Puparium The hardened skin of the final-instar larva (higher Diptera, Strepsiptera) or last nymphal instar (Aleyrodidae), in which the pupa forms.

Pupation Becoming a pupa.

Pylorus The anterior hindgut where the Malpighian tubules enter, sometimes indicated by a muscular valve.

Pyrethrin One of the insecticidal chemicals present in the plant pyrethrum.

Pyrethroids Synthetic chemicals with similarity in structure to pyrethrins.

Quasisocial Social behaviour in which individuals of the same generation co-operate and nest-share without division of labour.

Queen A female belonging to the reproductive caste in eusocial or semisocial insects (Fig. 11.7), called a **gyne** in social Hymenoptera.

Quiescence A slowing down of metabolism and development in response to adverse environmental conditions; *see also* **Diapause**.

Radius In wing venation, the fourth longitudinal vein, posterior to the subcosta; with a maximum of six branches (Fig. 2.21).

Rake The process on the distal apex of the tibia of a bee that gathers pollen into the press (Fig. 11.3).

Rank The level in a taxonomic hierarchy, e.g. species, genus, family, order.

Raptorial Adapted for capturing prey by grasping.

Rectal pad Thickened sections of the epithelium of the rectum involved in water uptake from the faeces (Figs 3.17, 3.18).

Rectum (*adj.* rectal) The posterior part of the hindgut (Figs 3.1, 3.13).

Reflex A simple response to a simple stimulus.

Refractory period The time interval during which a nerve will not initiate another impulse; in reproduction, the period during which a mated female will not re-mate.

Refuge A safe place.

Release Specifically, to stimulate a particular behaviour.

Releaser A particular stimulus whose signal stimulates a specific behaviour.

Remigium The anterior part of the wing, usually more rigid than the posterior **clavus** and with more veins (Fig. 2.20).

Reniform Kidney-shaped.

Replete An individual ant which is distended by liquid food.

Replicate Specifically of disease organisms, to increase in numbers.

Reservoir Of diseases, the natural host and geographic range.

Resilin A rubber-like or elastic protein in some insect cuticle.

Resistance The ability to withstand (e.g. insecticides).

Resource tracking A relationship, for example between parasite and host or plant and pollinator, in which the evolution of the association is based on ecology rather than phylogeny; *see also* **Coevolution; Phyletic tracking**.

Respiration A metabolic process in which substrates (food) are oxidized using molecular oxygen; also used inappropriately to mean breathing, as through spiracles or gas exchange across thin cuticle.

Retinaculum Particularly used of the specialized hooks or scales on the base of the fore wing that lock with the **frenulum** of lepidopteran hind wings in flight; also the retaining hook of the springtail **furcula** (spring) (Box 8.2).

Retinula cell A sensory cell of the light receptor organs (**ommatidium, stemma** or **ocellus**) comprising a rhabdom of rhabdomeres and connected by nerve axons to the optic lobe (Fig. 4.10).

Rhabdom The central zone of the retinula consisting of microvilli filled with visual pigment; comprising rhabdomeres belonging to several different retinula cells (Fig. 4.10).

Rhabdomere One of typically eight units comprising a rhabdom.

Rheophilic Liking running water.

Rhizosphere A zone surrounding the roots of plants, usually richer in fungi and bacteria than elsewhere in the soil.

Riparian Riverside.

Rostrum A facial extension that bears the mouthparts at the end (Boxes 10.8 and 14.4(a)); *see also* **Proboscis**.

Rotenone A particular legume-derived chemical with insecticidal and other toxic effects.

Round dance A communication dance of honey bees (Box 11.2).

Salivarium *or* **salivary reservoir** The cavity into which the salivary gland opens, between the hypopharynx and the labium (Figs 3.1, 3.14).

Salivary gland The gland that produces saliva (Fig. 3.1).

Saltatorial Adapted for jumping or springing.

Saprophage (*adj.* saprophagous) An eater of decaying organisms.

Sarcolemma The outer sheath of a striated muscle fibre.

Scale Flattened setae (unicellular outgrowths of the cuticle).

Scape The first segment of the antenna (Fig. 2.10).

Schizont The second stage in the asexual cycle of a malaria parasite (*Plasmodium* spp.), produced by the **pre-erythrocytic schizogonous cycle** or by division of a **trophozoite** (Fig. 14.1).

Schwann cell A cell surrounding the axon of a scolopidium (Fig. 4.3).

Sclerite A plate on the body wall surrounded by membrane or sutures.

Sclerotization Stiffening of the cuticle by cross-linkage of protein chains.

Scolopale cell In a chordotonal organ, the sheath cell that envelops the dendrite (Fig. 4.3).

Scolopidia In a chordotonal organ, the combination of three cells, the **cap, scolopale** and **dendrite** (Fig. 4.3).

Scopa A brush of thick hair on the hind tibia of adult bees.

Scraper The ridged surface drawn over a **file** to produce stridulatory sounds.

Scutellum The posterior third of the **alinotum** (either meso- or metanotum), lying behind the **scutum** (Fig. 2.18).

Scutum The middle third of the **alinotum** (either meso- or metanotum), in front of the **scutellum** (Fig. 2.18).

Secondary cycle In a disease, the cycle that

involves an atypical host; *see also* **Primary cycle**.

Secondary pest outbreak Previously harmless insects becoming pests following insecticide treatment for a primary pest.

Secondary plant compounds Plant chemicals assumed to be produced for defensive purposes.

Secondary segmentation Any segmentation that fails to match the embryonic segmentation; more specifically, the insect external skeleton in which each apparent segment includes the posterior (intersegmental) parts of the primary segment preceding it (Fig. 2.7).

Sector A major wing-vein branch and all of its subdivisions.

Semi-aquatic Living in saturated soils, but not immersed in free water.

Seminal vesicle Male sperm storage organs (Fig. 3.20(b)).

Semiochemical Any chemical used in intra- and interspecific communication.

Semisocial Social behaviour in which individuals of the same generation co-operate and nest-share with some division of reproductive labour.

Sensillum (*pl.* sensilla) A sense organ, either simple and isolated, or part of a more complex organ.

Sensory neurone A nerve cell that receives and transmits stimuli from the environment (Fig. 3.5).

Serial homology The occurrence of identically derived features on different segments of the same insect.

Serosa The membrane covering the embryo (Fig. 6.5).

Seta (*pl.* setae) A cuticular extension, a **trichoid sensillum**; also called a **hair** or **macrotrichium**.

Sheath cell Any cell enveloping another (Fig. 4.3).

Sibling Full brother or sister.

Single cycle disease A disease involving one species of host, one parasite and one insect vector.

Siphunculus (*pl.* siphunculi), *see* **Cornicle**.

Sister group The closest related group of the same rank as the group under study.

Smell The olfactory sense, the detection of air-borne chemicals.

Sociality The condition of living in an organised community.

Soldier In social insects, an individual worker belonging to a subcaste involved in colony defence.

Solitary Non-colonial, occurring singly or in pairs.

Species (*adj.* specific) A group of all individuals that can interbreed, mating within the group (sharing a gene pool) and producing fertile offspring, usually similar in appearance and behaviour – but *see also* **Polymorphic** – and sharing a common evolutionary history.

Specific (*or* **pair-wise**) **coevolution** Concerted evolutionary change that takes place between two species, in which the evolution of a trait in one leads to reciprocal development of a trait in a second organism in a feature that evolved initially in response to a trait of the first species; *see also* **Guild coevolution**.

Sperm competition In multi-mated females, the syndrome by which sperm from one mating compete with other sperm to fertilize the eggs.

Sperm precedence The preferential use by the female of the sperm of one mating over others.

Spermatheca The female receptacle for sperm deposited during mating (Fig. 3.20(a)).

Spermathecal gland A tubular gland off the spermatheca, producing nourishment for sperm stored in the spermatheca (Fig. 3.20(a)).

Spermatophore An encapsulated package of spermatozoa (Fig. 5.4).

Spermatophylax In katydids, a proteinaceous part of the spermatophore eaten by the female after mating (Box 5.2).

Sphecophily Plant pollination by wasps.

Spina An internal apodeme of the intersegmental sternal plate called the spinasternum.

Spinasternum An **intersternite** bearing a spina (Fig. 2.18), sometimes fused with the eusterna of the prothorax and mesothorax, but never the metathorax.

Spine A multicellular unjointed cuticular extension, often thorn-like (Fig. 2.6(a)).

Spiracle An external opening of the **tracheal system** (Fig. 3.10(a)).

Sporogony The asexual cycle of a malaria parasite (*Plasmodium* spp.) in the mosquito, in which an **oocyst** undergoes nuclear division to produce **sporozoites**.

Sporozoites In the asexual cycle of a malaria parasite (*Plasmodium* spp.) in the mosquito, the product of **sporogony**; stored in the salivary gland and passed to the vertebrate host during mosquito feeding (Fig. 14.1).

Spur An articulated spine.

Stadium The period between moults, the instar duration or intermoult period.

Startle display A display made by some cryptic insects upon discovery, involving exposure of a startling colour or pattern, such as eye spots.

Statary The sedentary, stationary phase of army ants.

Stemma (*pl.* stemmata) The 'simple' eye of many larval insects, sometimes aggregated into a more complex visual organ.

Stenogastrous Having a shortened or narrow abdomen.

Sterile male technique A means of controlling insects by swamping populations with large numbers of artificially sterilized males.

Sternellum The small sclerite of the eusternum, lying posterior to the basisternum (Fig. 2.18).

Sternum (*pl.* sterna; *adj.* sternal; *dim.* sternite) The ventral surface of a segment (Fig. 2.7).

Stipes The distal part of the maxilla, bearing **galea**, **lacinia** and **maxillary palp** (Fig. 2.10).

Stomach poison An insecticidal poison that acts after ingestion into the insect gut.

Stomatogastric nervous system The nerves associated with the fore- and midgut.

Stomodeal trophallaxis, *see* **Oral trophallaxis**.

Stomodeum (*adj.* stomodeal), *see* **Foregut**.

Striated muscle Muscles in which myosin and actin filaments overlap to give a striated effect.

Stridulation The production of sound by rubbing two rough or ridged surfaces together.

Style In apterygote insects, small appendages on abdominal segments, homologous to abdominal legs.

Stylet One of the elongate parts of piercing/sucking mouthparts (Figs 2.13, 2.14, 10.4), a needle-like structure.

Subalare (*pl.* subalaria) A small sclerite, one of the epipleurites that lies posterior to the pleural wing process, an attachment for the direct flight muscles (Fig. 2.18).

Subcosta In wing venation, the third longitudinal vein, posterior to the costa (Fig. 2.21).

Subgenual organ A chordotonal organ on the proximal tibia that detects substrate vibration (Fig. 4.4).

Subimago (*also* **subimaginal instar**) In Ephemeroptera, the winged penultimate instar; subadult.

Suboesophageal ganglion The fused ganglia of the mandibular, maxillary and labial segments, forming a ganglionic centre beneath the oesophagus (Figs 3.6, 3.14).

Subsocial A social system in which adults look after immature stages for a certain period.

Substitutional ovipositor An ovipositor formed from extensible posterior abdominal segments; *see also* **Appendicular ovipositor**.

Succession An ecological sequence.

Superficial layer The outermost of the three layers of epicuticle often bearing a lipid or wax and/or cement layer, with outer and inner epicuticle beneath it (Fig. 2.1); *see also* **Inner epicuticle**; **Outer epicuticle**.

Superlingua (*pl.* superlinguae) A lateral lobe of the hypopharynx (Fig. 2.10), the remnant of a leg appendage of the third head segment.

Superparasitism The occurrence of more parasitoids within a host than can complete their development within the host.

Supplementary reproductive In termites, a potential replacement reproductive within its natal nest, which does not become alate; also called a **neotenic** or **ergatoid**.

Suture An external groove that may show the fusion of two plates (Fig. 2.10).

Symbiont An organism that lives in **symbiosis** with another.

Symbiosis A long-lasting, close and dependent relationship between organisms of two different species.

Sympathetic nervous system, *see* **Visceral nervous system**.

Sympatric Overlapping geographic distributions of organisms or taxa; *see also* **Allopatric**.

Synanthropic Associated with humans or their dwellings.

Synapse The site of approximation of two nerve cells at which they may communicate.

Synchronous muscle A muscle that contracts once per nerve impulse.

Synergism The enhancement of the effects of two substances that is greater than the sum of their individual effects.

Synomone A communication chemical that benefits both the receiver and the producer; *see also* **Allomone**; **Kairomone**.

Systematics The science of biological classification.

Systemic insecticide An insecticide taken into the body of a host (plant or animal) that kills insects feeding on the host.

Taenidium (*pl.* taenidia) The spiral thickening of the tracheal wall that prevents collapse.

Tagma (*pl.* tagmata) The group of segments that form a major body unit (head, thorax, abdomen).

Tagmosis The organization of the body into major units (head, thorax, abdomen).

Tapetum A reflective layer at the back of the eye formed from small tracheae.

Target-site resistance Increased tolerance by an insect to an insecticide through reduced sensitivity at the target site.

Tarsomere A subdivision of the tarsus (Fig. 2.19).

Tarsus (*pl.* tarsi) The leg segment distal to the tibia, comprising one to five tarsomeres and apically bearing the pretarsus (Fig. 2.19).

Taste Chemoreception of chemicals in a liquid dissolved form.

Taxis (*pl.* taxes) An orientated movement of an organism.

Taxon (*pl.* taxa) Any taxonomic unit (species, genus, family, phylum, etc.).

Taxonomy (*adj.* taxonomic) The theory and practice of naming and classifying organisms.

Tegmen (*pl.* tegmina) A leathery, hardened fore wing (Fig. 2.22(c)).

Tegula One of the articular sclerites of the neopteran wing, lying at the base of the costa (Fig. 2.21); *see also* **Axillary sclerites**; **Humeral plate**.

Telotaxis Orientation and movement directly towards a visual stimulus.

Telotrophic (*or* **acrotrophic**) **ovariole** An ovariole in which the nurse cells are only within the germarium; the nurse cells remain connected to the oocytes by long filaments as the oocytes move down the ovariole; *see also* **Panoistic ovariole**; **Polytrophic ovariole**.

Teneral The condition of a newly eclosed adult insect, which is unsclerotized and unpigmented.

Tentorium The endoskeleton of the head, including anterior and posterior tentorial arms – cuticular invaginations.

Tergum (*pl.* terga; *adj.* tergal; *dim.* tergite) The dorsal surface of a segment (Fig. 2.7).

Terminalia The terminal abdominal segments involved in the formation of the genitalia.

Testis (*pl.* testes) One of (usually) a pair of male gonads (Figs 3.1(b), 3.20(b)).

Thanatosis Feigning death.

Thelytoky (*also* **thelytokous parthenogenesis**) Parthenogenesis producing only female offspring.

Thorax The middle of the three major divisions (**tagma**) of the body, comprising pro-, meso- and metathorax (Fig. 2.18).

Threshold The minimum level of stimulus required to initiate (release) a response.

Tibia (*pl.* tibiae) The fourth leg segment, following the femur (Fig. 2.19).

Tolerance The ability of a plant to withstand insect attack and recover.

Tonofibrillae Fibrils of cuticle that connect a muscle to the epidermis (Fig. 3.2).

Tormogen cell The socket-forming epidermal cell associated with a seta (Figs 2.6, 4.1).

Trachea (*pl.* tracheae) A tubular element of the insect gas exchange system, within which air moves (Figs 3.10, 3.11).

Tracheal system The insect gas exchange system, comprising tracheae and tracheoles (Figs 3.10, 3.11).

Tracheole The fine tubules of the insect gas exchange system (Fig. 3.10(b)).

Transovarial transmission The transmission of micro-organisms between generations via the eggs.

Transverse At right angles to the longitudinal axis.

Triad (*adj.* triadic) Triplets of long wing veins (paired main veins and an intercalated longitudinal vein).

Trichogen cell A hair-forming epidermal cell associated with a seta (Figs 2.6, 4.1).

Trichoid sensillum A hair-like cuticular projection; a seta, hair or macrotrichium (Figs 2.6(b), 3.5, 4.1).

Tritocerebrum The posterior (or posteroventral) paired lobes of the insect brain, the ganglia of the third segment, functioning in handling the signals from the body.

Triungulin An active, dispersive first-instar larva of insects including many that undergo hypermetamorphosis.

Trochanter The second leg segment, following the coxa (Fig. 2.19).

Trochantin A small sclerite anterior to the coxa (Figs 2.18, 2.19).

Troglobite (*also* **troglobiont**; **troglodyte**, *adj.* troglodytic) Cave-dweller; *see also* **Cavernicolous**.

Trophamnion In parasitoids, the enveloping membrane surrounding the poly-embryonically derived multiple individuals that arise from a single egg, derived from the host's haemolymph.

Trophic Relating to food; of an egg of a social insect, degenerate and used in feeding other members of the colony.

Trophocyte The dominant metabolic and storage cell of the fat body.

Trophogenesis (*adj.* trophogenic) In social insects, the determination of caste type by differential feeding of the immature stages (in contrast to genetic determination of caste).

Trophozoite The first stage in the asexual cycle of a malaria parasite (*Plasmodium* spp.), derived from a merozoite (either from a liver cell or an erythrocyte) (Fig. 14.1).

Trypanosomiasis A disease caused by *Trypanosoma* protozoans, transmitted to humans predominantly by reduviid bugs (Chagas' disease) or tsetse flies (sleeping sickness).

Tymbal A stretched elastic membrane capable of sound production when flexed.

Tympanum (*pl.* tympana) *or* **tympanal organ** Any organ sensitive to vibration, comprising a tympanic membrane (thin cuticle), an air sac and a chordotonal sensory organ attached to the tympanal membrane (Fig. 4.4).

Unguis (*pl.* ungues) Claw (Fig. 2.19).

Unguitractor plate The ventral sclerite of the pretarsus that articulates with the **claws** (Fig. 2.19).

Uniporous Having a single opening.

Univoltine Having one generation in one year; *see also* **Bivoltine**; **Multivoltine**.

Urea A minor component of insect nitrogenous excretion, $CO(NH_2)_2$ (Fig. 3.19).

Uric acid The main nitrogenous excretion product, $C_5H_4N_4O_3$ (Fig. 3.19).

Uricotelism An excretory system based on uric acid excretion.

Urocyte (*also* **urate cell**) A cell that acts as a temporary store for urate excretion products.

Vagina A pouch-like or tubular genital chamber of the female genitalia.

Valve Generally, any unidirectional opening flap or lid; in female genitalia, the blade-like structures comprising the ovipositor shaft (also called **gonapophysis**) (Fig. 2.23(b)).

Valvifer In female insect genitalia, derivations of **gonocoxites** 8 and 9, supporting the valves of the ovipositor (Fig. 2.23(b)).

Vannal fold, *see* **Anal fold**.

Vannus The anal area of the wing anterior to the jugal area (Fig. 2.20).

Vas deferens The ducts that carry sperm from the testes (Fig. 3.20(b)).

Vector Literally a bearer; specifically a host of a disease transmissible to another species of organism.

Vectorial capacity A mathematical expression of the probability of disease transmission by a particular vector.

Venter The lower surface of the body.

Ventilate To pass air or oxygenated water over a gas exchange surface.

Ventral Towards or at the lower surface (Fig. 2.8).

Ventral diaphragm A membrane lying horizontally above the nerve cord in the body cavity, separating the perineural sinus from the perivisceral sinus (Fig. 3.9).

Ventral nerve cord The chain of ventral ganglia.

Ventral tube In Collembola, a ventral sucker, also called collophore (Box 8.2).

Ventriculus The tubular part of the midgut, the main digestive section of the gut (Fig. 3.13).

Vertex The top of the head, posterior to the frons (Fig. 2.9).

Vibration dance A communication dance of honey bees (Box 11.2).

Vicariance The division of the range of a species by Earth history events (e.g. ocean or mountain formation).

Visceral (*also* **sympathetic**) **nervous system** The nerve system that innervates the gut, reproductive organs and tracheal system.

Vitellarium The structure within the ovariole in which **oocytes** develop and yolk is provided to them (Fig. 3.20(a)).

Vitelline membrane The outer layer of an **oocyte**, surrounding the yolk (Fig. 5.7).

Vitellogenesis The process by which oocytes grow by yolk deposition.

Viviparity The bearing of live young (i.e. post-egg hatching) by the female; *see also* **Adenotrophic viviparity**; **Haemocoelous viviparity**; **Ovoviviparity**; **Pseudoplacental viviparity**.

Voltinism The number of generations per year.

Vulva The external opening of the copulatory pouch (bursa copulatrix) or vagina of the female genitalia (Fig. 3.20(a)).

Waggle dance A communication dance of honey bees (Box 11.2).

Wasmannian mimicry A form of **mimicry** which allows an insect of another species to be accepted into a social insect colony.

Wax A complex lipid mixture giving water-proofing to the cuticle, providing covering, or building material.

Wax canals (*or* **filaments**) Fine tubules that transport lipids from the pore canals to the surface of the epicuticle (Fig. 2.1).

Wax layer The lipid or waxy layer outside the epicuticle (Fig. 2.1); may be absent.

Wax mirrors Overlapping plates on the venter of the fourth to seventh abdominal segments of social bees that serve to direct the wax flakes that are produced beneath each mirror.

Weed Any plant 'in the wrong place', particularly used of plants away from their natural range, or invading human monocultural crops.

Wood borer An insect that tunnels into live or dead wood.

Worker In social insects, a member of the sterile caste that assists the reproductives.

Xylophage (*adj.* xylophagous) An eater of wood.

Zoocecidia Plant galls induced by animals such as insects, mites and nematodes, as opposed to those formed by the plant response to micro-organisms.

Zoophilic Preferring other animals to humans, especially used of feeding preference of blood-feeding insects.

Zygote A fertilized egg; in malaria parasites (*Plasmodium* spp.) resulting from fusion of a microgamete and macrogamete (Fig. 14.1).

FURTHER READING

Torre-Bueno, J.R. de la (1989) *The Torre-Bueno Glossary of Entomology*, 2nd edn, The New York Entomological Society in cooperation with the American Museum of Natural History, New York.

REFERENCES

CREDITS FOR VIGNETTES

Chapter 1	After various sources; especially Michener, 1974
Chapter 2	After scanning electron micrograph by C.A.M. Reid
Chapter 3	After Uvarov, 1966
Chapter 4	After Blaney, 1976
Chapter 5	After Dobson, 1989
Chapter 6	After Johnson and Lyon, 1992
Chapter 7	After Anon., 1991; Henwood, 1993
Chapter 8	After Eisenbeis and Wichard, 1987
Chapter 9	After Wiggins, 1978
Chapters 10–12	After Blaney, 1976
Chapter 13	After Wickler, 1968
Chapter 14	After Anon., 1991
Chapter 15	After Burton and Burton, 1975

REFERENCES FOR FIGURES, TABLES AND BOXES

Alexander, B. A. (1992) An exploratory analysis of cladistic relationships within the superfamily Apoidea, with special reference to sphecid wasps (Hymenoptera). *Journal of Hymenoptera Research*, **1**, 25–61.

Ando, H. (ed.) (1982) *Biology of the Notoptera*, Kashiyo-Insatsu Co. Ltd, Nagano, Japan.

Anon. (1991) *Ladybirds and Lobsters, Scorpions and Centipedes*, British Museum (Natural History), London.

Anon. (1992) *The Indonesian National IPM Program*, Jakarta Pusat.

Askew, R.R. (1971) *Parasitic Insects*, Heinemann, London.

Atkins, M.D. (1980) *Introduction to Insect Behaviour*, Macmillan Publishing Co., Inc., New York.

Austin, A.D. and Browning, T.O. (1981) A mechanism for movement of eggs along insect ovipositors. *International Journal of Insect Morphology and Embryology*, **10**, 93–108.

Badonnel, A. (1951) Ordre des Psocoptères, in *Traité de Zoologie: Anatomie, Systématique, Biologie. Tome X. Insectes Supérieurs et Hémiptéroïdes, Fascicule II*, (ed. P.-P. Grassé), Masson, Paris, pp. 1301–40.

Bandsma, A.T. and Brandt, R.T. (1963) *The Amazing World of Insects*, George Allen and Unwin, London.

Bar-Zeev, M. (1958) The effect of temperature on the growth rate and survival of the immature stages of *Aedes aegypti* (L.). *Bulletin of Entomological Research*, **49**, 157–63.

Bartell, R J., Shorey, H.H. and Barton Browne, L. (1969) Pheromonal stimulation of the sexual activity of males of the sheep blowfly *Lucilia cuprina* (Calliphoridae) by the female. *Animal Behaviour*, **17**, 576–85.

Barton Browne, L., Smith, P.H., van Gerwen, A.C.M. and Gillott, C. (1990) Quantitative aspects of the effect of mating on readiness to lay in the Australian sheep blowfly, *Lucilia cuprina*. *Journal of Insect Behavior*, **3**, 637–46.

Beard, J. (1989) Viral protein knocks the guts out of caterpillars. *New Scientist*, **124(1696–1697)**, 21.

Beccari, O. (1877) Volume 1 Istituto Sordo-Muti, *Malesia, (Genova)*, **1**, 167–92.

Belwood, J.J. (1990) Anti-predator defences and ecology of neotropical forest katydids, especially the Pseudophyllinae, in *The Tettigoniidae: Biology, Systematics and Evolution*, (eds W.J. Bailey and D.C.F. Rentz), Crawford House Press, Bathurst, pp. 8–26.

Bennet-Clark, H.C. (1989) Songs and the physics of sound production, in *Cricket Behavior and Neurobiology*, (eds F. Huber, T.E. Moore and W. Loher), Comstock Publishing Associates, a division of Cornell University Press, Ithaca and London, pp. 227–61.

Binnington, K.C. (1993) Ultrastructure of the attachment of the bacteria *Serratia entomophila* to foregut cuticle of *Costelytra zealandica* (Coleoptera: Scarabaeidae) and a review of nomenclature for insect epicuticular layers. *International Journal of Insect Morphology and Embryology*, **22(2–4)**, 145–55.

Birch, M.C. and Haynes, K.F. (1982) *Insect*

Pheromones, Studies in Biology no. 147, Edward Arnold, London.

Blaney, W.M. (1976) *How Insects Live,* Elsevier-Phaidon, Oxford.

Bonhag, P.F. and Wick, J.R. (1953) The functional anatomy of the male and female reproductive systems of the milkweed bug, *Oncopeltus fasciatus* (Dallas) (Heteroptera: Lygaeidae). *Journal of Morphology,* **93,** 177–283.

Borror, D.J., Triplehorn, C.A. and Johnson, N.F. (1989) *An Introduction to the Study of Insects,* 6th edn, Saunders College Publishing, Philadelphia, etc.

Brackenbury, J. (1990) Origami in the insect world. *Australian Natural History,* **23(7),** 562–9.

Bradley, T.J. (1985) The excretory system: structure and physiology, in *Comprehensive Insect Physiology, Biochemistry, and Pharmacology, Volume 4, Regulation: Digestion, Nutrition, Excretion,* (eds G.A. Kerkut and L.I. Gilbert), Pergamon Press, Oxford, New York, Toronto, Sydney, Paris, Frankfurt, pp. 421–65.

Brower, J.V.Z. (1958) Experimental studies of mimicry in some North American butterflies. Part III. *Danaus gilippus berenice* and *Limenitis archippus floridensis. Evolution,* **12,** 273–85.

Brower, L.P., Brower, J.v.Z. and Cranston, F.P. (1965) Courtship behavior of the queen butterfly, *Danaus gilippus berenice* (Cramer). *Zoologica,* **50,** 1–39.

Burton, M. and Burton, R. (1975) *Encyclopedia of Insects & Arachnids,* Octopus Books, London.

Calder, A.A. and Sands, D.P.A. (1985) A new Brazilian *Cyrtobagous* Hustache (Coleoptera: Curculionidae) introduced into Australia to control salvinia. *Journal of the Entomological Society of Australia,* **24,** 57–64.

Caudell, A.N. (1920) Zoraptera not an apterous order. *Proceedings of the Entomological Society of Washington,* **22,** 84–97.

Chapman, R.F. (1982) *The Insects. Structure and Function,* 3rd edn, Hodder and Stoughton, London, Sydney, Auckland, Toronto.

Chapman, R.F. (1991) General anatomy and function, in *The Insects of Australia,* 2nd edn, (CSIRO), Melbourne University Press, Carlton, pp. 33–67.

Cherikoff, V. and Isaacs, J. (1989) *The Bush Food Handbook,* Ti Tree Press, Balmain, N.S.W.

Chu, H.F. (1949) *How to Know the Immature Insects,* Wm. C. Brown Company Publishers, Dubuque, Iowa.

Clements, A.N. (1992) *The Biology of Mosquitoes, Volume 1, Development, Nutrition and Reproduction,* Chapman & Hall, London.

Cohen, E. (1991) Chitin biochemistry, in *Physiology of the Insect Epidermis,* (eds K. Binnington and A. Retnakaran), CSIRO Publications, Melbourne, pp. 94–112.

Common, I.B.F. (1990) *Moths of Australia,* Melbourne University Press, Carlton, Victoria.

Common, I.F.B. and Waterhouse, D.F. (1972) *Butterflies of Australia,* Angus and Robertson, Sydney.

Corner, E.J.H. (1952) *Wayside Trees of Malaya, Volume II,* Government Printing Office, Singapore.

Cornwell, P.B. (1968) *The Cockroach,* Hutchinson & Co Ltd, London.

Cox, J.M. (1987) Pseudococcidae (Insecta: Hemiptera). *Fauna of New Zealand (number) 11,* Science Information Publishing Centre, DSIR.

Coyne, J.A. (1983) Genetic differences in genital morphology among three sibling species of *Drosophila. Evolution,* **37,** 1101–17.

CSIRO (1970) *The Insects of Australia,* 1st edn, Melbourne University Press, Carlton.

CSIRO (1991) *The Insects of Australia,* 2nd edn, Melbourne University Press, Carlton.

Daly, H.V., Doyen, J.T. and Ehrlich, P.R. (1978) *Introduction to Insect Biology and Diversity,* McGraw-Hill Book Company, New York.

Darlington, A. (1975) *The Pocket Encyclopaedia of Plant Galls in Colour,* 2nd edn, Blandford Press Ltd, Dorset.

De Foliart, G.R. (1989) The human use of insects as food and as animal feed. *Bulletin of the Entomological Society of America,* **35,** 22–35.

De Klerk, C.A., Ben-Dov, Y. and Giliomee, J.H. (1982) Redescriptions of four vine infesting species of *Margarodes* Guilding (Homoptera: Coccoidea: Margarodidae) from South Africa. *Phytophylactica,* **14,** 61–76.

Dean, J., Aneshansley, D.J., Edgerton, H.E. and Eisner, T. (1990) Defensive spray of the bombardier beetle: a biological pulse jet. *Science,* **248,** 1219–21.

Deligne, J., Quennedey, A. and Blum, M.S. (1981) The enemies and defense mechanisms of termites, in *Social Insects, Volume II,* (ed. H.R. Hermann), Academic Press, New York, London, Toronto, Sydney, San Francisco, pp. 1–76.

Devitt, J. (1989) Honeyants: a desert delicacy. *Australian Natural History,* **22(12),** 588–95.

Dodson, G. (1989) The horny antics of antlered flies. *Australian Natural History,* **22 (12),** 604–11.

Dow, J.A.T. (1986) Insect midgut function. *Advances in Insect Physiology,* **19,** 187–328.

Duffy, E.A.J. (1963) *A Monograph of the Immature*

Stages of Australasian Timber Beetles (Cerambycidae), British Museum (Natural History), London.

Eastham, L.E.S. and Eassa, Y.E.E. (1955) The feeding mechanism of the butterfly *Pieris brassicae* L. *Philosophical Transactions of the Royal Society of London,* **B239,** 1–43.

Eberhard, W.G. (1985) *Sexual Selection and Animal Genitalia,* Harvard University Press, Cambridge, Massachusetts, and London, England.

Eibl-Eibesfeldt, I. and Eibl-Eibesfeldt, E. (1967) Das Parasitenabwehren der Minima-Arbeiterinnen der Blattschneider-Ameise (*Atta cephalotes*). *Zeitschrift für Tierpsychologie,* **24,** 278–81.

Eidmann, H. (1929) Morphologische und physiologische Untersuchungen am weiblichen Genitalapparat der Lepidopteren. I. Morphologischer Teil. *Zeitschrift für angewandte Entomologie,* **15,** 1–66.

Eisenbeis, G. and Wichard, W. (1987) *Atlas on the Biology of Soil Arthropods,* 2nd edn, Springer-Verlag, Berlin, Heidelberg, New York, London, Paris, Tokyo.

Elliott, J.M. and Humpesch, U.H. (1983) A key to the adults of the British Ephemeroptera, *Freshwater Biological Association Scientific Publication,* **47,** 1–101.

Entwistle, P.F. and Evans, H.F. (1985) Viral control, in *Comprehensive Insect Physiology, Biochemistry and Pharmacology, Volume 12, Insect Control,* (eds G.A. Kerkut and Gilbert, L.I.), Pergamon Press, Oxford, New York, Toronto, Sydney, Paris, Frankfurt, pp. 347–412.

Evans, F.D. (1978) Megaloptera and aquatic Neuroptera, in *An Introduction to the Aquatic Insects of North America,* (eds R.W. Merritt and K.W. Cummins), Kendall/Hunt Publishing Company, Dubuque, Iowa, pp. 133–45.

Ferrar, P. (1987) *A Guide to the Breeding Habits and Immature Stages of Diptera Cyclorrhapha,* Part 2, Entomonograph volume 8, E.J. Brill/Scandinavian Science Press, Leiden, Copenhagen.

Filshie, B.K. (1982) Fine structure of the cuticle of insects and other arthropods, in *Insect Ultrastructure, Volume 1,* (eds R.C. King and H. Akai), Plenum Press, New York and London, pp. 281–312.

Fjellberg, A. (1980) *Identification Keys to Norwegian Collembola,* Utgitt av Norsk Entomologisk Forening.

Foldi, I. (1983) Structure et fonctions des glandes tégumentaires de Cochenilles Pseudococcines et de leurs sécrétions. *Annales de la Societé Entomologique de France (N.S.),* **19,** 155–66.

Freeman, W.H. and Bracegirdle. B. (1971) *An Atlas of Invertebrate Structure,* Heinemann Educational Books, London.

Frisch, K.v. (1967) *The Dance Language and Orientation of Bees,* The Belknap Press of Harvard University Press, Cambridge, Massachusetts.

Froggatt, W.W. (1907) *Australian Insects,* William Brooks & Co. Ltd, Sydney.

Frost, S.W. (1959) *Insect Life and Insect Natural History,* 2nd edn, Dover Publications, Inc., New York.

Fry, C.H., Fry, K. and Harris, A. (1992) *Kingfishers, Bee-eaters and Rollers,* Christopher Helm (Publishers) Ltd, London.

Galil, J. and Eisikowitch, D. (1968) Pollination ecology of *Ficus sycomorus* in East Africa. *Ecology,* **49,** 259–69.

Gauld, I. and Bolton, B. (eds) (1988) *The Hymenoptera,* British Museum (Natural History), London, in association with Oxford University Press, Oxford.

Georghiou, G.P. (1986) The magnitude of the resistance problem, in *Pesticide Resistance: Strategies and Tactics for Management,* (eds National Research Council), National Academy Press, Washington, D.C., pp. 14–43.

Gibbons, B. (1986) *Dragonflies and Damselflies of Britain and Northern Europe,* Country Life Books, Twickenham.

Gray, E.G. (1960) The fine structure of the insect ear. *Philosophical Transactions of the Royal Society of London,* **B243,** 75–94.

Greany, P.D., Vinson, S.B. and Lewis, W.J. (1984) Insect parasitoids: finding new opportunities for biological control. *BioScience,* **34,** 690–6.

Grimstone, A.V., Mullinger, A.M. and Ramsay, J.A. (1968) Further studies on the rectal complex of the mealworm *Tenebrio molitor* L. (Coleoptera, Tenebrionidae). *Philosophical Transactions of the Royal Society,* **B253,** 343–82.

Gutierrez, A.P. (1970) Studies on host selection and host specificity of the aphid hyperparasite *Charips victrix* (Hymenoptera: Cynipidae). 6. Description of sensory structures and a synopsis of host selection and host specificity. *Annals of the Entomological Society of America,* **63,** 1705–9.

Gwynne, D.T. (1981) Sexual difference theory: Mormon crickets show role reversal in mate choice. *Science,* **213,** 779–80.

Gwynne, D.T. (1990) The katydid spermatophore: evolution of a parental investment, in *The Tettigoniidae: Biology, Systematics and Evolution,* (eds W.J. Bailey and D.C.F. Rentz), Crawford House Press, Bathurst, pp. 27–40.

Hadley, N.F. (1986) The arthropod cuticle, *Scientific American*, **255 (1)**, 98–106.

Hadlington, P. (1987) *Australian Termites and Other Common Timber Pests*, New South Wales University Press, Kensington.

Harris, W.V. (1971) *Termites: Their Recognition and Control*, 2nd edn, Longman, London.

Haynes, K.F. and Birch, M.C. (1985). The role of other pheromones, allomones and kairomones in the behavioural responses of insects, in *Comprehensive Insect Physiology, Biochemistry, and Pharmacology, Volume 9, Behaviour*, (eds G.A. Kerkut and L.I. Gilbert), Pergamon Press, Oxford, New York, Toronto, Sydney, Paris, Frankfurt, pp. 225–55.

Hely, P.C., Pasfield, G. and Gellatley, J.G. (1982) *Insect Pests of Fruit and Vegetables in NSW*, Inkata Press, Melbourne, Sydney, London.

Henwood, A. (1993) Still life in amber. *New Scientist*, **137 (1859)**, 31–4.

Hepburn, H.R. (1985) Structure of the integument, in *Comprehensive Insect Physiology, Biochemistry and Pharmacology, Volume 3, Integument, Respiration and Circulation*, (eds G.A. Kerkut and L.I. Gilbert), Pergamon Press, Oxford, New York, Toronto, Sydney, Paris, Frankfurt, pp. 1–58.

Hermann, H.R. and Blum, M.S. (1981) Defensive mechanisms in the social Hymenoptera, in *Social Insects, Volume II*, (ed. H.R. Hermann), Academic Press, New York, London, Toronto, Sydney, San Francisco, pp. 77–197.

Herms, W.B. and James, M.T. (1961) *Medical Entomology*, 5th edn, The Macmillan Company, New York.

Hill, A.S. and Roelofs, W.L. (1975) Sex pheromone components of the omnivorous leafroller moth, *Platynota stultana*. *The Journal of Chemical Ecology*, **1**, 91–9.

Hölldobler, B. (1984) The wonderfully diverse ways of the ant. *National Geographic*, **165**, 778–813.

Hölldobler, B. and Wilson, E.O. (1990) *The Ants*, Springer-Verlag, Berlin, Heidelberg, London, Paris, Tokyo, Hong Kong.

Holman, G.M., Nachman, R.J. and Wright, M.S. (1990) Insect neuropeptides. *Annual Review of Entomology*, **35**, 201–17.

Horridge, G.A. (1965) Arthropoda: general anatomy, in *Structure and Function in the Nervous Systems of Invertebrates, Volume II*, (eds T.H. Bullock and G.A. Horridge), W.H. Freeman and Company, San Francisco and London, pp. 801–964.

Hungerford, H.B. (1954) The genus *Rheumatobates* Bergroth (Hemiptera-Gerridae). *The University of Kansas Science Bulletin*, **36**, 529–88.

Imms, A.D. (1913) Contributions to a knowledge of the structure and biology of some Indian Insects. II. On *Embia major, sp. nov.*, from the Himalayas. *The Transactions of the Linnean Society of London*, **11**, 167–95.

Jobling, B. (1976) On the fascicle of blood-sucking Diptera. *Journal of Natural History*, **10**, 457–61.

Johnson, W.T. and Lyon, H.H. (1992). *Insects that Feed on Trees and Shrubs*, 2nd edn, Comstock Publishing Associates of Cornell University Press, Ithaca and London.

Julien, M.H. (ed.) (1992) *Biological Control of Weeds: A World Catalogue of Agents*, CAB International, Wallingford.

Katz, M., Despommier, D.D. and Gwadz, R.W. (1989) *Parasitic Diseases*, 2nd edn, Springer-Verlag, New York, Berlin, Heidelberg, London, Paris, Tokyo.

Keeley, L.L. and Hayes, T.K. (1987) Speculations on biotechnology applications for insect neuroendocrine research. *Insect Biochemistry*, **17**, 639–61.

Kettle, D.S. (1984) *Medical and Veterinary Entomology*, Croom Helm, London and Sydney.

Kukalová, J. (1970) Revisional study of the order Palaeodictyoptera in the Upper Carboniferous shales of Commentry, France. Part III. *Psyche*, **77**, 1–44.

Kukalová-Peck, J. (1991) Fossil history and the evolution of hexapod structures, in *The Insects of Australia*, 2nd edn, (CSIRO), Melbourne University Press, Carlton, pp. 141–79.

Lewis, T. (1973) *Thrips: their Biology, Ecology and Economic Importance*, Academic Press, London and New York.

Lindauer, M. (1960) Time-compensated sun orientation in bees. *Cold Spring Harbor Symposia on Quantitative Biology*, **25**, 371–7.

Link, E. (1909) Uber die Stirnaugen der hemimetabolen Insecten. *Zoologische Jahrbucher*, **27**, 281–376, 4 pls.

Loudon, C. (1989) Tracheal hypertrophy in mealworms: design and plasticity in oxygen supply systems. *Journal of Experimental Biology*, **147**, 217–35.

Lubbock, J. (1873) *Monograph of the Collembola and Thysanura*, The Ray Society, London.

Lüscher, M. (1961) Air-conditioned termite nests. *Scientific American*, **205(1)**, 138–45.

Lyal, C.H.C. (1986) Coevolutionary relationships of lice and their hosts: a test of Fahrenholz's Rule, in *Coevolution and Systematics*, (eds A.R. Stone and D.L. Hawksworth), Systematics Association, Oxford, pp. 77–91.

Malcolm, S.B. (1990) Mimicry: status of a classical evolutionary paradigm. *Trends in Ecology and Evolution*, **5**, 57–62.

Masler, E.P., Kelly, T.J. and Menn, J.J. (1991) Biologically active insect peptides, in *Insect Neuropeptides: Chemistry, Biology, and Action*, (eds J.J. Menn, T.J. Kelly and E.P. Masler), American Chemical Society, Washington, pp. 6–19.

Matsuda, R. (1965) Morphology and evolution of the insect head. *Memoirs of the American Entomological Institute*, **4**, 1–334.

McAlpine, D.K. (1990) A new apterous micropezid fly (Diptera: Schizophora) from Western Australia. *Systematic Entomology*, **15**, 81–6.

McAlpine, J.F. (ed.) (1981, 1987) *Manual of Nearctic Diptera, Volumes 1 & 2,*, Research Branch, Agriculture Canada, Monograph Nos 27 & 28.

McIver, S.B. (1985) Mechanoreception, in *Comprehensive Insect Physiology, Biochemistry, and Pharmacology, Volume 6, Nervous System: Sensory*, (eds G.A. Kerkut and L.I. Gilbert), Pergamon Press, Oxford, New York, Toronto, Sydney, Paris, Frankfurt, pp. 71–132.

Mercer, W.F. (1900) The development of the wings in the Lepidoptera. *New York Entomological Society*, **8**, 1–20.

Merritt, R.W., Craig, D.A., Walker, E.D., Vanderploeg, H.A. and Wotton, R.S. (1992) Interfacial feeding behavior and particle flow patterns of *Anopheles quadrimaculatus* larvae (Diptera: Culicidae). *Journal of Insect Behavior*, **5**, 741–61.

Michelsen, A. and Larsen, O.N. (1985) Hearing and sound, in *Comprehensive Insect Physiology, Biochemistry, and Pharmacology, Volume 6, Nervous System: Sensory*, (eds G.A. Kerkut and L.I. Gilbert), Pergamon Press, Oxford, New York, Toronto, Sydney, Paris, Frankfurt, pp. 495–556.

Michener, C. D. (1974) *The Social Behavior of Bees*, The Belknap Press of Harvard University Press, Cambridge, Massachusetts.

Miyazaki, M. (1987) Forms and morphs of aphids, in *Aphids: their Biology, Natural Enemies and Control, Volume A*, (eds A.K. Minks and P. Harrewijn), Elsevier, Amsterdam, Oxford, New York, Tokyo, pp. 27–50.

Monteith, S. in Monteith, G. (1990) Life inside an ant-plant. *Wildlife Australia*, **27(4)**, 5.

Nosek, J. (1973) *The European Protura*, Muséum D'Histoire Naturelle, Genéve.

Novak, V.J.A. (1975) *Insect Hormones*, Chapman & Hall, London.

O'Connell, R.J. (1975) Olfactory receptor responses to sex pheromone components in the redbanded leafroller moth. *The Journal of General Physiology*, **65**, 179–205.

Oliveira, P.S. (1988) Ant-mimicry in some Brazilian salticid and clubionid spiders (Araeae: Salticidae, Clubionidae). *Biological Journal of the Linnean Society*, **33**, 1–15.

Palmer, M.A. (1914) Some notes on life history of ladybeetles. *Annals of the Entomological Society of America*, **7**, 213–38.

Poisson, R. (1951) Ordre des Hétéroptéres, in *Traité de Zoologie. Anatomie, Systématique, Biologie. Tome X. Insectes Supérieurs et Hémiptéroïdes, Fascicule II*, (ed. P.-P. Grassé), Masson, Paris, pp. 1657–803.

Preston-Mafham, K. (1990) *Grasshoppers and Mantids of the World*, Blandford, London.

Raabe, M. (1986) Insect reproduction: regulation of successive steps. *Advances in Insect Physiology*, **19**, 29–154.

Richards, A.G. and Richards, P.A. (1979) The cuticular protuberances of insects. *International Journal of Insect Morphology and Embryology*, **8**, 143–57.

Richards, G. (1981) Insect hormones in development. *Biological Reviews of the Cambridge Philosophical Society*, **56**, 501–549.

Richards, O.W. and Davies, R.G. (1959) *A. D. Imms' Outline of Entomology*, Methuen & Co. Ltd, London.

Richards, O.W. and Davies, R.G. (1977) *Imms' General Textbook of Entomology, Volume I: Structure, Physiology and Development*, 10th edn, Chapman & Hall, London.

Riddiford, L.M. (1991) Hormonal control of sequential gene expression in insect epidermis, in *Physiology of the Insect Epidermis*, (eds K. Binnington and A. Retnakaran), CSIRO Publications, Melbourne, pp. 46–54.

Roth, L.M. (1982) Blattaria, in *Synopsis and Classification of Living Organisms, Volume 2*, (ed. S.P. Parker), McGraw-Hill Book Company, New York, etc., pp. 346–9.

Rumbo, E.R. (1989) What can electrophysiology do for you?, in *Application of Pheromones to Pest Control*, Proceedings of a Joint CSIRO-DSIR Workshop, July 1988, (ed. T.E. Bellas), Division of Entomology, CSIRO, Canberra, pp. 28–31.

Sainty, G.R. and Jacobs, S.W.L. (1981) *Waterplants of New South Wales*, Water Resources Commission, New South Wales.

Salt, G. (1968) The resistance of insect parasitoids to the defence reactions of their hosts. *Biological Reviews of the Cambridge Philosophical Society*, **43**, 200–32.

Samson, P.R. and Blood, P.R.B. (1979) Biology and temperature relationships of *Chrysopa* sp., *Micromus tasmaniae* and *Nabis capsiformis*. *Entomologia Experimentalis et Applicata*, **25**, 253–9.

Santos Oliveira, J. F., Passos de Carvalho, J., Bruno de Sousa, R. F. X. and Madalena Simão, M. (1976) The nutritional value of four species of insects consumed in Angola. *Ecology of Food and Nutrition*, **5**, 91–7.

Schmutterer, H. (1990) Properties and potential of natural pesticides from the neem tree, *Azadirachta indica*. *Annual Review of Entomology*, **35**, 271–97.

Schwabe, J. (1906) Beitrege zur Morphologie und Histologie der tympanalen Sinnesapparate der Orthopteren. *Zoologica, Stuttgart*, **50**, 1–154.

Sheppard, A.W. (1992) Predicting biological weed control. *Trends in Ecology and Evolution*, **7**, 290–1.

Smith, P.H., Gillott, C., Barton Browne, L. and van Gerwen, A.C.M. (1990) The mating-induced refractoriness of *Lucilia cuprina* females: manipulating the male contribution. *Physiological Entomology*, **15**, 469–81.

Smithers, C.N. (1982) Psocoptera, in *Synopsis and Classification of Living Organisms, Volume 2*, (ed. S.P. Parker), McGraw-Hill Book Company, New York etc., pp. 394–406.

Snodgrass, R.E. (1935) *Principles of Insect Morphology*, McGraw-Hill Book Company, New York and London.

Snodgrass, R.E. (1946) The skeletal anatomy of fleas (Siphonaptera). *Smithsonian Miscellaneous Collections*, **104 (18)**, 1–89.

Snodgrass, R.E. (1956) *Anatomy of the Honey Bee*, Comstock Publishing Associates, Ithaca, New York.

Snodgrass, R.E. (1957) A revised interpretation of the external reproductive organs of male insects. *Smithsonian Miscellaneous Collections*, **135(6)**, 1–60.

Snodgrass, R.E. (1967) *Insects: Their Ways and Means of Living*, Dover Publications, Inc., New York.

Spencer, K.A. (1990) *Host Specialization in the World Agromyzidae (Diptera)*, Kluwer Academic Publishers, Dordrecht, Boston, London.

Spradbery, J.P. (1973) *Wasps: An Account of the Biology and Natural History of Solitary and Social Wasps*, Sidgwick and Jackson, London.

Stanek, V.J. (1969) *The Pictorial Encyclopedia of Insects*, Hamlyn, London, New York, Sydney, Toronto.

Stanek, V.J. (1977) *The Illustrated Encyclopedia of Butterflies & Moths*, Octopus Books Limited, London.

Stern, V.M., Smith, R.F., van den Bosch, R. and

Hagen, K.S. (1959) The integrated control concept. *Hilgardia*, **29**, 81–101.

Stoltz, D.B. and Vinson, S.B. (1979) Viruses and parasitism in insects. *Advances in Virus Research*, **24**, 125–71.

Struble, D.L. and Arn, H. (1984) Combined gas chromatography and electroantennogram recording of insect olfactory responses, in *Techniques in Pheromone Research*, (eds H.E. Hummel and T.A. Miller), Springer-Verlag, New York, Berlin, Heidelberg, Tokyo, pp. 161–78.

Sullivan, D.J. (1988) Hyperparasites, in *Aphids. Their Biology, Natural Enemies and Control, Volume B*, (eds A.K. Minks and P. Harrewjin), Elsevier, Amsterdam, Oxford, New York, Tokyo, pp. 189–203 .

Sutherst, R.W. and Maywald, G.F. (1985) A computerised system for matching climates in ecology. *Agriculture, Ecosystems and Environment*, **13**, 281–99.

Sutherst, R.W. and Maywald, G.F. (1991) Climate modelling and pest establishment. *Plant Protection Quarterly*, **6**, 3–7.

Suzuki, N. (1985) Embryonic development of the scorpionfly, *Panorpodes paradoxa* (Mecoptera, Panorpodidae) with special reference to the larval eye development, in *Recent Advances in Insect Embryology in Japan*, (eds H. Ando and K. Miya), ISEBU Co. Ltd, Tsukubo, pp. 231–8.

Szabó-Patay, J. (1928) A kapus-hangya. *Termeszettudományi Közlöny, Budapest*, pp. 215–19.

Tait, N. (1985) Snorkels and aqualungs. *Australian Natural History*, **21**, 489–93.

Terra, W.R. and Ferreira, C. (1981) The physiological role of the peritrophic membrane and trehalase: digestive enzymes in the midgut and excreta of starved larvae of *Rhynchosciara*. *Journal of Insect Physiology*, **27**, 325–31.

Trueman, J.W.H. (1991) Egg chorionic structures in Corduliidae and Libellulidae (Anisoptera). *Odonatologica*, **20**, 441–52.

Uvarov, B. (1966) *Grasshoppers and Locusts*, Cambridge University Press, Cambridge.

van den Bosch, R. and Hagen, K.S. (1966) Predaceous and parasitic arthropods in Californian cotton fields. *Californian Agricultural Experimental Station Bulletin 820*, 1–32.

Waage, J.K. (1986) Evidence for widespread sperm displacement ability among Zygoptera (Odonata) and the means for predicting its presence. *Biological Journal of the Linnean Society*, **28**, 285–300.

Waterhouse, D.F. (1974) The biological control of dung. *Scientific American*, **230(4)**, 100–9.

Watson, J.A.L. and Abbey, H.M. (1985) Seasonal cycles in *Nasutitermes exitiosus* (Hill) (Isoptera: Termitidae). *Sociobiology*, **10**, 73–92.

Wheeler, Q.D. (1990) Insect diversity and cladistic constraints. *Annals of the Entomological Society of America*, **83**, 1031–47.

Wheeler, W.M. (1910) *Ants: Their Structure, Development and Behavior*, Columbia University Press, New York.

White, D.S., Brigham, W.U. and Doyen, J.T. (1984) Aquatic Coleoptera, in *An Introduction to the Aquatic Insects of North America*, 2nd edn, (eds R.W. Merritt and K.W. Cummins), Kendall/Hunt Publishing Company, Dubuque, Iowa, pp. 361–437.

White, G.B. (1985) *Anopheles bwambae* sp.n., a malaria vector in the Semliki Valley, Uganda, and its relationships with other sibling species of the *An. gambiae* complex (Diptera: Culicidae). *Systematic Entomology*, **10**, 501–22.

Whitten, M.J. (1989) The relevance of molecular biology to pure and applied entomology. *Entomologia Experimentalis et Applicata*, **53**, 1–16.

Wickler, W. (1968) *Mimicry in Plants and Animals*, Weidenfeld and Nicolson, London.

Wiggins, G.B. (1978) Trichoptera, in *An Introduction to the Aquatic Insects of North America*, (eds. R.W. Merritt and K.W. Cummins), Kendall/Hunt Publishing Company, Iowa.

Wigglesworth, V.B. (1964) *The Life of Insects*, Weidenfeld and Nicolson, London.

Wigglesworth, V.B. (1972) *The Principles of Insect Physiology*, 7th edn, Chapman & Hall, London.

Williams, J.L. (1941) The relations of the spermatophore to the female reproductive ducts in Lepidoptera. *Entomological News*, **52**, 61–5.

Wilson, M.R. and Claridge, M.F. (1991) *Handbook for the Identification of Leafhoppers and Planthoppers of Rice*, CAB International, Wallingford.

Winston, M.L. (1987) *The Biology of the Honey Bee*, Harvard University Press, Cambridge, Massachusetts, London.

Womersley, H. (1939) *Primitive Insects of South Australia*, Government Printer, Adelaide.

Youdeowei, A. (1977) *A Laboratory Manual of Entomology*, Oxford University Press, Ibadan.

Zacharuk, R.Y. (1985) Antenna and sensilla, in *Comprehensive Insect Physiology, Biochemistry, and Pharmacology, Volume 6, Nervous System: Sensory*, (eds G.A. Kerkut and L.I. Gilbert), Pergamon Press, Oxford, New York, Toronto, Sydney, Paris, Frankfurt, pp. 1–69.

Zanetti, A. (1975) *The World of Insects*, Gallery Books, New York.

INDEX

Readers are referred also to the Glossary (pages 433–461): glossary entries are not cross referenced in this index. All *italicized* entries below are the scientific names of insects, except where the following abbreviations indicate:

Bact. – Bacterial name
Mam. – Mammal name
Vir. – Viral name

Bot. – Botanical name
Nem. – Nematode name

Fung. – Fungal name
Prot. – Protozoan name